INTERPRETING THE *NATIONAL* ELECTRICAL *CODE*®

P9-CBK-347

**Based on the 2005
NATIONAL ELECTRICAL CODE®**

INTERPRETING THE *NATIONAL* ELECTRICAL *CODE*®

Seventh Edition

**Truman C. Surbrook
Professor
Michigan State University**

**Jonathan R. Althouse
Instructor
Michigan State University**

THOMSON

DELMAR LEARNING

Australia Canada Mexico Singapore Spain United Kingdom United States

Interpreting the *National Electrical Code*®, 7th Edition

Truman C. Surbrook and Jonathan R. Althouse

Vice President, Technology and Trades, SBU:
Alar Elken

Editorial Director:
Sandy Clark

Senior Acquisitions Editor:
Stephen Helba

Development:
Dawn Daugherty

Marketing Director:
Dave Garza

Channel Manager:
Dennis Williams

Marketing Coordinator:
Stacey Wiktorek

Production Director:
Mary Ellen Black

Production Manager:
Larry Main

Production Editor:
Ruth Fisher

Library of Congress Cataloging-in-Publication Data:

Surbrook, Truman C.
 Interpreting the National Electrical Code / Truman C. Surbrook, Jonathan R. Althouse.—7th ed.
  p. cm.
 "Based on the 2005 National Electrical Code."
 Includes index.
 ISBN 1-4018-5213-0 (core text)—ISBN 1-4018-5214-9 (ig.)
 1. Electric engineering—Insurance requirements—United States. 2. National Fire Protection Association. National Electrical Code (2005) I. Althouse, Jonathan R. II. National Fire Protection Association. National Electrical Code (2005) III. Title

TK260.S87 2005
621.319'24'021873—dc22   2004061548

BRIEF CONTENTS

CONTENTS

PREFACE

INTERPRETING THE NATIONAL ELECTRICAL CODE® is a text and workbook designed as a self-contained study course for learning to read and interpret the meaning of the Code, and to find information about how to do wiring installations. This text is intended for use by personnel with some experience in electrical wiring and use of the *National Electrical Code®*. The text is organized into 15 units covering the entire Code. The individual units cover a particular subject and show how to find information about that subject in the entire Code.

PREREQUISITES

Electrical wiring experience and a basic familiarity with the *NEC®* are desirable for the student to maximize learning from this text and the course in which it is used. Basic arithmetic, written communication, and reading skills are required for the course. Students with an interest in electrical wiring and the Code will, after completion of this text and course, be able to find information from the Code needed to do residential, commercial, farm, and industrial wiring and to be successful with electrical examinations.

EACH ARTICLE DISCUSSED

A brief description of the type of information found in every article of the Code is included in the text. Code article discussions are included in the unit of the text where the article contains important information for the subject of that unit. The Contents provides a list of Code articles included in each unit to provide a quick reference for the student to find information in this text.

CODE CHANGES DISCUSSED

This text contains a discussion of the *NEC®* changes of significance since the last edition of the Code. The articles that pertain to the subject of the unit as listed in the Contents are contained in each unit. This discussion of changes is intended to explain the meaning and application of the change. Every article of the Code is included in this discussion of Code changes.

USEFUL WIRING INFORMATION INCLUDED

Each unit of this text contains useful information for understanding the meaning behind Code requirements, or information on the subject of the unit useful for applications in the field that is not included in the Code. This text can, therefore, be used as a handy reference for making wiring installations.

SAMPLE CALCULATIONS

Many types of calculations are needed for making electrical installations. This text includes a discussion of the Code requirements, and then provides sample calculations as would be typically encountered in everyday wiring. Methods have been developed in this text that help the electrician better understand the Code requirements for making electrical calculations.

ILLUSTRATIONS

The illustrations in this text provide numerous references to Code sections to help the reader find important information in the Code. The purpose of the illustrations is to help the reader gain an understanding of a typical application of a particular Code section. It is not possible for an illustration to provide all information contained in a particular Code section. It is important that the reader study the particular Code section referenced to gain a complete understanding of the meaning of the Code section.

PRACTICAL CODE QUESTIONS AND PROBLEMS

Learning to find information in the *NEC®* requires looking up answers to practical everyday wiring installation questions. Each unit of this text contains two sets of 15 practical questions or problems pertaining to the subject of the unit. The Beginning Worksheet contains questions that are at the journey electrician or first license level and are especially useful for persons studying to take the first level electrician examination. The Advanced Worksheet contains questions that are at the master or advanced electrician level and are especially useful for persons studying to take a master or advanced electrician examination. The questions are multiple choice and the answers to the questions are found in the Code articles listed for that particular unit. Space is provided to write the Code sections where the answer was found. When calculations are required to obtain an answer, space may not be adequate to write the solution in the text. The instructor will give the answer and work the calculation during the class session. An answer sheet to hand in for grading is provided at the end of the text for the student to circle the answer and write the Code reference.

METRICS (SI) AND THE *NEC®*

The electrical trade in the United States continues to be based upon the inch-pound system of measurements, while most of the remainder of the world uses SI metric (Systeme International d'Unites) dimensions. The 2005 edition of the Code will continue to provide dimensions in SI metric followed with the inch-pound equivalent in parentheses. This text will provide dimensions in inch-pound units followed with the SI metric equivalent in parentheses. Electrical materials and equipment used in the United States continue to be provided with inch-pound dimensions, although some plans and specifications may have dimensions given as SI metric. In this case, it will be necessary to make a conversion. *Annex A-2* in this text gives common conversions from SI metric to inch-pound units and from inch-pound units to SI metric.

In general, the SI metric dimensions in the Code have been rounded to an even number. For example, 12 in. is actually 305 mm. The Code rounds 12 in. to an even 300 mm. This rounding of units can result in a significant difference in measurements and in some calculations. *Annex A-1* at the end of this text gives common lengths used in the Code and the SI metric conversions. When these distances are rounded as in the Code, *Annex A-1* gives the difference. The Code in *90.9(D)* permits either inch-pound or SI metric measurement to be used for electrical installations, even though they may be slightly different.

Wire sizes continue to be given in the Code as AWG (American Wire Gage) and kcmil (thousands of circular mils). Standard SI metric wire sizes are in sq. millimeters and are different than standard AWG sizes. Conversions to sq. millimeters are given in the Code for standard AWG and kcmil sizes. These are considered soft conversions since the basis for sizing the wires is AWG and kcmil. A hard conversion is a switch to standard metric sizes in sq. millimeters.

Trade sizes of conduit and tubing continue to be used in the Code. Conduit threads continue to be specified as National Pipe Threads (NPT). Some equipment may be provided with metric pipe threads in which case adapters to NPT are provided. Since conduit and tubing of different types have different dimensions; standard trade sizes are used, which have approximately the same internal diameter. Actual dimensions of the different types of conduit and tubing are given in *Table 4* of the Code. These standard trade sizes for the purpose of SI metric conversion have been given a number called a metric designator. This metric designator is the approximate inside diameter in millimeters. Metric designators for standard conduit and tubing trade sizes are given in *Table 300.1(C)* and in *Annex A-3* of this text.

INTERNATIONAL RESIDENTIAL CODE®

The *International Residential Code®* (*IRC®*) applies to the construction of one- and two-family dwellings. It is a standard developed by the International Code Council, Inc. and revised every three years. The first edition carried the date 2000, and the second edition carried the date 2003. The next edition will have the date 2006. The *IRC®* has been adopted by a number of jurisdictions in the United States as the standard for construction of one- and two-family dwellings. The *IRC*® contains standards for building, mechanical, plumbing, and electrical installations. The electrical section of the 2003 *IRC*® is based on the 2002 edition of the *National Electrical Code*® (*NEC*®). The *IRC*® only covers 120/240-volt, single-phase, 3-wire services rated up to and including 400 amperes. Other sizes and types of service to a one- or two-family dwelling are permitted as covered in the *NEC*®. The *IRC*® has a numbering system different from the *NEC*®, and much of the text is written and organized differently in the two Codes. There are a few differences in the requirements of the *IRC*® as compared to the *NEC*®, but most rules in these two Codes are essentially equivalent.

It is important for persons learning the electrical trade to determine if the *International Residential Code*® applies in a particular jurisdiction. If the *IRC*® applies, it is recommended that persons studying this text practice looking up answers in the *IRC*® as well as in the *NEC*®. The *INSTRUCTOR'S GUIDE* for this text also provides Code references from the *IRC*® where applicable. In a few cases the answers are different, and those differences are explained. Electrical requirements are covered in *Chapter 33* through *Chapter 42* of the *IRC*®. The section numbering system for the *IRC*® is different than for the *NEC*®. This text does not discuss the *IRC*® or give references from the *IRC*®. It will be left up to the instructor to explain how to find answers to questions based on the *IRC*®.

INSTRUCTOR'S GUIDE

An *INSTRUCTOR'S GUIDE* is available that gives a detailed lesson plan for each of the 15 sessions. It is also possible to offer the course in segments of less than 15 sessions, and suggestions for a shorter format are provided. The *INSTRUCTOR'S GUIDE* contains all answers to the student homework questions and problems, as well as complete calculations. Code references are also given for each homework question or problem. Some questions or problems have multiple Code references. All possible Code references are provided with a background discussion to help the instructor explain the reasoning behind the answer and Code reference.

TO THE STUDENT

This is a text and workbook that can be used for self-study, but it was developed primarily for use with three to four hours of instruction for each unit. The greatest benefit will be derived from the text and course if the student will make a concerted effort to work all problems and answer all questions. The only way to learn to find information in the *NEC*® and to interpret the information is to use the Code to find answers to daily wiring questions. The class discussion of the answers to questions will be of greatest meaning when the student has made an effort to find the answer and Code reference.

The instructor will briefly review the main points of the Code material that will be used to answer the homework questions. These pointers may save time lost searching for answers in the Code. The instructor will review the changes in the Code of greatest significance since the last edition. This will help to avoid confusion with the previous Code.

The student should quickly read the Code discussion portion of the text that describes the type of information found in each article. This will help to make question look-up more efficient. Then the student should read carefully the portion of the text that discusses Code material and gives sample calculations. Most problems will be similar to these sample calculations. The student should work the homework problems. It will take two to four hours to answer the questions and work the homework problems. It usually works best to spend some time on several occasions rather than trying to work the homework all at one time.

Before returning to class, write the answer and the Code reference on the answer sheet to hand in at the beginning of class.

Important Note: Please note that this textbook was completed after all the normal steps in the *NFPA 70* review cycle—Proposals to Code-Making-Panels, review by Technical Correlating Committee, Report on Proposals, Comments to Code-Making-Panels, review by Technical Correlating Committee, Report on Comments, NFPA Annual Meeting, and ANSI Standards Council—and before the actual publication of the 2005 edition of the *NEC*®. Every effort has been made to be technically correct, but there is always the possibility of typographical errors, or appeals made to the NFPA Board of Directors after the normal review cycle that could change the appearance or substance of the Code.

If changes do occur after the printing of this textbook, these changes will be included in the *INSTRUCTOR'S GUIDE* and will be incorporated into the textbook upon its next printing.

Please note also that the Code has a standard method to introduce changes between review cycles, called "Tentative Interim Amendment," or TIA. These TIAs and typographical errors can be downloaded from the NFPA Web site, http://www.nfpa.org, to make your copy of the Code current.

ABOUT THE AUTHORS

One of the authors of *INTERPRETING THE NATIONAL ELECTRICAL CODE*®, Truman C. Surbrook, has extensive practical experience in electrical wiring, as well as many years of experience as an instructor in the electrical trade.

Truman C. Surbrook, Ph.D., is a Professor of Biosystems and Agricultural Engineering at Michigan State University, a registered Professional Engineer, and Master Electrician. Dr. Surbrook developed an Electrical Apprenticeship Program at Michigan State University, and later served as chair for curriculum development for a statewide electrical inspector training short course. Dr. Surbrook developed a highly successful and comprehensive *NEC*® training course through Outreach Programs of Michigan State University.

Dr. Surbrook has spent his professional career teaching, authoring textbooks, training bulletins, and papers, working with youths, electricians, contractors, and inspectors, and conducting research in the areas of electrical wiring and electronics. He has been an active member and officer of several professional organizations and has served on numerous technical committees. Organizations in which he has been most active are the Institute of Electrical and Electronic Engineers, American Society of Agricultural Engineers, International Association of Electrical Inspectors, the National Fire Protection Association and the Michigan Agricultural Electric Council. Dr. Surbrook has served on several Ad Hoc Technical Committees of the National Electrical Code, and he has been a member of National Electrical Code-Making Panels 13 and 19. For 15 years, Dr. Surbrook was writer and producer of a weekly radio program dealing with electrification. He has written many bulletins, articles, and technical publications on various subjects related to electrical wiring.

Jonathan R. Althouse, M.S., who is Dr. Surbrook's coauthor, is an Instructor of Biosystems and Agricultural Engineering at Michigan State University, a licenced Master Electrician and Electrical Contractor. Mr. Althouse is the Coordinator of the Electrical Technology Apprenticeship Program at Michigan State University.

Mr. Althouse has developed an electrical awareness program for high school students entitled *Teaching Electrification in Agribusiness Classes in High Schools* (T.E.A.C.H.S.). He has also developed and provided instruction for *NEC*® Update Courses and Journey and Master Electrician examination prep classes. Mr. Althouse has been involved with organizations such as the International Association of Electrical Inspectors, the American Society of Agricultural Engineers, and the Michigan Agricultural Electric Council. Mr. Althouse recently has developed a Michigan State University internet site that provides technical information to personnel in the electrical trade: http://www.egr.msu.edu/age/.

REVIEWERS

The following instructors provided recommendations during the writing and production of *INTERPRETING THE NATIONAL ELECTRICAL CODE*®:

Madeline Borthick, IEC, Houston, TX
Deanna Hanieski, Professor, Lansing Community College, Lansing, MI
Wayne Tanner, Franklin City VTS, Chambersburg, PA
Gerald W. Williams, Oxnard College, Oxnard, CA
Mike Forister, Cheyenne, WY
Frank Holcomb, Tennesee Technology Center, Paris, TN

ACKNOWLEDGMENTS

It is with sincere gratitude that the authors express appreciation to the following individuals who have either helped with some aspect of the development of this text or have provided field testing and critical review of the material to ensure completeness and accuracy:

Lori A. Althouse, Editing Assistant
  Eaton Rapids, MI
Matthew and Elizabeth Althouse
  Eaton Rapids, MI
Dennis Cassady, Electrical Inspector
  Kentwood, MI
John P. Donovan, Standard Electric Co.
  Lansing, MI
George Little, Electrical Inspector
  Farmington, MI

John Negri, Electrical Consultant
  Bloomingdale, MI
James O'Donnell, Electrical Inspector
  Marquette, MI
Mary Surbrook, Editing Assistant
  Williamston, MI
Bailey, Spencer, and Emma Surbrook
  Lansing, MI
James Worden, Electrical Inspector
  Port Huron, MI

UNIT 1

General Wiring and Fundamentals

OBJECTIVES

After completion of this unit, the student should be able to:

- name the sponsoring organization of the *National Electrical Code®*.
- know how and when to submit a proposal to change the *National Electrical Code®*.
- explain what is a tentative interim amendment.
- describe direct current and alternating current.
- calculate the unknown quantity if only two of these quantities are known: volts, current, or resistance.
- determine the current in a circuit or equipment if power, voltage, and power factor are known.
- determine the area of a conductor if the conductor diameter is given.
- calculate the total resistance of a circuit if the values of resistance in either series or in parallel are given.
- choose the correct ampacity table from *Article 310* to determine the minimum size of conductor required for a wiring application.
- answer wiring installation questions relating to *Articles 90, 100, 110, 200, 300, 310, 320, 324, 328, 330, 332, 334, 336, 338, 340, 382, 394, 590,* or *Chapter 9, Tables 8* or *9*.
- state at least five significant changes that occurred from the 2002 to the 2005 Code for *Articles 90, 100, 110, 200, 300, 310, 320, 324, 328, 330, 332, 334, 336, 338, 340, 382, 394, 590,* or *Chapter 9, Tables 8* or *9*.

ORIGIN OF THE *NATIONAL ELECTRICAL CODE®*

The first *National Electrical Code®* was developed in 1897. In 1911, the National Fire Protection Association (NFPA) became the sponsor, and the Code has been revised on numerous occasions since that date. Now it is revised every three years. The time schedule for revising the 2005 Code is listed in the back of the Code. The next revision will be the 2008 edition.

The *NEC®* is available for adoption as the electrical law in a governmental jurisdiction. That governmental jurisdiction may add one or more amendments to allow for local needs, preferences, or conditions. It is not the intent of this text to cover amendments to the Code made by local or state jurisdictions, but an instructor using this material for a course may cover these amendments as an addendum to this material.

Process of Revising the Code

The process of revising the Code begins with any interested person who submits a proposal for changing the Code not later than the closing date for accepting proposals. This closing date is given in each edition of the Code. Proposals are mailed to the Vice President of the Technical Council, National Fire Protection Association, Batterymarch Park, Quincy, MA 02269. A blank form for submitting proposals is available on the NFPA Web site, http://www.nfpa.org.

The proposals for changing a particular section or adding a new section or article are reviewed by the appropriate Code-Making Panel (listed in the front of the Code) at the date listed in the Code. The Code-Making Panel actions on proposals are published in a document titled the *Committee Report on Proposals*, which is usually available to the public in July of the year shown in the timetable of events in the Code. Any interested person can obtain a copy of this document and make public comment that must be received by the NFPA not later than the closing date listed for public comment. A comment form is in the front of the *Committee Report on Proposals*.

Code-Making Panels meet in December of every third year to act on public comments. The Panel is only permitted to take action on sections of the Code for which public comment was received. The Code-Making Panel's actions at these two meetings constitute the changes that occur in the Code. The *Committee Report on Comments* reports on the Panel action taken on public comment. The electrical section meeting of the National Fire Protection Association in May 2007 results in the adoption of the 2008 Code. Members present at that meeting are permitted to make motions to change Code-Making Panel action taken on a particular proposal published in the *Committee Report on Proposals* or change the action taken on a comment as published in the *Committee Report on Comments*. If the motion receives a majority vote of members present at the meeting, the change becomes a part of the new Code. A person who submitted a proposal that was rejected and made public comment that was rejected, can make a motion for adoption of the proposal at the annual meeting. Following this meeting of the electrical section of NFPA, the change process is completed, and NFPA publishes the Code. There is a brief period after the annual meeting when an appeal of panel action can be submitted to the *NEC® Correlating Committee*. The new edition of the Code is available for public sale by September of the year of the NFPA Annual Meeting when the Code was adopted. The *Committee Report on Proposals*, *Committee Report on Comments*, and a preliminary draft of the *NEC®* can be accessed on the NFPA web site at http://www.nfpa.org.

There is another means by which the Code may change in addition to the process previously described. This is by the issuing of a Tentative Interim Amendment or TIA. These are generally changes of such importance that they should not wait until the next Code. There is a process through which they are approved, and if approved, a TIA is added to the Code only for the duration of that edition. A TIA will automatically be submitted as a proposal for the next Code change, and it must go through the same process as all other proposals to remain as a part of the Code.

ELECTRICAL CALCULATIONS WITH A CALCULATOR

An electronic calculator is a valuable tool for making calculations required for electrical installations. It is important to become familiar with the proper use of your calculator. The procedures for making calculations may be different from one calculator to another. The following procedure is common for many calculators, but it may be different for your calculator. If this procedure does not work for your calculator, then check the calculator instructions for each example problem. Practice with your calculator until you become comfortable with its operation. This will help you with the problems in this text as well as eliminating errors in electrical calculations on the job. There are more efficient methods of making the following calculations on some calculators; however, the following method works on most calculators.

Example of String Multiplication

Calculate power (P) when the voltage (V), current (A), and power factor (pf) are known.

$$P = V \times A \times pf = 480\ V \times 22\ A \times 0.75 = 7920\ W$$

Step	Press	Display
1	480	480
2	×	480
3	22	22
4	=	10,560
5	×	10,560
6	.75	.75
7	=	7,920 ←

Example of Addition and Multiplication

Determine the cost of several runs of conduit. The runs are 22 ft, 74 ft, and 15 ft, and the cost is $1.15 per ft.

Conduit cost = (22 ft + 74 ft + 15 ft) × $1.15 = $127.65

Step	Press	Display
1	22	22
2	+	22
3	74	74
4	=	96
5	+	96
6	15	15
7	=	111
8	×	111
9	1.15	1.15
10	=	**127.65** ◄—

Example of String Multiplication and Division

Determine the full-load current of a 3-phase, 37.5-kVA transformer with a 208-volt secondary winding.

$$A = \frac{kVa \times 1000}{1.73 \times V} = \frac{37.5 \times 1000}{1.73 \times 208} = 104.21 \ A$$

Step	Press	Display
1	37.5	37.5
2	×	37.5
3	1000	1000
4	=	37,500
5	÷	37,500
6	1.73	1.73
7	=	21,676.3
8	÷	21,676.3
9	208	208
10	=	**104.21** ◄—

FUNDAMENTALS OF ELECTRICITY

An electrician must have a basic understanding of the fundamentals of electricity to perform most effectively in the field and also to pass electrical examinations. This section of the text is a brief review of minimum basic electrical fundamentals.

AC and DC

The Code refers to alternating current (ac) and direct current (dc). Direct current travels only in one direction in the circuit. A typical source of direct current is a battery or a rectifier. Figure 1.1 shows how the

Figure 1.1   A plot of voltage in a direct current circuit over a period of time is a straight line.

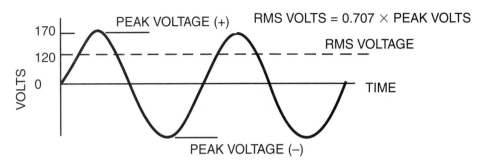

Figure 1.2   The plot of voltage in an alternating current circuit over a period of time is a sine wave, and the voltage that would show on a meter would be the rms voltage.

voltage in a direct current circuit would look if plotted on a graph over a period of time. An oscilloscope is an electrical instrument that can show the voltage and time plot of a direct current circuit. The voltage will be either a positive or a negative constant value. The level of voltage remains at a constant or a nearly constant magnitude. The polarity will remain either positive or negative. As long as the polarity does not change from positive to negative or from negative to positive, the electrical current will flow in one direction in the circuit. The polarity must change if the current is to reverse direction of flow in the circuit.

In the case of an alternating current circuit, the voltage is constantly changing. The voltage increases to a maximum, then decreases back to zero. Then the voltage builds to a maximum with opposite polarity, then decreases in magnitude back to zero. This completes one cycle. Normal electrical power operates at 60 cycles per second or hertz (Hz). A graph of an ac voltage varying with time is shown in Figure 1.2. This is what an ac voltage looks like if viewed on the screen of an oscilloscope, and the waveform is known as a sine wave from the mathematics of trigonometry. If an analog voltmeter with a needle or a digital voltmeter is connected to measure voltage across a component of an ac circuit, the voltmeter will indicate the root-mean-square (rms) voltage as indicated in Figure 1.2. The actual voltage is changing as shown in Figure 1.2, but the analog or digital voltmeter will register only a single value of voltage. The rms voltage is compared with the alternating voltage in Figure 1.2. The rms voltage is 0.707 times the peak of the varying voltage (sine wave) of the circuit. Or, the peak voltage of the sine wave is 1.414 times the rms voltage. Root-mean-square is used in the definition of voltage in *Article 100* of the Code. Alternating current frequencies other than 60 Hz are sometimes used for special applications in electrical equipment.

Many voltmeters are not true rms (root-mean-square) meters. The common hand-held voltmeter actually measures the average value of the alternating sine wave at 60 Hz, which is 0.637 times the peak value. The voltmeter output is calibrated to increase the average voltage value by 1.11 so the output will be equal to the rms voltage (0.637 × 1.11 = 0.707). If the voltage is something other than a sine wave or if the power source is being altered by solid-state switching, the voltage indicated by the meter will be in error. It is also possible in some electrical systems for more than one frequency to be present at the same time. This is the case when harmonic frequencies are present on the same 3-phase, 4-wire circuits as mentioned in *NEC® 220.61(C)*. Under these conditions, an averaging voltmeter will not indicate an accurate rms voltage. The amount of the error will depend on the amount of harmonic distortion of the 60-Hz sine wave. True rms reading voltmeters are available.

Metric Conversions

Metric units are provided in the Code followed by the approximate equivalent English units. In most cases, the dimensions are distances, which are given in meters (m), centimeters (cm), and millimeters (mm). It is helpful to visualize the differences between common distances used in the English and Metric system. Figure 1.3 compares 1000 ft with 1 kilometer and 1 mile. Figure 1.3 also gives some useful conversion units.

For the most part, distances, areas, volumes, and weights are the same as given in the previous edition of the Code, except the metric units are listed first. Frequently, the metric units have been rounded to even numbers. In some areas of the Code, the metric units have not been rounded. A table has been included in *Annex A* as a quick reference to some of the common metric length conversions. The table also gives the actual metric conversions. The longer distances may have a difference between the English and Metric units as much as several in. because of rounding of the Metric conversion.

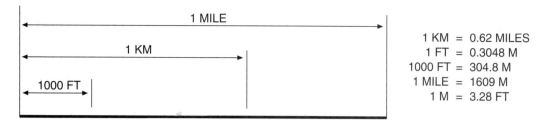

Figure 1.3   **A comparison of common distances, Metric and English, used in the electrical trade.**

Since many distance dimensions are given in the Code in millimeters, it is useful to have a comparison with the standard English units in fractions of an in. Figure 1.4 is a magnified version of the one-in. scale compared to millimeters.

Measurement units given in the Code are required to be SI units (Systeme International d'Unites) frequently called metric units. The SI units are to be followed by the units commonly used in the United States known in the Code as inch-pound units, and these units will be in parentheses. This requirement is found in *90.9(B)*, which calls for "hard conversions" from SI units to inch-pound units. "Soft conversions" for inch-pound units to SI units are permitted in some situations as given in *90.9(C)*. A soft conversion is an exact equivalent measurement in SI units and inch-pound units. Examples of hard conversions are switches to standard metric sizes such as metric conduit and tubing sizes and metric wire sizes. Another example of a hard conversion is the support of a cable within a certain distance of a box. The hard conversion may call for the support to be not more than 300 mm from the box. The SI dimension of 300 mm is a replacement for 12 in., which is actually 305 mm. For the moment, the support will be permitted to be 305 mm (12 in.) from the box, but sometime in the future the 12 in. will be deleted and the standard metric distance of 300 mm will be the maximum distance permitted. For now, either the standard SI units or the inch-pound equivalent units will be permitted according to *90.9(D)*. In this text, inch-pound units will be used followed by the SI units in parenthesis.

When trade sizes are used, these are no exact dimensions and a hard or soft conversion is not meaningful. In these cases, trade size designators are used. For example, in *300.1(C)* a trade size 1 in. conduit or tubing is given a trade size designator of 27. In this case, the 27 is the approximate inside diameter in mm. In some cases, an industry standard is to express a dimension as inch-pound units and there is no meaningful SI unit equivalent. An example is National Pipe Thread and thread taper of 3⁄4 in. per ft. The ratio of 1 to 16 given in *500.8(D)* is not a metric equivalent. *Annex A-1* of this text gives the rounded SI unit equivalent for typical dimensions used in the electrical trade.

Wire Dimensions

Conductor dimensions used in the electrical trade are overall conductor diameter, and the conductor cross-sectional area. Conductor diameter is given in millimeters (mm), and conductor cross-sectional area is given in square millimeters (mm²). Conductor diameter using in. and cross-sectional area using square in. is still provided in the Code. In the future, there will be standard conductor sizes in square millimeters (mm²) with an allowable ampacity listed for each standard size.

Wire dimensions are also given in American Wire Gauge (AWG) and in thousands of Circular Mils (kcmil). The **k** is the metric symbol for 1000. The **c** represents **circular** for Circular Mils. The **mil** is 1/1000. In the case of electrical wires, the mil represents 1/1000 of an in. The area of a wire is determined by taking the diameter of a wire using in., multiplying by 1000, and then squaring that number. Squaring a num-

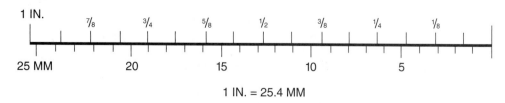

Figure 1.4   **An expanded one-in. scale compared to the equivalent dimensions in millimeters.**

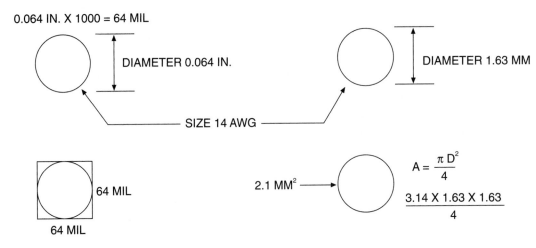

Figure 1.5   **In the inch-pound system, the diameter of a wire in mils is squared to get Circular Mils, where in the Metric system the area is in square millimeters found using the formula for the area of a circle.**

ber means the number is multiplied by itself. The old designation for thousands of Circular Mils is (MCM). The **M** is the Roman numeral for 1000. Some wires encountered in electrical wiring will be marked **kcmil** and some will be marked **MCM.** Both of these designations represent thousands of Circular Mils. Area of a wire is expressed as Circular Mils, not as a true area. Common wire sizes are listed in *Table 8, Chapter 9* of the Code. The area of the wire in square millimeters and Circular Mils is given in the table.

When the diameter of a wire is given using in. as shown in *Table 8, Chapter 9* of the Code, the diameter can be converted to mils by multiplying the diameter using in. by 1000. A mil is 0.001 in. A size 14 AWG solid wire has a diameter of 0.064 in., and therefore the diameter would be 0.064 times 1000, or 64 mils. This is illustrated in Figure 1.5.

The Circular Mil area of a wire is obtained by squaring the diameter of the wire in mils, or multiplying the diameter by itself. The area of a size 14 AWG solid wire would be 64 mils times 64 mils, or 4096 Circular Mils (4.096 kcmil) as shown in Figure 1.5. The values given in *Table 8, Chapter 9* of the Code are approximate, so they may not match exactly with a calculation.

In other parts of the world, wire sizes are in square millimeters (mm^2), but the Code does not give standard metric wire sizes or the metric equivalent sizes for standard AWG and kcmil sizes. Table 1.1 gives the metric equivalent sizes in mm^2 for standard AWG and kcmil sizes.

Table 1.1   **Metric equivalent conductor sizes for common AWG and kcmil wire sizes.**

AWG	mm²	kcmil	mm²
18	0.8	250	127
16	1.3	300	152
14	2.1	350	177
12	3.3	400	203
10	5.3	500	253
8	8.4	600	304
6	13.3	700	355
4	21.2	750	380
3	26.7	800	405
2	33.6	900	456
1	42.4	1000	507
1/0	53.5	1250	633
2/0	67.4	1500	760
3/0	85	1750	887
4/0	107	2000	1013

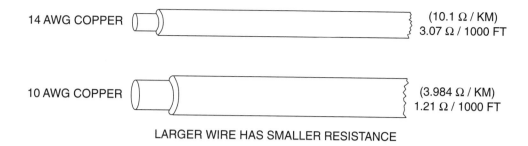

Figure 1.6   A wire with a larger cross-sectional area has a lower resistance for a given length.

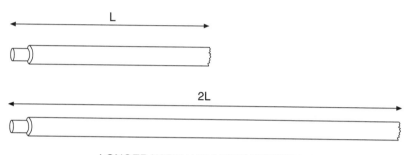

Figure 1.7   A long wire has a higher resistance than a shorter wire of the same size.

Wire Resistance

Conductor resistance is measured in ohms. The resistance of a conductor is different for different types of materials such as copper and aluminum. A conductor with a larger cross-sectional area than another conductor has a lower resistance for a given length. For example, the resistance listed for one kilometer of size 10 AWG solid uncoated copper conductor is 3.984 ohms (1.21 ohms per 1000 ft). A size 14 AWG copper conductor is smaller than a size 10 AWG, and a one kilometer (km) length of a size 14 AWG solid uncoated copper conductor has a resistance of 10.1 ohms (3.07 ohms per 1000 ft). The resistance of a conductor decreases as the cross-sectional area of the conductor increases. This is illustrated in Figure 1.6.

The resistance of a conductor increases as the length of the conductor increases. If one conductor of the same size is twice as long as another, the longer conductor will have twice the resistance of the shorter conductor. This is illustrated in Figure 1.7. The resistance of one kilometer (km) or 1000 ft of conductor at a temperature of 75°C is given in *Table 8, Chapter 9* of the Code. If the resistance of the length of conductor other than one kilometer or 1000 ft is desired, Equation 1.1 will be useful in determining the actual resistance of the conductor.

$$\text{Resistance}_{\text{Desired length}} = \frac{\text{Desired length (m)}}{1000 \text{ m}} \times \text{Resistance}_{1 \text{ km}} \qquad \textbf{Eq. 1.1}$$

$$\text{Resistance}_{\text{Desired length}} = \frac{\text{Desired length (ft)}}{1000 \text{ ft}} \times \text{Resistance}_{1000 \text{ ft}}$$

Temperature has an effect on the resistance of an electrical conductor. As the temperature of a copper or an aluminum wire increases, the resistance will also increase. The change in resistance of a wire as tem-

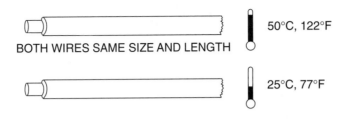

BOTH WIRES SAME SIZE AND LENGTH   50°C, 122°F

                                                   25°C, 77°F

HOTTER WIRE HAS HIGHER RESISTANCE

Figure 1.8   The resistance of a wire increases as the temperature of the wire increases.

perature increases is illustrated in Figure 1.8. A wire carrying electrical current for several hours will have a higher temperature and a higher resistance than a wire not carrying electrical current. The temperature of the area where a wire is to be installed is frequently different from one area to another. For copper and aluminum wires, and most other wires, the resistance increases as the temperature increases. The resistivity (unit value of resistance of a material) of copper and aluminum wire is different, and it increases as the temperature increases. Table 1.2 gives values of resistivity for copper and aluminum at three different temperatures. Note that the resistivity increases as the temperature increases.

Table 1.2   Resistivity of copper and aluminum conductors at different temperatures.

Conductor type	Resistivity, K		
	25°C	50°C	75°C
	(ohm mm² / m)		
Copper	0.0180	0.0197	0.0214
Aluminum	0.0294	0.0323	0.0352
	(ohm cmil / ft)		
Copper	10.79	11.83	12.87
Aluminum	17.69	19.43	21.18

The resistance of a given length of a particular size of copper or aluminum wire also can be determined at a particular temperature using Equation 1.2. The equation gives the relationship between total wire resistance, the resistivity of the wire, the cross-sectional area of the wire, and the length of the wire. The Circular Mil area of the wire is found in *Table 8, Chapter 9*.

$$R = K \times \frac{L}{A}$$ **Eq. 1.2**

R = resistance in ohms
K = resistivity of material in ohm cmil / ft (ohm mm² / m)
L = length of conductor in ft (meters)
A = area of conductor in cmil (mm²)

Example 1.1   Determine the approximate resistance of a 246 ft (75 m) length of size 3 AWG uncoated copper conductor which is at a temperature of 75°C.

Answer: Look up the area of a size 3 AWG conductor in *Table 8, Chapter 9* of the Code and find 52,620 cmil (26.67 mm²). Look up the resistivity, K, for a copper conductor in Table 1.2 at 75°C and find 12.87 ohm cmil/ft (0.0214 ohm mm² / m). Use these values in Equation 1.2 and find the resistance of 246 ft (75 m) of size 3 AWG uncoated copper conductor to be 0.060 ohms.

$$R = 12.87 \text{ ohm cmil} / \text{ft} \times \frac{246 \text{ ft}}{52,620 \text{ cmil}} = 0.060 \text{ ohm}$$

$$R = 0.0214 \text{ ohm mm}^2 / \text{m} \times \frac{75 \text{ m}}{26.67 \text{ mm}^2} = 0.060 \text{ ohm}$$

Since the resistance of 1000 ft and one kilometer of size 3 AWG uncoated copper conductor is given in *Table 8, Chapter 9*, another way to get the approximate resistance of a 246 ft (75 m) length is to multiply the resistance per one-thousand ft by 0.246 (resistance per kilometer by 0.075 km) to get 0.060 ohms (Equation 1.1).

Table 8 in *Chapter 9* gives the approximate cross-sectional area of the actual conductor in square millimeters in the second column and in circular mils in the third column. In the case of a stranded conductor, these are the approximate areas of a single strand multiplied by the number of strands. When determining the resistance of a conductor, the actual cross-sectional area of the conductor material is needed. There is an overall area column in *Table 8*. In the case of a stranded conductor, this is an area determined using the overall diameter of the strands bundled together. The overall diameter of a conductor made up of strands will be greater than the diameter of a solid conductor of the same material cross-sectional area. This can be seen by comparing the overall diameters for conductor sizes 8 AWG and smaller in *Table 8*. When determining fill of a raceway, the overall area of the conductor is needed. When determining resistance of a conductor, the actual cross-sectional area of the conductor material is needed. *Table 8* gives the actual area of conductors in both square millimeters and Circular Mils. Actual conductor area is needed for all sizes of conductors when making voltage drop calculations. Table 1.1 gives the conductor cross-sectional area in square millimeters for the complete range of conductor sizes.

Ohm's Law

It is important to understand the relationship between electrical current, voltage, and resistance of an electrical circuit. This relationship is known as Ohm's Law. Equation 1.3 is Ohm's Law arranged to solve for current. The current in amperes is equal to the voltage divided by resistance in ohms. Assume the resistance of a circuit is held constant but the voltage is changed. According to Equation 1.3, if the voltage is increased, the current will increase. If the voltage is decreased, the current will decrease. This is illustrated in Figure 1.9. Note when the voltage is doubled from 120 volts to 240 volts, the current in the circuit of Figure 1.9 doubles from 12 amperes to 24 amperes.

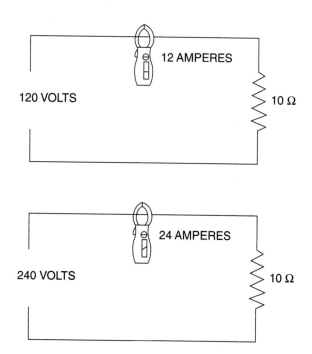

Figure 1.9 According to Ohm's Law, doubling the voltage will result in a doubling of the current in the circuit if the resistance remains constant.

$$\text{Current} = \frac{\text{Voltage}}{\text{Resistance}} \qquad\qquad I = \frac{E}{R} \qquad\qquad \text{Eq. 1.3}$$

Ohm's Law can be written to solve for the voltage when the current flow and resistance are known. Equation 1.4 is Ohm's Law arranged to solve for voltage. If the current flow through a wire increases, the voltage drop along the wire will increase. Figure 1.10. Ohm's Law for a 10-ampere flow through wires with two different resistances. The voltage drop will be doubled for the wire with twice as much resistance when the current flow remains constant.

$$\text{Voltage} = \text{Current} \times \text{Resistance} \qquad\qquad E = I \times R \qquad\qquad \text{Eq. 1.4}$$

The resistance of a circuit or a device can be determined using Equation 1.5. If the current flow and the voltage are known, the resistance can be calculated.

$$\text{Resistance} = \frac{\text{Voltage}}{\text{Current}} \qquad\qquad R = \frac{E}{I} \qquad\qquad \text{Eq. 1.5}$$

Example 1.2   The voltage drop across an incandescent lightbulb is 120 volts, and the current flow through the lightbulb is 0.8 amperes. Determine the resistance of the lightbulb filament when it is lighted.

Answer: The resistance is desired, and the voltage and current are known; therefore, use Equation 1.5.

$$\text{Resistance} = \frac{120 \text{ V}}{0.8 \text{ A}} = 150 \text{ } \Omega$$

Some equations for Ohm's Law may read impedance (Z) instead of resistance (R). Impedance is in ohms, and it is the combined effect of resistance, inductive reactance, and capacitive reactance in a circuit.

Circuits

Electricity must always travel in a circuit from a source of voltage back to the source of voltage. A simple circuit contains a voltage source, resistance of the wires and the load, and electrical current flow. The relationship that governs these quantities is Ohm's Law, which was just discussed. Most electrical circuits

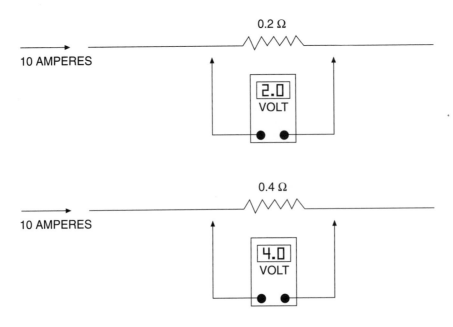

Figure 1.10   Increasing the resistance for a constant current flow will result in an increase in voltage along the circuit.

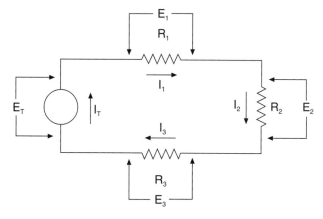

Figure 1.11 There is only one path through a circuit where the loads or resistances are arranged in series.

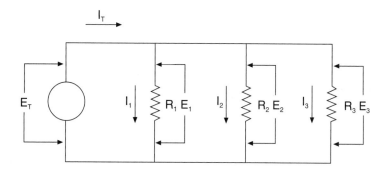

Figure 1.12 There are multiple paths through a circuit where the loads or resistances are arranged in parallel.

actually consist of several resistances in series or several resistances in parallel. Series and parallel circuits with three resistances are shown in Figure 1.11 and Figure 1.12. Troubleshooting electrical problems is much easier if the electrician has a basic understanding of series and parallel circuits.

A series circuit has all of the loads or resistances connected so there is only one path through the circuit. Therefore, if there is only one path, the electrical current is the same everywhere in the circuit, as illustrated with Equation 1.6. The combined resistance of the series circuit is the sum of the individual resistances, as represented by Equation 1.7. The voltage at the source of the circuit will drop across the individual series resistances. The sum of the voltage drops across all of the resistances in series will add up to the circuit source voltage, as indicated by Equation 1.8. Refer to the circuit of Figure 1.11 of a series circuit.

$$I_T = I_1 = I_2 = I_3 \qquad\qquad \textbf{Eq. 1.6}$$

$$R_T = R_1 + R_2 + R_3 \qquad\qquad \textbf{Eq. 1.7}$$

$$E_T = E_1 + E_2 + E_3 \qquad\qquad \textbf{Eq. 1.8}$$

Example 1.3 Assume the resistances in the series circuit of Figure 1.11 are 4 ohm, 8 ohm, and 12 ohm, and the circuit is powered at 120 volts. Determine: (1) the total circuit resistance, (2) the current flow in the circuit, and (3) the voltage drop across the 8-ohm resistor.

Answer: (1) The resistors are in series; therefore, use Equation 1.7 to find the total resistance.

$$R_T = 4\ \Omega + 8\ \Omega + 12\ \Omega = 24\ \Omega$$

(2) Total voltage and total resistance are known; therefore, use Ohm's Law (Equation 1.3) to solve for current.

$$\text{Current} = \frac{120\ V}{24\ \Omega} = 5\ A$$

(3) Ohm's Law can be used to determine the voltage drop across the 8-ohm resistor (Equation 1.4). The current determined in (2) flows through each resistor because this is a series circuit.

$$\text{Voltage} = 5\ \text{A} \times 8\ \Omega = 40\ \text{V}$$

This is the voltage drop across the 8-ohm resistor. The voltage drop across the 4-ohm resistor is 20 volts, and across the 12-ohm resistor is 60 volts. The voltage drop across all series resistors adds to 120 volts.

A parallel circuit has all of the loads or resistances arranged so there are as many paths for the electrical current to follow as there are resistances in parallel, as shown in Figure 1.12. The total electrical current flowing in the circuit is the sum of the electrical current in each path of the circuit, as indicated by Equation 1.9. The total circuit resistance is the most difficult quantity to determine. The reciprocal of the total resistance is the sum of the reciprocals of each resistance in parallel, as indicated in Equation 1.10. The total resistance of the circuit will always be less than the value of the smallest resistance in parallel. The voltage across each parallel resistance will be equal and the same as the source voltage, as illustrated in Equation 1.11.

$$I_T = I_1 + I_2 + I_3 \qquad \text{Eq. 1.9}$$

$$\frac{1}{R_T} = \frac{1}{R_1} + \frac{1}{R_2} + \frac{1}{R_3} \qquad \text{Eq. 1.10}$$

$$E_T = E_1 = E_2 = E_3 \qquad \text{Eq. 1.11}$$

Example 1.4   Assume the parallel resistances in Figure 1.12 have the values of 4 ohm, 6 ohm, and 12 ohm, and the circuit is operated at 120 volts. Determine the following for the circuit: (1) total circuit resistance, (2) total circuit current flow, and (3) the current flow through the 6-ohm resistor.

Answer: (1) This is a circuit with all resistances in parallel. The total resistance of the circuit can be determined using Equation 1.10.

$$\frac{1}{R_T} = \frac{1}{4} + \frac{1}{6} + \frac{1}{12} = 0.250 + 0.167 + 0.083 = 0.500$$

$$R_T = \frac{1}{0.500} = 2\ \Omega$$

(2) Ohm's Law can be used to determine the total current flow in the circuit (Equation 1.3).

$$\text{Current} = \frac{120\ \text{V}}{2\ \Omega} = 60\ \text{A}$$

(3) All of the resistances are in parallel, and there are 120 volts across each resistor. The current flowing through the 6-ohm resistor is found by using Equation 1.3, and dividing 120 volts by 6 ohms to get 20 amperes. Using Equation 1.3, the current through the 4-ohm resistor is 30 amperes, and 10 amperes through the 12-ohm resistor. Note the current through each resistor adds to the total current, which was 60 amperes (20 A + 30 A + 10 A = 60 A). Another way to determine the total resistance of the parallel circuit is to first find the total current. For this circuit, the total current was 60 amperes. Then use Ohm's Law Equation 1.5 to find the total circuit resistance of 2 ohms (120 V / 60 A = 2 ohm).

Power and Work Formulas

Power is the rate of doing work, as represented by Equation 1.12. It will take twice as much power to lift a weight a given number of ft in one minute as it will to lift the same weight the distance in two minutes. Horsepower (hp) is a common unit of measure of power, but watts (W) is also a unit of measure of power. The conversion from horsepower to several other units of power is shown as follows:

$$\text{Power} = \frac{\text{Work}}{\text{Time}} \qquad \text{Eq. 1.12}$$

$$\begin{aligned} 1\ \text{hp} &= 33{,}000\ \text{ft-lb/min} \\ &= 44{,}760\ \text{J/min} \\ &= 746\ \text{W} \end{aligned}$$

Electrical power in watts can be determined if the volts, amperes, and power factor of the circuit are known. For a direct current circuit or an alternating current circuit where the loads are resistance type, such as incandescent lights or electric resistance heaters, the power factor is 1.0. In the case of electric discharge lights, motors, and many other types of equipment, the power factor will be less than 1.0. A power factor meter can be used to determine the power factor of a particular circuit. Equation 1.13 is used to determine the single-phase power of a circuit, and Equation 1.14 is used to determine the power of a 3-phase circuit.

$$\text{Power}_{\text{Single-phase}} = \text{Volts} \times \text{Amperes} \times \text{Power Factor} \qquad \text{Eq. 1.13}$$

$$\text{Power}_{\text{3-phase}} = 1.73 \times \text{Volts} \times \text{Amperes} \times \text{Power Factor} \qquad \text{Eq. 1.14}$$

Example 1.5   A circuit of fluorescent luminaires (lighting fixtures) draws 15.4 amperes at 120 volts with a circuit power factor of 0.72. Determine the single-phase power in watts drawn by the circuit.

Answer: This is a single-phase power problem. Therefore, Equation 1.13 is used.

$$\text{Power} = 120\ \text{V} \times 15.4\ \text{A} \times 0.72 = 1331\ \text{W}$$

It is often useful to determine the current drawn by a particular load. This can be done by rearranging the power (Equation 1.13) to solve for current. The new form of the equation to determine current when the watts and voltage are known is given by Equation 1.15.

$$\text{Amperes} = \frac{\text{Power}}{\text{Volts} \times \text{Power Factor}} \qquad \text{Eq. 1.15}$$

The power factor is 1.0 for dc loads and resistance ac loads such as incandescent lamps and resistance heating elements. The use of Equation 1.15 is illustrated by an example.

Example 1.6   Determine the current drawn by a 100-watt incandescent lightbulb operating at its rated 120 volts.

Answer: Use Equation 1.15 and note that the power factor is 1.0 for an incandescent lightbulb.

$$\text{Amperes} = \frac{100\ \text{W}}{120\ \text{V} \times 1.0} = 0.83\ \text{A}$$

Heat is a form of work, and work is power multiplied by time. Heat is produced as electrical current flows through an electrical conductor. Equation 1.16 shows the relationship between heat (in watt-seconds), electrical current (in amperes), resistance (in ohms), and time (in seconds).

$$\text{Heat} = I^2 \times R \times \text{Time} \qquad \text{Eq. 1.16}$$

The kilowatt-hour (kWh) is a unit of measure of electrical work. A smaller unit of work is the watt-second. The watt-second is equal to one joule (J), which is the metric unit of measure of heat. The British thermal unit (Btu) is also a unit of measure of heat. The following are conversion factors for different units of measure of work or heat:

1 kWh	= 3413 Btu = 3,600,000 joules
1 watt-sec.	= 1 joule
1 watt-hr.	= 3600 joules
1 kilojoule	= 0.948 Btu

Example 1.7   A wire supplying power to an electric motor has a resistance of 0.2 ohm. Assume that a short circuit occurs at the electric motor controller, and the short-circuit current of 3500 amperes flows for one second. Determine the heat produced in the wire by the short-circuit.

Answer: The heat produced is determined by using Equation 1.16. For simplicity, assume the resistance of the wire remains constant during the short circuit.

$$\text{Heat} = 3500 \text{ A} \times 3500 \text{ A} \times 0.2 \ \Omega \times 1 \text{ s} = 2,450,000 \text{ J}$$

This is 2450 kilojoules or 2323 Btu.

Efficiency

The efficiency of a system or equipment is the output divided by the input. Efficiency of an electric motor can be calculated using Equation 1.17 provided it is known exactly how much power is being developed by the motor. The watts drawn by the motor can be determined by using Equation 1.13 for a single-phase circuit, or Equation 1.14 for a 3-phase circuit.

$$\textbf{Efficiency} = \frac{\textbf{Power Out}}{\textbf{Power In}} = \frac{\textbf{Horsepower Developed} \times \textbf{746}}{\textbf{Watts}} \qquad \textbf{Eq. 1.17}$$

Example 1.8   A single-phase electric motor draws 9.8 amperes at 115 volts with a power factor of 0.82. The motor is developing $1/2$ horsepower. Determine the efficiency of the motor.

Answer: Efficiency is determined by dividing the power developed by the input power to operate the motor using Equation 1.17. The power out of the motor is in horsepower; therefore, it must first be converted to electrical power.

$$\textbf{Efficiency} = \frac{\textbf{0.5 hp} \times \textbf{746 W/hp}}{\textbf{9.8 A} \times \textbf{115 V} \times \textbf{0.82}} = \textbf{0.40 or 40\%}$$

Student Practice Problems

These practice problems on electrical fundamentals will help improve skills at working with the basic concepts of electricity previously discussed. Look for the equation from the previous discussion that relates to the problem. It is helpful to list the known information and the quantity desired. Then try to find an equation using that information. Refer to the example problems for help.

1. The resistance of 328 ft (100 m) of size 4 AWG uncoated copper wire with resistance as given in Code *Table 8, Chapter 9* is:
    A. 0.050 ohms.   B. 0.101 ohms.   C. 0.257 ohms.   D. 1.250 ohms.

2. You observe a technician using an oscilloscope to measure the voltage at an outlet. The value of the peak voltage of the sine wave on the screen is 167 volts. If you were to measure this same voltage with your rms meter, the rms voltage would be:
    A. 95 volts.   B. 110 volts.   C. 118 volts.   D. 130 volts.

3. A baseboard electric heater operates at 240 volts and draws 3.33 amperes. The resistance of the heating element is:
    A. 72 ohms.   B. 95 ohms.   C. 110 ohms.   D. 240 ohms.

4. The current drawn by a 200-watt incandescent lamp operating at 120 volts is:
    A. 0.50 ampere.   B. 1.00 ampere.   C. 1.36 amperes.   D. 1.67 amperes.

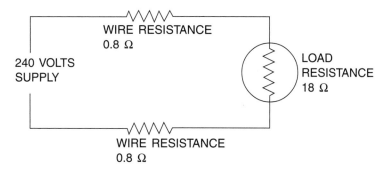

Figure 1.13   A typical series circuit is one in which the resistance of the wire supplying the load is in series with the resistance of the load.

5. An 18-ohm load is connected to a 240-volt power source using wires that each have a resistance of 0.8 ohm. The circuit is shown in Figure 1.13. The voltage across the terminals of the load is:
    A. 120 volts.   B. 130 volts.   C. 220 volts.   D. 240 volts.

6. A 120-volt circuit supplies two loads connected in parallel. One load has a resistance of 20 ohms, and the other load has a current flow of 4 amperes. The circuit is shown in Figure 1.14. The total current flowing in the circuit is:
    A. 4 amperes.   B. 5 amperes.   C. 7.5 amperes.   D. 10 amperes.

7. The total resistance of the circuit with the parallel loads shown in Figure 1.14 is:
    A. 6 ohms.   B. 12 ohms.   C. 18 ohms.   D. 24 ohms.

8. A 3-phase electric motor draws 68 amperes and operates at 230 volts with a power factor of 0.78. The power in watts drawn by the motor is:
    A. 11,200 watts.   B. 16,350 watts.   C. 21,105 watts.   D. 28,654 watts.

9. The 230-volt, 3-phase electric motor of Problem 8 is developing 25 horsepower. The efficiency of the motor is:
    A. 56%.   B. 78%.   C. 81.6%.   D. 88.4%.

10. A house is operating six 100-watt incandescent lamps, four 60-watt lamps, and four 40-watt lamps for six hours. The amount of electrical energy used during the six hours is:
    A. 6 kWh.   B. 8.4 kWh.   C. 10 kWh.   D. 12.5 kWh.

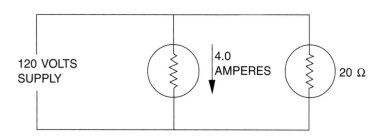

Figure 1.14   A typical parallel circuit is one in which the loads are all connected to the same voltage source, such as the luminaires (light fixtures) in a room.

CODE DISCUSSION

A person in the electrical trade who makes proper use of the Code must understand the background of the Code. Important information about the use of the Code is contained on the inside front cover and inside back cover. The first page of the Code and the last several pages of the Code also contain important information about the use of the Code, as well as the process by which it is maintained. The following information is from the front of the Code: *A vertical line in the margin indicates the text was reworded from the previous edition of the Code or the text is completely new. A bullet in the margin means that one or more paragraphs was deleted from that location. Sometimes that material was moved to another location in the Code. A superscript "x" indicates the material was extracted from another document. In order for that material to be changed, the change must be acted upon by the group that is responsible for the original document. A time table for revising the Code is given in the back of the Code along with a form for submitting a proposal to make a change in any section or article.*

Article 80 is a set of rules available for adoption by a governmental jurisdiction for administering and enforcing the electrical code in that jurisdiction. *Article 80* is not to be considered to be a part of the Code unless specifically approved by a governmental jurisdiction. A similar set of administrative rules was available in the past as document *NFPA 70L*. *NEC® 80.9* states to which installations the new electrical code applies. For example, the new code applies to additions, alterations, and repairs, but not to the existing building unless the new work causes the existing building wiring to be unsafe. *NEC® 80.13* states what authority is placed upon the enforcing agency. For example, in *80.13(1)*, the enforcing agency is permitted to make interpretations where a section of the Code may not be clear. This same rule is covered in *90.4* and is not new. *NEC® 80.13(13)* gives the enforcing agency the authority to order wiring or equipment be made exposed if it was concealed before an inspection could be performed. The makeup and duties of an electrical board are covered in *80.15*. The permit process is covered in *80.19*, and plan review requirements are covered in *80.21*. Qualifications of electrical inspectors are covered in *80.27*. Key points are that the inspector have experience in the electrical trade and the inspector be certified by a nationally recognized inspector certification program. There is no requirement in *Article 80* that the inspector hold an electrical license.

Article 90 sets the basic ground rules for understanding the remainder of the Code. The entire Code is directed at meeting the purpose stated in *90.1* which is the "practical safeguarding of persons and property from the hazards arising from the use of electricity." Following the rules of the Code will mean the installation is "essentially free from hazard." *NEC® 90.2* lists the installations covered by the Code and those that are not covered. Installations made on private or public property by a utility, such as area lighting, and is under the complete control of that utility for installation and maintenance, is generally not considered to be covered by the Code according to *90.2(A)(4)* and *90.2(B)(5)*. *NEC® 90.4* gives the local inspection agency authority to make interpretations when there seems to be a debate as to the meaning of a Code section. *NEC® 90.5(C)* states that fine print notes (FPN) are only explanatory and not enforceable. *NEC® 90.9* gives the Code rules relative to metric units. The Code will permit trade size designators rather than true metric conversions, such as in the case of raceway sizes. In some cases, English units only will be given where it is an established industry practice. It is very important to note that *90.9(D)* authorizes the use of approximate metric conversions. See the table of distances in *Annex A* of this text. In some cases, the Metric dimension and the English dimension may differ by several in. For example, "in sight from" is defined as a maximum distance of 50 ft (15 m). As illustrated in Figure 1.15, the difference between 15 m and 50 ft is $9^7/16$ in. Whether this be used as a "not greater than" or a "not less than" requirement, either the English or Metric dimension is considered meeting the intent of the Code. A disconnect located 15.24 m from a motor controller is still considered to be located within sight of the controller according to *90.9(D)*.

Article 100 provides definitions of terms that are generally used in more than one article of the Code. If the term is used only in a specific article, the definition is usually provided in that article. There are several definitions in particular that will be helpful in understanding various sections throughout the Code. *NEC® 110.2* requires that conductors and equipment installed according to the Code shall be approved. **Approved** means acceptable to the authority having jurisdiction. *NEC® 110.3* gives a list of criteria the inspector can use to evaluate an installation or equipment for approval. **Listing** and **labeling** are given as criteria for judging suitability of equipment. Those definitions are in *Article 100*. Another important definition is **continuous load**. Any load that is expected to operate at a maximum for three hours or more is considered to be a continuous load.

A multiwire branch-circuit is common and there are several specific rules that apply to this type of circuit. A **multiwire branch-circuit** (branch-circuit, multiwire) has two or more ungrounded conductors with

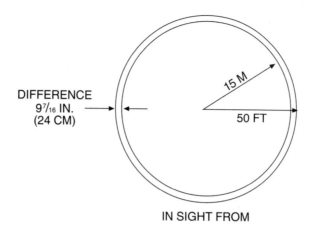

Figure 1.15 Approximate Metric conversions are permitted in the Code, and where there is a difference such as 50 ft (15 m), both dimensions are considered as meeting the intent of the Code.

a voltage between them, and share a common grounded (neutral) conductor as illustrated later in Figure 1.19. Two terms that can be confusing are receptacle and receptacle outlet. A **receptacle** is defined as a single contact device for cord and plug connected equipment. A **receptacle outlet** is a point on the electrical system where there may be one or more receptacles or other devices.

Article 110 contains general information about electrical installations. A section that is important for the sizing of conductors for a branch-circuit or feeder is *110.14(C)*. When determining the minimum size of conductor for a circuit, the insulation on the conductors frequently has a higher temperature rating than the conductor terminations. Conductor terminations may be marked with the maximum temperature rating, but often they are not marked. It is in the case when the termination temperature is not marked that *110.14(C)* becomes important. The basic rules are illustrated in Figure 1.16. According to *110.14(C)(1)(a)*, if a termination is not marked otherwise, it has a 60°C rating if the overcurrent device protecting the circuit is rated 100 amperes or less or if the conductor is size 1 AWG or smaller. *NEC® 110.14(C)(1)(b)* specifies that if the overcurrent device protecting the circuit is rated greater than 100 amperes or if the conductor is size 1/0 AWG or larger, the termination is required to have a minimum rating of 75°C unless otherwise marked.

Article 110 also gives the minimum work space requirements of electrical equipment. For electrical equipment operating at not over 600 volts, the minimum work space requirements are provided in *110.26*.

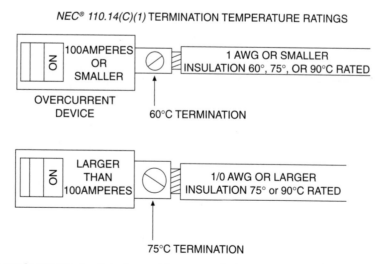

Figure 1.16 If a conductor termination does not have a marked temperature rating, *110.14(C)(1)* sets the temperature rating based upon the rating of overcurrent device or conductor size.

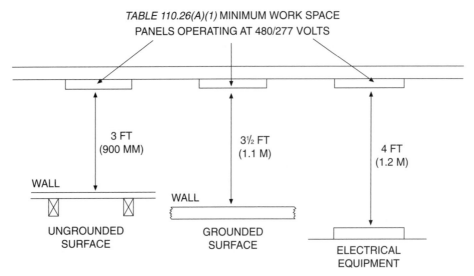

Figure 1.17 **A minimum clear working space is required in front of electrical equipment based upon the voltage to ground in the equipment and whether it is facing an ungrounded surface, a grounded surface or equipment, or other equipment with exposed live parts.**

Figure 1.17 illustrates minimum working space clear distances required to be provided in front of electrical equipment operating at over 150 volts to ground. For equipment operating over 600 volts, the minimum required working clearances are given in *110.34*.

   Article 200 primarily covers the means of identification of the grounded-circuit conductor (usually the neutral), and grounded-circuit conductor terminations on devices. *NEC® 200.6(B)* permits a conductor with insulation with a color other than white or green to be identified as a grounded conductor by adding a white marking, which completely encircles the conductor, at each termination at the time of installation. In the case of switch loops, where multiconductor cable is used, *200.7(C)(2)* permits a white or gray conductor to be re-identified as an ungrounded conductor at all locations where the conductor is visible. The re-identified white conductor is also required to be the feed, supplying the source of power to the switch. Under no circumstances is it permitted to be the switched return conductor to the load. This rule is illustrated in Figure 1.18.

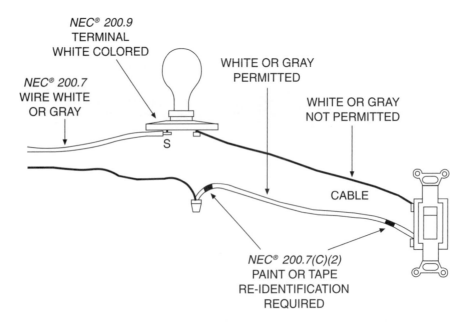

Figure 1.18 **Grounded-circuit conductors shall be white or gray, but a white or gray conductor when part of a cable assembly is permitted to be re-identified as an ungrounded conductor when used as the feed to the switch.**

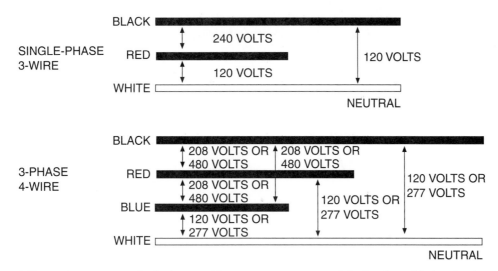

Figure 1.19 A common neutral wire is run with two or more ungrounded wires in multiwire circuits.

Article 300 is another section that covers general wiring installation methods. *NEC® 300.4* covers the protection of conductors passing through structural members. There are also minimum clearance requirements for the installation of cables and some raceways run on the surface of studs and joists. The issue is the protection of the conductors from damage from nails and other fasteners. Underground installations are covered in *300.5*. *Table 300.5* gives the minimum depth of burial permitted for different types of occupancies and different conditions. For farm and some commercial installations, *300.7* is particularly important for conditions in which a raceway is exposed to different temperatures. There can be problems of condensation in the raceway system, or there can be problems with damage to the conduit or separation of the conduit due to expansion and contraction.

A multiwire branch-circuit is defined in *Article 100* as a circuit that has two or more ungrounded conductors with a potential between them and sharing a common grounded conductor. Typical single-phase and 3-phase multiwire branch-circuits are illustrated in Figure 1.19. A requirement that deals with multiwire circuits is *300.13(B)*. Removing a device in a multiwire circuit is not permitted to interrupt the neutral wire. This is illustrated in Figure 1.20. Numerous other miscellaneous, but important, requirements are covered such as supporting conduit, preventing induced currents, traversing ducts, and preventing fire spread.

A minimum free length of 6 in. (150 mm) of conductor is required to be provided at each box to make up splices and to connect to devices and luminaires (fixtures). This rule is found in *300.14*. If the box has an opening dimension of less than 8 in. (200 mm), then there shall be at least 3 in. (75 mm) of conductor

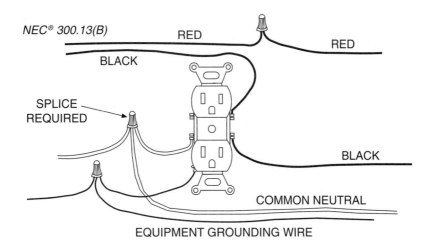

Figure 1.20 Device removal in multiwire circuits is not permitted to open the neutral conductor.

NEC® 300.14 MINIMUM LENGTH OF FREE CONDUCTOR;
MINIMUM OF 3 IN. (75 MM) OUTSIDE OF BOX OPENING

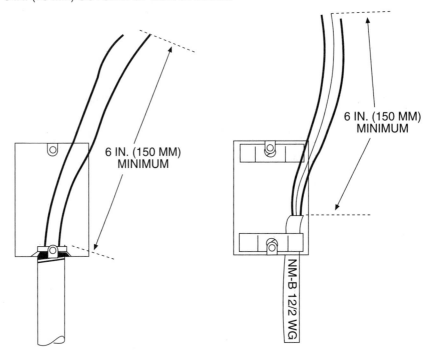

6 IN. (150 MM)
MINIMUM

6 IN. (150 MM)
MINIMUM

NM-B 12/2 WG

Figure 1.21   The 6 in. (150 mm) minimum length of free conductor is measured from the point where the conductor emerges from the raceway or cable sheath, and there must be a minimum of 3 in. (75 mm) of conductor extending outside of boxes with an opening less than 8 in. (200 mm).

extending outside the box. This would mean that in the case of a box that is more than 3 in. (75 mm) deep, the actual free length of conductor may be required to be more than 6 in. (150 mm). The measurement of the free length of conductor is from the point where the conductors emerge from the raceway, or the cable jacket as illustrated in Figure 1.21.

Article 310 is one of the most frequently used articles of the Code. This article contains the ampacity tables used to determine the minimum size of wire permitted for a circuit or load. Several methods can be used to determine the ampacity of electrical conductors, as stated in *310.15*. Generally, *Table 310.16, Table 310.17, Table 310.18, Table 310.19,* and *Table 310.20* will be used to determine the minimum size of wire for a particular circuit. *Table 310.16* can be used for all situations of not more than three wires in cable or raceway or in the earth. *Table 310.17* applies to single insulated wires in free air. This would be the case for single overhead conductors mounted on insulators. *Table 310.20* is for insulated conductors supported on a messenger such as triplex cable. *NEC® 310.15(C)* provides a method permitted to be used to determine the ampacity of a conductor when the calculations are performed by or checked by an engineer. A number of ampacity tables are contained in *Annex B* of the Code. These tables are not generally permitted to be used to determine the ampacity of a conductor unless the governmental jurisdiction specifically permits the *Annex B* ampacity tables to be used. Several examples will help in understanding which table is to be used for which situation. Keep in mind that the ampacity tables only apply directly to raceways and cables that contain not more than three current-carrying conductors. If there are more than three wires, or if the wires are exposed to a temperature greater than 30°C or 87°F, the ampacity as found in *Table 310.16* or *Table310.17* shall be reduced. Conductor derating for ambient temperature and number of wires in raceway or cable is covered in *Unit 2*.

The allowable ampacity of a conductor depends upon the size of the conductor, the type of insulation, and the temperature rating of the splices and terminations. The allowable current must be limited to a value that will not allow the temperature of the conductor to exceed the withstand temperature rating of the insulation. At points where conductors are spliced or terminated into a lug or under a screw terminal, the conductor operating temperature is not permitted to be higher than the maximum temperature limit of the termination. If a termination operates at more than the maximum temperature rating, it most likely will deteriorate and eventually fail. A fire may result. If the conductor terminations in a circuit are rated 75°C, and the conductor insulation has a different temperature rating, such as THHN at 90°C, generally the column of the table with a temperature matching the lowest temperature rating is used to find the allowable ampacity of the conductor for that circuit. That is not necessarily the case when applying adjustment and correction factors.

Another factor that determines the actual ampere rating of a conductor is the rating of the overcurrent device (fuse or circuit breaker) used to protect the conductor from overheating. The actual load or calculated load for a circuit is not permitted to exceed the rating of the overcurrent device no matter what size of wire is used. If a 20-ampere circuit breaker protects a circuit where the conductor is size 10 AWG copper with Type THHN insulation, the calculated load is not permitted to exceed 20 amperes, and if the actual load is continuous, it is not permitted to exceed 16 amperes for this circuit breaker. All of these factors must be considered when selecting the minimum size conductor for a particular circuit.

Copper conductor sizes 14, 12, and 10 AWG, and aluminum conductor sizes 12 and 10 AWG, when used for most circuits have an overcurrent device maximum limit. In a previous edition of the Code, this limit was given in a footnote at the bottom of *Table 310.16* and *Table 310.17*. Those overcurrent device limits are now in *240.4(D)*. The overcurrent device rating for size 14 AWG copper wire is 15 amperes, for size 12 AWG it is 20 amperes, and for size 10 AWG it is 30 amperes.

Example 1.9   Three size 10 AWG, Type THWN insulated copper conductors are contained in a conduit in a building. Determine the ampacity of the conductors if the ambient temperature does not exceed 30°C and all terminations are rated 75°C.

Answer: *Table 310.16* applies to this situation. Find the type of insulation in the copper conductor part of the table (75°C column), and find the ampacity that corresponds to size 10 AWG wire, which is 35 amperes.

Example 1.10   Type USE Aluminum Cable, size 2 AWG, consisting of three current-carrying conductors, is directly buried in the earth at a depth of 24 in. (600 mm). Determine the ampacity of the cable if all terminations are rated 75°C.

Answer: *Table 310.16* applies in this situation. Find the ampacity of size 2 AWG aluminum in the column headed with conductor insulation Type USE. The cable is rated at 90 amperes.

Example 1.11   Type XHHW aluminum wire, size 4 AWG, is installed overhead in free air as single conductors on insulators. Assume that the ambient temperature does not exceed 30°C and that at times this would be considered a wet location. Assume the terminations at the drip loop are rated at 90°C. Determine the ampacity of the aluminum wire in this situation.

Answer: *Table 310.17* is used in the case of single conductors in free air. Note that there are two columns of aluminum wire with Type XHHW insulation listed, the 75°C column and the 90°C column. It will be necessary to refer to Type XHHW insulation in *Table 310.13* for an explanation of when to use the two columns. Because this is considered at times to be a wet location, the 75°C column is used. The size 4 AWG aluminum wire is rated at 100 amperes.

When there are more than three current-carrying conductors in a raceway, cable, or trench in earth, the allowable ampacities in *Table 310.16* require an adjustment. *NEC® 310.15(B)(2)* gives the adjustment factors depending upon the number of current-carrying conductors. *NEC® 310.15(B)(4)* explains when the neutral is required to be counted as a current-carrying conductor. This subject will be discussed in detail in the next unit. *NEC® 310.60* provides directions for determination of ampacity of conductors operating at 2001 to 35,000 volts.

Article 320 covers Type AC Armored Cable. Type AC Cable generally consists of two or three insulated conductors and an equipment grounding conductor within a flexible metallic covering. The uses not permitted are covered in *320.12*. The cable shall be supported within 12 in. (300 mm) of the end of each run and at intervals not to exceed $4^{1}/2$ ft (1.4 m) as illustrated in Figure 1.22. The cable is subject to damage similar to other cables, therefore, *320.15* requires that exposed runs of Type AC Cable closely follow the surface of the building, or that running boards be provided. In the case of an accessible ceiling space, Type AC Cable is permitted to be secured by support wires that are provided for the sole support of electrical wiring. As for exposed work, the cable in an accessible ceiling is required to be supported within 12 in. (300 mm) of each termination, and at intervals not to exceed $4^1/2$ ft (1.4 m). Type AC Cable is permitted to be run through bored or punched holes in wood or metal framing members as well as notches in wood members. For other than vertical runs through framing members, the maximum support distance is $4^1/2$ ft (1.4 m). The cable is not required to be secured to the framing members. The maximum distance from a termination to a secure means of support is not permitted to exceed 12 in. (300 mm).

Article 324 provides specifications for the materials and installation for flat conductor cable to be placed under carpet squares. This is a technique for extending power to work stations in a large open room. Definitions are given in 324.2. Uses permitted and not permitted are covered in 324.10 and 324.12. The maximum voltage permitted for flat conductor cable is 300 volts between ungrounded conductors and a maximum of 150 volts from any ungrounded conductor to the grounded conductor. General use circuits are not permitted to be rated in excess of 20 amperes. These voltage and current requirements are stated in 324.12. Flat Conductor Cable is permitted to be installed only under carpet squares defined in 324.41 as having a maximum dimension of 36 in. (900 mm) on each side.

Article 328 provides basic specifications and uses permitted for medium-voltage cable that is intended for use where energized at from 2001 to 35,000 volts. Type MV Cable is available as a single-conductor

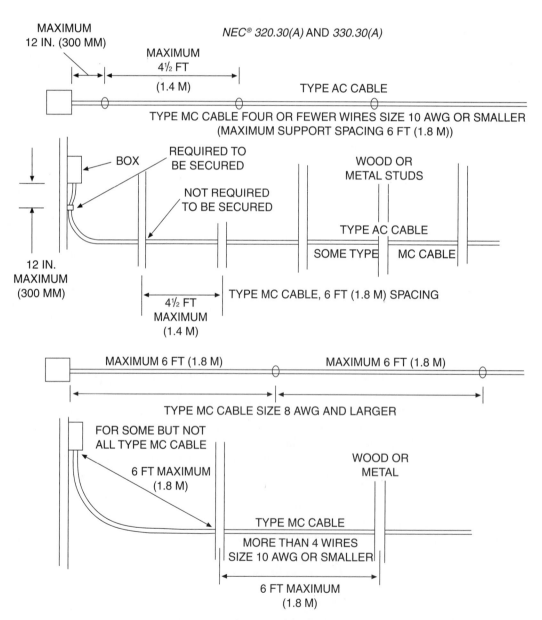

Figure 1.22   Type AC Armored Cable is required to be supported within a distance of 12 in. (300 mm) of a box, and Type MC Metal-Clad Cable is required to be supported within a distance of 6 ft (1.8 m) of a box.

cable or as a multiconductor cable. The cable is permitted to be installed in raceway, cable trays, direct burial according to *310.7*, and supported by a messenger in air as specified in *Article 396*. When installed in tunnels, the rules for installation are found in *Article 110, Part IV*. General installation requirements and clearances from live parts and equipment are given in *Part III* of *Article 110*. The method of determining the ampacity of the conductors rated 2001 to 35,000 volts is specified in *310.60(B)*. It is permitted to use *Table 310.67* through *Table 310.86*, or the formula in *310.60(D)* under engineering supervision. *Annex B* gives examples of the use of the formula under engineering supervision.

Article 330 provides information on the uses permitted, installation, and construction of metal-clad Type MC Cable. The conductors are contained within a Flexible Metallic Sheath. It is permitted to be used for services, feeders, and branch-circuits. Type MC Cable is available with conductors operating at not more than 600 volts and also with conductors operating at more than 600 volts. *NEC® 330.30* requires that Type MC Cable be supported at intervals not to exceed 6 ft (1.8 m). If the conductors are size 10 AWG and smaller and there are four or fewer conductors, Type MC Cable is required to be supported within 12 in. (300 mm) of terminations. This is illustrated in Figure 1.22. If there are five or more conductors in the cable, or if conductors are size 8 AWG and larger, the distance from the termination to the first support is permitted to be up to 6 ft (1.8 m). Type MC Cable is permitted to be run through bored or punched holes in wood or metal framing members, as well as notches in wood members. For other than vertical runs through framing members, the maximum support distance is 6 ft (1.8 m). The cable is not required to be secured to the framing members, and the distance from a termination to the first framing member is permitted to be up to 6 ft (1.8 m). It is also permitted to be used as a luminaire (lighting fixture) or equipment whip in lengths up to 6 ft (1.8 m) within accessible ceiling spaces. In this case, the Type MC Cable is secured only at the box and at the luminaire (lighting fixture) or equipment supplied.

Article 332 covers the construction and installation of Type MI Mineral-Insulated, Metal-Sheathed Cable. Type MI Cable has the insulated conductors contained within a liquidtight and gastight continuous sheath with a densely packed mineral insulation filling the space between the conductors and the copper or alloy steel sheath. The cable can be cut to length with special fittings applied at the terminations. Type MI Cable has a wide variety of applications, and it is permitted to be fished into existing building spaces.

In the case of Type MI Cable, the termination fittings have a maximum temperature rating, generally 90°C. The cable itself is capable of operating at a much higher temperature. Rules for determining the allowable ampacity of Type MI Mineral-Insulated, Metal-Sheathed Cable are found in *332.80*. Temperature adjustment of the cable itself is generally not considered a problem unless heat is conducted to the terminal fittings such that their maximum temperature rating would be exceeded. *NEC® 332.80* permits Single-Conductor MI Cable assembled into a bundle to have the allowable ampacity to be determined from *Table 310.17*, even though that table is for single-conductors in free air. This is permitted as long as the cable is installed so that the maximum terminal fitting temperature is not exceeded.

Article 334 deals with Nonmetallic-Sheathed Cable, usually Type NM. *NEC® 334.112* specifies that the conductors within the nonmetallic sheath shall have insulation rated at 90°C. These cables are marked Type NM-B. *NEC® 334.80* requires the allowable ampacity of a cable to be determined using the 60°C column of *Table 310.16*. Uses not permitted are given in *334.12*, and locations where Type NM-B cable is permitted to be installed are not directly stated. It is necessary to understand the basic types of construction in order to determine where Nonmetallic-Sheathed Cable is permitted to be installed. It is not permitted to be installed in buildings that are of Type I or Type II construction. Type I and Type II structures are required to be of noncombustible construction materials. Type III construction is a combination of combustible and noncombustible materials, Type IV is permitted to have heavy timber construction, and Type V is permitted to be wood frame construction.

Nonmetallic-Sheathed Cable is permitted to be used as a wiring method for one-family, two-family, and multifamily dwellings that are permitted to be of Type III, IV, or V construction. Type NM-B cable is permitted to be used as a wiring method concealed in walls, floors, and ceilings with a 15-minute fire rating for nondwelling buildings that are of Type III, IV, or V construction. It is not permitted to be installed as an exposed wiring method above suspended or dropped ceilings except for one-family, two-family, or multifamily dwellings of Type III, IV, or V construction. In *334.30(B)(2)* luminaires (lighting fixtures) and equipment installed in suspended ceilings of one-family, two-family, and multifamily dwellings are permitted to be supplied using Type NM-B cable provided the distance from the last point of support to the luminaire (fixture) or equipment does not exceed 4½ ft (1.4m).

It is important to protect Nonmetallic-Sheathed Cable from damage. One form of damage is bending the cable in a small radius. The minimum radius of bend is five times the diameter of the cable, as stated in *334.24*. *NEC® 334.30* requires Type NM-B Cable to be secured in place at intervals not exceeding 4¹/₂ ft (1.4 m) and within 12 in. (300 mm) of terminations. According to *314.17(C) Exception*, the case of a non-

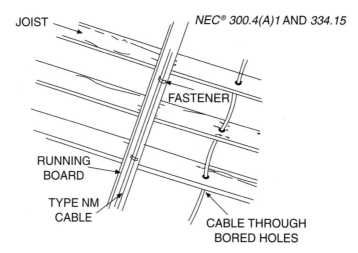

Figure 1.23 Small sizes of Type NM-B Cable are required to be attached to running boards or run through bored holes in joists.

metallic single-gang box, where the cable is not secured directly to the box, the Nonmetallic-Sheathed Cable is required to be secured at a distance from the box of not more than 8 in. (200 mm) measured along the cable. Cables run through holes in metal or wood framing members are considered to meet the requirement of being supported and secured. Nonmetallic-Sheathed Cable is permitted to be attached to the underside of joists in basements for 2-wire size 6 AWG, or 3-wire size 8 AWG. Cables are permitted to be installed directly to the surface of joists and rafters in attics and crawl spaces, but protection is required near access openings as specified in *334.23,* and shown in Figure 1.23. *NEC® 334.30 Exception 3* permits Nonmetallic-Sheathed Cable to be run unsupported except at the terminations in lengths up to $4^1/2$ ft (1.4 m) to supply power to a luminaire (fixture) or equipment in an accessible ceiling.

Article 336 covers the installation of Type TC Power and Control Tray Cable. The uses not permitted are the main emphasis of this article. The ampacity of Type TC Cable is specified in *336.80.* If the cable is smaller than size 14 AWG, the ampacity is determined using the rules of *392.11.* Determination of conductor allowable ampacity when installed in cable trays will be discussed in *Unit 12.* Type TC Cable is available as a multiconductor cable and as single-conductor cable. Most single-conductor cables size 1/0 AWG and larger have a Type TC rating in addition to the normal conductor insulation rating.

Article 338 covers the use of Type SE Service-Entrance Cable and Type USE Underground Service-Entrance Cable. Type SE Cable is permitted for use as circuits and feeders within a building. The common types of Service-Entrance Cable in use are Type SE style U with two insulated conductors and a bare conductor within the nonmetallic covering, and Type SE style R, which has three insulated conductors and a bare equipment grounding conductor within the nonmetallic outer covering. These cable types are shown in Figure 1.24. According to *338.12(A)(3)(b),* Type SE Cable when used for interior wiring is installed according to the same rules as for Type NM Cable.

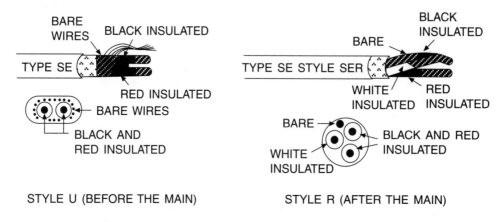

Figure 1.24 Type SE Service-Entrance Cable is available as style U or style R.

Article 340 covers Type UF Underground Feeder and Branch-Circuit Cable for use underground and within buildings. The installation of this cable within buildings is required to follow the requirements of Type NM-B Nonmetallic-Sheathed Cable in *Article 334*. It is important to note that there will be two grades of this cable in use. When Type UF Cable is installed within a building, the individual insulated wires in the cable are required to have a 90°C insulation temperature rating. This requirement on the 90°C rated insulation is stated in *340.12(1)* where the conductor requirements for Type UF Cable are to be the same as for Type NM-B Cable. Cables are available with wires that have an insulation temperature rating of only 60°C. Both are permitted to be installed underground.

Article 382 deals with nonmetallic extensions, typically a Nonmetallic-Sheathed Cable installed on the surface of walls or ceilings. It is permitted to be attached as a nonmetallic extension to the exposed surface of walls and ceilings to extend power from an existing outlet of a 15- or 20-ampere circuit. This technique provides a means of extending wiring without having to enter the wall or ceiling cavity. This type of material is permitted to be messenger supported as an aerial cable for industrial buildings where the nature of the application requires a flexible means of connecting power to equipment.

Article 394 covers a type of wiring called Concealed Knob-and-Tube Wiring. This type of wiring was used in buildings in the early years of wiring. Single-insulated conductors were held in place with insulators called knobs and cleats. The conductor then passed through a structural member and ran inside a porcelain tube. This type is encountered in existing older buildings.

Article 590 provides rules for the installation of wiring for temporary electrical power or lighting. This temporary power may be used for construction, remodeling, and similar activities. It may also be used to supply decorative lighting not generally intended for use for more than 90 days. *NEC® 590.4(A)* requires a service meeting the requirements of *Article 230* to supply a temporary wiring system. For a temporary electrical system at a construction site, *590.4(D)* does not permit receptacles to be installed on the same circuit with temporary lighting. In the case of a multiwire branch-circuit, all ungrounded conductors are required to be simultaneously disconnected according to *590.4(E)*. This means single-pole overcurrent devices as disconnects are not permitted for multiwire branch-circuits. *NEC® 590.6(A)* requires ground-fault circuit-interrupter (GFCI) protection for all receptacles on 15-, 20-, or 30-ampere rated 125-volt circuits that are not part of the permanent wiring system. GFCI protection is also required for receptacles on circuits of the permanent wiring system that are used for temporary electrical power. This same section also permits listed GFCI cord sets to suffice as the required protection.

Table 8, in *Chapter 9*, is useful in that it provides a comparison of conductor sizes in AWG or kcmil and the equivalent area in square millimeters. The table also gives the cross-sectional area of the conductors using square in. This information is necessary when determining the minimum size of raceway when one or more of the conductors are bare. The table also gives the approximate resistance of copper and aluminum conductors. This information is useful especially when determining the voltage drop caused by conductors. Voltage drop will be covered later in the text. Resistance is affected by the temperature of the conductor. The temperature of a conductor will rise when current is flowing. *Table 8* resistances are based upon an operating temperature of 75°C. A formula is provided in the footnote to the table to adjust the resistance for another operating temperature.

Table 9 in *Chapter 9* provides information about the inductive reactance of conductors when supplying loads where the power factor is less than one. This information is used to determine the approximate voltage drop on conductors. When conductors are larger than size 1/0 AWG supplying ac loads with a power factor less than one, the inductive reactance of the conductor begins to have a significant effect on the voltage drop in addition to the resistance of the conductor. Even the type of raceway has an effect on the inductive reactance. The effect is greater in Rigid Steel Conduit as compared to Rigid Nonmetallic Conduit. A copper feeder conductor supplying an ac load with a power factor of 0.85 using size 500 kcmil conductors run in Rigid Steel Conduit will have as much voltage drop due to inductive reactance as due to resistance of the conductor. In this case, using the values from *Table 8* would give an inaccurate approximation of voltage drop. Generally, *Table 8* resistances are adequate for calculations of voltage drop for conductor sizes up to 1/0 AWG. It is recommended that impedance values from *Table 9* be used for voltage drop calculations for larger conductor sizes. There is a formula in the footnote of *Table 9* that can be used to determine the approximate impedance of conductors when supplying loads with a power factor other than 0.85. Some experience with using trigonometry is needed to apply that formula.

SIZING CONDUCTORS FOR A CIRCUIT

The size of conductors for a branch-circuit or a feeder depends upon the load to be served and the rating of overcurrent device chosen for the circuit. The rule for determining the minimum size overcurrent device permitted for a branch-circuit is found in *210.20(A)*. In the case of a feeder the minimum overcurrent device rating is determined according to *215.3*. In either case, the rule is the same. The overcurrent device shall have a rating not less than the noncontinuous load plus 1.25 times the continuous load. *NEC® 240.6(A)* lists the standard ratings of overcurrent devices. Choose an overcurrent device rating that is larger than the calculation of *210.20(A)* or *215.3*. The following example will illustrate the point:

Example 1.12 A feeder supplies 92 amperes of continuous load and 70 amperes of noncontinuous load. Determine the minimum rating of overcurrent device permitted for this feeder. The feeder consists of three Type THHN copper current-carrying conductors in raceway.

Answer: This is a feeder, so use the rule of *215.3*. Multiply the 92 amperes of continuous load by 1.25 and add to the noncontinuous load of 70 amperes to get 185 amperes.

$$92 \text{ A} \times 1.25 = 115 \text{ A}$$
$$70 \text{ A} \times 1.00 = \underline{70 \text{ A}}$$
$$185 \text{ A}$$

The overcurrent device is required to have a rating not less than 185 amperes. From *240.6(A)*, the next higher standard rating is 200 amperes.

The next step is to determine the minimum size conductor for the feeder. *NEC® 240.4* requires the conductor be protected in accordance with the conductor ampacity as given in *310.15*. *NEC® 215.2(A)* specifies that for a feeder the minimum conductor size is not permitted to be smaller than the noncontinuous load plus 1.25 times the continuous load. This minimum size is determined without any consideration of adjustment or correction factors. Because the wiring method is conductors in raceway, *Table 310.16* will be used to determine conductor ampacity. The conductor insulation is 90°C rated, but it is necessary in this case to use the 75°C column of *Table 310.16*. The reason is that conductor termination temperature has not been specified for the circuit, in which case it is necessary to proceed according to the rules in *110.14(C)(1)(b)*. If the overcurrent device rating is greater than 100 amperes or the conductor is larger than size 1 AWG, the terminations are rated 75°C unless otherwise specified. The minimum size copper THHN conductor permitted for this feeder with a 200-ampere overcurrent device is 3/0 AWG. If this had been a branch-circuit rather than a feeder, the method of determining the minimum conductor size would have been the same, except the rule for branch-circuits is found in *210.19(A)*.

Table 310.16 gives the allowable ampacity of conductors when the conductors are run in raceway, in cable, or directly buried in the earth. If circuit conductors operate continuously carrying the listed allowable ampacity, the conductor insulation temperature rating given at the top of the column will not be exceeded. This holds true only if the temperature around the conductors does not exceed 86°F (30°C), and there are not more than three current-carrying conductors in the cable, raceway, or trench. If these limitations are exceeded, the values shown in *Table 310.16* must be adjusted. There are temperature correction factors at the bottom of *Table 310.16*. The following example will illustrate how the temperature correction factors are applied.

Example 1.13 A raceway containing copper conductors runs through an area of a building where the temperature is likely to reach 120° F (49°C) as shown in Figure 1.25. Determine the allowable ampacity of a size 3 AWG copper conductor with THWN insulation run through this area.

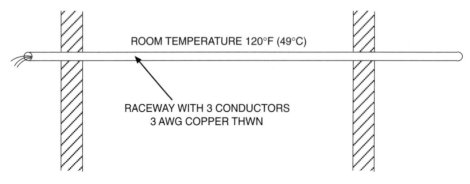

ROOM TEMPERATURE 120°F (49°C)

RACEWAY WITH 3 CONDUCTORS
3 AWG COPPER THWN

Figure 1.25 A raceway containing three size 3 AWG copper THWN conductors runs through a room with a typical temperature of 120°F.

Answer: Look up the allowable ampacity of a size 3 AWG copper conductor in the 75°C column of *Table 310.16* and find 100 amperes. Now continue down that same column into the temperature correction section of the table. Fahrenheit temperatures are on the right-hand side, so continue down to the row that includes 120°F. Find the correction factor 0.75. Now multiply the 100 amperes by 0.75 to get 75 amperes, which is the allowable ampacity of a size 3 AWG copper THWN conductor under these conditions.

If there are more than three current-carrying conductors in the cable, raceway, or trench in earth, then the allowable ampacity given in *Table 310.16* must be adjusted. Each conductor produces some heat, therefore, the maximum insulation temperature is likely to be exceeded if the ampacity in the table is used without adjustment. *NEC® 310.15(B)(2)(a)* gives the rules for applying adjustment factors when there are more than three current-carrying conductors in a raceway or cable. This subject will be discussed in more detail in the next unit. The following example will illustrate how the adjustment factors are applied.

Example 1.14   A raceway contains nine size 8 AWG copper current-carrying conductors with THHN insulation, illustrated in Figure 1.26. Determine the allowable ampacity of these conductors in this raceway.

Answer: Look up the allowable ampacity of a size 8 AWG copper conductor in the 90°C column of *Table 310.16* and find 55 amperes. Look up the adjustment factor for nine current-carrying conductors in raceway from *Table 310.15(B)(2)(a)* and find 0.7. Now multiply 55 amperes by 0.7 to get 38.5 amperes, which is the allowable ampacity of a size 8 AWG copper THHN conductor under these conditions.

On occasion, a raceway or cable containing more than three current-carrying conductors runs through an area with a high ambient (surrounding) temperature. In this case, both a temperature correction factor and an adjustment factor for more than three current-carrying conductors must be applied to the allowable ampacity found in the table. The next example will illustrate the point.

Example 1.15   A raceway containing six size 10 AWG copper current-carrying conductors with THHN insulation passes through an area of a building where the temperature is expected to be 110°F for long periods of time. Determine the allowable ampacity of these conductors under these conditions.

Answer: Look up the allowable ampacity of a size 10 AWG copper Type THHN conductor in the 90°C column of *Table 310.16* and find 40 amperes. Next, continue down the ampacity column to the temperature correction section of the table and find the correction factor for 110°F which is 0.87. Now look up the adjustment factor for six current-carrying conductors in *Table 310.15(B)(2)(a)* and find 0.8. Multiply 40 amperes from *Table 310.16* by 0.87 and 0.8 to get 27.8 amperes, which is the allowable ampacity of a size 10 AWG copper THHN conductor under these conditions

Determining the allowable ampacity of a conductor is as simple as looking up the value in the appropriate table and applying adjustment or correction factors if necessary. Determining the minimum size con-

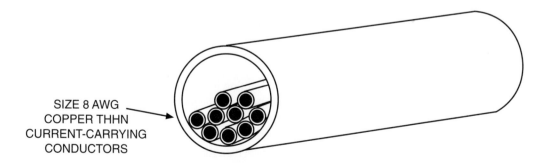

SIZE 8 AWG
COPPER THHN
CURRENT-CARRYING
CONDUCTORS

Figure 1.26   A raceway contains nine size 8 AWG copper THHN current-carrying conductors.

ductor for a specific load and circuit involves the application of *240.4*. The conductor usually is required to be protected by a fuse or circuit breaker with a rating not higher than the allowable ampacity of the conductor. This subject will be discussed in more detail in the next unit when correction and adjustment factors are involved.

Example 1.16   A circuit supplies a continuous load of 38 amperes and is protected by a 50-ampere circuit breaker. All conductor terminations in the circuit are rated at 75°C. If the circuit conductors are THWN copper, determine the minimum size permitted for the circuit.

Answer: Look up a conductor in this case in the 75°C column of *Table 310.16* which has an allowable ampacity of not less than 50 amperes, which would be size 8 AWG.

The allowable ampacity of the conductor and the rating of the overcurrent device do not always match up exactly like the previous example. Consider a circuit supplying a continuous load of 50 amperes, which is required to be protected with an overcurrent device rated not less than 70 amperes. With the terminations rated at 75°C and copper THWN conductors, the allowable ampacity is found in the 75°C column of *Table 310.16*. According to *210.19(A)*, the minimum size conductor permitted for this circuit must have an allowable ampacity not less than 1.25 times the continuous load, which is 62.5 amperes (1.25 × 50 A = 62.5 A). The choices are size 6 AWG at 65 amperes and size 4 AWG at 85 amperes. One is a little too small for the overcurrent device, and the other is much too large. The size 6 AWG is permitted to be used in this case according to *240.4(B)*. If the overcurrent device protecting the circuit is rated not over 800 amperes, it is permitted to choose a conductor adequate to supply the load (62.5 amperes) and protect that conductor with the next higher standard rating of overcurrent device. The standard ratings of overcurrent devices are listed in *240.6(A)*. The allowable ampacity of the size 6 AWG copper THWN conductor does not match a standard rating of overcurrent device. It is then permitted to round up to the next standard rating, which in this case is 70 amperes. If this is a branch-circuit that supplies multiple receptacles for cord- and plug-connected loads, then the allowable ampacity of the conductor must be equal to or higher than the rating of the overcurrent device.

Example 1.17   A feeder conductor in a building supplies a 140-ampere load with aluminum Type THWN conductors in raceway. The feeder overcurrent device is only required to be rated 150 amperes. In this case, the installer decided to protect the feeder at 200 amperes. Determine the minimum size conductors permitted for the feeder.

Answer: *Table 310.16* is used because the wiring method is conductors in raceway. There is no mention of termination temperature, so the rule in *110.14(C)(1)(b)* will apply. The terminations will be 75°C rated because the overcurrent device is larger than 100 amperes. From the 75°C aluminum column of *Table 310.16*, find size 4/0 AWG rated 180 amperes and size 250 kcmil rated 205 amperes. Neither conductor corresponds to a standard rating of overcurrent device listed in *240.6(A)*. The size 4/0 AWG copper conductor has an allowable ampacity greater than the load, and the next higher rated overcurrent device is 200 amperes.

There is a rule in *240.4(D)* that restricts the size of conductor for some circuits. In the past, this rule was a footnote to some of the allowable ampacity tables. This rule sets the maximum rating of overcurrent device for a size 14 copper conductor at 15 amperes. For a size 12 copper conductor, the maximum overcurrent device rating is 20 amperes. And for size 10 copper, the maximum is 30 amperes. There are some exceptions to this rule, which will be discussed later in the text. The following example will illustrate the application of this rule:

Example 1.18   A circuit supplies 15 amperes of lighting load. The circuit is required by *210.20(A)* to be protected by a 20-ampere rated circuit breaker. Determine the minimum size copper THHN conductors permitted for this circuit if there are only two current-carrying conductors in raceway.

Answer: The termination temperature is not specified, therefore, the rule of *110.14(C)(1)(a)* will apply. The terminations will be assumed to be rated 60°C. The allowable ampacity will be found in the 60°C column of *Table 310.16*. This is a continuous load, therefore, the minimum conductor size, according to

210.19(A), is required to be not less than 1.25 times the load, which is 18.8 amperes (15 A × 1.25 = 18.8 A). According to *Table 310.16*, a size 14 AWG copper conductor is rated 20 amperes. The rule in *240.4(D)* only permits a 15-ampere overcurrent device on a size 14 AWG copper conductor. In this case, a size 12 AWG copper conductor is the minimum permitted because the overcurrent device is rated 20 amperes.

USING THE CODE TO ANSWER QUESTIONS

The numerous people who write the Code attempt the difficult task of choosing words that prohibit unsafe installations while not excluding any of the numerous acceptable installation methods. They try to achieve this task with words that we can understand. The Code is constantly changing, however, because of the introduction of new materials and techniques and because unacceptable confusion and loopholes are discovered.

One point that has led to confusion and misunderstanding in the field is the phrase **approved for the purpose. Approved** is defined in the Code as meaning approved by the authority having jurisdiction (the inspector). In many cases, it simply means that a judgment call must be made based on the inspector's knowledge and background and local conditions. The Code has made an attempt to place more responsibility on the manufacturer to indicate the suitability of various products for a specific purpose. The Code now uses the phrase **listed for the purpose** rather than **approved for the purpose.** The following points will help in understanding the meaning of the Code.

1. Read each section carefully and think about what the section is saying. Try not to let personal bias obscure the true meaning of the section.

2. Keep the purpose of the Code in mind as you read the Code, *90.1.* All sections of the Code are directed to this purpose.

3. Look for the word "shall" and how it is used in the section. For example, "shall not be permitted" is a prohibitive statement. "Shall be permitted" is a permissive statement. Another similar type of statement is "shall not be less than," which fixes a lower limit.

4. Fine print notes (FPNs) are scattered throughout the Code to either act as an advisory or clarify the previous Code material. These FPNs are not considered a legally binding part of the Code.

5. Be sure to read the **scope** at the beginning of each article. The specific sections apply only under the conditions specified in the scope.

6. Definitions are frequently provided in an article to add important clarity. Read these definitions, as they are often necessary to the understanding of a particular section. Most definitions are grouped in *Article 100* if they are used in two or more articles. If used only in a specific article, the definitions will be given only in that article. If a definition is not given in the Code, then it is permitted to use the definition from a dictionary.

7. Do not confuse **wiring specifications** with Code requirements. Just because in your experience you have seen only one particular wiring method for an application, do not automatically assume this is a Code requirement. It is the option of the owner, architect, or another code to specify a particular method or material providing it is not in conflict with the Code.

8. Footnotes to the tables are considered part of the table and are a binding part of the Code.

9. Local jurisdictions have the right to adopt amendments to the Code. Sometimes the electric utility will have a requirement not covered in the Code. The utility standards apply to the installation of service equipment.

10. It is important to remember the limitations placed on manufacturers' materials and equipment by the testing process and the physical behavior of materials and equipment in a particular environment. For example, Type THHN wire ampacity cannot be based on 90°C if the wire is connected to a circuit-breaker with a maximum temperature rating of 75°C. Some organizations providing testing services are Underwriters Laboratories (UL), Electrical Testing Laboratories (ETL), and Canadian Standards Association (CSA).

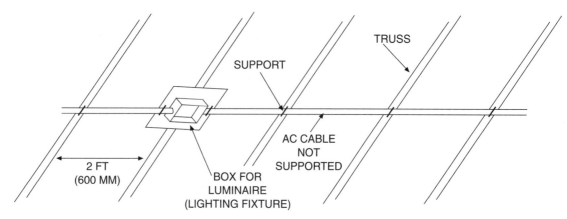

Figure 1.27 **Type AC Armored Cable is run as exposed wiring on the under side of room trusses supported only by attachment to each truss.**

The following example question will help illustrate the technique of finding information in the Code. Look for key terms that state the subject of the question, then look them up in the Code index. After using the Code for a period of time, you will remember many of the article numbers.

Example 1.19 Type AC Armored Cable is to be installed as open exposed wiring to supply luminaires (lighting fixtures) in a building of truss construction with an open ceiling. The trusses are spaced 2 ft (600 mm) on centers with a luminaire (lighting fixture) attached to the bottom of every fourth truss. Is it permitted to run the Type AC Cable perpendicular to the trusses without any means of support between trusses as shown in Figure 1.27, or is it required to be attached to running boards?

Answer: The subject of this question is Type AC Armored Cable, open exposed wiring, and supports. If you know the article number, then it probably would be most efficient to go directly to that article. Otherwise, look up Armored Cable, Type AC, in the index. Look down the list of subtopics and exposed work is in *320.15*, and supports are covered in *320.30*. Type AC Cable is not a separate listing, although under cable, there is a listing for Armored Cable, Type AC. According to *320.15*, Armored Cable is permitted to be on the underside of each truss (joist) provided the support spacing is not exceeded and the cable is supported at each truss. In *320.30*, Armored Cable run in the manner described, is required to be supported within 12 in. (300 mm) of each box, and at intervals not to exceed 4^1/2 ft (1.4 m). If these requirements are also met, the Type AC Armored Cable is permitted to be run as shown in Figure 1.27.

STUDENT CODE PRACTICE

Answer the following wiring questions and give the Code reference where the answer is found. The answer will be found in the Code articles listed in the objectives at the beginning of this unit. Pick out the key words that describe the subject of each question and look them up in the Code index. When you are finished, check your skill by comparing with the key words and answers at the end.

1. A conductor with black insulation is permitted to be re-identified as a grounded conductor at the time of installation with a distinctive white marking that completely encircles the conductor at the conductor terminations for:
    A.   any size conductor.
    B.   sizes 6 AWG and larger.
    C.   sizes 8 AWG and larger.
    D.   sizes smaller than 6 AWG.
    E.   sizes larger than 6 AWG.

    Key words _____   Code reference _____

2. The maximum branch-circuit voltage permitted between conductors supplying luminaires (lighting fixtures) in a dwelling unit is:
    A.   120 volts.
    B.   208 volts.
    C.   240 volts.
    D.   277 volts.
    E.   480 volts.

    Key words _____   Code reference _____

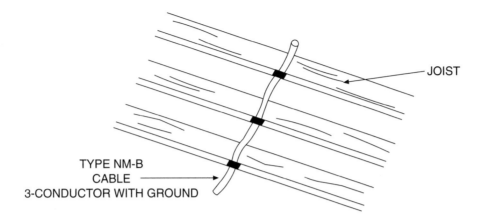

Figure 1.28 Type NM-B Nonmetallic-Sheathed Cable with three conductors and an equipment ground is run attached to the bottom of floor joists in an unfinished basement.

3. The largest solid conductor generally permitted to be installed in raceway is size:

   A. 12 AWG.
   B. 10 AWG.
   C. 8 AWG.
   D. 6 AWG.
   E. 4 AWG.

   Key words _____ Code reference _____

4. The minimum size copper conductor permitted for a 120-volt general illumination branch-circuit in a dwelling is:

   A. not specified in the Code, but depends upon the load to be served.
   B. 18 AWG.
   C. 16 AWG.
   D. 14 AWG.
   E. 12 AWG.

   Key words _____ Code reference _____

5. A 3-conductor Nonmetallic-Sheathed Cable with ground run exposed at an angle to the joists of an unfinished basement and fastened to the lower edge of the joists, as shown in Figure 1.28, is permitted for sizes not smaller than:

   A. 14 AWG.
   B. 12 AWG.
   C. 10 AWG.
   D. 8 AWG.
   E. 6 AWG.

   Key words _____ Code reference _____

Answers to Code Practice: The first question is about grounded conductor identification. Look up grounded conductors, then identification. This can lead to the answer. Now look up conductors, grounded, then identification. This one leads to *200.6,* which gives the answer. Other ways to find the answer in the index are by looking under identification, grounded conductors or conductor identification. The answer to the first question is **E** and the Code reference is *200.6(B).*

The second question is about maximum branch-circuit voltages for dwellings. Start by looking up voltage, maximum. There is no reference for this subject, but there is a listing for branch-circuit limits, under voltage, which does lead to the answer. Another approach is to look up branch-circuits, voltage. This lists voltage limitations which leads to the answer **A** and the Code reference is *210.6(A).*

The third question is about the largest solid conductor in raceway. If a conductor is not solid, it is stranded. One approach is to look up conductors, raceway, solid or stranded. There is an entry for stranded, which leads to the answer. Another approach is to look under raceway, conductors, solid or stranded, but this leads to a dead end. The answer is **B** and the Code reference is *310.3.*

Question four is about minimum size branch-circuit conductors in dwellings. Start by looking up conductors, minimum size, or conductors, branch-circuits, minimum size. The first reference leads to a reference that gives the answer. Another approach is to look under branch-circuits, conductors, minimum size. This gives a reference, but it is not the correct subject, so it is another dead end. The answer is **D** and the Code reference is *310.5* and *Table 310.5*.

Question five is about Nonmetallic-Sheathed Cable exposed in unfinished basements. Look under Nonmetallic-Sheathed Cable, exposed, basements. The reference is Nonmetallic-Sheathed Cable, exposed work, which leads to the answer. Another approach is to look up basements, Nonmetallic-Sheathed Cable, which will also lead to the answer. The answer is **D** and the Code reference is *334.15(C)*.

MAJOR CHANGES TO THE 2005 CODE

These are the changes to the 2005 *NEC®* that correspond to the Code sections studied in this unit. The following analysis explains the significance of the changes from the 2002 to the 2005 Code only, and this analysis is not intended to be used in place of the Code. Refer to the actual section of the 2005 Code for the exact wording and meaning of each section discussed. Changes are indicated in the Code with a vertical line in the margin. If material was deleted or moved to another location in the Code, the location of the deletion is indicated with a dark dot in the margin. Any article covered in this unit, for which significant changes occurred, are included in the following discussion.

Article 100   **Definitions**

There is a new definition of a **system bonding jumper,** which is intended to apply to a separately derived system. A system bonding jumper is a connection between a grounded-circuit conductor and the equipment grounding conductor of a separately derived system as shown in Figure 1.29. At a service this is called the main bonding jumper. The system bonding jumper is permitted to be installed at the source of the separately derived system or at the first disconnecting means supplied by the system.

The definition of a **device** now includes an apparatus that only controls electrical energy; it is not required to carry electrical energy. This definition was limited in the past and did not seem to include electronic control devices.

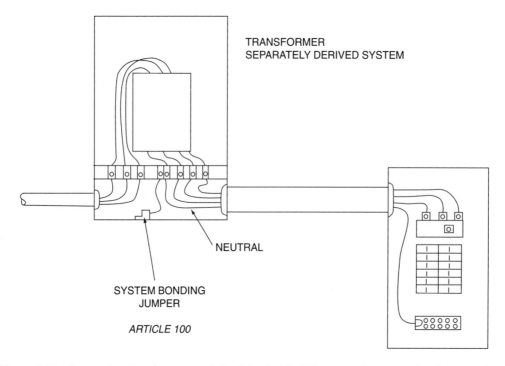

Figure 1.29   System bonding jumper as defined in *Article 100* means the connection between the grounded-circuit conductor and the equipment grounding conductor of a separately derived system.

A **dwelling unit** is now required to provide permanent provisions for living, cooking, sleeping, and sanitation. The previous definition only required space to be provided for eating, living, and sleeping. There was no requirement for permanent provisions for these functions. Space provided for a microwave oven and similar cooking equipment is not considered permanent provisions for cooking.

Solidly grounded is now defined as the connection to ground without using a resistor or an impedance device.

The term **grounding electrode** is now defined as a device making an electrical connection to the earth. In the past there was no definition, only a description as to what was considered acceptable as a grounding electrode.

The term **guest room** was added. It means a space providing facilities for living, sleeping, sanitary, and storage. Permanent provisions for cooking is not provided. Providing space for a microwave is not considered as providing permanent provisions for cooking.

There is a new definition of **handhole,** which is an enclosure into which a person can reach, but not enter, for the purpose of accessing electrical wiring and components. A handhole is associated with equipment of an underground wiring installation. The handhole can be at the base of a pole supporting a luminaire (lighting fixture) where the wiring is emerging from the earth into the pole.

A fine print note was added to the definition of **qualified person** that makes reference to NFPA 70E-2004 for electrical safety training requirements.

Supplementary overcurrent protection is used in numerous articles of the Code, but there is no common definition of the term. It is an overcurrent device that is intended for the protection of equipment or appliances and not for the protection of the circuit. This is not a change of intent from past use of supplementary overcurrent protection.

Article 110   Requirements for Electrical Installations

110.15: This section requires that the phase conductor with the higher voltage to ground for a 4-wire delta, 240/120-volt 3-phase system be marked orange or identified by some other effective means at every point where connections are made and the grounded-circuit conductor is present. The change is that only the phase with the higher voltage to ground is permitted to be labeled orange. This is illustrated in Figure 1.30. Other conductors, if identified with a color, must be identified with a color, other than

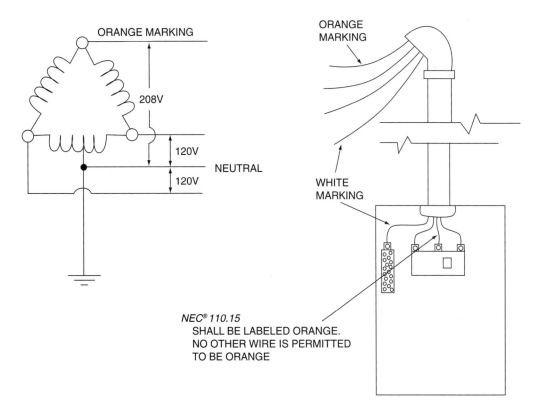

Figure 1.30   Only the phase conductor with the higher voltage to ground of a 3-phase, 4-wire, delta, 240/120 volt electrical system is permitted to be labeled orange at all locations where the grounded conductor is present.

orange. This rule does not prohibit an orange marking on any ungrounded conductor of a different electrical system.

110.26: Enclosures that can be locked and containing electrical equipment operating at not over 600 volts are considered to be accessible to qualified persons. The previous edition of the Code specified lock and key, which was sometimes interpreted as only permitting a lock that required a key to open.

110.26(C)(2): For electrical equipment containing overcurrent, switching, or control devices and rated 1200 amperes or more and with a width of more than 6 ft (1.8 m), an exit is required at each end of the working space unless equipment is arranged so that the exit route is unobstructed from all parts of the room. These exit doors are also required to open out with pressure release door latches or panic bars. The change in this section is that the width requirement was deleted. It makes no difference how wide the equipment, if it is rated 1200 amperes or more, an exit at each end of the room is required. This is illustrated in Figure 1.31.

110.31: Vaults, rooms, closets, and areas surrounded by fences or screens that can be locked and containing electrical equipment operating at over 600 volts are considered to be accessible to qualified persons. The previous edition of the Code specified lock and key, which was sometimes interpreted as only permitting a lock that required a key to open.

Part V: The entire *Part IV* of *Article 314* was moved to *Article 110*. This material provides requirements for manholes and other electric enclosures intended for personnel entry. There were essentially no changes made to *Part V*.

Article 200   Use and Identification of Grounded Conductors

200.6(B): A grounded conductor with other than white or green insulation and larger than size 6 AWG is now permitted to be re-identified using gray tape as well as white tape that completely encircles the conductor. In the past, only white tape was permitted to re-identify a grounded conductor that did not have white or gray insulation.

200.6(D): If a premises is supplied by more than one wiring system, each with a grounded conductor, the grounded conductors of each system are required to be identified in such a manner that they can be easily recognized when run in the same raceway, cable, box, or other enclosure. One method was

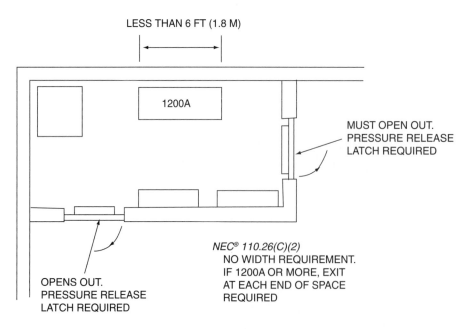

Figure 1.31   The requirement for a personnel exit at each end of an electrical room is now only based upon equipment rated 1200 amperes or more, and there is no longer a minimum equipment width.

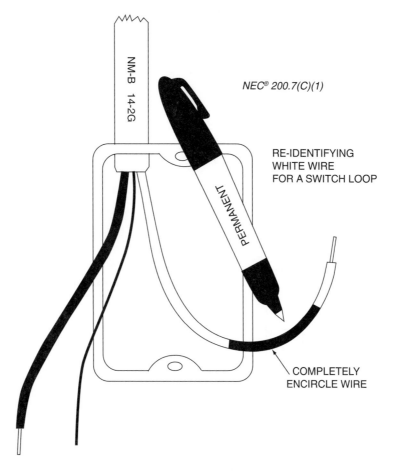

NEC® 200.7(C)(1)

RE-IDENTIFYING
WHITE WIRE
FOR A SWITCH LOOP

COMPLETELY
ENCIRCLE WIRE

Figure 1.32 **When the white insulated wire of cable used in a switch loop as an ungrounded conductor is re-identified, the marking is now required to completely encircle the conductor.**

to have a colored strip on a white insulated wire. Now it is permitted to have a colored stripe on a gray insulated wire. The other change is that the means of identifying the grounded conductors is required to be permanently posted at each branch-circuit panelboard. An example may be where a 208/120-volt 4-wire wye system supplies receptacles and lighting, and a 480/277-volt 4-wire system supplies electric discharge lighting. The neutral conductors, if run in the same raceway or enclosure, are required to be uniquely distinguishable from each other, and the means of identification is now required to be posted at each panelboard.

200.7(C)(1): When cable is installed as a switch loop for lighting, the white or gray wire is required to be re-identified to indicate it is no longer a grounded conductor. The change is that this re-identification is required to be a color other than white, gray, or green, and that the identification shall encircle the conductor in a manner similar to Figure 1.32. In the past the identification was not required to completely encircle the conductor.

Article 300 **Wiring Methods**

300.4(A)(1) Exception 2: When cables and some raceways are installed through bored holes in wood members, a steel plate with a minimum thickness of $^1/_{16}$ in. (1.6 mm) is required when the edge of the hole is less than $1^1/_4$ in. (32 mm) from the outer edge of the wood member. A listed steel plate of lesser thickness is now permitted.

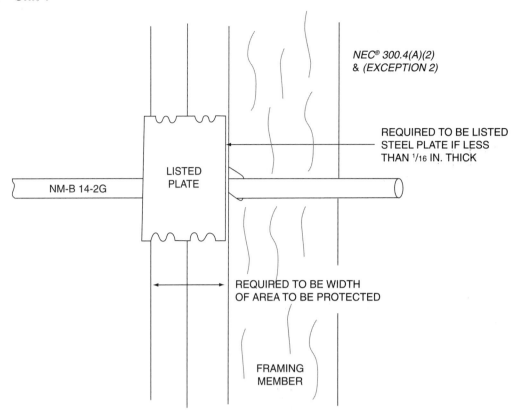

NEC® 300.4(A)(2)
& (EXCEPTION 2)

REQUIRED TO BE LISTED
STEEL PLATE IF LESS
THAN ¹/₁₆ IN. THICK

LISTED
PLATE

NM-B 14-2G

REQUIRED TO BE WIDTH
OF AREA TO BE PROTECTED

FRAMING
MEMBER

Figure 1.33 A metal plate used to protect cables or raceway from damage from nails or screws is permitted to be less than ¹/₁₆ in. thick if listed, and it must be wide enough to completely cover the area requiring protection.

300.4(A)(2): When cables or some raceways are installed through notched holes in wood members, the steel plate is now required to have sufficient length and width to completely cover the notched area of the wood member as shown in Figure 1.33.

300.4(A)(2) Exception 2: When cables and some raceways are installed through a notch in wood members, a steel plate with a minimum thickness of ¹/₁₆ in. (1.6 mm) is required to protect the cable or raceway from damage caused by nails or screws. A listed steel plate of lesser thickness is now permitted.

300.4(B)(2) Exception: Nonmetallic-Sheathed Cable and Electrical Nonmetallic Tubing passing through metal building members in such a way they may be damaged by nails or screws are required to be protected with a steel plate, sleeve, or clip with a minimum thickness of ¹/₁₆ in. (1.6 mm). A listed steel plate, sleeve, or clip of lesser thickness is now permitted.

300.4(D) Exception 3: When cable or some raceways are installed lengthwise along a framing member or concealed with furring strips such that the cable is less than 1¹/₄ in. (32 mm) from the edge, a steel plate or sleeve with a minimum thickness of ¹/₁₆ in. is required to be installed to protect the cable or raceway from damage due to nails or screws. This new exception permits a listed steel plate or sleeve with a lesser thickness to be installed to protect the cable or raceway.

300.4(E) Exception 2: Cables and some raceway installed in shallow grooves of walls, floors, or ceilings is required to be protected from damage from nails or screws with a steel plate with a minimum thickness of ¹/₁₆ in. (1.6 mm). Now a listed steel plate is permitted to be of a lesser thickness.

300.6(B): Nonferrous wiring materials, including support hardware embedded or encased in concrete or in direct contact with the earth, are now required to be provided with supplementary corrosion protection. The previous edition of the Code permitted the wiring materials to be of a nonferrous material installed without corrosion protection if the material was judged to be suitable for the conditions.

300.6(C): Nonmetallic boxes, equipment, and wiring materials are now required to be listed as sunlight resistant where installed exposed to sunlight. If nonmetallic materials are exposed to chemical deterioration, the equipment shall be inherently resistant to the particular chemical agent or be identified for use with the particular chemical agent.

300.11(A)(1): When wiring is installed in the space above a fire-rated suspended ceiling, the wiring is permitted to be supported by separate support wires that are identified to distinguish them form the ceil-

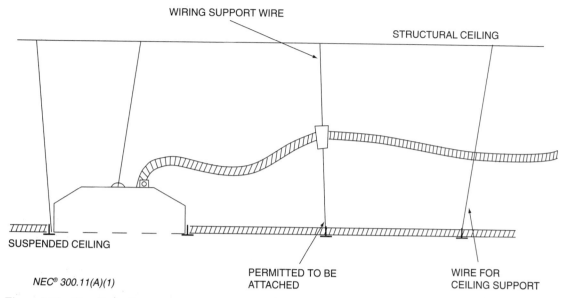

WIRING SUPPORT WIRE

STRUCTURAL CEILING

SUSPENDED CEILING

NEC® 300.11(A)(1)

PERMITTED TO BE
ATTACHED

WIRE FOR
CEILING SUPPORT

Figure 1.34 Now it clearly states that wires installed to support wiring above a suspended ceiling are permitted to be secured to the ceiling grid.

ing support wires. The change is that it is now made clear that one end of the support wires for the electrical equipment is permitted to be attached to the ceiling grid such as shown in Figure 1.34.

300.22(B): Liquidtight Flexible Metal Conduit is no longer permitted to be installed in a duct or plenum for connection to equipment. The previous edition of the Code permitted lengths not to exceed 4 ft of Liquidtight Flexible Metal Conduit to be installed for the termination to equipment in ducts or plenums.

Table 300.50: There was a revision of this table, which gives minimum cover requirements for high-voltage cables. There were several exceptions to this section, which were added into *Table 300.50* as a new column or a footnote. *Exception 2* of the previous edition of the Code was deleted that allowed the installation depth of burial to be reduced by 6 in. (150 mm) for every 2 in. (50 mm) of concrete cover over the installation. This exception no longer applies.

Article 310   Conductors for General Wiring

310.6 Exception: This section requires solid dielectric-insulated conductors operating at more than 2000 volts in permanent installations to be shielded. The exception did not require a metallic shield for conductors operating up to 8000 volts under the conditions specified, and with minimum insulation thicknesses as given in *Table 310.63*. This exception now only permits nonshielded cables for use at up to 2400 volts.

310.8(D)(3): Conductors and cables exposed to direct rays of the sun shall be listed as sunlight resistant. The change is that now included are coverings such as tape and sleeving. These materials shall be listed as sunlight resistant when installed on conductors exposed to direct rays of the sun. These materials shall also be listed for the application such as water resistant when installed outside and exposed to the weather.

310.10 FPN No. 2: Conductors installed in raceway outdoors in close proximity to a rooftop and exposed to direct rays of the sun are likely to experience a temperature in excess of the ambient temperature of the air. This fine print note indicates the conductors could experience a temperature rise of up to 30°F (17°C) above ambient. The implication, but not a requirement, is that 30°F (17°C) should be added to the expected highest ambient temperature for the purpose of applying the temperature adjustment factors at the bottom of the ampacity tables.

Table 310.16 is based upon an ambient temperature of 86°F (30°C). When the conductor is expected to operate in an area where the ambient temperature is above 86°F, an adjustment is required to be applied. Wiring supplying rooftop environmental air equipment may in particular be exposed to high ambient temperatures. For example, size 12 AWG copper THHN wire run in raceway is rated at 30 amperes. If the same circuit is run outside where the ambient temperature is likely to reach 110°F, the temperature adjustment factor for THHN insulation is 0.87, which means the size 12 AWG copper wires are rated at 26.1 amperes. If this same circuit is run outside in direct sunlight in close proximity to the rooftop, it may be desirable to make an adjustment for an additional rise in temperature. If an additional

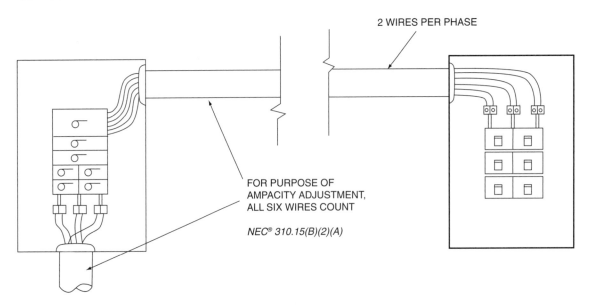

2 WIRES PER PHASE

FOR PURPOSE OF
AMPACITY ADJUSTMENT,
ALL SIX WIRES COUNT

NEC® 310.15(B)(2)(A)

Figure 1.35   **When conductors are installed with several parallel conductors per phase in the same raceway, each wire of the parallel set is to be counted as a current-carrying conductor for the purpose of applying the ampacity adjustment factors of** *Table 310.15(B)(2)(a).*

30°F is added to the 110°F ambient temperature, the total temperature adjustment would be based on 140°F. Based upon this temperature adjustment, size 12 AWG copper wire is rated at 21.3 amperes.

310.15(B)(2)(a): In the past there was a debate as to whether parallel conductors of the same phase installed in the same raceway were counted as one conductor or as multiple conductors for the purpose of applying the ampacity adjustment factors of *Table 310.15(B)(2)(a).* It was the intent that each conductor was to be counted as a current-carrying conductor. This is illustrated in Figure 1.35. Now there is a new final sentence that clearly states that each conductor of a set is to be counted for the purpose of applying the adjustment factors of *Table 210.15(B)(2)(a).*

Table 310.63: Minimum solid dielectric-insulation thickness and jacket thickness for non-shielded conductors is given in this table. The table no longer recognizes cables operating at 5001 to 8000 volts. This table is related to *310.6 Exception (c)* where a nonmetallic jacket over the insulation was only required for conductors operating at 5001 to 8000 volts. Now those conductors are required to have a metallic sheath, and conductors operating at 2400 volts are required to have a nonmetallic jacket over the insulation.

Article 320   **Armored Cable: Type AC**

320.12: This section stating uses not permitted for Type AC cable was revised and specific locations were deleted. The general statement at the beginning of *320.10* makes it clear that elsewhere in the Code use of Type AC cable may be prohibited. There was no need to repeat those uses not permitted in this section. Instead, this section now states the conditions that would be used to judge if Type AC cable is not permitted to be installed. In particular those conditions are: exposed to physical damage, in damp or wet locations, exposed to corrosive fumes or vapors, installed in hollow spaces of masonry, or embedded in masonry where damp or wet conditions exist.

Article 324   **Flat Conductor Cable: Type FCC**

324.6: This is a new section that requires Type FCC Flat Conductor Cable to be listed, although *324.100(A)* also requires this cable to be listed.

Article 330   **Metal-Clad Cable: Type MC**

330.12: The section on uses not permitted was revised and is quite specific and complete as to what conditions are considered to be adverse for Type MC cable. This new revised section should be useful in determining the conditions under which Type MC cable can be installed. There is a new fine print note that states that Type MC cable suitable for direct burial is also considered suitable for concrete encasement.

Article 332   **Mineral-Insulated, Metal-Sheathed Cable: Type MC**

332.12: The list of uses not permitted for Type MI cable now includes providing protection from physical damage for underground runs where necessary. The change is that protection of the underground cable is not required everywhere.

Article 334   **Nonmetallic-Sheathed Cable: Types NM, NMC, and NMS**

334.2: Definitions of Type NM, Type NMC, and Type NMS cables were added to this section. Type NMC has a corrosion-resistant nonmetallic jacket. Type NMS is permitted to contain control, communications, data, and signaling conductors along with the power conductors.

334.10(4): Nonmetallic-Sheathed Cables are not permitted to be installed in Cable Trays except in buildings of Type III, Type IV, or Type V construction. Type III construction is a combination of combustible and noncombustible building materials. Type IV is heavy timber construction. Type V is construction of combustible materials such as wood frame construction with plasterboard surfaces. For Type I and II construction, the main building materials are required to be noncombustible.

334.80: A new paragraph was added calling attention to an abnormal heating condition that can be encountered when more than two Nonmetallic-Sheathed Cables are bundled together and pass through a hole in wood frame materials where the hole is sealed to prevent fire spread. The ampacity adjustment factors of *Table 310.15(B)(2)(a)* are required to be applied. Since the insulation on the conductors is rated at 90°C, the ampacity in the 90°C column of *Table 310.16* can be used for the purpose of derating. The actual load on the conductors is to be used for the calculation. For a dwelling, the actual load may be difficult to determine since the 180 VA does not apply to receptacles in a dwelling and many loads only operate for short periods of time. For a nondwelling application, the load on the conductors can be determined.

Article 336   **Power and Control Tray Cable: Type TC**

336.10(7): Type TC cable that is permitted to be installed between a cable tray and utilization equipment or a device is now required to be marked Type TC-ER. This cable meets the impact and crush standards of Type MC cable.

Article 338   **Service-Entrance Cable: Types SE and USE**

338.10(B)(2) Exception: The neutral in Type SE cable is permitted to be uninsulated for the purpose of supplying feeders from one building to another as permitted in *250.32* and *Part II* of *Article 225*. This means that Type SE Style U with an uninsulated conductor is permitted to be used for a feeder from one building to another. This is illustrated in Figure 1.36 where Type SE cable, Style U is installed from the service panel in one building to the overhead triplex cable. Type SE cable, Style U is then

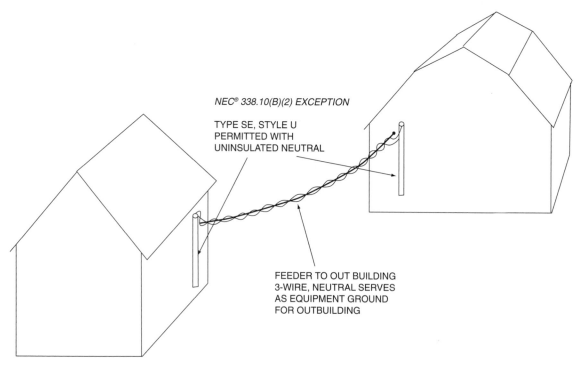

NEC® 338.10(B)(2) EXCEPTION

TYPE SE, STYLE U
PERMITTED WITH
UNINSULATED NEUTRAL

FEEDER TO OUT BUILDING
3-WIRE, NEUTRAL SERVES
AS EQUIPMENT GROUND
FOR OUTBUILDING

Figure 1.36   When Type SE cable is used as a part of a feeder or branch-circuit from one building to another, the neutral wire protected by an outer cable jacket is not required to be insulated, thus permitting the use of Style U cable for this application.

used at the second building from the overhead conductor to the panelboard in the second building. According to the previous edition of the Code, the grounded conductor (neutral) was required to be insulated for this application, thus requiring Style R when Type UF cable was used as a part of the feeder run. Style R has three insulated conductors and a bare equipment grounding conductor.

338.10(B)(4)(b): Type USE cable is not required to have a flame-retardant outer covering and is not permitted to enter a building even to terminate. The previous edition of the Code permitted Type USE cable to emerge from the ground outside and enter a building for termination provided it was terminated within 6 ft (1.8 m) after entering the building.

Article 340   Underground Feeder and Branch-Circuit Cable: Type UF

340.112: It now states clearly that when installed as a nonmetallic-sheathed cable, the conductors are required to have 90EC insulation. This is also covered by *340.10(4)*, which requires Type UF cable installed as a nonmetallic-sheathed cable to meet the requirements of *Part III* of *Article 334*.

Article 590   Temporary Installations

Article 527 on temporary installations was considered to be out of place and was moved to the end of *Chapter 5* as new *Article 590*. The actual changes are as follows:

590.4(B): Use of Type NM and Type NMC cables for temporary light and power feeders during construction of a building is permitted in a building of any type of construction and a building of any height. It is now made clear that NM and NMC cables are not required to be concealed within fire-rated construction when used as temporary wiring in any type of building of any height.

590.4(C): Use of Type NM and Type NMC cables for temporary light or power branch-circuits during construction of a building is permitted in a building of any type of construction and a building of any height. It is now made clear that NM and NMC cables are not required to be concealed within fire-rated construction when used as temporary wiring in any type of building of any height.

590.4(J) Exception: It is not permitted to support overhead spans of temporary branch-circuits and feeders with live vegetation according to *590.4(D)*. This new exception does permit live vegetation as support for overhead spans of feeders and branch-circuits if for a period of not more than 90 days for holiday decorative lighting and similar purposes. Strain relief is required to be installed to protect the cable from movement of the vegetation. The exception also applies to overhead spans for light and power during emergencies and for testing.

590.5: This new section requires decorative lighting and accessories used for 90 day temporary holiday lighting to be listed.

Table 8, Chapter 9: Conductor Properties

Table 8: The metric cross-sectional area of conductors was added to the table for sizes 250 kcmil through 2000 kcmil. This is the actual cross-sectional area of the metal conductor, which is used for determination of conductor ampacity. The overall cross-sectional area of the conductors was given in the previous edition of the Code, and that value is used for conductor fill in raceways and conductor spacing in cable trays. An example of where this new data is used is when adjusting the size of an equipment grounding conductor because the ungrounded conductor size is increased to compensate for voltage drop.

WORKSHEET NO. 1—BEGINNING GENERAL WIRING AND FUNDAMENTALS

Mark the single answer that most accurately completes the statement based upon the 2005 Code. Also provide, where indicated, the Code reference that gives the answer or indicates where the answer is found, as well as the Code reference where the answer is found.

1.   The current drawn by a 250-watt incandescent lamp operating at 120 volts is:
    A.   0.40 amperes.   C.   1.56 amperes.   E.   3.00 amperes.
    B.   1.04 amperes.   D.   2.08 amperes.

2.   A copper conductor that is 150 meters in length has a resistance of 0.14 ohms. If the current flow through the conductor is 50 amperes, the voltage drop along the conductor is:
    A.   1.4 volts.   C.   7 volts.   E.   357 volts.
    B.   3.7 volts.   D.   14 volts.

3.   A 208/120-volt electrical panelboard is mounted on a concrete block wall and the panelboard faces metal equipment in the room that is mounted to the concrete floor as shown in Figure 1.37. The minimum distance from the front of the panelboard to the metal equipment is not permitted to be less than:
    A.   2 ft (600 mm).   C.   3.5 ft (1 m).   E.   10 ft (3 m).
    B.   3 ft (900 mm).   D.   4 ft (1.2 m).

Code reference _____

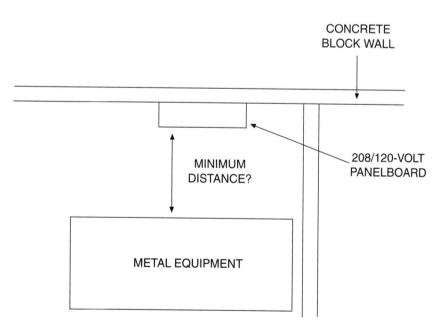

Figure 1.37   A 208/120-volt panelboard is mounted on a concrete block wall, and metal equipment is located in front of the panelboard.

4. A branch-circuit consists of three copper conductors with THHN insulation run in Electrical Metallic Tubing with no other conductors. The conductor terminations are rated for a maximum of 75°C. The ambient temperature in the area of this installation is not expected to exceed 30°C. If the conductors for this circuit are size 8 AWG, the maximum ampacity of the conductors permitted to be used to determine the minimum conductor size is:

   A. 40 amperes.   C. 55 amperes.   E. 65 amperes.
   B. 50 amperes.   D. 60 amperes.

   Code reference_____

5. A 15-ampere, 120-volt branch-circuit consisting of Type UF Cable is supplied from the service panel in a dwelling, and the entire circuit is protected with a GFCI. The direct burial cable supplies a luminaire (lighting fixture) on a post in the yard, and is run under an unpaved drive used only for dwelling-related vehicles, shown in Figure 1.38. The cable is required to be buried to a minimum depth of:

   A. 1 ft (300 mm).   D. 3 ft (900 mm).
   B. 1.5 ft (450 mm).   E. 4 ft (1.2 m).
   C. 2 ft (600 mm).

   Code reference_____

6. Type MC Cable with three size 8 AWG copper conductors is run in a commercial building through metal studs in a wall. When terminating at a metal box, the cable is required to be supported a distance from the box of not more than:

   A. 1 ft (300 mm).   C. 3 ft (900 mm).   E. 6 ft (1.8 m).
   B. 1.5 ft (450 mm).   D. 4.5 ft (1.4 m).

   Code reference_____

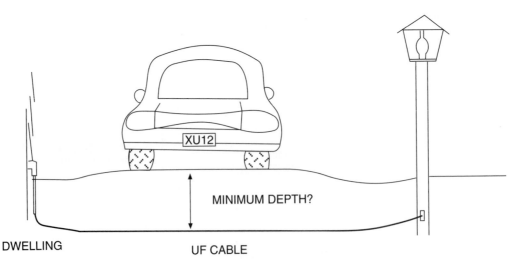

Figure 1.38 Determine the minimum depth of burial required for Type UF Cable under a dwelling drive with circuit rated 15 amperes and GFCI protected.

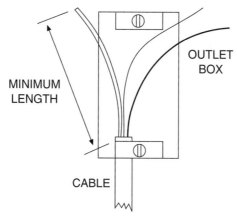

Figure 1.39   There is a minimum length of free conductor required at an outlet or junction box for making up splices and connections.

7. Type NM-B Cable enters a metal device box and is secured by a cable clamp at the bottom of the box. The cable is installed so the cable jacket extends $\frac{1}{4}$ in. (6 mm) beyond the cable clamp and free conductors begin at the end of the cable jacket as shown in Figure 1.39. If the device box has a depth of 3.5 in. (90 mm), the minimum permitted length of free conductor in this box is required to be not less than:

    A. 3 in. (75 mm).   D. 8 in. (200 mm).

    B. 6 in. (150 mm).   E. 12 in. (300 mm).

    C. 6.5 in. (163 mm).

Code reference _____

8. Type NM-B Cable is used in a dwelling as a switch loop from a ceiling box at a lighting outlet to a single-pole switch on the wall. A 120-volt, black, insulated conductor and a white, insulated neutral conductor are run using Type NM-B Cable to the ceiling box as shown in Figure 1.40. The white, insulated conductor from the ceiling box to the switch box:

    A. is permitted to be either the 120-volt supply to the switch or the return to the light.

    B. is only permitted to be the return to the light.

    C. if marked to identify it as an ungrounded conductor, is permitted to be the return to the light.

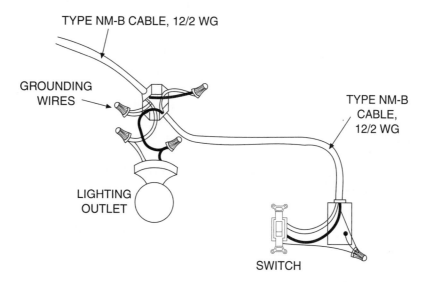

Figure 1.40   Nonmetallic-Sheathed Cable is used to wire a circuit containing a ceiling luminaire (lighting fixture) controlled by a single-pole wall switch.

D.   is only permitted to be the return to the light and must be marked to identify it as an ungrounded conductor.

E.   is only permitted to be the 120-volt feed to the switch and must be permanently marked to identify it as an ungrounded conductor.

Code reference_____

9.   If a person wishes to submit a proposal to amend a section of the 2005 *NEC®*, which may become a part of the 2008 *NEC®*, the proposal must be received either by mail or fax at the NFPA office in Quincy, MA, not later than 5:00 PM EST on:
A.   November 4, 2005.   D.   May 15, 2007.
B.   October 24, 2006.   E.   July 22, 2008.
C.   November 1, 2006.

Code reference_____

10.   The cross-sectional area of a size 8 AWG solid copper conductor is:
A.   4110 cmil (2.08 mm²).   D.   16,510 cmil (8.367 mm²).
B.   6530 cmil (3.31 mm²).   E.   26,240 cmil (13.30 mm²).
C.   10,380 cmil (5.261 mm²).

Code reference_____

11.   A point on the wiring system at which current is taken to supply utilization equipment is called:
A.   a tap.   D.   an appliance.
B.   a service point.   E.   an outlet.
C.   a circuit.

Code reference_____

12.   In a dwelling, for the purpose of supplying cord- and plug-connected loads 1440 volt-amperes or less, or less than 0.25 horsepower, the nominal voltage between conductors supplying device terminals shall not exceed:
A.   120 volts.   C.   240 volts.   E.   480 volts.
B.   208 volts.   D.   277 volts.

Code reference_____

13.   Type NM-B Nonmetallic-Sheathed Cable is required to be secured by staples, cable ties, straps, or similar fittings at intervals along the cable, shown in Figure 1.41, not exceeding:
A.   3 ft (900 mm).   C.   4.5 ft (1.4 m).   E.   8 ft (2.5 m).
B.   4 ft (1.2 m).   D.   6 ft (1.8 m).

Code reference_____

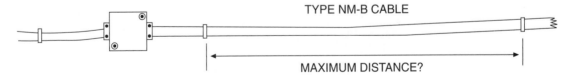

TYPE NM-B CABLE

MAXIMUM DISTANCE?

Figure 1.41   Determine the maximum support spacing for Nonmetallic-Sheathed Cable run along the surface of a building.

14. The ampacity of a size 8 AWG copper Type UF Cable installed within a building with 90°C insulation on the conductors and not in conditions requiring ampacity adjustment is:

    A. 35 amperes.   C. 45 amperes.   E. 55 amperes.
    B. 40 amperes.   D. 50 amperes.

                                        Code reference_____

15. Type AC Armored Cable entering a box is required to be fastened in place by an approved means within a distance of the box of not more than:

    A.   6 ft (1.8 m).   D. 12 in. (300 mm).
    B. 24 in. (600 mm).   E.   8 in. (200 mm).
    C. 18 in. (450 mm).

                                        Code reference_____

WORKSHEET NO. 1—ADVANCED GENERAL WIRING AND FUNDAMENTALS

Mark the single answer that most accurately completes the statement based upon the 2002 Code. Also provide, where indicated, the Code reference that gives the answer or indicates where the answer is found, as well as the Code reference where the answer is found.

1. A 3-phase, 480-volt electric motor drawing 21 amperes with a power factor of 0.58 will be drawing power of approximately:
   A. 5260 watts.   C. 7308 watts.   E. 11,190 watts.
   B. 5846 watts.   D. 10,114 watts.

2. Three resistors are connected in parallel, as shown in Figure 1.42, and they have resistances of 6 ohms, 9 ohms, and 18 ohms. The total resistance of the circuit is:
   A. 2 ohms.   C. 4 ohms.   E. 33 ohms.
   B. 3 ohms.   D. 11 ohms.

3. The terminal for a size 1/0 AWG copper conductor without any markings to indicate the temperature rating of the terminal is assumed to have a minimum terminal temperature rating of:
   A. 30°C.   C. 60°C.   E. 90°C.
   B. 40°C.   D. 75°C.

   Code reference_____

4. An aluminum overhead triplex feeder conductor supplies a building with a calculated load of 135 amperes and the feeder is protected at the supply end with a 150-ampere circuit breaker. If the conductor insulation has a rating of 75°C, the smallest conductor permitted for this feeder is size:
   A. 1 AWG.   C. 2/0 AWG.   E. 4/0 AWG.
   B. 1/0 AWG.   D. 3/0 AWG.

   Code reference_____

5. A 20-ampere rated branch-circuit supplies fluorescent luminaires (lighting fixtures) with a total load of 14.2 amperes. All terminations are 75°C rated, there are only three current-carrying conductors in EMT, and the conductor insulation is THHW as shown in Figure 1.43. The minimum size copper conductor permitted for this branch-circuit is:
   A. 16 AWG.   C. 12 AWG.   E. 8 AWG.
   B. 14 AWG.   D. 10 AWG.

   Code reference_____

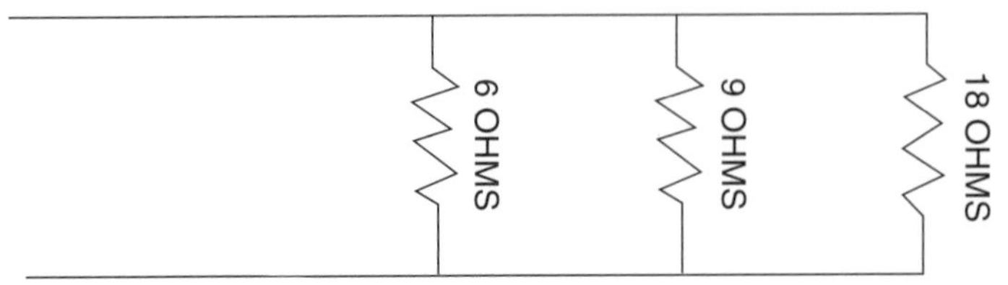

Figure 1.42   Three resistors are arranged in parallel with values of 6 ohms, 9 ohms, and 18 ohms.

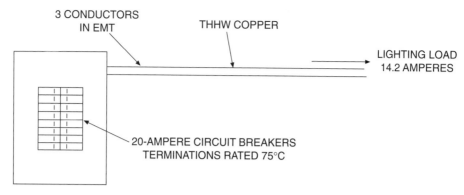

Figure 1.43 Three Type THHW copper conductors are run in Electrical Metallic Tubing to supply a lighting load of 14.2 amperes on a 20-ampere rated branch-circuit with all terminations rated 75°C.

6. Receptacles rated 120 volts and 15- or 20-amperes installed on construction sites for the purpose of supplying power for portable equipment are not permitted to be:
   A. of the grounding type.
   B. GFCI protected.
   C. supplied with Type NM-B Cable if the building is more than three floors in height.
   D. supplied with Type NM-B Nonmetallic-Sheathed Cable for other than dwellings.
   E. installed on branch-circuits that also supply temporary lighting.

   Code reference_____

7. An electrical panel is installed on a wall in a room of a commercial building where the distance from the floor to the structural ceiling is 15 ft (4.5 m). The distance from the floor to the top of the panelboard is 6 ft (1.8 m). An air-handling duct that will not cause dripping due to condensation is:
   A. not permitted to be installed above the panelboard.
   B. permitted to be installed above the panelboard.
   C. permitted to be installed not less than 3 ft (900 m) above the panelboard.
   D. permitted to be installed not less than 4 ft (1.2 m) above the panelboard.
   E. permitted to be installed not less than 6 ft (1.8 m) above the panelboard.
   Code reference_____

8. Type FCC Flat Conductor Cable is to be installed on the surface of a concrete floor and covered with carpet squares to supply general-purpose branch-circuits for work stations in a large room. The maximum rating permitted for the branch-circuits is:
   A. not specified in the Code.   D. 30 amperes.
   B. 15 amperes.   E. 40 amperes.
   C. 20 amperes.

   Code reference_____

9. Type MI Mineral-Insulated Cable with three size 8 AWG copper conductors is installed as concealed wiring inside a building in a dry location. The end seal fittings for the cable and terminals have a 90°C rating. The maximum permitted allowable ampacity of the cable is:
   A. 40 amperes.   C. 30 amperes.   E. 55 amperes.
   B. 35 amperes.   D. 50 amperes.

   Code reference_____

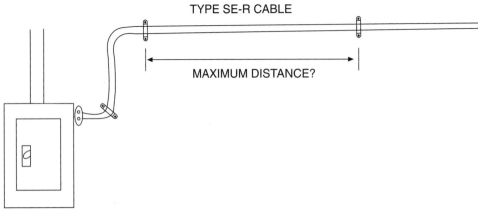

TYPE SE-R CABLE

MAXIMUM DISTANCE?

Figure 1.44   Type SE style R Service-Entrance Cable is run along the flat surface of the interior of a building as a feeder to another location in the same building.

10. Type SE-R Service-Entrance Cable with three insulated conductors and a bare equipment grounding conductor contained within the outer nonmetallic sheath is used as a feeder to provide power from the service panel to a panelboard located in another part of the building as shown in Figure 1.44. The cable run along the flat surface of structural materials is required to be supported at intervals not to exceed:
A. 6 ft (1.8 m).   D. 24 in. (600 mm).
B. 4^1/2 ft (1.4 m).   E. 12 in. (300 mm).
C. 3 ft (900 mm).

Code reference_____

11. In an industrial building with a staff of maintenance electricians, Type TC Power and Control Tray Cable with the same crush and impact requirements as Type MC Cable is permitted to be installed as open wiring between a cable tray and a machine and protected from physical damage provided the maximum distance between supports does not exceed:
A. 50 ft (15 m).   D. 15 ft (4.5 m).
B. 25 ft (7.5 m).   E. 6 ft (1.8 m).
C. 20 ft (6 m).

Code reference_____

12. The *National Electrical Code*® does not apply to wiring installations in:
A. underground mines.   D. recreational vehicles.
B. floating buildings.   E. public buildings.
C. carnivals.

Code reference_____

13. Type UF Cable is installed as surface wiring in a building where damp conditions are likely to exist. The cable is used with nonmetallic boxes and fittings that prevent the entrance of moisture. The cable is required to be supported at intervals not to exceed:
A. 10 ft (3 m).   D. 3 ft (900 mm).
B. 6 ft (1.8 m).   E. 24 in. (600 mm).
C. 4^1/2 ft (1.4 m).

Code reference_____

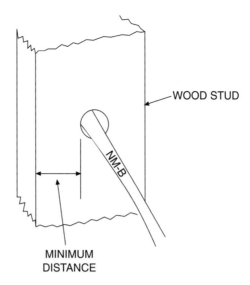

Figure 1.45   Type NM-B Nonmetallic-Sheathed Cable is run through bored holes in wood studs.

14.   Nonmetallic-Sheathed Cable, Type NM-B, run through bored holes in wood studs in a dwelling, are only permitted to be installed without a metal plate protecting the cable from penetration by screws or nails if the distance from the edge of the hole, shown in Figure 1.45, to the nearest edge of the stud is not less than:
A. $3/4$ in. (19 mm).   D. $1^1/4$ in. (32 mm).
B. $7/8$ in. (22 mm).   E.   2 in. (50 mm).
C.   1 in. (25 mm).

Code reference _____

15.   A dwelling is wired with Type NM-B Nonmetallic-Sheathed Cable. In the case of a lighting outlet controlled from two locations, a switch loop from the lighting outlet runs to the first 3-way switch with a conductor with two insulated conductors and an equipment ground, and between the 3-way switches with cable containing three insulated conductors and an equipment ground. The ungrounded conductor that originates at the lighting outlet runs to the common terminal of the second 3-way switch as shown in Figure 1.46. For this particular installation, the traveler conductors connecting the 3-way switches are required to be the red conductor and:
A.   either the white conductor with a permanent marking at each end or the black conductor.
B.   only the white conductor with a permanent marking at each end.
C.   only the black conductor with a red marking at each end.
D.   only the black conductor with a white marking at each end.
E.   only the black conductor.

Code reference _____

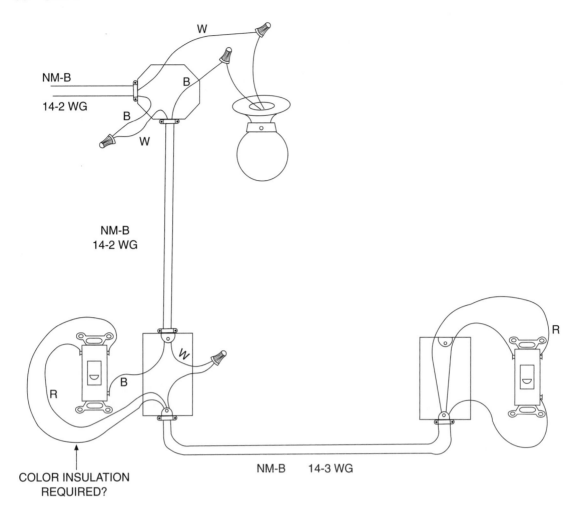

W

NM-B

B

14-2 WG

B

W

NM-B
14-2 WG

R

W

B

R

COLOR INSULATION
REQUIRED?

NM-B   14-3 WG

Figure 1.46   Two 3-way switches are used to control a luminaire (lighting fixture) using Type NM-B Nonmetallic-Sheathed Cable. Note the colors of conductors connected to the switches and luminaire (lighting fixture).

UNIT 2

Wire, Raceway, and Box Sizing

OBJECTIVES

After completion of this unit, the student should be able to:

- determine the size of a conductor for a circuit considering ambient temperature and more than three conductors in the raceway, cord, or cable.

- determine the minimum size conduit permitted when the conductors are all the same size and type of insulation.

- determine the minimum size conduit permitted when the conductors are different sizes and different types of insulation.

- determine the minimum size wireway and conduit nipples permitted for conductors.

- determine the minimum size junction box or device box permitted to take conductor fill into consideration.

- determine the minimum dimensions for pull boxes for straight pulls and angle pulls permitted.

- determine the minimum dimensions permitted for conduit bodies for various applications.

- answer wiring installation questions relating to *Articles 312, 314, 342, 344, 348, 350, 352, 353, 354, 356, 358, 360, 362, 366, 376, 378, 386, 388, Chapter 9 Tables 1, 2, 4, 5,* and *5A,* and *Annex C.*

- state at least five significant changes that occurred from the 2002 to the 2005 Code for *Articles 312, 314, 342, 344, 348, 350, 352, 353, 354, 356, 358, 360, 362, 366, 376, 378, 386, 388,* or *Chapter 9 Tables 1, 2, 4, 5,* and *5A,* or *Annex C.*

CODE DISCUSSION

The emphasis of this unit is to determine the minimum size of conductors for specific circuit and feeder applications if the actual or calculated load current is known. The ampacity tables in *Article 310* were discussed in *Unit 1.* Emphasis of this unit is on determination of the size and the installation of conductors, raceway systems, and boxes. A brief discussion of some key points made in the article dealing with raceways, cabinets, and boxes follows, with example calculations later in this unit.

Article 312 is on cabinets and cutout boxes used to enclose electrical equipment. Wire bending space and space requirements for wires in gutters within the enclosures are covered in *312.6.* When a wire or wires leave a lug or terminal and leave the enclosure through the wall opposite the lug, the distance from the lug to that enclosure wall is determined from *Table 312.6(B).* When the conductors leave an enclosure wall adjacent to the lug, the distance from the lug or terminal to the opposite wall is found in *Table 312.6(A).* These wire bending space requirements are illustrated in Figure 2.1.

Article 314 applies to outlet devices and junction boxes, pull boxes, and conduit bodies. It also covers requirements for handhole enclosures. Fittings permitted to contain splices or devices as permitted elsewhere in the code are required to meet the requirements of this article. Boxes, conduit bodies, and handhole enclosures shall be installed such that, after installation, they are accessible without removing any part of the building, or excavating, according to *314.29.* Boxes are permitted to be installed behind easily removable panels such as ceiling tiles in suspended ceilings where the ceiling tiles are easily removed.

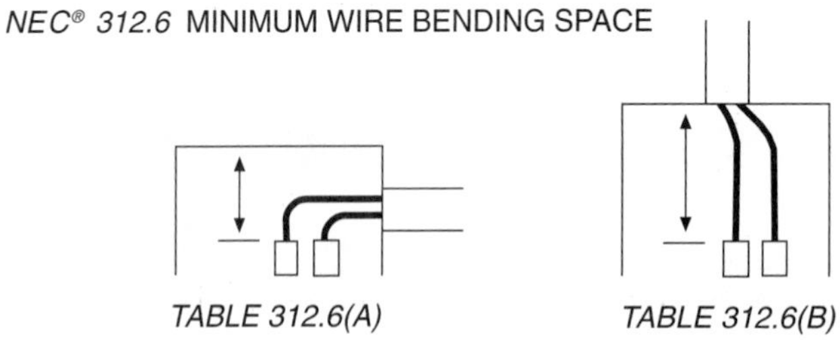

NEC® 312.6 MINIMUM WIRE BENDING SPACE

TABLE 312.6(A)   TABLE 312.6(B)

Figure 2.1   Minimum wire bending space is required from a lug or terminal to the opposite wall of the enclosure.

Minimum requirements for volume for the wires, devices, and fittings are given in *314.16*. In the case of standard device boxes, the maximum number of wires permitted in a box is given in *Table 314.16(A)*. The issue is that there is adequate physical space to prevent damage to conductors, prevent unnecessary pressure on splices and terminations, and prevent excessive heat produced within the box from current flowing in the wires and devices. If different sizes of wire enter a box or if a standard box is not used, then the minimum permitted volume of the box is determined using *Table 314.16(B)*. When wire sizes 4 AWG and larger are contained in a box, the minimum size permitted is determined on the basis of physical length and width of the box, according to the rules of *314.28*. Now the box is known as a pull box.

The rules for supporting boxes are given in *314.23*. Cable or raceway is required to be secured to metal boxes according to *314.17(B)*. Nonmetallic-Sheathed Cable and Type UF Cable are permitted to enter a single-gang nonmetallic box with dimensions not exceeding $2^1/4$ in. by 4 in. (57 mm by 100 mm) without being secured to the box, according to the *Exception* to *314.17(C)*. The cable is required to be fastened at a distance along the cable of not more than 8 in. (200 mm) from the box, and the cable sheath is required to extend into the box opening a distance of not less than $^1/4$ in. (6 mm), as shown in Figure 2.2. This only applies for single-gang boxes in walls and ceilings.

Article 342 is on Intermediate Metal Conduit (IMC). It is a metal raceway that is permitted to be threaded. It is permitted to be used in essentially the same applications as Rigid Metal Conduit. This type of conduit has a smaller wall thickness than Rigid Metal Conduit. Intermediate Metal Conduit is available in trade sizes $^1/2$ through 4 (16 through 103). It shall be supported within 3 ft (900 mm) of a box, fitting, or cabinet, and at intervals not more than 10 ft (3 m) unless threaded couplings are used. The 3 ft (900 mm) spacing requirement for supports at IMC terminations is permitted to be increased to not more than 5 ft (1.5 m) when building structural supports do not permit supporting the IMC within 3 ft (900 mm) of the termination according to *342.30(A)*. When threaded couplings are used, straight runs of IMC are permitted to be supported with the same maximum intervals as Rigid Metal Conduit, which are given in *Table 342.30(B)(2)*. Exposed vertical risers with threaded couplings are permitted to be supported at intervals not to exceed 20 ft (6 m), provided no other means of support is available for industrial machinery and fixed equipment.

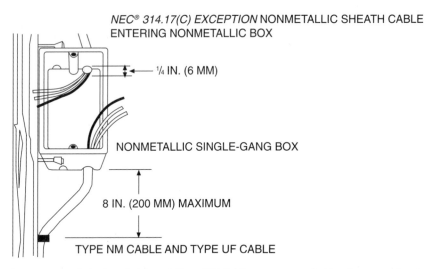

NEC® 314.17(C) EXCEPTION NONMETALLIC SHEATH CABLE
ENTERING NONMETALLIC BOX

$^1/4$ IN. (6 MM)

NONMETALLIC SINGLE-GANG BOX

8 IN. (200 MM) MAXIMUM

TYPE NM CABLE AND TYPE UF CABLE

Figure 2.2   Nonmetallic-Sheathed Cable and Type UF Cable are not required to be secured to a single-gang non-metallic box if it is secured with 8 in. (200 mm) of the box and extends into the box at least $^1/4$ in. (6 mm).

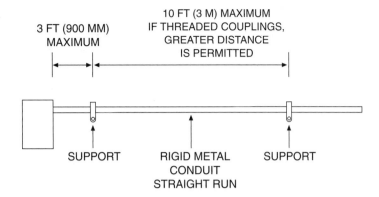

NEC® 344.30(B)(2) MAXIMUM DISTANCE BETWEEN SUPPORTS
FOR RIGID METAL CONDUIT IS GIVEN IN TABLE 344.30(B)(2) FOR
STRAIGHT RUNS WITH THREADED COUPLINGS.

Figure 2.3 **Maximum support spacing for Rigid Metal Conduit is 10 ft (3 m) unless threaded couplings are used; then maximum spacing is found in *Table 344.30(B)(2)*.**

Article 344 covers Rigid Metal Conduit (RMC) that has thicker walls than other types of metal conduit and tubing, and it is permitted to be threaded. Galvanized Rigid Steel Conduit is generally used where high mechanical strength is needed, and rigid aluminum conduit is often used where weight is required to be minimized. Rigid Metal Conduit is available in trade sizes from $1/2$ through 6 (16 through 155). There is a minimum radius of bend permitted for all field bends. The minimum bend radius depends on the trade diameter (metric designator) of the conduit as given in *Table 2, Chapter 9*. There is a minimum radius of bend for one-shot and full shoe benders, and another minimum radius required if other methods of bending are used such as a hickey style bender. The minimum radius is measured to the centerline of the conduit.

Rigid Metal Conduit shall be supported within 3 ft (900 mm) of a box, fitting, or cabinet, and at intervals not to exceed 10 ft (3 m). For straight runs of RMC with threaded couplings, the maximum support spacing is permitted to be increased to the distances given in *Table 344.30(B)(2)*. This is illustrated in Figure 2.3. The 3 ft (900 mm) spacing requirement for supports at conduit terminations is permitted to be increased to not more than 5 ft (1.5 m) when building structural supports do not permit supporting the conduit at a lesser distance, according to *344.30(A)*. Exposed vertical risers with threaded couplings are permitted to be supported at intervals not to exceed 20 ft (6 m) if no means of intermediate support is available for connection to industrial machinery and fixed equipment.

Article 348 is on Flexible Metal Conduit (FMC). It is of a spiral metal construction to provide flexibility and mechanical strength. This type of raceway is permitted for use in dry locations, and it is popular for use where flexibility is needed to connect raceway wiring systems to luminaires (lighting fixtures) and equipment. The minimum trade diameter generally permitted to be used is trade size $1/2$ (16); however, there are several applications given in *348.20(A)* where trade size $3/8$ (12) is permitted.

Flexible Metal Conduit shall be supported within 12 in. (300 mm) of a box, fitting, or enclosure with some exceptions. Where limited flexibility is necessary, as shown in Figure 2.4, lengths up to 3 ft. (900 mm)

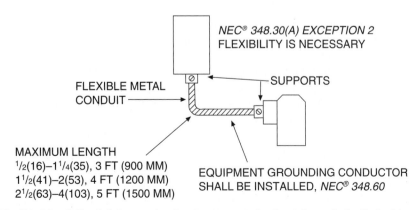

Figure 2.4 **Flexible Metal Conduit is permitted to be supported only at the ends for limited lengths when flexibility is necessary.**

are permitted to be supported only at the end connectors. Flexible Metal Conduit taps to luminaires (lighting fixtures) are permitted in lengths up to 6 ft (1.8 m), supported only at the terminals. An equipment grounding conductor is generally required to be installed because the Flexible Metal Conduit is usually not considered to be an acceptable equipment grounding conductor. The only condition is that, where the fittings are listed for grounding, circuit conductors within the listed Flexible Metal Conduit, not more than 6 ft (1.8 m) in length, are protected from overcurrent at not more than 20 amperes as provided in *250.118*.

Article 350 and *Article 356* describe Liquidtight Flexible Metal Conduit (LFMC) and Liquidtight Flexible Nonmetallic Conduit (LFNC) as having a nonmetallic liquidtight outer covering. They are available in trade sizes up to 4 (103). Both Liquidtight Flexible Metal Conduit and Liquidtight Flexible Nonmetallic Conduit are permitted for use for exposed and concealed wiring when the conditions of installation, operation, or maintenance require flexibility or require protection from liquids, vapors, or solid materials. Liquidtight Flexible Nonmetallic Conduit shall be listed and marked for the purpose when installed outdoors or for direct burial. Liquidtight Flexible Nonmetallic Conduit Type LFNC-B is permitted to be installed in lengths greater than 6 ft (1.8 m).

Liquidtight Flexible Conduit installed as a fixed wiring system is required to be supported at a distance of not more than 12 in. (300mm) from a box, fitting, or enclosure. Liquidtight Flexible Metal Conduit (LFMC) as a fixed wiring system is permitted to be supported at intervals not to exceed 4¹/₂ ft. (1.4 m). Liquidtight Flexible Nonmetallic Conduits, Type LFNC-A and Type LFNC-C, are not permitted to be installed in lengths greater than 6 ft (1.8 m), and are permitted to be supported only by the connectors. Liquidtight Flexible Nonmetallic Conduit, Type LFNC-B is permitted to be installed in lengths greater than 6 ft (1.8 m) as a fixed wiring system with a maximum support spacing of 3 ft (900 mm). Type LFNC-B is only permitted to be installed as a flexible connection to equipment supported only by the connectors in lengths not greater than 3 ft (900 mm). It is permitted to be installed in lengths up to 6 ft (1.8 m) supported only by the connectors to luminaires (lighting fixtures) or to other equipment in accessible ceilings.

An equipment grounding conductor is required to be installed through the Liquidtight Flexible Nonmetallic Conduit if the equipment or circuit supplied is required to be grounded. In the case of Liquidtight Flexible Metal Conduit, it is not permitted to be used as an equipment grounding conductor unless both the conduit and the fittings are listed for equipment grounding, and the length does not exceed 6 ft (1.8 m). If these requirements are met, then trade sizes of ³/₈ and ¹/₂ (12 and 16), are not required to have a supplemental equipment grounding conductor, provided the circuit conductors are protected from overcurrent at not more than 20 amperes. Trade sizes of ³/₄ through 1¹/₄ (21 through 35) are not required to have an equipment grounding conductor if the circuit overcurrent protection is not more than 60 amperes. Where necessary because flexibility is required, an equipment grounding conductor must be installed. These rules are found in *250.118*.

Article 352 concerns Rigid Nonmetallic Conduit (RNC), which is resistant to corrosion from moisture and most chemicals. It is available made from polyvinyl chloride (PVC) or from a reinforced thermosetting resin material (RTRC). The reinforced thermosetting resin conduit is stiffer than PVC and is permitted to have support spacings increased from those listed in *Table 352.30*. RNC is available as Schedule 40, which is the standard wall thickness, and as Schedule 80, which has a thicker wall and thus will generally withstand greater impact before damage will occur. RNC is available in trade sizes ¹/₂ through 6 (16 through 155). Standard lengths are 10 ft (3.048 m), although it is available as a continuous length from reels. The conduit is not threaded. It is joined to fittings by brushing an adhesive solvent on the conduit and then placing the conduit into the fitting. If done properly, this will form a watertight seal at the fitting. The conduit is bent by applying heat to the area to be bent until the conduit softens. Then the conduit is placed in a form to make the desired bend and allowed to cool and harden. Factory-made bends are available.

Rigid Nonmetallic Conduit changes length when it is exposed to changes in temperature. If an installation will be subject to a large temperature variation during normal use, then it may be necessary to install expansion fittings to allow for the thermal expansion and contraction. It is important to remember thermal expansion and contraction when installing Rigid Nonmetallic Conduit, especially when installing conduit supports. If RNC is installed in a location where it will experience a change in temperature, supports must be of a type that will allow the RNC to move as expansion and contraction occur. *Table 352.44(A)* gives the change in length of PVC when exposed to different temperature changes. The expansion rate for PVC Rigid Nonmetallic Conduit in *Table 352.44(A)* is 0.04056 in./100 ft °F or 0.06084 mm/m °C. *Table 352.44(B)* gives the change in length of RTRC Rigid Nonmetallic Conduit, and the expansion rate for that material is 0.0180 in./100 ft °F or 0.0270 mm/m °C. The temperatures in *Tables 352.44(A)* and *(B)* are temperature differences. One column is in degrees Celsius and the other is in degrees Fahrenheit. If the temperature difference is known in Fahrenheit degrees, multiply by 5 and divide by 9 to convert to Celsius degrees. A temperature difference of 45°F is equal to 25°C (45°F × 5 / 9 = 25°C). If the temperature difference is known in Celsius

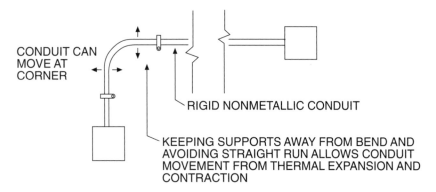

NEC® 352.44 ALLOW FOR THERMAL EXPANSION AND CONTRACTION OF RIGID NONMETALLIC CONDUIT

Figure 2.5   Install Rigid Nonmetallic Conduit so it can move due to thermal expansion and contraction if it will be exposed to a large temperature variation.

degrees, multiply by 9 and divide by 5 to get Fahrenheit degrees. A temperature difference of 75°C is equal to 135°F (75°C × 9 / 5 = 135°F).

If Rigid Nonmetallic Conduit supports are installed in the correct locations, the conduit can move when the temperature changes without damage. If possible, avoid confining the conduit so expansion and contraction will occur without applying stress to boxes, fittings, cabinets, and supports, as shown in Figure 2.5. If a straight run of Rigid Nonmetallic Conduit is installed so it cannot expand and contract with a change in temperature, then an expansion fitting is required if the length due to temperature change will be more than 1/4 in. (6 mm) according to *352.44*. A 120 ft (36.58 m) long straight section of PVC type Rigid Nonmetallic Conduit will change in length 4.87 in. (124 mm) if the temperature change is 100°F (55.6°C). Look up the expansion factor of 4.06 in. per 100 ft (3.38 mm/m) for PVC in *Table 352.44(A)*. Then multiply the expansion factor by hundreds of feet (120/100 = 1.2 hundreds) to get 4.87 in. (124 mm) of length change. Table 2.1 gives the change in length of PVC Rigid Nonmetallic Conduit for both English and metric units. Verify the numbers in the previous example by looking up the values in Table 2.1 on the next page.

A disadvantage of using Rigid Nonmetallic Conduit is that it tends to sag between supports; therefore, supports are required to be closer together than for comparable sizes of metal conduit and tubing. The maximum permitted support spacing for different trade diameters is given in *Table 352.30(B)*. The maximum support spacing permitted for Rigid Nonmetallic Conduit is shown in Figure 2.6. For trade sizes 1/2 through 1 (16 through 27), the maximum permitted spacing is 3 ft (900 mm). Rigid Nonmetallic Conduit of all trade diameters shall be supported within 3 ft (900 mm) of a box, fitting, or cabinet except support spacing for reinforced thermosetting resin conduit may be greater.

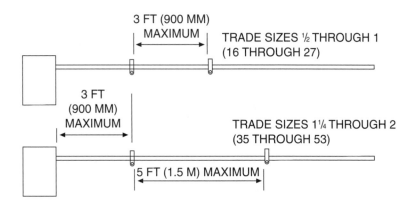

NEC® 352.30 SUPPORT SPACING OF RIGID NONMETALLIC CONDUIT DEPENDS ON THE TRADE DIAMETER

Figure 2.6   Support spacing for Rigid Nonmetallic Conduit depends on the trade diameter of the conduit.

Table 2.1 Change in length of PVC Rigid Nonmetallic Conduit with change in temperature.

Temperature change °C	Length change (mm/m)	Temperature change °F	Length change (in./100ft)	Temperature change °C	Length change (mm/m)	Temperature change °F	Length change (in./100ft)
2.8	0.17	5.0	0.20	47.2	2.87	85.0	3.45
5.0	0.30	9.0	0.37	48.0	2.92	86.4	3.50
5.6	0.34	10.0	0.41	49.0	2.98	88.2	3.58
6.0	0.37	10.8	0.44	50.0	3.04	90.0	3.65
7.0	0.43	12.6	0.51	51.0	3.10	91.8	3.72
8.0	0.49	14.4	0.58	52.0	3.16	93.6	3.80
8.3	0.51	15.0	0.61	52.8	3.21	95.0	3.85
9.0	0.55	16.2	0.66	53.0	3.22	95.4	3.87
10.0	0.61	18.0	0.73	54.0	3.29	97.2	3.94
11.0	0.67	19.8	0.80	55.0	3.35	99.0	4.02
11.1	0.68	20.0	0.81	55.6	3.38	100.0	4.06
12.0	0.73	21.6	0.88	56.0	3.41	100.8	4.09
13.0	0.79	23.4	0.95	57.0	3.47	102.6	4.16
13.9	0.85	25.0	1.01	58.0	3.53	104.4	4.23
14.0	0.85	25.2	1.02	58.3	3.55	105.0	4.26
15.0	0.91	27.0	1.10	59.0	3.59	106.2	4.31
16.0	0.97	28.8	1.17	60.0	3.65	108.0	4.38
16.7	1.01	30.0	1.22	61.0	3.71	109.8	4.45
17.0	1.03	30.6	1.24	61.1	3.72	110.0	4.46
18.0	1.10	32.4	1.31	62.0	3.77	111.6	4.53
19.0	1.16	34.2	1.39	63.0	3.83	113.4	4.60
19.4	1.18	35.0	1.42	63.9	3.89	115.0	4.66
20.0	1.22	36.0	1.46	64.0	3.89	115.2	4.67
21.0	1.28	37.8	1.53	65.0	3.95	117.0	4.75
22.0	1.34	39.6	1.61	66.0	4.02	118.8	4.82
22.2	1.35	40.0	1.62	66.7	4.06	120.0	4.87
23.0	1.40	41.4	1.68	67.0	4.08	120.6	4.89
24.0	1.46	43.2	1.75	68.0	4.14	122.4	4.96
25.0	1.52	45.0	1.83	69.0	4.20	124.2	5.04
26.0	1.58	46.8	1.90	69.4	4.22	125.0	5.07
27.0	1.64	48.6	1.97	70.0	4.26	126.0	5.11
27.8	1.69	50.0	2.03	71.0	4.32	127.8	5.18
28.0	1.70	50.4	2.04	72.0	4.38	129.6	5.26
29.0	1.76	52.2	2.12	72.2	4.39	130.0	5.27
30.0	1.83	54.0	2.19	73.0	4.44	131.4	5.33
30.6	1.86	55.0	2.23	74.0	4.50	133.2	5.40
31.0	1.89	55.8	2.26	75.0	4.56	135.0	5.48
32.0	1.95	57.6	2.34	76.0	4.62	136.8	5.55
33.0	2.01	59.4	2.41	77.0	4.68	138.6	5.62
33.3	2.03	60.0	2.43	77.8	4.73	140.0	5.68
34.0	2.07	61.2	2.48	78.0	4.75	140.4	5.69
35.0	2.13	63.0	2.56	79.0	4.81	142.2	5.77
36.0	2.19	64.8	2.63	80.0	4.87	144.0	5.84
36.1	2.20	65.0	2.64	80.6	4.90	145.0	5.88
37.0	2.25	66.6	2.70	81.0	4.93	145.8	5.91
38.0	2.31	68.4	2.77	82.0	4.99	147.6	5.99
38.9	2.37	70.0	2.84	83.0	5.05	149.4	6.06
39.0	2.37	70.2	2.85	83.3	5.07	150.0	6.08
40.0	2.43	72.0	2.92	84.0	5.11	151.2	6.13
41.0	2.49	73.8	2.99	85.0	5.17	153.0	6.21
41.7	2.54	75.0	3.04	86.0	5.23	154.8	6.28
42.0	2.56	75.6	3.07	86.1	5.24	155.0	6.29
43.0	2.62	77.4	3.14	87.0	5.29	156.6	6.35
44.0	2.68	79.2	3.21	88.0	5.35	158.4	6.42
44.4	2.70	80.0	3.24	88.9	5.41	160.0	6.49
45.0	2.74	81.0	3.29	89.0	5.41	160.2	6.50
46.0	2.80	82.8	3.36	90.0	5.48	162.0	6.57
47.0	2.86	84.6	3.43	91.0	5.54	163.8	6.64

Several types of Rigid Nonmetallic Conduit are intended only for underground installations. Type A is a thin-walled Rigid Nonmetallic Conduit that must be installed underground embedded in concrete. Type EB is also a thin-walled Rigid Nonmetallic Conduit that has a stiffer wall thickness than Type A. It too must be installed underground encased in concrete. Type HDPE schedule 40 Rigid Nonmetallic Conduit is permitted to be installed without concrete encasement, but only for underground direct burial applications. Rules for installing High Density Polyethylene Conduit, Type HDPF, are given in *Article 353*. It is also important to note the designation markings on reinforced thermosetting resin conduit. If it is marked underground, it is only permitted to be installed underground.

Article 354 describes the construction, use, and installation of a preassembled cable in nonmetallic underground conduit. This is a smooth outer surface nonmetallic conduit, which comes in continuous lengths usually on reels. The cable or conductors are already installed in the conduit. It is permitted to be used only for underground installations except for terminating in a building. This type is designated NUCC. The purpose of this product is the ease of installation of underground circuits to minimize the possibility of damage to the conductors during installation. This product is available in trade sizes from $^1/_2$ to 4 (16 through 103). The same fill requirements apply as for other conduit installations. It is permissible to remove the conductors at a later time and replace them with new conductors.

Article 358 deals with Electrical Metallic Tubing (EMT), and it is frequently used where the raceway is not exposed to physical damage. The tubing is not permitted to be threaded. This type of raceway is popular because it is easy to cut, bend, and install. Electrical Metallic Tubing is available in trade sizes $^1/_2$ through 4 (16 through 103). Minimum bending radius to the centerline of the tubing is the same as for Rigid Metal Conduit. If the bends are made in the field, the minimum bending radius is found in *Table 2 of Chapter 9*.

The maximum support spacing for Electrical Metallic Tubing is 10 ft (3 m). The tubing shall also be supported within 3 ft (900 mm) of a box, fitting, or enclosure. There is an exception that permits the support to be up to 5 ft (1.5 m) from the box, fitting, or enclosure if a practical means of support is not available a lesser distance.

Article 360 deals with a Flexible Metallic Tubing (FMT) that is liquidtight. It is important to read the list of uses permitted and uses not permitted covered in *360.10* and *360.12*. This material is not permitted to be used in lengths to exceed 6 ft (1.8 m) as stated in *360.12(6)*. The use of Flexible Metallic Tubing as an equipment grounding conductor is covered in *250.118(7)*. An equipment grounding conductor is required to be installed unless the circuit conductors in the tubing are protected from overcurrent at not more than 20 amperes, the fittings are listed for grounding, and the total length of Flexible Metallic Tubing in any grounding path is not more than 6 ft (1.8 m). Radius of bends is given in *360.24*, and the radius depends on whether the bend is fixed or may be infrequently flexed after installation.

Article 362 deals with Electrical Nonmetallic Tubing (ENT), which is a pliable corrugated raceway that can be bent by hand. It is available in trade sizes $^1/_2$ through 2 (16 through 53). The uses permitted are discussed in the article; however, in general, it is intended for use as exposed wiring, or as a concealed wiring method in walls, floors, and ceilings, and above suspended ceilings. If the building is more than three floors above grade, then walls, ceilings, floors, and suspended ceilings must have a 15-minute finish fire rating. An ENT installation is illustrated in Figure 2.7. Electrical Nonmetallic Tubing is permitted to be installed as sur-

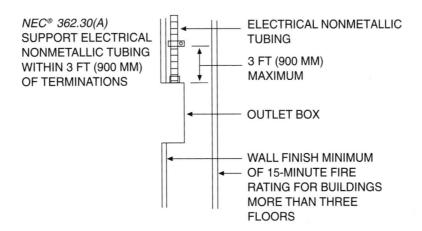

Figure 2.7   A minimum 15-minute finish fire rating is required when Electrical Nonmetallic Tubing is installed as concealed wiring in buildings of more than three floors above grade level.

face wiring, provided it is not subjected to physical abuse and provided the building is not over three floors in height. Electrical Nonmetallic Tubing shall be supported at intervals not exceeding 3 ft (900 mm), and it shall be supported within 3 ft (900 mm) of a termination at a box, fitting, or enclosure.

Article 366 is about auxiliary gutters, which are the same material as wireway except the purpose is different. Auxiliary gutters are limited to 30 ft (9.14 m) in length, and their purpose is to contain wiring between enclosures and devices such as at a motor control center or to connect the wiring and make taps for a group of enclosures making up a service entrance to a building. Auxiliary gutters constructed of metal are to be supported at intervals not to exceed 5 ft (1.5 m). *NEC® 366.22(A)* permits up to 30 current-carrying wires in a metal auxiliary gutter without applying the derating factors of *310.15(B)(2)(a)*, provided that the conductor cross section does not exceed 20% of the cross-sectional area of the metal auxiliary gutter. For nonmetallic auxiliary gutter, there is no 30 current-carrying conductor limit, but the derating factors of *310.15(B)(2)(a)* must be applied whenever there are more than three current-carrying conductors at any one cross section. *NEC® 366.100(A)* requires adequate electrical and mechanical continuity of the complete auxiliary gutter system. This requirement would indicate that the auxiliary gutter would be considered to be acceptable to serve as an equipment grounding conductor. Conductors entering or leaving an auxiliary gutter are required to have a wire bending space determined according to *Table 312.6(A)*. When conductors size 4 AWG and larger enter and leave an auxiliary gutter, the distance between cable or conduit entries are not permitted to be less than given in *314.28* as illustrated for wireway in Figure 2.8.

Article 376 and *Article 378* cover wireways, which are raceways of square cross section with a removable or hinged cover along one side. These are used as a raceway for conductors from one location to another. Metallic and nonmetallic wireway is available. Change in length due to change in temperature must be considered when nonmetallic wireway is installed in locations where it will be exposed to a change in temperature. The cross-sectional area of the wire is not permitted to exceed 20% of the cross-sectional area of the wireway. The derating factors of *310.15(B)(2)(a)* do not apply if there are not more than 30 current-carrying wires in the metal wireway and the fill is not over 20%. For nonmetallic wireway, there is no 30 current-carrying conductor limit, but the derating factors of *310.15(B)(2)(a)* must be applied whenever there are more than three current-carrying conductors at any one cross section. Wireways are not permitted to serve as equipment grounding conductors unless they are listed for the purpose.

Metallic wireway mounted horizontally is required to be supported at each end and at intervals not to exceed 5 ft (1.5 m) unless listed for greater support spacings. If the wireway is manufactured as one solid length of more than 5 ft (1.5 m), the support spacing is permitted to be at the ends but at intervals not to exceed 10 ft (3 m). Vertical runs of wireway are permitted to be supported at intervals of not more than 15 ft (4.5 m) with not more than one joint between supports. Nonmetallic wireway is required to be supported at terminations and at intervals not to exceed 3 ft (900 mm) unless listed for greater support intervals up to 10 ft (3 m). For a vertical run, the maximum support spacing is every 4 ft (1.2 m).

The minimum size of wireway for an application will depend upon not only the size, type, and number of conductors at any cross-section, but also how the wireway is used. The total cross-sectional area of the conductors is not permitted to exceed 20% of the inside cross-sectional area of the wireway. Nonmetallic wireways are required to have the inside cross-sectional area marked. Conductors frequently enter perpendicular to the wireway as shown in Figure 2.8. The minimum distance from the raceway entry to the opposite wall of the wireway is required to be not less than the wire bending requirements of *Table 312.6(A)* for one conductor per terminal. When conductors size 4 AWG and larger enter and leave a wireway, the distance between cable or conduit entries is not permitted to be less than given in *314.28* as illustrated in Figure 2.8. The distance is six times the trade diameter of the largest raceway when the entries are offset, and eight times the trade diameter of the largest raceway when the entries are straight across from each other. Another way to deal with a straight pull through a wireway is to run the raceway straight through without the conductors actually entering the wireway. This will avoid having to install a wireway with a dimension greater than necessary for the conductors run in the wireway.

Article 386 and *Article 388* are about surface metal raceways and surface nonmetallic raceways. These are raceways attached to the surface of walls or ceilings. A typical application is where the raceway is run on the surface to extend from an existing outlet to a new outlet location. Equipment grounding for surface metal raceway is covered in *386.60*. It is not permitted to be used as an equipment grounding conductor unless the surface metal raceway is specifically listed for the purpose. Raceway support requirements are not stated in this article, but they are required to meet the general support requirements of *300.11(A)*. *NEC® 300.11(A)* simply states that raceway shall be securely fastened in place, and there are no specific Code requirements for surface raceway support.

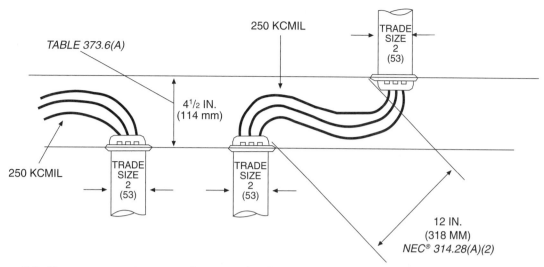

Figure 2.8 For a raceway entry or a conductor passing through an Auxiliary Gutter or Wireway with offset raceway or cable entries, the distance between raceway or cable entries shall not be less than given in *314.28*.

Chapter 9, Table 1 Notes provide information necessary for the use of the tables. *Notes 3* and *4* are of particular interest. *Note 3* states that equipment grounding conductors, if present, are to be counted when determining conduit or tubing fill. *Note 4* covers conduit and tubing nipples. As illustrated in Figure 2.9, conduit or tubing not more than 24 in. (600 mm) in length is considered to be a nipple. The wire fill is permitted to be 60% of the conduit or tubing total cross-sectional area. Also, the derating factors of *310.15(B)(2)(a)* do not apply in the case of a conduit or tubing nipple.

Table 2 of *Chapter 9* gives the minimum bending radius permitted for conduit and tubing. For the case of factory bends or field bends made with a one-piece bending tool, the first column of *Table 2* is used. When bends are made with a tool that makes bends in multiple steps, the column marked "other bends" is used. In all cases, the minimum radius of bend is made to the centerline of the conduit or tubing. In the case of Flexible Metal Conduit (FMC), Liquidtight Flexible Metal Conduit (LFMC), Liquidtight Flexible Nonmetallic Conduit (LFNC), and Flexible Nonmetallic Tubing (FNT), the minimum bending radius is measured to the centerline of the conduit, but the other bend column is used. In the case of Rigid Nonmetallic Conduit, where field bends can be made by various methods, there is no specification of which column of *Table 2* is to be used. As long as the cross-section of the conduit is not distorted, the first column of *Table 2* can be used.

Tables in *Chapter 9* or *Annex C* are used to determine the maximum number of wires and cables permitted to be installed in raceway. The total cross-sectional area of the conductors including the insulation is not permitted to exceed a maximum percentage of the cross-sectional area of the conduit or tubing. For most applications, the maximum is 40% fill. *Table 1* gives the maximum permitted percentage of cross-sectional area the wires are permitted to fill in conduit and tubing. *Table 4* gives the internal diameter and cross-

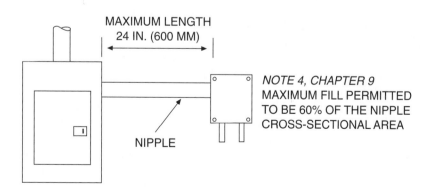

Figure 2.9 Conduit or tubing between enclosures that is not more than 24 in. (600 mm) in length is considered to be a nipple.

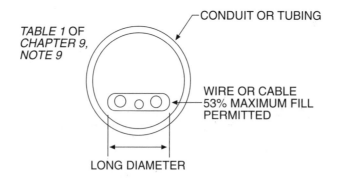

Figure 2.10   A Type NM-B Cable is installed in conduit or tubing.

sectional area for the common types of conduit and tubing. The available trade diameter for the different types of conduit and tubing are provided. The table also gives the maximum cross-sectional area permitted for each size and type of conduit or tubing for one, two, and three or more conductors. Specific directions are provided in the notes of this table that tell how to determine the minimum conduit or tubing diameter permitted for specific wires or cable. *Note 9* tells how to determine the cross-sectional area of conduit or tubing for a multiconductor cable with an elliptical cross section.

> **Example 2.1**   A Type NM-B Cable with two insulated size 14 AWG conductors and one bare conductor has a maximum dimension of $^3/_8$ in. (0.375 in. or 9.5 mm). Determine the cross-sectional area of the cable to find the minimum trade diameter Rigid Metal Conduit permitted to be installed, as shown in Figure 2.10.
>
> **Answer:**   *Note 9* of *Table 1, Chapter 9* requires that for the purpose of determining conduit or tubing fill, the maximum dimension of the cable is to be used as though it was the diameter of a circular cable. The minimum trade diameter Rigid Metal Conduit required is trade size $^1/_2$ (16). Look up the area in the "one conductor" column of *Table 4*. The area of a circle is as follows:
>
> $$\text{Area of a circle} = \frac{3.14 \times \text{Diameter} \times \text{Diameter}}{4}$$
>
> $$= \frac{3.14 \times 0.375 \times 0.375}{4} = 0.110 \text{ in.}^2$$
>
> $$= \frac{3.14 \times 9.5 \text{ mm} \times 9.5 \text{ mm}}{4} = 70.8 \text{ mm}^2$$

When the wires are all of the same size and type of insulation, the cross-sectional area of the wires will be identical. In this case, *Note 1, Chapter 9*, specifies that the appropriate table in *Annex C* is permitted to be used to look up the minimum size of conduit or tubing for the wires. There are separate tables in *Annex C* for each type of conduit or tubing. Conductor strands may be made compact with no void space, or they may be round. The conductors with compact strands have a smaller cross-sectional area than conductors of the same wire gage with round strands. The tables in *Annex C* for compact wires are designated with the letter A such as *Table C1(A)* through *Table C12(A)*.

If the wires are different sizes and different insulation types, the cross-sectional areas of each size or type will be different. In this case, it will be necessary to calculate the total cross-sectional area of the conductors. *Tables 5* and *5A* give the diameter and cross-sectional area of conductors and their insulation. For aluminum conductors with compact conductor configuration, the values are found in *Table 5A*. Then *Table 4* is used to determine the minimum permitted size of conduit or tubing for the wires. *Table 4* provides data on trade sizes (metric designator) of conduit and tubing, such as internal diameter and internal cross-sectional area. The table gives values of area that are percentages of the total area, for example, 1.342 sq. in. (866 mm²) is 40% of the total cross-sectional area of trade size 2 (53) Electrical Metallic Tubing.

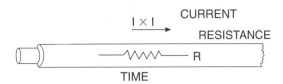

INCREASE IN ANY OF THESE INCREASES HEAT PRODUCED IN WIRE

Figure 2.11 **Heat produced in a conductor is proportional to the square of the current times the resistance of the conductor times the amount of time the current flows through the conductor.**

SAMPLE CALCULATIONS

Methods for making the calculations to select the minimum size permitted for conductors, conduit, and boxes for specific installations are discussed. These methods are the same as those used in the Code, but, as the Code often uses the trial and error method for minimum size determination, the methods presented in this unit use direct calculations from which the minimum permitted size may be determined. It is suggested that the student copy the various formulas in the margin of the Code page where the size is to be determined for easy reference in the future.

Protecting Conductors

Electrical wires are required to be protected to prevent insulation damage. There are three common ways that insulation damage can occur, and insulated conductors are required to be protected from such damage.

- Protect insulation from excessive temperature (*310.10*).
- Protect insulation and wire from physical damage (*300.4*).
- Prevent deterioration of insulation by environmental factors such as chemicals and moisture (*110.11*).

The ampere rating of a conductor will have an effect on the operating temperature of a conductor. A conductor heats as electrical current flows through the conductor. The amount of heat produced by current flow can be calculated using Equation 1.16. Figure 2.11 represents current flow through a conductor.

$$\text{Heat} = I^2 \times R \times t$$

Where: I = current, in amperes
R = resistance, in ohms
If **t** is in hours, heat is in watt-hours
If **t** is in seconds, heat is in joules

Watt-hours can be converted to British thermal units (Btu) by multiplying by 3.413. As electrical current flows through a conductor, the temperature will rise. It takes approximately three hours of steady current flow for the temperature of the wiring system to reach a maximum operating temperature. This is why, in the Code, continuous load is considered a load operating for three hours or longer. The Code is concerned about heat produced under various conditions, such as 100% of the rated maximum conductor ampacity, 80% maximum conductor ampacity, and 50% conductor ampacity. Table 2.2 compares the joules of heat produced in 100 ft of size 12 AWG copper with a wire resistance of 0.16 ohm.

Compare the amounts of heat produced by the wire, as shown in Table 2.2. Conductors produce only one-quarter as much heat when operating at 50% load as they do at 100% of the rated maximum conductor ampacity. The conductors produce only 64% as much heat at 80% of conductor ampacity as at 100%. These percentages were determined based on the heating of 100 ft of size 12 AWG wire using Equation 1.16.

Table 2.2 **Approximate heat produced by current flow along 100 ft of size 12 AWG copper wire.**

Circuit rating	Amperes	Heat produced
50%	10	16 J/s
80%	16	41 J/s
100%	20	64 J/s

Ambient or surrounding temperature has an effect on the operating temperature of a wire. The ampacity of conductors for conditions specified in the tables of *Article 310* is determined at a specified ambient temperature. Temperature correction factors are provided at the bottom of the ampacity tables of *Article 310,* which are used to adjust the ampacity of the table for ambient temperatures other than those specified for the table. For example, consider a copper wire, size 3 AWG, which is in conduit in free air with 60°C insulation with an ambient temperature of 120°F (49°C). The ampere rating of the wire is found in *Table 310.16* as 85 amperes. The ampacity correction factor at the bottom of the table for 120°F (49°C) is 0.58, which results in an allowable ampere rating of the wire under these conditions of 49 amperes.

$$\text{Allowable wire ampacity} = 85 \text{ A} \times 0.58 = 49 \text{ A}$$

The environment around a conductor affects the rate at which heat produced by current flow is removed from the conductor. This is the reason why several ampacity tables are in *Article 310* of the Code. Heat is removed from the conductors at different rates if the conductors are overhead in free air, in cable, or directly buried in the earth. If the conductors are wet or dry, the rate at which heat is removed from the conductor changes under certain conditions. For example, referring to *Table 310.16,* conductors with insulation Type XHHW are permitted to have the ampacity determined as a 90°C rated conductor, but if the conditions are wet, it is considered a 75°C conductor.

When there is more than one conductor in cable or raceway carrying electrical current, the overall temperature of the conductors builds more rapidly because each conductor is producing heat. The ampacity tables of *Article 310* were based on a maximum of three current-carrying conductors. When there are more than three current-carrying conductors in a cable or raceway, an adjustment factor is used to determine the allowable ampere rating for a conductor. These adjustment factors are found in *310.15(B)(2)(a)*. When there are ten current-carrying conductors in cable or raceway, the adjustment factor drops to 50%. Because of the significant drop in the adjustment factor from nine to ten conductors in a single run of cable or conduit, there is a real incentive to limit the number of wires to not more than nine.

There may be a situation where all of the conductors will not be energized at the same time. This is called load diversity. If the load diversity is not more than 50%, then other adjustment factors can be used. A 50% load diversity would mean that in any given cable or run of conduit or tubing, only 50% of the conductors would be energized at any time. The fine print note in *310.15(B)(2)(a)* calls attention to *Table B-310-11,* which is in *Annex B* at the end of the Code. These adjustment factors are permitted to be used only when there is a load diversity of at least 50% and approved by the authority having jurisdiction.

Neutral as a Current-Carrying Conductor

Consider the case of more than three conductors in conduit, tubing, cord, or cable. The first step is to actually determine the number of current-carrying conductors. A neutral conductor, when serving as a common conductor to more than one ungrounded conductor, may carry only the unbalanced load between the ungrounded conductors. *NEC® 310.15(B)(4)* discusses this issue of when to count the neutral. Frequently, in the case of multiwire feeders and branch-circuits, the neutral conductor is not considered a current-carrying conductor. Therefore, it is not counted for the purpose of derating. The following discussion explains when to count the neutral conductor for the purpose of derating for more than three conductors in raceway or cable.

The single-phase, 120/240-volt 3-wire feeder, or multiwire branch-circuit, does not require the neutral to be counted as a current-carrying conductor. This is true even if the major portion of the load is electric discharge lighting or data processing equipment. This type of circuit or feeder is shown in Figure 2.12.

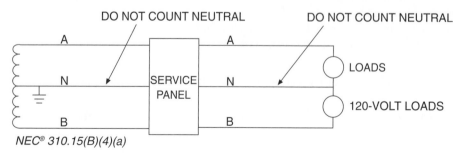

Figure 2.12   The neutral is not counted as a current-carrying conductor for a single-phase, 120/240-volt feeder or multiwire branch-circuit.

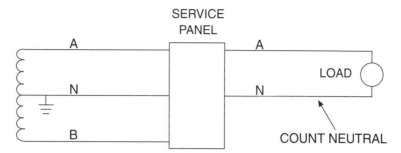

Figure 2.13   The neutral is counted as a current-carrying conductor for a 2-wire, single-phase, 120-volt branch-circuit or feeder.

For the case of the 120-volt 2-wire circuit, the neutral is counted as a current-carrying conductor, as illustrated in Figure 2.13. It does not matter if the neutral is from a 3-wire 120/240-volt single-phase system, or from a 208/120-volt 3-phase system, the neutral is counted as a current-carrying conductor. The neutral also is counted as a current-carrying conductor of a 2-wire 277-volt circuit.

A 3-wire feeder, or multiwire branch-circuit, can be obtained from a 3-phase, 4-wire wye electrical system with the feeder or branch-circuit operating at 208/120 volts. Balancing the 120-volt loads will not reduce the current on the neutral *(310.15(B)(4)(b))*. The neutral will always carry significant current, and the neutral is counted as a current-carrying conductor. Figure 2.14 shows this 3-wire circuit derived from a wye electrical system.

The 3-phase, 4-wire delta 240/120-volt feeder has a neutral conductor that serves as a common conductor for two of the ungrounded conductors, just like the 120/240-volt 3-wire single-phase system of Figure 2.12. The 4-wire, 240/120-volt delta system is shown in Figure 2.15. The neutral is not counted as a current-carrying conductor because the neutral only carries current due to the unbalance between the 120-volt loads on each of the ungrounded conductors (ungrounded conductors A and C). The 120-volt loads are not required to be balanced. The heat produced by current in the 120-volt circuit conductors is not greater in the unbalanced condition than when in the balanced condition.

For the 3-phase, 4-wire wye, 208/120-volt and 480/277-volt electrical systems, the neutral is not counted unless the major portion of the load consists of electronic computers, data processing equipment, or electric discharge lighting. In the case of a 4-wire set of branch-circuits with a common neutral serving electric discharge lights in a room, the neutral is required to be counted as stated in *310.15(B)(4)(c)*. If less

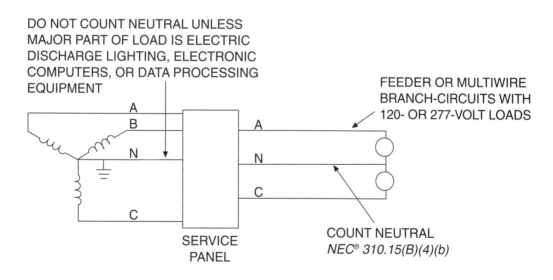

Figure 2.14   The neutral is counted as a current-carrying conductor for a 3-wire feeder or branch-circuit derived from a wye system.

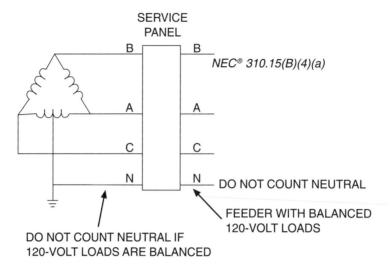

Figure 2.15 **The neutral is not counted as a current-carrying conductor for a 4-wire feeder derived from a 240/120-volt delta electrical system with reasonably balanced 120-volt loads.**

than half of the line-to-neutral load in a building was electric discharge lighting, electronic computers, or data processing equipment, then the feeder neutral most likely would not be required to be counted as a current-carrying wire. The feeder and multiwire branch-circuits of a 4-wire wye system are shown in Figure 2.16.

Sizing Conductors

The Code in *Article 310* requires that the allowable ampacity permitted for a conductor be determined based on the ampacity derating factors for ambient temperatures higher than that for which the ampacity table was developed. When there are more than three current-carrying conductors in cable or raceway, the derating factors of *310.15(B)(2)(a)* shall also apply. A conductor is chosen from the proper ampacity table in the Code by choosing the wire that seems to be adequate, and then multiplying the allowable ampacity in the table by the temperature correction factor if it applies, and by the derating factor from *310.15(B)(2)(a)* if it applies. This derated allowable ampacity value is not permitted to be smaller than the rating of the circuit subject to the provisions of *240.4*. If it is smaller than the circuit rating, then it will be necessary to choose a larger wire and repeat this calculation. The allowable ampacity of a conductor if a temperature derating factor and a derating factor for more than three conductors in a cable or conduit apply can be determined using Equation 2.1.

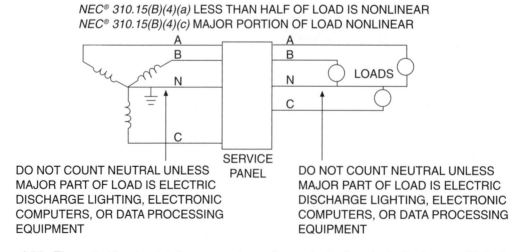

Figure 2.16 **The neutral is not counted as a current-carrying conductor for a 4-wire feeder or multiwire branch-circuit unless the major portion of the load is electric discharge lighting, electronic computers, or data-processing equipment.**

Adjusted Allowable Ampacity   =   Table Allowable Ampacity ×

Derating Factor (*310.15(B)(2)*) ×

Temperature Correction Factor   **Eq. 2.1**

The rules for sizing conductors for a branch-circuit are in *210.19(A)(1)* and for feeders in *215.2(A)(1)*. These rules are the same, but before they can be applied, first determine the size of overcurrent device protecting the branch-circuit or feeder. The minimum size overcurrent device permitted for a branch-circuit is specified in *210.20(A)* and for feeders in *215.3*. The rules are the same. The minimum overcurrent device rating is to be not less than 1.25 times the continuous load plus any noncontinuous load. Look up the standard ratings of overcurrent devices in *240.6(A)*.

The next step in sizing a conductor for a branch-circuit or a feeder is to determine the minimum permitted conductor size assuming no correction factors or adjustment factors apply. The minimum conductor size is required to have an allowable ampacity not less than 1.25 times the continuous load plus any noncontinuous load. The column of the ampacity table to use depends upon the lowest temperature rating of a component of the circuit. It is necessary to know the temperature rating of splicing devices and terminations of the circuit. It is also necessary to know the insulation temperature rating of the conductors. If no information is known about termination temperature, then the rule in *110.14(C)(1)* will apply. The lowest temperature rating of a component in a circuit will determine the column of *Table 310.16* that will be used to determine the minimum size of the conductor for the circuit or feeder. Assume some terminations in a circuit are rated 60°C and some are rated 75°C. Assume the conductor insulation is THHN with a 90°C rating. For this circuit it will be necessary to look up the minimum conductor size using the 60°C column of *Table 310.16*.

Next, determine if any adjustment or correction factors will apply. Is there a portion of the circuit where there are more than three current-carrying conductors? If the answer is yes, then determine the lowest temperature rating of any component in that portion of the circuit or feeder. For example, assume the conductors being sized have THHN insulation, but other conductors in that same run of raceway have THWN insulation. Then the ampacity to which the adjustment factor will be applied will be from the 75°C column. If all conductors in that portion of the raceway had 90°C rated insulation, then the ampacity to which the adjustment factor will be applied will be from the 90°C column of *Table 310.16*. This is permitted even though in another part of the circuit there are terminations rated at only 60°C. Look up the allowable ampacity in *Table 310.16* and apply any temperature correction factor and adjustment factor from *310.15(B)(2)(a)* that applies to the section of the circuit in question.

NEC® 210.19(A)(1) states in the first sentence that the conductors for a branch-circuit shall have an ampacity not less than the maximum load to be served. The conductor ampacity is the value determined after applying the adjustment and correction factors. That value must not be less than the load. That means 1.0 times the continuous load plus 1.0 times the noncontinuous load. This will determine if the minimum conductor size of the first step is still adequate to handle the load when adjustment and correction factors are applied. The following examples will illustrate the process of conductor size selection.

Example 2.2   The home runs for eight 120-volt lighting circuits supplied from a 120/240-volt, 3-wire, single-phase electrical system are contained in the same conduit, which runs for 30 ft (9.14 m) through an area of a building where the ambient temperature typically runs to 120°F (49°C). There are no fixtures or conductor terminations in the room that may operate up to 120°F (49°C). Each circuit supplies eight fluorescent luminaires (lighting fixtures), each of which draws 1.6 amperes for a total of 12.8 amperes per circuit. Multiwire branch-circuits are used with a common neutral for two ungrounded conductors. The total conductors within the conduit are eight ungrounded conductors and four neutral conductors. This circuit is illustrated in Figure 2.17. All circuits are protected with 20-ampere circuit breakers, and all terminations in the circuits are 75°C rated. Determine the minimum size of Type THHN copper conductors for the circuits.

Answer: The overcurrent device for each circuit has already been selected at 20 amperes. The next step is to determine the minimum size conductor permitted without considering any adjustment or correction factors according the *210.19(A)(1)*. The circuit supplies a continuous load, therefore, the minimum size conductor must have an allowable ampacity of not less than 1.25 × 12.8 amperes, which is 16 amperes. The 75°C column of *Table 310.16* will be used because of the conductor terminations. It would appear that a size 14 AWG copper conductor is acceptable, but *240.4(D)* only permits a 15-ampere

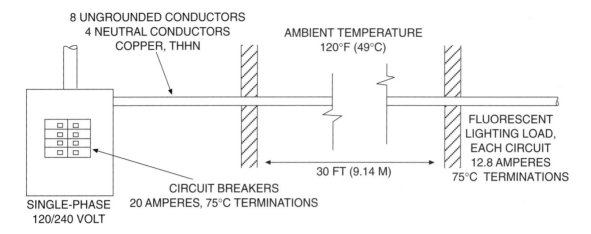

8 UNGROUNDED CONDUCTORS
4 NEUTRAL CONDUCTORS
COPPER, THHN

AMBIENT TEMPERATURE
120°F (49°C)

FLUORESCENT
LIGHTING LOAD,
EACH CIRCUIT
12.8 AMPERES
75°C TERMINATIONS

30 FT (9.14 M)

CIRCUIT BREAKERS
20 AMPERES, 75°C TERMINATIONS

SINGLE-PHASE
120/240 VOLT

Figure 2.17   Eight circuits of a 120/240-volt, 3-wire electrical system supply fluorescent fixtures using multiwire branch-circuits that run through a room with an ambient temperature of 120°F (49°C).

overcurrent device on a size 14 AWG copper wire. These circuits have 20-ampere overcurrent devices, therefore, a size 12 AWG copper wire is required.

Next, apply any adjustment factors to a size 12 AWG copper Type THHN conductor to determine if it is still acceptable. The most severe conditions are in the room with the 120°F (49°C) ambient temperature. The 90°C column of *Table 310.16* can be used in this area to apply the temperature correction and other adjustment factors. For an ambient temperature of 120°F (49°C), the temperature correction factor is 0.82.

Now determine the adjustment factor from *310.15(B)(2)(a)*. First, it is necessary to determine if the neutral conductors are counted as current-carrying conductors. According to *310.15(B)(4)(a)*, the neutrals are not counted, therefore, the total number of current carrying conductors for this circuit is eight. From *Table 310.15(B)(2)(a)*, the adjustment factor is 0.7. Use Equation 2.1 to determine the adjusted allowable ampacity of THHN copper conductors under these conditions.

Adjusted Allowable Ampacity = 30 A × 0.82 × 0.7 = 17.2 A

Finally, compare the adjusted allowable ampacity of the conductors with the actual load which, in this case, is 12.8 amperes. The adjusted allowable ampacity is greater than the load, but it is less than the rating of the overcurrent device. There is a rule in *240.4(B)* that permits a circuit conductor that does not match a standard rating of an overcurrent device to be protected by the next larger rating of overcurrent device which, in this case, is 20 amperes. The standard ratings of overcurrent devices are listed in *240.6(A)*. The conclusion is that the size 12 AWG copper THHN conductor are adequate for these circuits under these conditions.

Example 2.3 Fluorescent luminaires (lighting fixtures) are installed on 120-volt circuits each of which draws 14.4 amperes. The electrical supply in the building is 3-phase, 4-wire, 208/120-volt service. Six 20-ampere lighting circuits are run in the same raceway to an area of the building where the luminaires (lighting fixtures) are installed. These are multiwire branch-circuits with six ungrounded conductors and two neutrals. The conductors are copper with THWN insulation, and all circuit terminations are 75°C rated. Determine the minimum size conductors for the circuits.

Answer: The overcurrent devices for the circuits have already been selected at 20 amperes, which require a size 12 AWG copper conductor. The load on the circuit is 14.4 amperes which, multiplied by 1.25, gives 18.0 amperes as the minimum conductor allowable ampacity without adjustment factors applied. Fluorescent luminaires (lighting fixtures) are generally considered to be nonlinear loads. In that case, *310.15(B)(4)(c)* requires the neutrals to be counted as current-carrying conductors. For this raceway run, there are eight current-carrying conductors which, according to *Table 310.15(B)(2)(a)*, gives an adjustment factor of 0.7. The 75°C column of *Table 310.16* is used to determine the unadjusted allowable ampacity of the conductors in the raceway. Multiply the 25 amperes by 0.7 to get an adjusted allowable ampacity of 17.5 amperes. The allowable ampacity of the conductor is greater than the load of 14.4 amperes. *NEC® 240.4(B)* permits this 17.5-ampere conductor to be protected with the next larger standard rating of overcurrent device as listed in *240.6(A)*, which is 20 amperes. The conclusion is that a size 12 AWG copper THWN conductor is the minimum size permitted for the circuit.

Example 2.4   Two 3-phase, 3-wire circuits supply a 48-ampere continuous load. The circuit conductors are copper with THHN insulation, and all circuit-conductor terminations are 75°C rated. In route to supplying the load, the circuits run through a room for a distance of 25 ft (7.62 m) that typically has an ambient temperature of 125°F (51.6°C), as shown in Figure 2.18. Determine the minimum size conductors for these circuits.

Answer: First, determine the minimum rating of overcurrent device for the circuits using the rule of *210.20(A)*. The overcurrent device is required to have a rating not less than 1.25 times the continuous load, plus the noncontinuous load. For this circuit, the minimum rating is required to be 60 amperes (1.25 × 48 A = 60 A). A 60-ampere circuit breaker or set of fuses is the minimum required for this circuit.

Next, determine the minimum size conductor for the circuit assuming adjustment and correction factors do not apply. The rule is found in *210.19(A)*. The minimum conductor allowable ampacity, not considering adjustment or correction factors, is 1.25 times the continuous load plus the noncontinuous load. For this circuit, the minimum conductor size must have an allowable ampacity of 60 amperes (1.25 × 48 A = 60 A). The termination temperature rating of 75°C determines the column of *Table 310.16* that is used to find the minimum conductor size for this circuit which is size 6 AWG.

Finally, the temperature-correction factor and the adjustment for more than three conductors in the raceway are applied. The conductor insulation is THHN, therefore, it is permitted to use the 90°C column of *Table 310.16* to find the allowable ampacity to which the correction and adjustment factors are applied. Use Equation 2.1 to determine the adjusted allowable ampacity of the conductors. Start with size 6 AWG to see if after adjustment it has an allowable ampacity not less than the 48-ampere load. There are six current-carrying conductors in the raceway.

$$\text{Adjusted Allowable Ampacity} = 75\ \text{A} \times 0.8 \times 0.76 = 45.6\ \text{A}$$

The adjusted allowable ampacity of the size 6 AWG copper THHN conductors is only 45.6 amperes, which is less than the load of 48 amperes. It will be necessary to try the next larger size conductor. The adjusted allowable ampacity of a size 4 AWG copper THHN conductor is 57.8 amperes. Using the rule of *240.4(B)*, the minimum size conductor permitted for these circuits is size 4 AWG copper Type THHN.

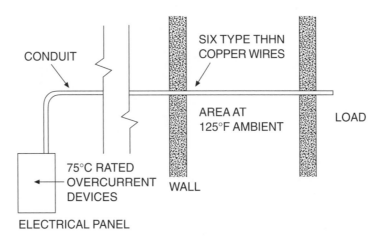

Figure 2.18   A run of conduit passes 25 ft (7.62 m) through a room with a high ambient temperature.

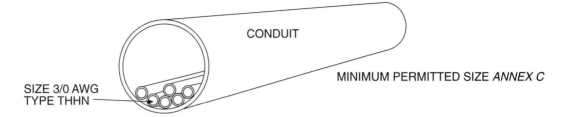

Figure 2.19   Six size 3/0 AWG, Type THHN wires are run in a Rigid Metal Conduit.

Sizing Conduit and Tubing

If all conductors in the conduit are the same size with the same type of insulation, the minimum trade diameter conduit or tubing permitted can be determined in accordance with *Note 1* of *Chapter 9. Note 1* makes reference to the tables in *Annex C.* There are separate tables for each of twelve types of conduit or tubing. There is one set of tables for conductors with round strands and a separate set of tables for conductors with compact strands. If a type of conduit or tubing is used that is not included in *Annex C,* the maximum number of conductors permitted in the conduit or tubing is determined by using the cross-sectional area provided by the manufacturer and calculating the conductor fill.

When the conductors in a conduit or tubing are of different sizes or types of insulation, the cross-sectional area of the conductors must be determined to calculate the minimum trade diameter of the conduit or tubing to be used. The diameter and cross-sectional area of common types of conductors are given in *Table 5* or *Table 5A* in *Chapter 9.* The total cross-sectional area of all conductors is first determined. Then, the minimum trade diameter conduit or tubing is found in *Table 4* for the type of conduit or tubing to be used. In the case of three or more conductors in conduit or tubing, the 40% fill column is used to determine the minimum trade diameter. Some examples will help illustrate how conduit or tubing is sized for a particular situation.

Example 2.5   Two feeders, each consisting of three size 3/0 AWG, Type THHN conductors, are run in the same Rigid Metal Conduit, as shown in Figure 2.19. Determine the minimum trade diameter Rigid Metal Conduit permitted for the six conductors.

Answer: Look up the minimum permitted Rigid Metal Conduit trade diameter directly in *Annex C, Table C8.* The minimum permitted trade diameter for wire with round strands is trade size $2^1/2$ (63). If the wire was of compact strand construction, the minimum permitted trade diameter conduit would be trade size 2 (53) from *Table C8A.*

Example 2.6   Determine the minimum trade diameter, Schedule 40 PVC Rigid Nonmetallic Conduit permitted to contain three size 2/0 AWG, Type THW wires, six size 12 AWG, Type THWN wires, and one bare size 6 AWG copper equipment grounding wire. This conduit is shown in Figure 2.20.

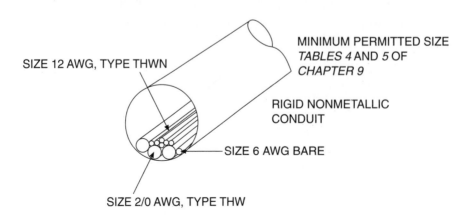

Figure 2.20   Insulated wires and bare copper equipment grounding wires are run in Rigid Nonmetallic Conduit.

Answer: It is necessary to determine the total cross-sectional area of the wires. Wire cross-sectional area is not permitted to exceed 40% of the inside cross-sectional area of the conduit. Minimum trade size 2 (53) is found in *Table 4*.

size 2/0 AWG, Type THW, *Table 5*
$3 \times 0.2624 = 0.7872$ in.² $3 \times 169.3 = 507.9$ mm²
size 12 AWG, Type THWN, *Table 5*
$6 \times 0.0133 = 0.0798$ in.² $6 \times 8.581 = 51.5$ mm²
size 6 AWG, bare, *Table 8*
$1 \times 0.027 = \underline{0.027}$ in.² $1 \times 17.09 = \underline{17.1}$ mm²
0.8940 in.²   576.5 mm²

Sizing Conduit and Tubing Nipples

A conduit or tubing nipple is defined in *Note 4* of *Chapter 9* as a length of conduit or tubing that does not exceed 24 in. (600 mm) in length. Figure 2.9 illustrates an application of a nipple. The cross-sectional area of the conductors is not permitted to exceed 60% of the cross-sectional area of the nipple. The nominal inside cross-sectional area of trade diameter conduit and tubing is given in the 100% column of *Table 4, Chapter 9*. For example, the total cross-sectional area of a trade size 2 (53) Rigid Metal Conduit nipple is 3.408 sq. in. (2199 mm²).

The minimum trade diameter of a conduit or tubing nipple permitted is determined by first finding the total cross-sectional area of all conductors contained within the nipple. This was done for the conductors in Figure 2.20, and the total cross-sectional area of the conductors was determined in Example 2.6 to be 0.8940 sq. in. (576.5 mm²). Assume the conduit of Example 2.6 is a nipple. Using the cross-sectional area, select a trade size (metric designator) from *Table 4* that is expected to be the correct size, and multiply the total cross-sectional area of that nipple by 0.6 (60%). The resulting area from this calculation is required to be equal to or larger than the cross-sectional area of the conductors. Consider a trade size 1¹/₄ (35) Rigid Metal Conduit nipple. The conductor cross-sectional area of the conductors is permitted to be 0.9156 sq. in. (591 mm²).

$$1.526 \text{ sq. in.} \times 0.6 = 0.9156 \text{ sq. in.}$$

$$(985 \text{ mm}^2 \times 0.6 = 591 \text{ mm}^2)$$

This value is larger than the cross-sectional area of the wire; therefore, trade size 1¹/₄ (35) Rigid Metal Conduit nipple is large enough for the conductors. Equation 2.2 can be used to determine the area of a nipple directly that has a cross-sectional area sufficient for the conductors. The total cross-sectional area of the conductors is divided by 0.6 (60%) to determine the minimum nipple cross-sectional area required. Then, just proceed down the 100% column of *Table 4* until an area is found that is equal to or greater than the value calculated. *Table 4* for each type of conduit and tubing now has a 60% fill column. The calculation method used in the past is no longer necessary once the conductor total cross-sectional area has been determined. For the previous example, look up a conduit or tubing trade size that has a 60% area not less than 0.8940 sq. in. (576.5 mm²). If the nipple is Rigid Metal Conduit, then trade size 1¹/₄ (35) is adequate.

$$\textbf{Actual Inside Area} = \frac{\textbf{Wire Cross-sectional Area}}{\textbf{0.6}} \qquad \textbf{Eq. 2.2}$$

Example 2.7 Assume the conductors form Figure 2.20 with a total cross-sectional area of 0.8940 sq. in. (576.5 mm²) are to be run through a Schedule 40 PVC Rigid Nonmetallic Conduit nipple between two panels. Determine the minimum trade diameter nipple permitted.

Answer: Now the method is real simple. Find *Table 4* for Schedule 40 Rigid Nonmetallic Conduit and look up a conduit in the 60% area column that is not less than 0.8940 sq. in. (576.5 mm²). Choose a trade size 1¹/₂ (41) nipple.

Wireway Sizing

The total cross-sectional area of the conductors is not permitted to exceed 20% of the cross-sectional area of the wireway, except at splices, *376.22* and *376.56*. The derating rules of *310.15(B)(2)(a)* for more than three conductors in raceway do not apply to metal wireway when there are not more than 30 current-carrying wires in the metal wireway. Wireway is square in cross-section; therefore, the cross-sectional area is determined by squaring the trade dimension of the wireway. For example, a 4-in. by 4-in. (10-cm × 10-cm) wireway has a cross-sectional area of 16 sq. in. (100 cm²). The interior corss-sectional area of nonmetallic wireway is marked on the wireway by the manufacturer, *378.12*. Equation 2.3 can be used to determine the minimum permitted dimension of wireway for electrical conductors. First determine the total cross-sectional area of the conductors, and then determine the minimum permitted cross-sectional area required with Equation 2.3.

$$\text{Actual Inside Area} = \frac{\text{Wire Cross-sectional Area}}{0.2} \qquad \text{Eq. 2.3}$$

Example 2.8 Determine the minimum wireway dimensions permitted to contain nine size 2/0 AWG, Type THW wires, and twelve size 12 AWG, Type THWN wires. This wireway is shown in Figure 2.21.

Answer: First, determine the total cross-sectional area of the conductors in the wireway. Equation 2.3 can be used to determine the minimum permitted wireway cross-sectional area. A 4-in. by 4-in. (10-cm × 10-cm) wireway has a cross-sectional area of 16.0 sq. in. (100 cm²). Therefore, a 4-in. (10-cm) square wireway is adequate for the conductors of this example.

Size 2/0 AWG, Type THW, *Table 5*

$$9 \times 0.2624 = 2.3616 \text{ in.}^2 \qquad\qquad 9 \times 169.3 = 1524 \text{ mm}^2$$

Size 12 AWG, Type THWN, *Table 5*

$$12 \times 0.0133 = \underline{0.1596 \text{ in.}^2} \qquad\qquad 12 \times 8.581 = \underline{103 \text{ mm}^2}$$

$$\text{Total Wire Area} \qquad 2.5212 \text{ in.}^2 \qquad\qquad\qquad\qquad 1626 \text{ mm}^2$$

$$\text{Actual Inside Area} = \frac{2.5212 \text{ in.}^2}{0.2} = 12.606 \text{ in}^2$$

$$\text{Actual Inside Area} = \frac{1626 \text{ mm}^2}{0.2} = 8130 \text{ mm}^2 = 81.3 \text{ cm}^2$$

Sizing Boxes

The easiest method to size standard boxes is to count the total conductors and conductor equivalents, then look up the size in *Table 314.16(A)*. This can be done only if the wires entering the box are all the same

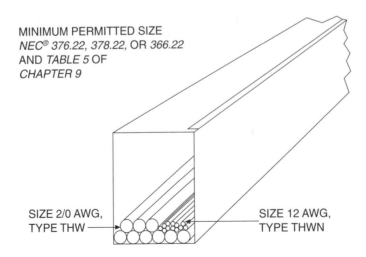

MINIMUM PERMITTED SIZE
NEC® 376.22, 378.22, OR *366.22*
AND *TABLE 5* OF
CHAPTER 9

SIZE 2/0 AWG,
TYPE THW

SIZE 12 AWG,
TYPE THWN

Figure 2.21 Minimum size of Wireway or Auxiliary Gutter based on 20% fill.

size. If they are not the same size, then the volume of the box must be determined using the volume require-ments of *Table 314.16(B)*. The rules for determining the minimum size device or junction box permitted are given in *314.16(B)*. The following steps are used when all wires in the box are the same size:

1. Count each circuit wire entering the box. An unbroken wire only counts as one. (Pigtails do not count.)
2. Where one or more internal cable clamps are used, one conductor equivalent shall be counted based upon the wire size secured by the clamp.
3. Fittings such as fixture studs and hickeys count as one conductor equivalent for each type of fitting.
4. Each device yoke counts as two wire equivalents based upon the wire size terminating at the device.
5. All grounding wires count as one wire equivalent unless the box contains equipment grounding wires for the circuit as well as for an isolated ground receptacle, such as may be used with computer equipment.

If wires of different sizes are in the same box, the equipment grounds are counted as a wire of the largest size. A device in the box is counted as a wire of the largest size connected to that device if more than one size of wire is in the box. Cable clamps are counted as a conductor equivalent to the size secured by the clamp.

Example 2.9   Determine the dimensions of the smallest size metal 3-in. by 2-in. (75-cm $\times$ 50-cm) standard device box required for the outlet shown in Figure 2.22. Both cables are size 14 AWG cop-per, with two insulated wires and a bare equipment grounding wire. There is a single-pole switch in this box with cable clamps.

Answer: Determine the number of conductors in the device box and the conductor equivalents for non-wire parts that take up space in the box. The procedure is described in *314.16(B)*. A 3-in. by 2-in. by 3^1/2-in. deep device box (75-cm $\times$ 50-cm $\times$ 90-cm) is found to be the minimum permitted from *Table 314.16(A)*.

Wires	4
Cable clamps	1
Other fittings	0
Device	2
Grounds	1
Total	8

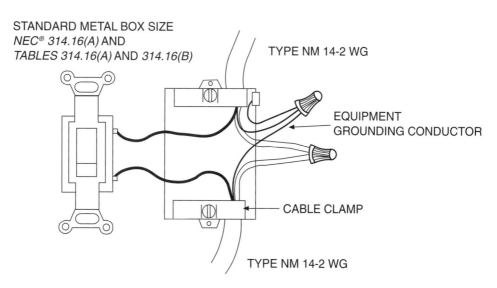

Figure 2.22   A switch is installed in a device box with size 14 AWG electrical cable with an equipment ground-ing wire.

Example 2.10   Consider another 3-in. by 2-in. device box similar to the one shown in Figure 2.22, except the conductors in the box are of different sizes. Conductors entering the box are a Type NM-B Cable with two size 14 AWG insulated and one bare equipment grounding conductor, and a Type NM-B Cable with two size 12 AWG insulated and one bare equipment grounding conductor. There are cable clamps in the box, and one single-pole switch. Determine the minimum depth standard device box permitted for this application.

Answer: To determine the minimum size box, it will be necessary to determine the volume of the box. First, do a conductor count and determine the number of conductors and conductor equivalents for each wire size based on the rules in *314.16(B)*. The cable clamp, device, and equipment grounds are counted as though they were a conductor of the largest size in the box. In this case the largest conductor is size 12 AWG. The minimum permitted capacity for the box is computed based upon the number of conductors of different sizes using the values from *Table 314.16(B)*. In this case, the minimum box volume permitted is 17.5 in.^{3} (287 cm^3). The volumes of standard boxes are listed in *Table 314.16(A)*. A 3-in. by 2-in. by 3^1/2-in. standard device box is adequate for this application and has a volume of 18 in.^{3} (295 cm^3).

	Size 14 AWG	Size 12 AWG
Wires	2	2
Equipment grounds	0	1
Device	0	2
Cable clamps	0	1
Other fittings	0	0
Totals	2	6

Size 14 AWG	2 × 2.0 in.^{3}	= 4.0 in.^{3}	2 × 32.8 cm^3	= 65.6 cm^3
Size 12 AWG	6 × 2.25 in.^{3}	= 13.5 in.^{3}	6 × 36.9 cm^3	= 221.4 cm^3
Totals		17.5 in.^{3}		287.0 cm^3

Nonmetallic boxes such as the one shown in Figure 2.23 do not have their volumes listed in *Table 314.16(A)*. The volume of the box is visibly displayed inside the box. Manufacturers can provide catalogs that give the dimensions and volumes of boxes available.

Example 2.11   Type NM-B Cable is installed into a single-gang nonmetallic device box where the cable is fastened such that cable clamps in the box are not necessary. The cable entering and leaving the box has size 14 AWG conductors, two of which are insulated and one is a bare equipment grounding conductor as shown in Figure 2.23. There is also a single-pole switch in the box. Determine the minimum volume required for this box.

Answer: If the nonmetallic box does not contain cable clamps, then the equivalent conductor count is seven, as tabulated below. A nonmetallic device box will be marked with the box volume. Look up the minimum volume requirement for each of the size 14 AWG conductors in *Table 370.16(B)*. Then multiply the seven conductor count of this device box by 2.0 in.^{3} (32.8 cm^3) for each wire in the box to get the minimum permitted box area of 14 in.^{3} (230 cm^3).

Wires	4
Cable clamps	0
Other fittings	0
Device	2
Grounds	1
Total	7

Size 14 AWG	7 × 2.0 in.^{3} = 14.0 in^3	7 × 32.8 cm^3 = 230 cm^3

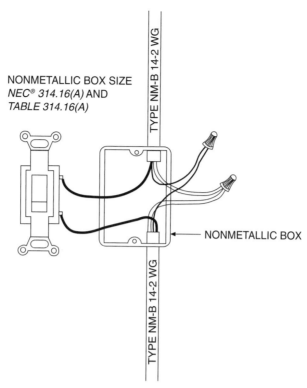

NONMETALLIC BOX SIZE
NEC® 314.16(A) AND
TABLE 314.16(A)

TYPE NM-B 14-2 WG

TYPE NM-B 14-2 WG

NONMETALLIC BOX

Figure 2.23   A nonmetallic box contains one device and has a Type NM Cable with two size 14 AWG insulated wires and a bare equipment grounding conductor entering the box.

Fixture wires, up to four including equipment grounding conductors smaller than size 14 AWG, entering a box from a canopy fixture are not counted for the purpose of determining the minimum size box. In some cases, fixture wires smaller than size 14 AWG may be required to be counted, *314.16(B)(1) Exception.* Be sure to count fixture studs and similar fittings, and if a fixture cover does not actually take up space inside a box, it is not counted. Plaster rings and raised covers for devices add volume to a box. Trade literature for device covers will list the amount of volume added for the various types and sizes. The volume of raised covers is usually marked on the cover.

Pull Boxes

Minimum dimensions are required for pull boxes and conduit bodies when the wires are size 4 AWG and larger. There are straight pulls in which the conductors enter one end of the box and leave the opposite end. For a straight pull, the minimum distance between openings is eight times the largest trade size conduit entering the pull box. The rule is found in *314.28(A)(1).*

Example 2.12   A trade size 2 (53) raceway enters and leaves opposite ends of a pull box as shown in Figure 2.24. The conductors in the raceway are size 2 AWG. Determine the minimum length of pull box permitted.

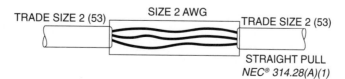

TRADE SIZE 2 (53)   SIZE 2 AWG   TRADE SIZE 2 (53)

STRAIGHT PULL
NEC® 314.28(A)(1)

Figure 2.24   A straight pull box in a run of trade size 2 (53) conduit where the conductors are size 4 AWG or larger.

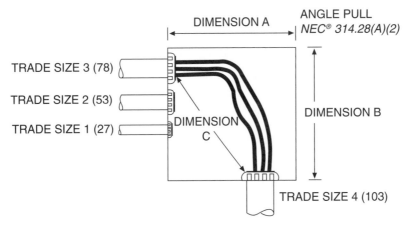

Figure 2.25   A pull box is used to make an angle pull for several conduits where the conductors are size 4 AWG or larger.

Answer: The minimum permitted length of pull box for a straight pull is determined by multiplying the largest trade diameter conduit by eight. The minimum length permitted for this installation is 16 in. (424 mm).

$$8 \times 2 \text{ in.} = 16 \text{ in.} \qquad 8 \times 53 = 424 \text{ mm}$$

For an angle pull, the distance to the opposite wall is not permitted to be less than six times the largest diameter conduit plus the sum of the trades sizes of other conduits or tubing on the same side in the same row. What is being sized is the distance to the opposite wall. This is confusing, and it will be necessary to read *314.28(A)(2)* carefully. An example of an angle pull box is shown in Figure 2.25. These required dimensions apply only when the wire size is 4 AWG and larger. If the wires are size 6 AWG and smaller, there is no minimum dimension requirement. In that case, the box is sized according to *314.16*.

Example 2.13   A pull box has three raceways entering one side, trades sizes 3 (78), 2 (53), and 1 (27), and a trade size 4 (103) raceway entering the adjacent side. Conductors passing from the trade size 4 (103) to the trade size 3 (78) raceways are size 4 AWG or larger. The pull box is shown in Figure 2.25. Determine the minimum dimensions of the pull box.

Answer: Dimension A is determined by multiplying the trade size 3 (78) diameter conduit by six and then adding to it the diameters of the other conduits on that same side in the same row. The minimum permitted length of dimension B is determined by multiplying the 4-in. diameter conduit (103) by six. The pull box is required to have a size not less than 21 in. (548 mm) by 24 in. (618 mm).

Minimum length of dimension A of the pull box:
$$6 \times 4 \text{ in.} \quad = 24 \text{ in.} \qquad\qquad 6 \times 103 = 618 \text{ mm}$$
Minimum length of dimension B of the pull box:

6×3 in. $= 18$ in.	$6 \times \;\;78 = 468$ mm
$+ \;2$ in.	$+ \;53$ mm
$+ \;1$ in.	$+ \;27$ mm
21 in.	548 mm

The purpose of specifying a minimum size of pull box is to provide an adequate amount of space within the pull box to install conductors size 4 AWG and larger without injuring the insulation on the conductors. In addition to specifying the minimum size of pull box, there is an additional requirement that any conduit entry containing the same conductors must be spaced a minimum distance apart of six times the diameter of the largest raceway involved. Referring to the pull box of Figure 2.25, the conductors pass from a trade size 4 (103) raceway to a trade size 3 (78) raceway. The distance between these two raceways is required to be not less than 24 in. (618 mm).

Dimension C = 6×4 in. = 24 in. $\qquad\qquad = 6 \times 103 = 618$ mm

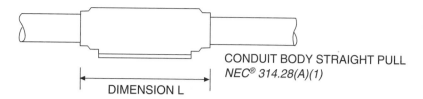

CONDUIT BODY STRAIGHT PULL
NEC® 314.28(A)(1)

DIMENSION L

Figure 2.26   A Conduit Body is used to provide access to a conduit to make a straight wire pull.

Conduit Bodies

The rules for determination of the minimum dimensions of pull boxes also apply to conduit bodies with one exception, which is the **LB**-type conduit body. The different situations considered are the straight run and the angle. These minimum dimension requirements apply only when the conduit body contains conductors size 4 AWG and larger.

The rules of *314.28(A)(1)* apply for a straight conduit body, and the length of the conduit body shall not be permitted to be less than eight times the trade diameter of the largest conduit entering the conduit body.

Example 2.14   A conduit body is used for wire pulling access in a straight run of conduit, as shown in Figure 2.26. There are four size 4 AWG, Type THHN wires installed in a trade size 1 (27) conduit. Determine the minimum permitted length for this straight conduit body.

Answer: This is a straight run, and the conduit body shall have a length not less than eight times the trade diameter of the largest conduit entering the conduit body.

$$8 \times 1 \text{ in.} = 8 \text{ in.} \qquad 8 \times 27 = 216 \text{ mm}$$

The rule for an angle conduit body is found in *314.28(A)(2)*. The dimension from a conduit entry to the opposite wall of the conduit body shall be required to be not less than six times the trade diameter of the conduit entering the conduit body. There is an exception that may be used for the dimension from the conduit entry to a removable cover. This same exception can be used when a conduit enters the back of a pull box opposite the removable cover. This exception allows the dimension to be determined by the one conductor per terminal column of *Table 312.6(A)*.

Example 2.15   A Type LB conduit body is installed in a run of conduit to make a right angle access through a wall, as shown in Figure 2.27. If four size 4 AWG Type THHN wires are run in trade size 1 (27) conduit, determine the minimum dimensions of a Type LB conduit body.

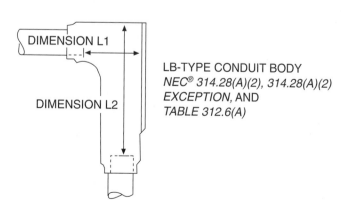

DIMENSION L1

LB-TYPE CONDUIT BODY
NEC® 314.28(A)(2), 314.28(A)(2)
EXCEPTION, AND
TABLE 312.6(A)

DIMENSION L2

Figure 2.27   Minimum dimensions are required for a Type LB Conduit Body when the wires are size 4 AWG and larger.

Answer: First, determine the minimum permitted length of dimension L2 that is required to be six times the trade diameter of the conduit to the opposite wall.

Dimension L2:   6 × 1 in. = 6 in.   6 × 27 = 162 mm

Dimension L1 is from the conduit entry to the removable cover. An exception permits this dimension to be determined using the "one conductor per terminal" column of *Table 312.6(A)*. Dimension L1 is a minimum of 2 in. (51 mm).

If the wires in the conduit body are size 6 AWG and smaller, then the rules of *314.28* do not apply. The conduit body is then required to have a minimum cross-sectional area of not less than twice the cross-sectional area of the largest conduit or tubing entering the conduit body, *314.16(C)*. If a conduit body is to contain splices, taps, or a device, *314.16(C)* requires the volume to be marked in the conduit body, and the fill shall not exceed the requirement of *314.16(B)*.

Sometimes cable assemblies with conductors larger than size 6 AWG join in a box for splicing or terminating. The rules in *314.28(A)* also apply, but there is no conduit or tubing size to use to determine the minimum dimensions of the box. The method to be used is explained in *314.28(A)*. First, determine the minimum size conduit or tubing that would have been needed if the conductors had been run in conduit or tubing. Then use that minimum conduit or tubing size to determine the dimensions of the straight or angle box. The method is explained in Example 2.16.

Example 2.16   A cable assembly containing four conductors size 2 AWG copper with Type THWN insulation enter one side of a box and leave the adjacent side of the box. Determine the minimum size of box required for splicing these conductors.

Answer: Because the conductors are larger than size 6 AWG, this is to be treated as an angle pull and sized according to *314.28(A)(2)*. First determine the minimum trade diameter conduit or tubing that would have been required if the conductors had been run in conduit or tubing rather than as a cable assembly. Look up the cross-sectional area of the size 2 AWG Type THWN conductors in *Table 5* and find the value 0.1158 sq. in. (74.71 mm²). Multiply this value by 4 conductors to get 0.4632 sq. in. (0.1158 in.² × 4 = 0.4632 in.²) (74.71 mm² × 4 = 299 mm²). Then look up the minimum size conduit or tubing from *Table 4* in the 40% column. The minimum is trade size 1¼ (35). If two different sizes are found, use whichever is smaller. Now use the rule of *314.28(A)(2)* and multiply by 6 to get the minimum dimension of 7.5 in. (1.25 in. × 6 = 7.5 in.) (35 mm × 6 = 210 mm). Another method of finding the minimum trade size conduit or tubing is to look up the size in *Annex C*. Usually *Table C4* for IMC will give the minimum trade size.

Sizing and Installing Parallel Conductor Sets

In the case of high-ampacity feeders, one conductor per phase may not be practical. Under these circumstances, two or more conductors per phase may be desirable or even necessary. The objective is to keep the resistance of each parallel path equal so the load current will divide equally on each wire of the set. The wires and terminations heat up when they carry current; therefore, the resistance changes when the temperature of the wire and terminations changes. For these reasons, special requirements are necessary when installing parallel sets of wires. The following is a summary of some special installation rules of *310.4* for sets of multiple conductors for each ungrounded conductor and the grounded conductor:

1.   Power conductors are permitted to be paralleled only for size 1/0 AWG and larger, except for special applications.
2.   All parallel conductors of a phase or neutral set shall be the same length.
3.   All conductors of the phase or neutral set shall be of the same material, cross-sectional area, and the same insulation type.
4.   All conductors of a phase or neutral set shall be terminated in the same manner.
5.   If run in more than one raceway, the raceways shall have the same physical properties, and the same length, and shall be installed in the same manner.
6.   If more than one raceway is used, make sure each phase wire and neutral, if present, is placed in each of the raceways to prevent eddy currents, as shown in Figure 2.28. This requirement is found in *300.20(A)*.
7.   If an equipment grounding wire is present and there is more than one raceway, an equipment grounding wire shall be in each raceway. The size of each equipment grounding conductor is determined

NEC® 310.4 CONDUCTORS ARE PERMITTED TO BE RUN IN PARALLEL

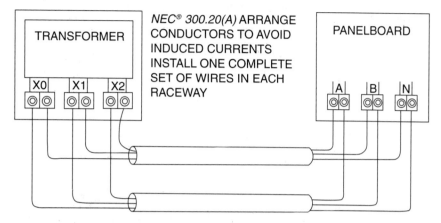

Figure 2.28 Care must be taken when paralleling sets of conductors, particularly when more than one raceway is used.

according to *250.122(F)*, which means the overcurrent rating of the feeder determines the minimum permitted size of the equipment grounding conductor in each raceway. See also Figure 5.10.

These rules for the installation of parallel sets of conductors apply to conductors of the same phase or neutral. Phase A is not required to be identical to phase B, for example. But all of the wires of phase A are required to be identical, as discussed in the previous points.

Determining the minimum parallel conductor size for a particular application can be confusing. In the case where there is a single main disconnecting means, the minimum permitted conductor size is determined by dividing the rating of the feeder overcurrent device by the number of parallel sets of conductors. If there are more than three current-carrying conductors in a single raceway, then a derating factor according to *310.15(B)(2)(a)* must be applied. Equation 2.4 can be used to determine the minimum permitted wire size for a feeder consisting of parallel sets of conductors. If the feeder rating is not over 800 amperes, then the next wire size smaller may be permitted to be used by *240.4(B)*. For some services, calculated load may be used.

Minimum Parallel Conductor Ampacity = $\dfrac{\textbf{Feeder Overcurrent Rating}}{\textbf{No. of Sets} \times \textbf{Derate Factor}}$ **Eq. 2.4**

Example 2.17 Determine the minimum permitted size of copper Type THWN conductors when three parallel sets are used for a 3-phase, 3-wire feeder with a 500-ampere overcurrent device that is not rated for 100% continuous operation and when the actual load current does not exceed 80% of the rating of the overcurrent device. All nine conductors are run in the same raceway.

Answer: The result of the calculation of Equation 2.4 gives a minimum feeder conductor rating of 238 amperes when there are three parallel sets. The minimum Type THWN copper wire size from *Table 310.16* is 250 kcmil with a rating of 255 amperes. This feeder has a rating of less than 800 amperes; therefore, *240.4(B)* will apply. In this case, a size 4/0 AWG conductor rated at 230 amperes is adequate for the load and will satisfy the requirement of *240.4(B)*.

$$\text{Minimum Parallel Conductor Ampacity} = \frac{500\ A}{3 \times 0.7} = 238\ A$$

Size 250-kcmil copper, Type THWN: 255 A × 0.7 × 3 = 536 A

Size 4/0 AWG copper, Type THWN: 230 A × 0.7 × 3 = 483 A

MAJOR CHANGES TO THE 2005 CODE

These are the changes to the 2005 *NEC®* that correspond to the Code sections studied in this unit. The following analysis explains the significance of the changes from the 2002 to the 2005 Code only and this analysis is not intended to be used in place of the Code. Refer to the actual section of the 2005 Code for the exact wording and meaning of each section discussed. Changes are indicated in the Code with a vertical line in the margin. If material was deleted or moved to another location in the Code, the location of the deletion is indicated with a dark dot in the margin. Only the articles with significant changes are listed in the following discussion.

Article 312   Cabinets, Cutout Boxes, and Meter Socket Enclosures

312.2(A): A new last sentence was added that now states, when raceway or cable enters an enclosure above the level of uninsulated live parts, in wet locations the fitting is required to be listed for use in wet locations.

312.4: This is a new section as it relates to flush-mounted cabinets and cutout boxes. This is essentially the same language as in *314.21* for boxes. If a cabinet or cutout box is flush mounted in a wall, the maximum space permitted between the cabinet and the edge of the plaster is ⅛ in. (3 mm). Larger gaps are required to be filled. This rule applies in the case of cabinets or cutout boxes that are to be equipped with flush covers.

Article 314   Outlet, Device, Pull, and Junction Boxes; Conduits Bodies; Fittings; and Handhole Enclosures

314.16(B)(1): A new sentence was added that deals with conductors passing through a box without being spliced. If the conductor is looped so that the length of conductor within the box is at least two times the minimum free conductor length required for attachment to a device (usually 12 in.), then the conductor is required to be counted as two conductors for the purpose of determining conductor fill. This is illustrated in Figure 2.29.

NEC® 314.16(B)(1)

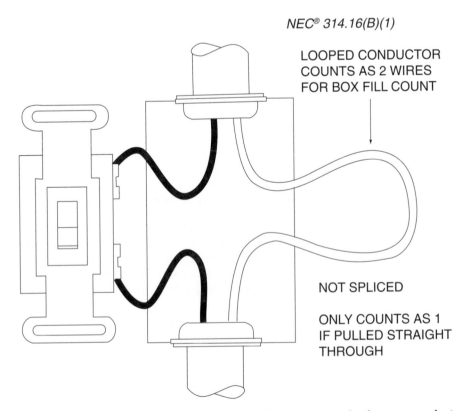

LOOPED CONDUCTOR
COUNTS AS 2 WIRES
FOR BOX FILL COUNT

NOT SPLICED

ONLY COUNTS AS 1
IF PULLED STRAIGHT
THROUGH

Figure 2.29   A wire that is not spliced and is pulled through a box is counted only as one conductor for the box fill count unless the conductor is looped equivalent in length to conductors that would be terminated, in which case the conductor counts as two for the purpose of figuring box fill.

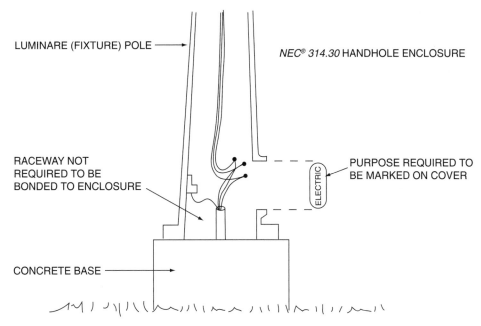

Figure 2.30   The cover to handhole enclosure is required to be marked as to the purpose, and an entering metal conduit or metal sheathed cable is not required to be bonded to the enclosure.

314.21: Repairing plaster or drywall around a box where a gap is greater than ¹/₈ in. only applies when the box is installed where flush-type covers or faceplates will be installed on the box.

314.23(B)(1): When screws are used to secure a box to a surface and the screws pass through the interior of the box, an approved method shall be used to prevent the exposed threads of the screw from causing abrasion of the conductor insulation.

314.27(D): This section specifies the conditions when a box is permitted to support a paddle fan. The requirements that were in *422.18* of the previous edition of the Code are now in this paragraph. The change is that when a box is listed to support a paddle fan weighing more than 35 lbs. (16 kg), but not more than 70 lbs. (32 kg), the maximum weight the box will support is required to be marked on the box.

314.30: This is a new section that provides requirements for handhole enclosures and enclosure covers. Paragraph (A) specifies the method of determining the minimum size of enclosure. Paragraph (B) does not require a conduit or cable entering a handhole enclosure to be secured to the handhole enclosure. Paragraph (C) requires the purpose of the handhole, such as electric, to be marked on the outside of the handhole cover, and if the cover weighs less than 100 lbs. (45 kg), it must be of a type that requires tools to remove. There are similar requirements that apply to manhole covers, *110.75(D)*. Handhole rules are illustrated in Figure 2.30.

Article 342   **Intermediate Metal Conduit: Type IMC**

342.22: In the second paragraph that makes reference to cables installed in IMC, it now states where "not prohibited" by the respective cable article. The words not prohibited replace the word permitted. If the cable article did not specifically permit the installation of cable in IMC, then some interpreted this as meaning it was not permitted. With the new wording, the cable article must specifically prohibit the use of the cable in IMC. Unless specifically prohibited, cables are permitted to be installed in any type of raceway.

    This same rule applies in the case of Rigid Metal Conduit *(344.22)*, Flexible Metal Conduit *(348.22)*, Liquidtight Flexible Metal Conduit *(350.22)*, Rigid Nonmetallic Conduit *(352.22)*, High Density Polyethylene Conduit *(353.33)*, Liquidtight Flexible Nonmetallic Conduit *(356.22)*, Electrical Metallic Tubing *(358.22)*, Flexible Metallic Tubing *(360.22)*, Electrical Nonmetallic Tubing *(362.22)*, Surface Metal Raceway *(386.22)*, and Surface Nonmetallic Raceway *(388.22)*.

342.30: There is now a specific reference to *300.18* relative to the installation of IMC, which means the raceway system must be installed as a complete system between outlets. There is a new raceway exception in *300.18* that permits isolated sections of IMC to be installed for the purpose of protecting cables from physical damage. This same rule applies in the case of Rigid Metal Conduit *(344.30)* and Electrical Metallic Tubing *(352.30)*.

342.30(B)(3): This section permits a vertical riser of IMC with threaded connections from equipment or industrial machinery to be supported with a spacing of not more than 20 ft. The top and bottom of the riser are now required to be supported and securely fastened. The previous edition of the code only required the top and bottom to be firmly supported. The intent did not change, but the meaning of these two references is different. Now it is required that the riser be supported in a manner in which it will not move vertically. Securely fastened means it is not permitted to move about horizontally, but be fixed in place. The previous edition of the Code could be interpreted as only requiring that the riser not be able to move vertically.

342.42(A): A new sentence was added that requires IMC threadless connectors and couplings installed in wet locations to meet the requirements of *300.15(A)*. This means that threadless connectors and couplings are to be installed in such a manner that moisture does not accumulate into boxes and enclosures from outside. This does not necessarily mean that threadless connectors are required to be listed for use in wet locations.

Article 344   **Rigid Metal Conduit: Type RMC**

344.30(B)(3): This section permits a vertical riser of Rigid Metal Conduit with threaded connections from equipment or industrial machinery to be supported with a spacing of not more than 20 ft. The top and bottom of the riser are now required to be supported and securely fastened. The previous edition of the code only required the top and bottom to be firmly supported. The intent did not change, but the meaning of these two references is different. Now it is required that the riser be supported in a manner in which it will not move vertically. Securely fastened means it is not permitted to move about horizontally, but be fixed in place. The previous edition of the Code could be interpreted as only requiring that the riser not be able to move vertically.

344.42(A): A new sentence was added that requires Rigid Metal Conduit threadless connectors and couplings installed in wet locations to meet the requirements of *300.15(A)*. This means that threadless connectors and couplings are to be installed in such a manner that moisture does not accumulate into boxes and enclosures from outside. This does not necessarily mean that threadless connectors are required to be listed for use in wet locations.

Article 348   **Flexible Metal Conduit: Type FMC**

348.30(A) Exception 2: Installations of Flexible Metal Conduit are to be secured within 12 in. (300 mm) of terminations and at intervals not to exceed $4^1/_2$ ft (1.4 m). This exception permits short lengths to be installed supported only by the connectors for termination at equipment such as electric motors. In the past, the maximum length to be installed supported only by the connectors was 3 ft (900 mm). The distances involved with large electric motors in particular makes 3 ft (900 mm) lengths of flexible metal conduit impractical. Now it is permitted to install larger trades sizes of Flexible Metal Conduit in longer lengths as shown in Figure 2.31. For trade sizes $^1/_2$ (16) through $1^1/_4$ (35), the maximum length is still 3 ft (900 mm). Now trade sizes $1^1/_2$ (41) through 2 (53) are permitted to be installed in lengths up to 4 ft (1200 mm) supported only by the connectors. For trade sizes $2^1/_2$ (63) through 4 (103) the maximum length permitted is 5 ft (1500 mm).

348.30(A) Exception 4: *Exception 3* permitted 6 ft (1.8 mm) lengths of Flexible Metal Conduit supported only by the connectors to be installed in a suspended ceiling for connecting to a luminaire. This rule did not apply for connections to other equipment installed in suspended ceilings. This new exception permits Flexible Metal Conduit to be installed in lengths up to 6 ft (1.8 m) from the last point of support to the termination at the luminaire or equipment. This was generally the practice in the past, but because it was not specifically stated, in some areas this practice was not permitted.

348.60: This section specifies when Flexible Metal Conduit is permitted to serve as an equipment grounding conductor and when it is not permitted. It also specifies the installation method to be used for equipment grounding conductors and equipment bonding jumpers. There is actually no change in this section, but a new sentence was added to make it clear that when Flexible Metal Conduit is not installed because

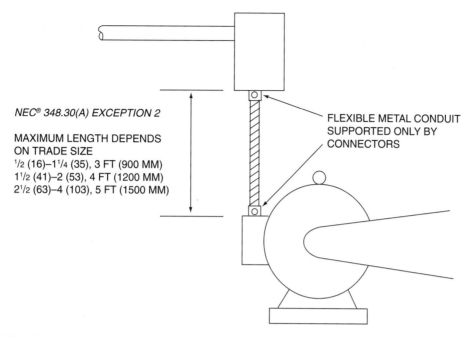

NEC® 348.30(A) EXCEPTION 2

MAXIMUM LENGTH DEPENDS
ON TRADE SIZE
¹/₂ (16)–1¹/₄ (35), 3 FT (900 MM)
1¹/₂ (41)–2 (53), 4 FT (1200 MM)
2¹/₂ (63)–4 (103), 5 FT (1500 MM)

FLEXIBLE METAL CONDUIT
SUPPORTED ONLY BY
CONNECTORS

**Figure 2.31   The length of flexible metal conduit that is permitted to be installed and supported only by the con-
nectors is at least 3 ft (900 mm), but may be up to 5 ft (1500 mm) depending upon the trade size of the conduit.**

flexibility is required, it is permitted to serve as an equipment grounding conductor as specified in
250.118(5).

Article 350   Liquidtight Flexible Metal Conduit: Type LFMC

350.30(A) Exception 4: *Exception 3* permitted 6 ft (1.8 m) lengths of Liquidtight Flexible Metal Conduit supported only by the connectors to be installed in a suspended ceiling for connecting to a luminaire. This
rule did not apply for connections to other equipment installed in suspended ceilings. This new exception permits Liquidtight Flexible Metal Conduit to be installed in lengths up to 6 ft (1.8 m) from the last
point of support to the termination at the luminaire or equipment. This was generally the practice in the
past, but because it was not specifically stated, in some areas this practice was not permitted. Liquidtight
Flexible Metal Conduit is not permitted to be installed in ducts, plenums, or suspended ceiling spaces
used for environmental air.

350.60: This section specifies when Liquidtight Flexible Metal Conduit is permitted to serve as an equipment
grounding conductor and when it is not permitted. It also specifies the installation method to be used for
equipment grounding conductors and equipment bonding jumpers. There is actually no change in this
section, but a new sentence was added to make it clear that when Liquidtight Flexible Metal Conduit is
not installed because flexibility is required, it is permitted to serve as an equipment grounding conductor
as specified in *250.118(6).*

Article 353   High Density Polyethylene Conduit: Type HDPE

This is a new Article covering an existing product that has special installation requirements, but was not specifically covered in the previous edition of the Code. High Density Polyethylene Conduit is a rigid nonmetallic conduit that is flexible enough to be handled as a coil and is often placed on a reel. It is intended for direct burial in the earth. It can be placed directly into a trench as a continuous length from a reel.
It is not permitted to be installed where exposed, or within a building. The cross-sectional area of HDPE
Conduit is close enough to PVC Rigid Nonmetallic Conduit that they are both included in the same conductor fill tables, *Table 4, Table C10,* and *Table C10(A).*

Article 356   LiquidTight Flexible Nonmetallic Conduit: Type LFNC

356.12(5): This is a new paragraph, but the intent did not change. Liquidtight Flexible Nonmetallic Conduit is not permitted to be installed in classified hazardous locations except as specifically permitted in *Chapter 5*.

356.30(1): This paragraph only applies when Liquidtight Flexible Nonmetallic Conduit is installed in lengths longer than 6 ft (1.8 m). It is required to be secured at intervals not to exceed 3 ft (900 mm) and within 12 in. (300 mm) of a box, enclosure, or fitting. If a piece of LFNC is not more than 6 ft (1.8 m) in length, then it is not required to be supported except by the connectors. This could be a 6 ft (1.8 m) piece of LFNC in a conduit run where the flexible section is needed to traverse an irregularity on a wall.

356.30(4): It is now permitted to install a section of Liquidtight Flexible Nonmetallic Conduit, Type B, for termination to a luminaire or equipment in an accessible ceiling and supported only by the connectors. This rule did not apply to terminations at equipment in the previous edition of the Code.

256.42: Couplings and connectors used with Liquidtight Flexible Nonmetallic Conduit are now required to be listed for use with LFCN. It now states specifically that straight LFNC fittings are permitted to be installed as direct burial in the earth or encased in concrete.

Article 358   Electrical Metallic Tubing: Type EMT

358.42: When installed in a wet location, EMT couplings and connectors are required to meet the requirements of *300.15(A)*. The previous edition of the Code required connectors and couplings installed in wet locations to be raintight. Now, by referencing *300.15(A)*, the requirement is that moisture not be permitted to enter the wiring system. This means that connectors and couplings that are not raintight are permitted to be used in locations such as on the underside of an enclosure.

Article 376   Metal Wireways

376.23(A): When conductors enter a metal wireway and bend to run through the wireway, there is a minimum distance required from the raceway or cable entry to the opposite side of the wireway based upon the size of wire involved. The distance is determined from *Table 312.6(A)* for wires connected to terminals. It is now specified that the minimum distance is determined using the one conductor per terminal column of *Table 312,6(A)*.

376.23(B): When a wire, size 4 AWG and larger, passes through a metal wireway where the wireway is used in a manner similar to a pull box, the raceway entries are required to be spaced apart a distance not closer than the distance determined in *314.18(A)* for pull boxes. If the wires enter the wireway as a cable rather than as raceway, there was no rule on how to determine the distances. A new last sentence was added to make it clear that the cable is to be assumed to be of the same trade size as the equivalent raceway required for the conductors.

376.56(B): Now there are rules for the installation of a power distribution block in a metal wireway. A power distribution block is a terminal block for making splices and taps to conductors and it is specifically designed and listed for installation in wireways. The rules require the power distribution block to be listed and installed to allow the minimum space requirements stated in the installation instructions. Wire bending space is required to be not less than that given in Table *312.6(A)*, and there are to be no exposed live parts after installations.

Article 378   Nonmetallic Wireways

378.23(A): When conductors enter a nonmetallic wireway and bend to run through the wireway, there is a minimum distance required from the raceway or cable entry to the opposite side of the wireway based upon the size of wire involved. The distance is determined from *Table 312.6(A)* for wires connected to terminals. it is now specified that the minimum distance is determined using the one conductor per terminal column of *Table 312.6(A)*.

378.23(B): When a wire, size 4 AWG and larger, passes through a nonmetallic wireway where the wireway is used in a manner similar to a pull box, the raceway entries are required to be spaced apart a distance not closer than the distance determined in *314.18(A)* for pull boxes. If the wires enter the wireway as a cable rather than as raceway, there was no rule on how to determine the distances. A new last sentence was added to make it clear that the cable is to be assumed to be of the same trade size as the equivalent raceway required for the conductors.

Article 386 **Surface Metal Raceways**

386.30: This is a new section requiring Surface Metal Raceway to be supported at intervals in accordance with the manufacturer's instructions. It is recognized that different types of Surface Metal Raceway will require support at different spacings, and it is the responsibility of the manufacturer to provide those support requirements. In cases where Surface Metal Raceway does not seem to be adequately supported, now there is a requirement that the installation must be in conformity with the manufacturer's instructions.

386.70: This section places requirements on Surface Metal Raceway that contains both signaling and light and power conductors. The signaling wires must be kept separated from the light and power wires, and each is run through the raceway in a separate compartment. In the past, the different compartments were required to be color coded. Now they are permitted to be color coded, stamped or imprinted with an identification.

It was also required in the past that the compartment for signaling wires and the compartment for light and power wires be maintained in the same relative position with respect to each other for the complete Surface Metal Raceway system. That requirement was deleted.

Article 388 **Surface Nonmetallic Raceways**

388.70: This section places requirements on Surface Nonmetallic Raceway that contains both signaling and light and power conductors. The signaling wires must be kept separated from the light and power wires and each is run through the raceway in a separate compartment. In the past the different compartments were required to be color coded. Now they are permitted to be color coded, stamped or imprinted with an identification.

Chapter 9 **Tables**

Table 2: This is a new table that gives radius of bends for conduit and tubing. It was *Table 344.24* in the previous edition of the Code. This table was moved to *Chapter 9* since other conduit and tubing articles use this table for determining the minimum radius of bends.

Table 4: The columns in this table give the cross-sectional area for various percentages of fill as listed at the top of the column. The columns were rearranged, but there were no changes made in the numbers within the column.

Table 5A: This table gives the diameter and cross-sectional areas for compact stranded aluminum conductors. Size 900 kcmil wire is now listed in the table, and the number of strands column was deleted. In the past, the listing of number of strands sometimes led the installer to believe that conductors with other numbers of strands might not be permitted. Conductors are not required to have a specific number of strands.

Annex C **Conduit and Tubing Fill Tables**

Note 2: The tables in *Annex C* are used to determine the minimum trade size conduit and tubing for conductors all of which are of the same size and have the same insulation. There is a two-hour fire-rated Type RHH insulation available for conductors, but it is thicker than other RHH insulation. Two-hour fire-rated RHH insulated conductors have a larger cross-sectional area than standard RHH insulated conductors. This means the tables in *Annex C* cannot be used to determine minimum conduit and tubing size for these conductors. A new note was added at the bottom of all tables that show RHH insulation to point out that the cross-sectional area must be obtained from the manufacturer for these conductors. The minimum size conduit or tubing is then determined by calculation using *Table 4* in *Chapter 9*.

Compact Strand Tables: All of the A tables in *Annex C* for conductors with compact strands now includes size 900 kcmil conductors.

WORKSHEET NO. 2—BEGINNING WIRE, RACEWAY, AND BOX SIZING

Mark the single answer that most accurately completes the statement based upon the 2005 Code. Also provide, where indicated, the Code reference that gives the answer or indicates where the answer is found, as well as the Code reference where the answer is found.

1. A metallic 3 in. by 2 in. (75 mm by 50 mm) device box contains one 3-way switch. The cables entering the box are one Type NM-B size 14 AWG copper 2-conductor with equipment ground, and one Type NM-B size 14 AWG copper 3-conductor with equipment ground, as shown in Figure 2.32. The device box contains cable clamps. The minimum permitted box depth for this installation is:
   A. 2$\frac{1}{4}$ in. (57 mm).
   B. 2$\frac{1}{2}$ in. (65 mm).
   C. 2$\frac{3}{4}$ in. (70 mm).
   D. 3$\frac{1}{2}$ in. (90 mm).
   E. 4 in. (100 mm).

   Code reference_____

2. A single-gang nonmetallic box without cable clamps is used instead of the metallic box shown in Figure 2.32. The cables entering the box are one Type NM-B size 14 AWG copper 2-conductor with equipment ground, and one Type NM-B size 14 AWG copper 3-conductor with equipment ground. There is also a 3-way switch in the box. The minimum box volume permitted is:
   A. 14 in.^{3} (229 cm^3).
   B. 15.75 in.^{3} (258 cm^3).
   C. 16 in.^{3} (262 cm^3).
   D. 18 in.^{3} (295 cm^3).
   E. 20.25 in.^{3} (332 cm^3).

   Code reference_____

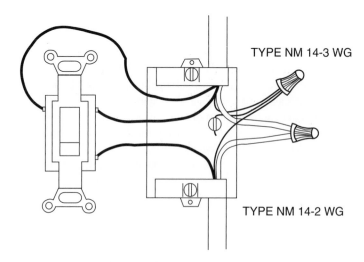

Figure 2.32   A 3-way switch is installed in a metal device box with size 14 AWG Type NM-B Cable entering both ends of the box.

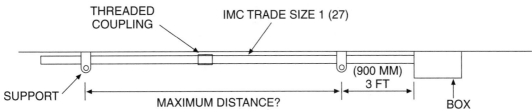

Figure 2.33   Intermediate Metal Conduit is run exposed on the ceiling of a room.

3. A straight run of trade size 1 (27) Intermediate Metal Conduit (IMC) is joined with threaded couplings and is supported within 3 ft (900 mm) of a box, as shown in Figure 2.33. The maximum distance permitted between supports of this straight run of IMC is:

    A. 10 ft (3 m).   D. 16 ft (4.9 m).
    B. 12 ft (3.7 m).   E. 20 ft (6.1 m).
    C. 14 ft (4.3 m).

    Code reference_____

4. A feeder consists of four size 4/0 AWG copper Type THWN conductors run in *Schedule* 40 PVC Type Rigid Nonmetallic Conduit. The minimum trade size conduit permitted for this feeder is:

    A. trade size 1 (27).   D. trade size 2 (53).
    B. trade size 1¹/4 (35).   E. trade size 2¹/2 (63).
    C. trade size 1¹/2 (41).

    Code reference_____

5. The conductor cross-sectional area is less than 20% of the cross-sectional area of a metal wireway. The adjustment factors for determining conductor allowable ampacity of *310.15(B)(2)(a)* do not apply to conductors in this wireway if the number of current-carrying conductors in the wireway does not exceed:

    A. thirty.   C. nine.   E. three.
    B. twelve.   D. six.

    Code reference_____

6. A nonmetallic single-gang trade size 2 by 3 (50 by 75) box without cable clamps is installed in a wall of a new dwelling. The wiring method is Nonmetallic-Sheathed Cable, Type NM-B with a minimum of ¹/4 in. of the cable sheath extending into the box. It is not required to secure the cable to this box if the cable is fastened within a distance from the box measured along the cable of not more than:

    A. 3 in. (75 mm).   D. 8 in. (200 mm).
    B. 4 in. (100 mm).   E. 12 in. (300 mm).
    C. 6 in. (150 mm).

    Code reference_____

7. Fluorescent lay-in luminaires (lighting fixtures) are mounted in an accessible suspended ceiling and supplied from 20-ampere branch-circuits. Flexible Metal Conduit supported by only a listed connector at a solidly mounted junction box and at the luminaire (lighting fixture) is permitted to be installed to supply a luminaire (fixture) in lengths not to exceed:

    A. 18 in. (450 mm).   D. 4 ft (1.2 m).
    B. 24 in. (600 mm).   E. 6 ft (1.8 m).
    C. 3 ft (900 mm).

    Code reference_____

8. A pull box is installed with a trade size 4 (103) Electrical Metallic Tubing entering one end and leaving the opposite end. The conductors in the run of raceway are size 500-kcmil copper Type RHW. The minimum length of a pull box permitted for this installation is:

   A. 12 in. (305 mm).
   B. 18 in. (457 mm).
   C. 24 in. (610 mm).
   D. 30 in. (762 mm).
   E. 32 in. (812 mm).

   Code reference_____

9. Trade size 1 (27) PVC Type Rigid Nonmetallic Conduit is run as exposed surface wiring in a building and securely fastened within a distance of not more than 3 ft (900 mm) of each box and cabinet. The maximum distance between supports in a run of this RNC is not permitted to be greater than:

   A. 3 ft (900 mm).
   B. 5 ft (1.5 m).
   C. 6 ft (1.8 m).
   D. 8 ft (2.5 m).
   E. 10 ft (3 m).

   Code reference_____

10. Size 500-kcmil Type THWN copper conductors are permitted to enter the top of a panelboard and terminate into lugs, as shown in Figure 2.34, provided the distance from the top of the lugs to the top of the panelboard enclosure is not less than:

    A. 6 in. (152 mm).
    B. 10 in. (305 mm).
    C. 11 in. (279 mm).
    D. 14 in. (356 mm).
    E. 15 in. (381 mm).

    Code reference_____

11. A 44-ampere continuous load is supplied by a circuit consisting of three copper Type THHN conductors run in Electrical Metallic Tubing as illustrated in Figure 2.35 on the next page. There are no other conductors in the EMT and the circuit is protected with a 60-ampere rated circuit breaker. All conductor terminations in the circuit are rated 75°C. The minimum size conductor permitted for this circuit is:

    A. 8 AWG.
    B. 6 AWG.
    C. 4 AWG.
    D. 3 AWG.
    E. 2 AWG.

    Code reference_____

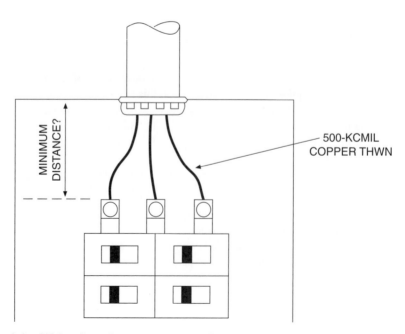

MINIMUM DISTANCE?

500-KCMIL COPPER THWN

Figure 2.34 A set of size 500-kcmil conductors enter a panelboard and terminate at the supply lugs.

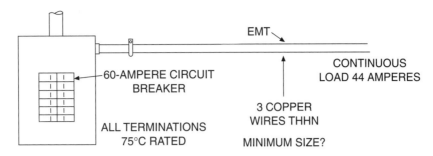

Figure 2.35   Three copper conductors with THHN insulation are run in EMT for a 60-ampere rated circuit to supply a 44-ampere load.

12. A 90° bend made in trade size 2 (53) PVC Rigid Nonmetallic Conduit in the field is not permitted to have a radius measured to the centerline of the conduit of less than:
    A. 9¹/₂ in. (241 mm).   D. 14 in. (356 mm).
    B. 10 in. (254 mm).   E. 15 in. (381 mm).
    C. 12 in. (305 mm).

    Code reference _____

13. Conductors are run to an electric motor in raceway with the final connection of Liquidtight Flexible Nonmetallic Conduit, Type LFNC-B, to provide for some flexibility at the motor for adjustments and to prevent transmission of vibration. The installation is shown in Figure 2.36. An equipment bonding jumper is included. The Liquidtight Flexible Nonmetallic Conduit supported only at its terminations is permitted to be installed for this purpose in lengths not exceeding:
    A. 18 in. (450 mm).   D. 4 ft (1.4 m).
    B. 2 ft (600 mm).   E. 6 ft (1.8 m).
    C. 3 ft (900 mm).

    Code reference _____

14. A metallic auxiliary gutter is installed at a motor control location for running circuit conductors from a panelboard to the various controllers. The auxiliary gutter is made mechanically and electrically continuous and is required to be supported at intervals not to exceed:
    A. 5 ft (1.5 m).   D. 10 ft (3 m).
    B. 6 ft (1.8 m).   E. 12 ft (3.7 m).
    C. 8 ft (2.5 m).

    Code reference _____

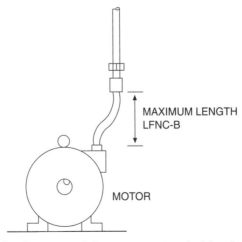

Figure 2.36   The final connection from a conduit to a motor terminal box is with a short length of LFNC-B Liquidtight Flexible Nonmetallic Conduit.

15. A service in a building has a main overcurrent device rated 1600 amperes. The service-entrance conductors are copper Type THWN run as four parallel sets each in a separate Rigid Metal Conduit, shown in Figure 2.37. All conductor terminations are rated 75°C, and there are only three current-carrying conductors in each service raceway. The minimum size conductors permitted for this service are size:

A.   500 kcmil.   C.   700 kcmil.   E.   800 kcmil.
B.   600 kcmil.   D.   750 kcmil.

Code reference _____

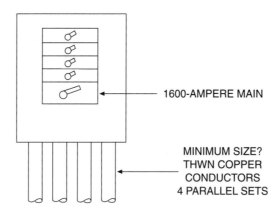

Figure 2.37   A panelboard serving as the main service for a building has a 1600-ampere main overcurrent device and is supplied by four parallel sets of copper THWN conductors.

WORKSHEET NO. 2—ADVANCED WIRE, RACEWAY, AND BOX SIZING

Mark the single answer that most accurately completes the statement based upon the 2005 Code. Also provide, where indicated, the code reference that gives the answer or indicates where the answer is found, as well as the Code reference where the answer is found.

1. A trade size 1 (27) Electrical Metallic Tubing is run across a ceiling perpendicular to the trusses and attached to the bottom of the trusses as shown in Figure 2.38. The EMT is supported at each luminaire (lighting fixture) and at each truss. The maximum permitted distance from the luminaire (lighting fixture) to the first support is:

   A. 18 in. (450 mm).   C. 4 ft (1.2 m).   E. 10 ft (3 m).
   B. 3 ft (900 mm).   D. 5 ft (1.5 m).

   Code reference _____

2. A metal junction box has one size 8 AWG 3-conductor Type NM-B Cable with equipment ground entering and two size 10 AWG 3-conductor Type NM-B Cables with equipment ground entering, as shown in Figure 2.39 on the next page. The junction box is required to have a minimum volume of:

   A. 18 in.^{3} (295 cm^3).   D. 32 in.^{3} (524 cm^3).
   B. 22.5 in.^{3} (369 cm^3).   E. 36 in.^{3} (590 cm^3).
   C. 27 in.^{3} (443 cm^3).

   Code reference _____

3. A circuit consisting of three copper Type THHN conductors is run in Electrical Metallic Tubing (EMT) serving a continuous load of 44 amperes. There are six other conductors with THHN insulation in the same run of EMT for a total of nine current-

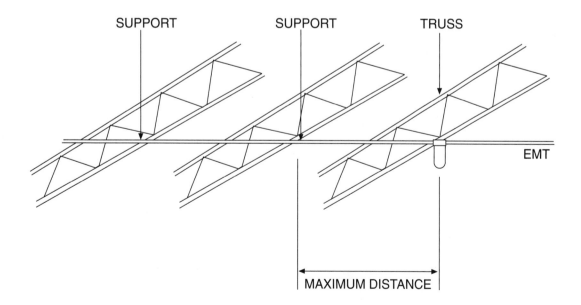

Figure 2.38 Electrical Metallic Tubing of the trade size 1 (27) is attached to the underside of ceiling trusses at the boxes and at the truss between the luminaires (lighting fixtures).

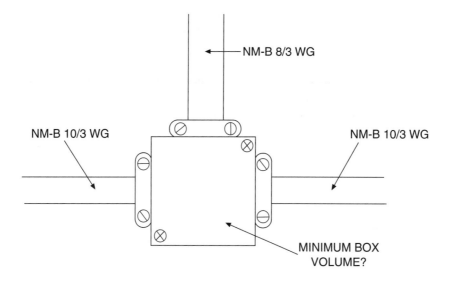

NM-B 8/3 WG

NM-B 10/3 WG

NM-B 10/3 WG

MINIMUM BOX
VOLUME?

Figure 2.39   A square junction box has two 3-conductor with equipment ground size 10 AWG cables and one 3-conductor with equipment ground size 8 AWG cable entering.

carrying conductors in the raceway, illustrated in Figure 2.40. If a 60-ampere over-current device protects the circuit, and all terminations are 75°C rated, the minimum size conductor permitted for the circuit is:

A. 12 AWG.   C. 8 AWG.   E. 4 AWG.

B. 10 AWG.   D. 6 AWG.

Code reference_____

4.   Several circuits are run in *Schedule* 40 PVC Rigid Nonmetallic Conduit as shown in Figure 2.41 on the next page. There are four conductors size 3/0 AWG copper Type THHN, six conductors size 10 AWG copper Type THHW, and one size 6 AWG bare copper equipment grounding conductor. The minimum trade size PVC conduit permitted for these conductors is:

A.   trade size 2 (53).   D. trade size $3^1/2$ (91).

B.   trade size $2^1/2$ (63).   E. trade size 4 (103)

C.   trade size 3 (78).

Code reference_____

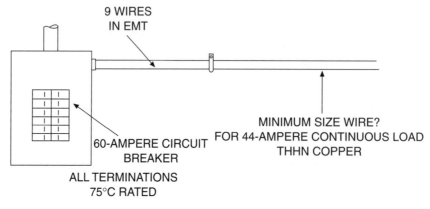

9 WIRES
IN EMT

MINIMUM SIZE WIRE?
FOR 44-AMPERE CONTINUOUS LOAD
THHN COPPER

60-AMPERE CIRCUIT
BREAKER

ALL TERMINATIONS
75°C RATED

Figure 2.40   A circuit protected at 60 amperes and with a 44-ampere load is run in a raceway with a total of nine current-carrying conductors.

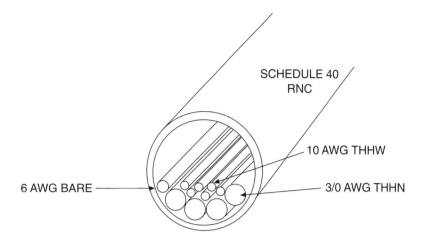

Figure 2.41   A *Schedule* 40 Rigid Nonmetallic Conduit contains one bare size 6 AWG conductor, six size 10 AWG Type THHW conductors, and four size 3/0 AWG Type THHN conductors.

5. Two enclosures are on back-to-back on opposite sides of a solid concrete wall and connected with a Rigid Metal Conduit that is only 14 in. (356 mm) in length. Four size 1 AWG copper Type THWN conductors are run through the conduit which, in this case, is required to have a minimum trade size of:
   A.   trade size ³/4 (21).
   B.   trade size 1 (27).
   C.   trade size 1¹/4 (35).
   D.   trade size 1¹/2 (41).
   E.   trade size 2 (53).

   Code reference _____

6. The general use 125-volt receptacles in a commercial building are supplied by 20-ampere rated multiwire branch-circuits with three ungrounded conductors and a common neutral from a 208/120-volt, 4-wire, 3-phase electrical system. The conductors for nine circuits are run in a single raceway to an area of the building including nine ungrounded conductors and three neutral conductors. For the purpose of applying adjustment factors to determine the allowable ampacity of the conductors, the number of current-carrying conductors for this set of circuits is:
   A.   six.
   B.   nine.
   C.   ten.
   D.   twelve.
   E.   sixteen.

   Code reference _____

7. Four Type THHN copper current-carrying conductors are run in EMT and supply two 20-ampere incandescent lighting circuits where all conductor terminations are rated 75°C. The load on the circuits is 14.8 amperes. In route to the area where the luminaires (lighting fixtures) are installed, the raceway passes through an area for about 20 ft (6.1 m) where the room temperature is 125°F (52°C). The wiring is illustrated in Figure 2.42. The minimum size conductors permitted for these circuits is:
   A.   not possible to determine with the information provided.
   B.   14 AWG.
   C.   12 AWG.
   D.   10 AWG.
   E.   8 AWG.

   Code reference _____

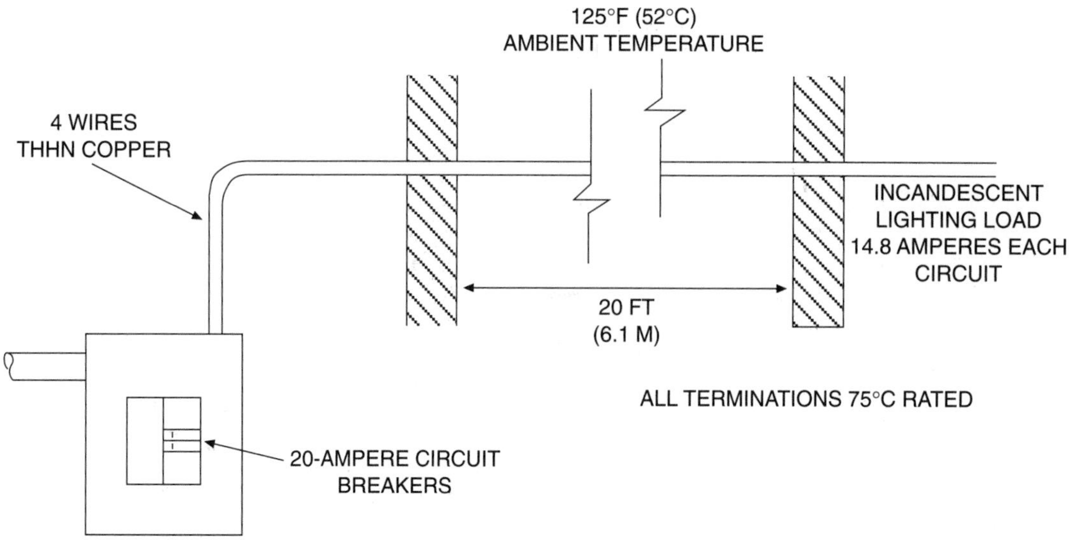

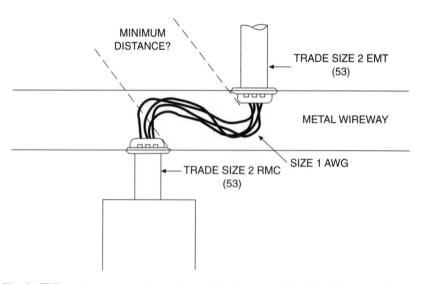

125°F (52°C)
AMBIENT TEMPERATURE

4 WIRES
THHN COPPER

INCANDESCENT
LIGHTING LOAD
14.8 AMPERES EACH
CIRCUIT

20 FT
(6.1 M)

ALL TERMINATIONS 75°C RATED

20-AMPERE CIRCUIT
BREAKERS

Figure 2.42   Four copper conductors with Type THHN insulation for 20-ampere circuits are run in the same raceway and pass through a room with a 125°F (52°C) ambient temperature.

8. A trade size 2 (53) EMT enters the top of a metal wireway, and out the bottom of the wireway is a trade size 2 (53) Rigid Metal Conduit connecting to a panelboard. Four size 1 AWG copper Type THHN conductors enter the wireway from the EMT and pass directly to the Rigid Metal Conduit as shown in Figure 2.43. The minimum permitted offset distance between the two raceways containing the same conductors is:
    A.   not specified in the Code for this application.
    B.   6 in. (152 mm).
    C.   8 in. (203 mm).
    D.   10 in. (254 mm).
    E.   12 in. (318 mm).

    Code reference_____

MINIMUM
DISTANCE?

TRADE SIZE 2 EMT
(53)

METAL WIREWAY

SIZE 1 AWG

TRADE SIZE 2 RMC
(53)

Figure 2.43   Size 1 AWG conductors pass through a metal wireway with a short distance offset.

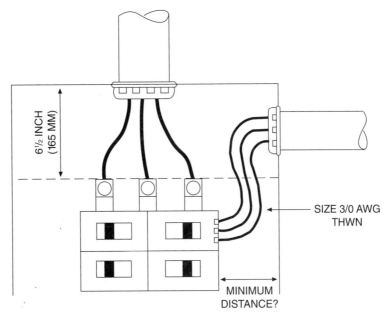

Figure 2.44 Size 3/0 AWG conductors connected to a disconnect make a 90° bend, travel to the top of the panel and then exit the wall opposite the disconnect terminals.

9. Conductors size 3/0 AWG Type THWN copper connected to a disconnect immediately make a 90° bend travel to the top of the panelboard then make another 90° bend and leave through the side of the panelboard as shown in Figure 2.44. The gutter space at the top of the panelboard is 6^1/2 in. (165 mm) deep. The installation is permitted provided the distance from the disconnect lugs to the side of the panelboard is not less than:

   A. 3 in. (76 mm).
   B. 4 in. (102 mm).
   C. 6 in. (152 mm).
   D. 6^1/2 in. (165 mm)
   E. 8 in. (203 mm).

   Code reference _____

10. Electrical Nonmetallic Tubing is permitted to be installed:

   A. for exposed work in buildings of any height.
   B. as concealed wiring only if the walls, floors, and ceilings provide a thermal barrier that has at least a 15-minute finish rating.
   C. as concealed wiring in buildings of any height with no requirement that the walls, floors, or ceiling have a fire rating.
   D. as concealed wiring in buildings of not more than three floors unless the walls, floors, and ceilings provide a thermal barrier that has at least a 15-minute finish rating.
   E. as concealed wiring in buildings of not more than three floors unless the walls, floors, and ceilings provide a thermal barrier that has at least a 1-hour finish rating.

   Code reference _____

11. A 140 ft (42.7 m) straight run of trade size 1 (27) PVC Type Rigid Nonmetallic Conduit is installed in an unheated parking structure where the temperature throughout the year is likely to change up to 120°F (66.7°C). The change in length for this run of PVC Conduit over a season will be:

   A. 3.48 in. (88 mm).
   B. 4.87 in. (124 mm).
   C. 5.12 in. (130 mm).
   D. 5.84 in. (148 mm).
   E. 6.82 in. (173 mm).

   Code reference _____

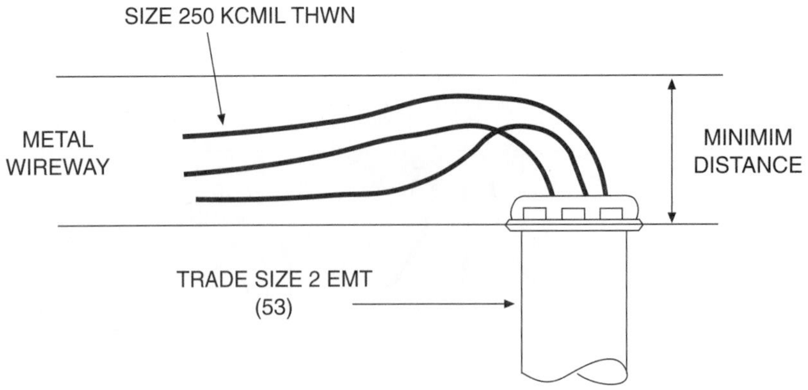

Figure 2.45 Size 250 kcmil conductors enter a metal wireway from a trade size 2 (53) EMT and make a 90° bend.

12. All conductors run within a metallic auxiliary gutter have insulation rated 75°C. All splices, taps and terminations of conductors within the auxiliary gutter or elsewhere in the circuits are rated 75°C. A size 8 AWG copper Type THWN conductor runs the entire length of a metallic auxiliary gutter and at one point along the gutter there are a total of 21 current-carrying conductors. The allowable ampacity of the size 8 AWG copper conductor for this application is:

   A. 22.5 amperes.   D. 40 amperes.
   B. 25 amperes.   E. 50 amperes.
   C. 35 amperes.

   Code reference _____

13. A trade size 2 (53) Electrical Metallic Tubing enters a metal wireway as shown in Figure 2.45, containing three Type THWN copper conductors size 250 kcmil. This metal wireway is required to have a width from the EMT entry to the opposite side of not less than:

   A. 4^1/2 in. (114 mm).   D. 7 in. (178 mm).
   B. 5 in. (127 mm).   E. 9 in. (229 mm).
   C. 5^1/2 in. (140 mm).

   Code reference _____

14. A pull box has a trade size 4 (103) conduit entering one side and trade sizes 2 (53) and 3 (78) conduit entering the adjacent side as shown in Figure 2.46 on the next page. Three size 4/0 AWG conductors run from the trade size 4 (103) to the trade size 3 (78) conduits. Three size 6 AWG and eight size 12 AWG conductors run from the trade size 4 (103) to the trade size 2 (53) conduits. The minimum distance permitted from the trade size 3 (78) conduit to the opposite wall (dimension A) is:

   A. 20 in. (521 mm).   D. 26 in. (660 mm).
   B. 22 in. (559 mm).   E. 30 in. (762 mm).
   C. 24 in. (618 mm).

   Code reference _____

15. A 3-phase feeder consists of two parallel sets of Type THHW copper conductors run in a single trade size 6 (155) Rigid Metal Conduit. There is a total of six current-carrying conductors in the conduit. The calculated load for the feeder is 560 amperes and the feeder conductors are protected with a 700-ampere rated overcurrent device. The minimum size conductors permitted for this feeder is:

   A. 400 kcmil.   C. 600 kcmil.   E. 750 kcmil.
   B. 500 kcmil.   D. 700 kcmil.

   Code reference _____

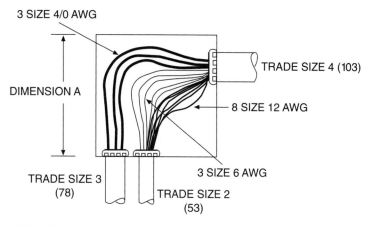

Figure 2.46   A pull box is installed for the purpose of making angle pulls.

UNIT 3

Outlets, Lighting, Appliances, and Heating

OBJECTIVES

After completion of this unit, the student should be able to:

- determine the location of electrical outlets in a dwelling.

- determine the minimum number of general lighting, small appliance, and laundry circuits permitted in a dwelling.

- state the clearance requirements for outside aerial feeders and branch-circuits.

- determine the minimum size flexible cord required for an application.

- state the installation requirements for lighting outlets in clothes closets.

- state the installation requirements for recessed lighting fixtures.

- determine the minimum size conductor for a storage-type electric water heater.

- determine the maximum number of baseboard electric heaters permitted to be installed on a 15- or 20-ampere branch-circuit.

- state if it is permitted to power a room air conditioner from an existing general lighting branch-circuit of a dwelling.

- state the cable type required to be installed for a push button of a door chime.

- answer wiring installation questions relating to *Articles 210, 220, Parts I and II, 225, 396, 398, 400, 402, 404, 406, 410, 411, 422, 424, 426, 720,* and *725.*

- state at least five significant changes that occurred from the 2002 to the 2005 Code for *Articles 210, 220, Parts I and II, 225, 404, 406, 410, 422, 424,* or *725.*

CODE DISCUSSION

The emphasis of this unit is on circuits and installation of outlets, appliances, and equipment. Methods are covered for determining the minimum number of general illumination, receptacle, and other circuits required for various types of buildings. Outside feeders and branch-circuit installations are also covered. Special circuits such as low-voltage, power-limited-control, and signaling circuit installation are covered in this unit.

Article 210, Part I contains specifications about branch-circuits such as circuit ratings (*210.3*) and voltage limitations for different types of occupancies and types of loads to be served (*210.6*). For example, in a dwelling, the maximum permitted voltage between conductors for luminaires (lighting fixtures) and general-use receptacles is 120 volts nominal. Multiwire branch-circuits are discussed in *210.4*. A multiwire branch-circuit is actually two or more circuits that share a common neutral conductor. In the case of a single-phase, 120/240-volt system, three wires are used to supply two circuits with a common shared neutral and 240 volts between the ungrounded conductors. In the case of a 3-phase 208/120-volt system, three circuits can be supplied with only four conductors. Generally it is assumed that when the loads on the multiwire circuits are the same and all are operating at the same time, the current on the common neutral conductor will be zero or nearly zero. This is not necessarily the case with a multiwire circuit derived from a 3-phase wye electrical system. Sometimes a multiwire branch-circuit will supply multiple devices on the same yoke or strap. A common example is a duplex receptacle where the tab between the ungrounded screw terminal for the two receptacles is removed. Each of the receptacles on the same strap is supplied from a separate circuit and uses a common neutral. See the definition of a multiwire branch-circuit in *Article 100*. It is required by *210.4(B)* that when multiple devices, such as receptacles, on the same yoke or strap are supplied by separate circuits, those circuits are required to originate from an overcurrent device that will de-energize both circuits at the same time. This can be accomplished using a 2-pole circuit breaker or two single-pole circuit breakers with a handle tie.

Ground-fault circuit-interrupter requirements for receptacles are covered in *210.8*. There are several general requirements that apply to any type of facility. All 15- and 20-ampere, 125-volt receptacles in bathrooms, kitchens, outdoor locations accessible to the public, and on rooftops of any type of facility are required to be GFCI protected. This can be accomplished with a ground-fault detecting receptacle, or a ground-fault detecting circuit breaker. Both the ungrounded conductor and the neutral pass through a current-sensing coil. If the current measured in those two wires is different by more than 0.006 amperes (6 milliamperes), the interrupter will trip. This means the current is finding a return path other than the neutral wire, which possibly could be a person. The dwelling GFCI requirements for protection of 125-volt, 15- and 20-ampere receptacles are summarized below:

- Receptacles serving *kitchen countertop* surfaces
- *Bathroom* receptacles
- *Outside* receptacles (see exception for snow-melting equipment)
- *Crawlspace* receptacles
- *Garage* receptacles (see exceptions for dedicated equipment receptacles)
- Receptacles in *accessory buildings* at or below grade used for storage or work areas (See Figure 3.1)
- Receptacles in *unfinished basement* or unfinished portion of a basement (see exceptions)
- Receptacles serving countertop and within 6 ft (1.8 m) of a *wet bar,* laundry, or utility sink
- Receptacles in a *boat house* or for a boat hoist

The minimum number of branch-circuits required for a building are specified in *210.11*. Calculating the minimum number of branch-circuits is covered later under *Sample Calculation.* Four 20-ampere branch-

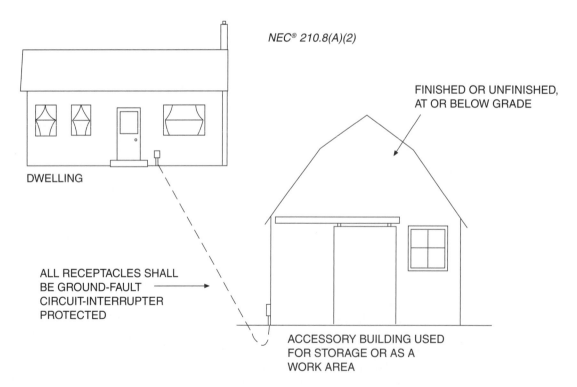

NEC® 210.8(A)(2)

FINISHED OR UNFINISHED, AT OR BELOW GRADE

DWELLING

ALL RECEPTACLES SHALL BE GROUND-FAULT CIRCUIT-INTERRUPTER PROTECTED

ACCESSORY BUILDING USED FOR STORAGE OR AS A WORK AREA

Figure 3.1   Receptacles in accessory buildings of dwellings, whether finished or unfinished, at or below grade, if used for storage or as work areas, are required to be GFCI protected for personnel.

circuits are required as a minimum in a dwelling. Two 20-ampere branch-circuits are required to supply the receptacles in the kitchen, dining room, and similar rooms. One 20-ampere branch-circuit is required to supply laundry equipment such as an electric washer and possibly a gas dryer. Another 20-ampere branch-circuit is required to supply the receptacle outlets in the bathroom or bathrooms of the dwelling. The minimum number of remaining general purpose 15- and 20-ampere circuits is calculated according to the method of *210.11(A)* and *220.12*.

Arcing faults can occur particularly when Nonmetallic-Sheathed Cable run concealed in walls, floors, and ceilings is damaged such as being penetrated by a nail or screw. In this case, current flows intermittently between conductors, or between a circuit conductor and the equipment grounding conductor or some other grounded structural member. This is known as a parallel arcing fault. An example of a series-arcing fault is when a loose connection occurs in either the ungrounded- or grounded-circuit conductor such as a loose screw at a switch or receptacle. Overcurrent devices such as circuit breakers and fuses require a specified amount of energy to open depending upon their rating. An arcing fault can create enough heat at the fault location to start a fire without drawing enough energy at the overcurrent device to cause it to open the circuit. It has been reported that dwelling fires caused by arcing faults have become prevalent enough to justify action to be taken to require a device on dwelling circuits that can detect an arcing fault in a circuit and disconnect power to the circuit. The device that is capable of detecting an arcing condition in a circuit is called an Arc-Fault Circuit-Interrupter (AFCI) and is defined in *210.12(A)*. The purpose of an AFCI is to detect a potentially fire-producing arcing condition anywhere in the circuit wiring of the dwelling or in equipment attached to the circuit such as a damaged cord to a floor lamp. An Arc-Fault Circuit-Interrupter is required to be installed in the circuit-panel in place of the circuit breaker, or adjacent to the circuit panel. There is more discussion of circuit-arcing conditions in *Unit 6,* and there is a diagram of an AFCI that replaces a normal circuit breaker in Figure 6.6.

Arc-Fault Circuit-Interrupters are required by *210.12(B)* for the protection of all branch-circuits that supply 125-volt, 15- and 20-ampere outlets in dwelling bedrooms. The definition of dwelling unit is in *Article 100* and includes one-family, two-family, and multifamily living units. Mobile homes and manufactured homes that also meet this definition are included.

Arc-Fault Circuit-Interrupters that are installed in a circuit panel to replace a typical circuit breaker are referred to as the "branch/feeder" type AFCI. A more recent development is the "combination" type AFCI, which is a receptacle outlet with arc-fault detection capability. Similar to a GFCI-type receptacle, the combination-type AFCI is of the feed-through type so the remainder of the circuit is also arc-fault protected. The branch/feeder AFCI and the combination AFCI provide different levels of arc-fault protection. Both types detect parallel arc faults where the current flows to the equipment grounding wire or to grounded structural metal. Sensitivity of an AFCI to series-arcing from the ungrounded wire to the neutral, or the opening of the circuit is different. The branch/feeder type AFCI sensitivity to series arcing is usually set at an arcing level of 50 to 75 amperes, which prevents nuisance tripping during normal operation, but it is not sensitive enough to detect some series arcing conditions that could result in a fire. The combination-type AFCI sensitivity is set at a 5-ampere arcing level for a period of not more than 1 second. This type is sensitive enough to detect a series-arcing condition and de-energize the circuit before a fire is likely to occur. Nuisance tripping of a combination AFCI device under normal conditions may occur, such as the opening of a switch or normal arcing in appliances such as variable speed equipment with an internal commutator and brushes. For situations where circuit reliability is of high importance, such as life-support equipment, the circuit can be protected using a branch/feeder type AFCI in the panelboard to protect the entire circuit, and a combination-type AFCI to protect the remainder of the circuit where high reliability for continuous power is not as important. By utilizing such a technique, the highest level of arc-fault protection can be obtained without jeopardizing reliability for a specific function.

Arc-Fault Circuit-Interrupter protection is intended for the entire circuit, and the device is intended to be installed in the service panelboard in place of a normal circuit breaker. The *Exception* to *210.12(B)* does permit the AFCI device to be installed up to 6 ft (1.8 m) from the panelboard as measured along the circuit conductor. This is a necessary exception if the combination-type AFCI is to be installed since it is a receptacle-type device. There is an additional requirement that the wiring between the service panelboard and the AFCI device be installed in metallic raceway such as RMC, IMC, EMT, FMC, or LFMC or be installed in a metal-sheathed cable such as Type AC Cable or Type MC Cable.

Article 210, Part II covers the ratings of branch-circuits, including those in dwellings. *NEC® 210.19(A)* states that the branch-circuit conductors shall have an ampacity of not less than the maximum load to be served. The rating of a branch-circuit is the rating of the overcurrent device protecting the circuit. The actual rating of any one branch-circuit is determined by the method in *210.20(A)*. The overcurrent device protecting a circuit is required to have a rating not less than 1.25 times any continuous load served plus 1.0 times any noncontinuous load served.

The minimum circuit rating of 40 amperes for a dwelling electric range is given in *210.19(A)(3)*. The neutral conductor of the range circuit is permitted to be of a smaller size than the ungrounded conductors, but it is not permitted to have an ampacity less than 70% of the branch-circuit rating, and it shall be no smaller than size 10 AWG copper (*210.19(A)(3) Exception 2*).

A tap is a conductor that connects to a branch-circuit or feeder conductor to serve a specific load. The tap conductor shall be of sufficient ampacity to supply the load. The tap conductor is permitted to be smaller than the branch-circuit or feeder conductor, but the Code sets a minimum size based on the rating of the branch-circuit or feeder. Additional discussion of taps is in *Units 6, 7, and 8. Exception 1* to *210.19(A)(3)* permits a tap with sufficient ampacity to serve the load but not less than 20 amperes for an electric wall-mounted oven or counter-mounted cooking unit in a dwelling when the circuit is protected at not more than 50 amperes. This is illustrated in Figure 3.2. There is another tap rule in this article. *NEC® 210.19(A)(4) Exception 1* permits taps to branch-circuits provided the tap conductor is to individual lampholders or luminaires (lighting fixtures) and is not more than 18 in. (450 mm) in length. For a branch-circuit rated up to 30 amperes, the tap shall have an ampere rating sufficient for the load, and not less than 15 amperes. For a branch-circuit rated up to 50 amperes, the tap shall have an ampere rating sufficient for the load, and not less than 20 amperes.

FPN 4 to *210.19(A)* recommends a maximum of 3% voltage drop be permitted on branch-circuits. There is a complete discussion of voltage drop in *Unit 6* with examples to show how to size a conductor to maintain voltage drop within the desired level. *NEC® 210.21(B)(1)* states that when a single receptacle is installed on a branch-circuit, the receptacle is not permitted to have a rating less than that of the branch-circuit rating. A single receptacle is defined in *Article 100*. It has provisions for the connection of only one cord-connected device. A duplex receptacle outlet is considered two receptacles. Therefore, if a single receptacle on a yoke is installed on a 20-ampere circuit, the minimum receptacle rating permitted is 20 amperes. If a duplex receptacle outlet is installed on a 20-ampere branch-circuit, the receptacle rating is permitted to be either 15 or 20 amperes. *NEC® 210.23* covers the permissible loads on branch-circuits of various ratings. For example, *210.23(A)* permits 15- and 20-ampere branch-circuits to supply typical residential and commercial lighting outlets. It is necessary to read the remainder of this section to determine the restrictions on the type of outlets and equipment permitted to be supplied from branch-circuits of various ratings.

The minimum requirements for providing outlets for lighting and receptacles are given in *Article 210, Part III*. The maximum permitted spacing of receptacle outlets in dwellings is given in *210.52*. A receptacle outlet shall be installed in listed rooms of a dwelling such that any point measured along the wall is not greater than 6 ft (1.8 m) from the outlet. Wall spaces 2 ft (600 mm) wide or wider shall have a receptacle outlet. These rules are illustrated in Figure 3.3. A minimum of two 125-volt, 20-ampere small appliance branch-circuits are required in a dwelling to serve the kitchen, dining room and similar rooms, *210.11(C)(1)*. Not fewer than two small-appliance branch-circuits are required to supply the receptacles serving the kitchen

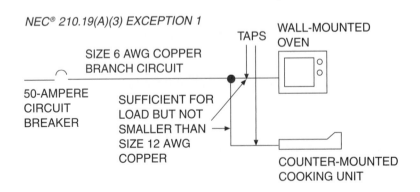

Figure 3.2   Taps in a dwelling range circuit shall have sufficient ampacity for the load, and shall not be less than 20 amperes.

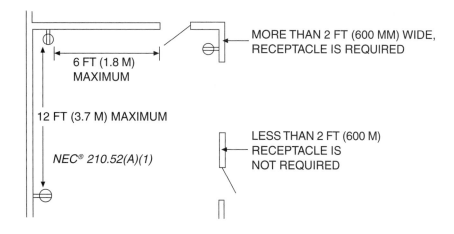

Figure 3.3   Required maximum spacing of receptacle outlets in a dwelling.

counters, *210.52(B)(3)*. All of the receptacles in the kitchen, dining room, and similar rooms are required to be supplied by small-appliance branch-circuits. It is permitted to have a wall switch-controlled receptacle in the dining room, but not the kitchen, for the purpose of supplying lighting served by a general-purpose branch-circuit, *210.70(A)(1) Exception 1*. Generally, the small appliance branch-circuits are only permitted to serve the receptacles in the kitchen, dining room, and similar rooms, *210.52(B)(2)*. The refrigerator is permitted to be supplied by a small-appliance branch-circuit, however, it is permitted to provide a dedicated 15-ampere rated branch-circuit for the refrigerator, *210.52(B)(1) Exception 2*. Receptacle outlets of the small appliance branch-circuits serving kitchen counters along a wall are required to be spaced such that no point along the wall line is more than 24 in. (600 mm) from a receptacle outlet, *210.52(C)(1)*. This is illustrated in Figure 3.4. A receptacle outlet is required to serve a peninsular counter that extends out into the room a distance from the connecting edge of at least 12 in. (300 mm) if at least 24 in. wide (600 mm), or a distance

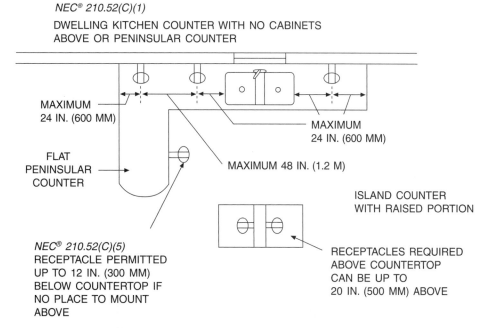

Figure 3.4   When a dwelling kitchen countertop is flat with no raised portions, and there is no other practical means to mount the receptacles within 18 in. (450 mm) above the surface, the receptacles are then permitted to be mounted no less than 12 in. (300 mm) below the countertop surface.

of 24 in. (600 mm) if less than 24 in. (600 mm) wide as shown in Figure 3.5. A receptacle is not required to be installed on a wall behind a range or sink unless the sink is set out from the wall more than 12 in. (300 mm) or out from a corner more than 18 in. (450 mm).

At least one receptacle is required to be installed at the front and one at the back of a one-family dwelling or at each unit of a two-family dwelling. It is stated in *210.52(E)* that this receptacle is to be accessible at grade level and not more than 6½ ft (2 m) above grade level. When there is a deck attached to be dwelling, sometimes a receptacle installed to serve the deck is not considered to be accessible to grade level. In some areas, another receptacle accessible at grade level may be required to be installed. A similar problem can arise with respect to a receptacle available for service of a central air-conditioning unit. In *210.63*, a receptacle on a 125-volt, single-phase, 15- or 20-ampere circuit is required to be located not more than 25 ft (7.5 m) from an air-conditioning unit. In the case of an air-conditioning unit located on the outside of a dwelling, this is another consideration when placing the required outside receptacle.

It is not uncommon for a dwelling basement to have unfinished portions separated by finished portions. It is made clear in *210.52(G)* that a receptacle is required to be installed in each unfinished portion of a dwelling basement if the unfinished portions are not adjoining such as shown in Figure 3.6.

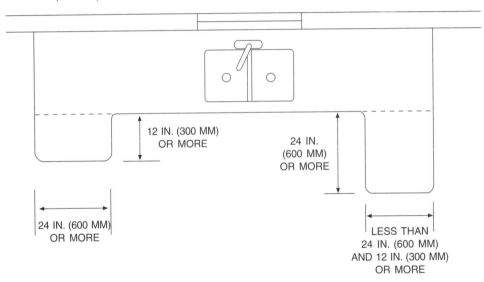

NEC® 210.52(C)(3) ONE RECEPTACLE OUTLET REQUIRED
TO SERVE PENINSULAR COUNTERS 24 IN. (600 MM)
BY 12 IN. (300 MM) OR LARGER

12 IN. (300 MM)
OR MORE

24 IN.
(600 MM)
OR MORE

24 IN. (600 MM)
OR MORE

LESS THAN
24 IN. (600 MM)
AND 12 IN. (300 MM)
OR MORE

Figure 3.5   One receptacle outlet is required to serve a peninsular counter with minimum dimensions of 12 in. (300 mm) by 24 in. (600 mm) or greater.

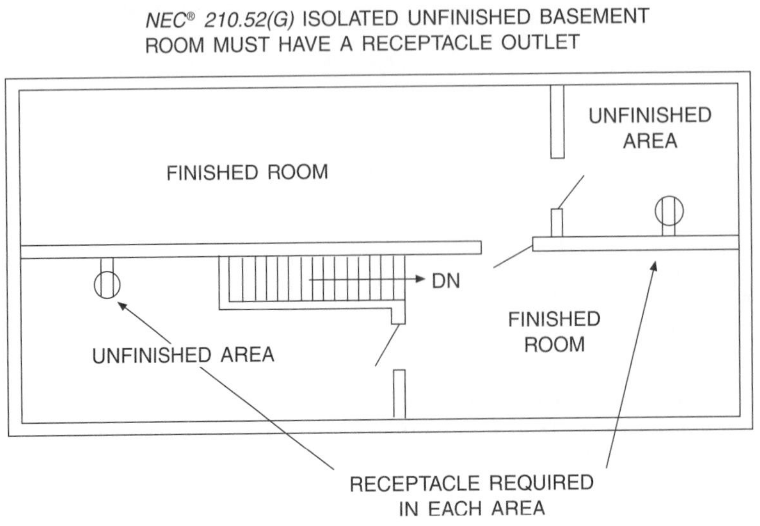

NEC® 210.52(G) ISOLATED UNFINISHED BASEMENT
ROOM MUST HAVE A RECEPTACLE OUTLET

UNFINISHED
AREA

FINISHED ROOM

DN

UNFINISHED AREA

FINISHED
ROOM

RECEPTACLE REQUIRED
IN EACH AREA

Figure 3.6   Unfinished areas of a basement in a dwelling that are isolated from each other by a finished area are each required to have a receptacle outlet.

NEC® 210.70(A) requires that every habitable room of a dwelling, including other listed areas, shall have a lighting outlet that is wall switch-controlled. A switched receptacle outlet may be used in place of an actual luminaire (lighting fixture) in some rooms, as shown in Figure 3.7. A lighting outlet is also required to be installed in hallways, stairways, attached garages, and detached garages with power. In the case of interior stairways, a switch controlling the lighting is required at each level when there are six steps or more. At least one lighting outlet containing a switch, or that is wall switch-controlled, is required for attics, underfloor spaces, basements, and utility rooms used for storage or containing equipment requiring servicing. The switch is required at the normal point of entry to the space, and the lighting outlet is required to be located at any equipment requiring servicing. Automatic control of lighting is permitted for hallways, stairways, and outdoor entrances.

Article 220, Parts I and *II* provides requirements for the determination of the number of outlets of various types permitted on a branch-circuit. *NEC® 220.12* provides important information for the determination of the number of branch-circuits required for receptacles and general lighting in buildings. *Table 220.12* gives the minimum unit load in VA/ft^2 or VA/m^2 of building area considered in determining the minimum number of branch-circuits for lighting and for determining the minimum lighting load for feeder calculation purposes. If the actual lighting load is known to be of a greater value, that load shall be considered the general lighting load.

NEC® 220.14(J) permits the general use receptacles in a dwelling to be included in the general illumination calculation. Small appliance receptacle loads and laundry receptacle load are specified elsewhere in the Code and are not a part of this general illumination load calculation. This means that the 3 volt-amperes per square ft (33 VA/m^2) from *Table 220.12* is used for a dwelling to determine the minimum number of general illumination branch-circuits that will supply the receptacle outlets and lighting fixtures. The significance of this is that the 180-volt-amperes requirement of *220.14(J)* does not apply to dwelling general-use receptacles. The number of outlets permitted on a circuit depends upon the loads to be served, and according to *220.18,* the total load is not permitted to exceed the rating of the circuit. In the case of general illumination in a dwelling, the load is determined based on the method described in *220.12,* which is the area of the dwelling times 3 volt-amperes per square ft (33 VA/m^2). This is converted into the number of circuits according to the method of *210.11(A)* where the total load in volt-amperes is divided by 120 volts and then divided into 15- or 20-ampere circuits.

Article 225 gives the requirements for the installation of branch-circuit and feeder conductors outside. Minimum size of conductors for overhead spans, protection of conductors, and overhead conductor clearances is specified in this article. Clearances above ground of aerial conductors are covered in *225.18.* Similar requirements for service conductors are covered in *230.24.* The conductor clearance over areas accessible only to pedestrians is a minimum of 10 ft (3.0 m). For a residential driveway and other driveways not subject to truck traffic, the minimum clearance is 12 ft (3.7 m), provided the conductors do not exceed 300 volts-to-ground. This clearance is increased to 15 ft (4.5 m) for conductors operating at more than 300 volts-

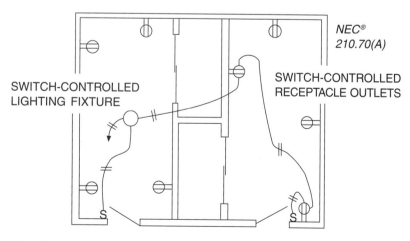

Figure 3.7   Wall switch-controlled lighting is required in every habitable room of a swelling.

to-ground. An 18-ft minimum conductor clearance is required when the driveway or road is subject to truck traffic. Branch-circuit and feeder conductor clearances above roofs are covered in *225.19*. The minimum overhead conductor clearances for roofs accessible to only pedestrians are summarized in Figure 3.8, and covered in *Exception 1 to 225.19*.

Part II of Article 225 gives the rules for supplying power from a building or structure on the property to another building or structure on the same property. *NEC® 250.32* deals with the neutral conductor and equipment grounding when supplying power to another building or structure on the same property. These rules deal with the disconnecting means for power to a building supplied by feeders or in some cases branch-circuits. In the case where qualified personnel are on duty at all times to facilitate safe disconnecting procedures, it is permitted for the disconnecting means to be located at a remote location on the property, *225.32 Exception 1*. The disconnecting means in general is located either inside or outside the building, *225.32*. Whether located inside or outside the disconnect is required to be at the closest practical location to the point where the conductors enter the building. Up to six disconnecting means are permitted as covered in *225.33*. These rules are summarized in Figure 3.9. The disconnecting means is required to be rated as suitable for use as service equipment.

Article 396 deals with Messenger-Supported Wiring, such as aerial triplex or quadruplex cable where a bare conductor supports the insulated conductors. It is permitted to field construct a support messenger and suitably attach the conductors to the messenger. There are no minimum conductor size requirements for overhead spans using Messenger-Supported Wiring. The ampacity of messenger-supported conductors is to be determined using the methods of *310.15,* which basically means the ampacity is determined in most cases using *Table 310.16* or *Table 310.20.*

Article 398 covers the situation in which conductors are run within a building or on the outside of a building where the individual conductors are supported by open wires on insulating devices. The wiring covered by this article is required to be exposed except where passing through a wall or floor. Open Wiring on Insulators is not permitted to be concealed except where passing through structural barriers such as walls. This wiring method is only permitted for industrial or agricultural installations.

Article 400 provides information and requirements on the use of flexible cords and cables. *Table 400.4* gives information about the various types of cords and cables and states the uses permitted. *Table 400.5* gives the allowable ampacity and ampacity derating factors applicable when more than three current-carrying conductors are in the cable of flexible cord. Refer to *Unit 2* for examples of how to determine the minimum size wire permitted when more than three current-carrying conductors are in flexible cord or cable. Markings on the flexible cord or cable with uses permitted and installation requirements are covered in this article. Temperature adjustment factors are also required to be applied to flexible-cord ampacity when the cord will

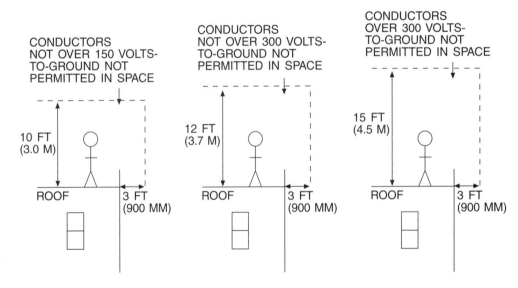

Figure 3.8   Minimum overhead conductor clearances for roofs accessible only to pedestrians.

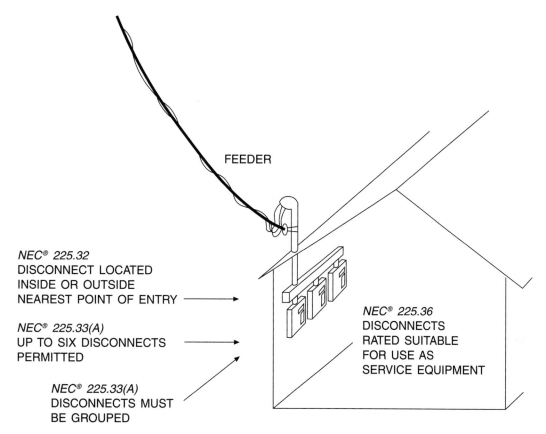

FEEDER

NEC® 225.32
DISCONNECT LOCATED
INSIDE OR OUTSIDE
NEAREST POINT OF ENTRY ———————▶

NEC® 225.33(A)
UP TO SIX DISCONNECTS ———————▶
PERMITTED

NEC® 225.33(A)
DISCONNECTS MUST
BE GROUPED

NEC® 225.36
DISCONNECTS
RATED SUITABLE
FOR USE AS
SERVICE EQUIPMENT

Figure 3.9   When one building is supplied power from another building or structure on the same property, the requirements for the disconnecting means for the building are generally the same as for installing a service in the building

be used in an area where the ambient temperature is above 30°C (86°F). Maximum temperature ratings are printed on some flexible cords. The temperature rating is likely different depending upon whether the cord is to be used in a dry or wet location. Temperature correction factors for flexible cords, as stated in *400.5*, are found at the bottom of *Table 310.16*.

Article 402 lists the markings on fixture wires, the types available and their uses, and general installation requirements. *Table 402.5* gives the permitted ampacity of the fixture wires of sizes 18 to 10 AWG.

Article 404 covers the installation, rating, and use of switches of various types, including knife switches and circuit breakers used as switches. All switches and circuit breakers used as switches are not permitted to be installed such that the center of the handle is more than 6 ft 7 in. (2 m) above the floor when in the "on" position, as stated in *404.8(A)*. *NEC® 240.83(D)* specifies that when a circuit breaker is used as a switch for 120- and 277-volt fluorescent lighting, the circuit breaker shall be marked "SWD" or "HID." A circuit breaker used as a switch for high-intensity discharge lighting is required to be marked "HID." The "SWD" marking may be on the side of the circuit breaker, or it may be on the small label next to the circuit-conductor terminal screw. If a circuit breaker is approved for use as a switch for high-intensity discharge lighting, the "HID" label is usually on the side of the circuit breaker.

Article 406 covers the ratings, types, and installation of receptacles, cord connectors, and attachment plugs. *NEC® 406.2(C)* deals with the situation in which aluminum conductors are attached directly to a receptacle outlet. The receptacle outlet is required to be marked CO/ALR if it is suitable for use with aluminum terminations. Do not confuse this marking with the usual marking of cu/al, which is frequently used for other types of terminations suitable for both copper and aluminum conductors. *NEC® 406.2(D)* covers receptacles with the equipment grounding terminal isolated from the yoke. Other rules on installation of isolated ground receptacles are found in *250.96(B)* and *250.146(D)*. These receptacle types are permitted to supply electronic equipment where electrical noise may be a problem. These receptacles are identified by an orange triangle. Permitted means of identification of receptacles with isolated grounds are illustrated in

Figure 3.10. When receptacles are installed outside exposed to weather, or in damp locations, a cover must be installed that will guard against the entrance of water under the typical operating conditions. *NEC® 406.8* specifies the means of protecting receptacles from the entrance of water. Any 15- or 20-ampere, 125- or 250-volt receptacle in a wet location is required to be provided with a cover that will prevent the entrance of water with or without the plug inserted into the receptacle as shown in Figure 3.11. *NEC® 406.10* specifies the dif-

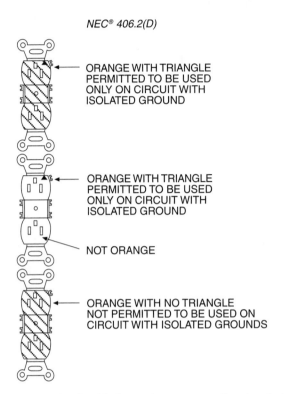

NEC® 406.2(D)

ORANGE WITH TRIANGLE PERMITTED TO BE USED ONLY ON CIRCUIT WITH ISOLATED GROUND

ORANGE WITH TRIANGLE PERMITTED TO BE USED ONLY ON CIRCUIT WITH ISOLATED GROUND

NOT ORANGE

ORANGE WITH NO TRIANGLE NOT PERMITTED TO BE USED ON CIRCUIT WITH ISOLATED GROUNDS

Figure 3.10   Receptacles installed on a circuit with the equipment grounding terminals isolated from the other circuit equipment grounds, as permitted in *NEC® 250.146(D)*, shall be identified with an orange triangle on the face of the receptacle.

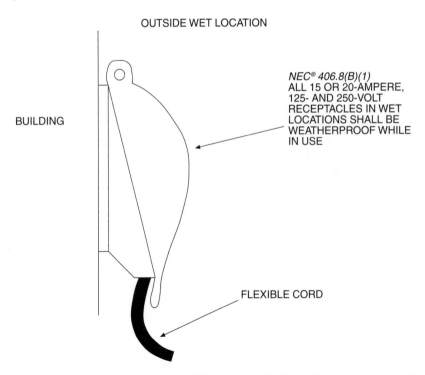

OUTSIDE WET LOCATION

BUILDING

NEC® 406.8(B)(1)
ALL 15 OR 20-AMPERE, 125- AND 250-VOLT RECEPTACLES IN WET LOCATIONS SHALL BE WEATHERPROOF WHILE IN USE

FLEXIBLE CORD

Figure 3.11   Receptacles rated 15- or 20-amperes, 125 volts installed in wet locations are required to have covers that are weatherproof regardless of the intended use of the receptacle.

ferent means of grounding a receptacle. The main rule requires a bonding jumper to be connected between the receptacle grounding terminal and a grounded metal box or equipment grounding wire if a nonmetallic box is used. Alternatives are direct metal-to-metal contact between a metal receptacle yoke and a metal box when the box is surface mounted. Fastening to a raised cover is not considered to be an acceptable means of grounding the receptacle unless the raised cover is listed for grounding. In the case of a flush-mounted metal box, a self-grounding receptacle with a contact device on the mounting screw to maintain electrical contact between the mounting screw and the receptacle yoke is considered an acceptable means of grounding the receptacle.

Article 410 gives requirements on the installation, location, grounding, support, and wiring of luminaires (lighting fixtures) and associated auxiliary equipment. *NEC® 410.8(A)* gives a definition of storage space in clothes closets. This is important because a luminaire (lighting fixture) is required to be installed such that a minimum clearance is maintained between the luminaire (lighting fixture) and the storage space. *Figure 410.8* gives the dimensions of the storage space in a clothes closet. *NEC® 410.8(C)* states that incandescent luminaires (lighting fixtures) with exposed or partially-exposed lamps are not permitted to be installed in a clothes closet. This means porcelain lamp receptacles with bare lamps are not permitted to be installed in clothes closets.

Fire is a danger if excessive heat is produced by improper installation or use of luminaires (lighting fixtures). Incandescent luminaires (fixtures) require a higher wattage to obtain the same amount of light as electric discharge luminaires (fixtures). Therefore, the heat produced by incandescent lamps is usually greater than for electric discharge luminaires (fixtures). Electric discharge luminaires (fixtures) usually have a ballast that is a source of heat in addition to the lamp. Ballasts for most fluorescent luminaires (fixtures) are required by *410.73(E)* to be thermally protected, and ballasts for recessed high-intensity discharge luminaires (fixtures) are required by *410.73(F)* to be thermally protected or inherently protected.

Recessed luminaires (lighting fixtures), if not installed properly, can create a fire hazard. *Part XI* of *Article 410* provides installation requirements for recessed luminaires (fixtures). For most applications, recessed incandescent luminaires (fixtures) are required to be thermally protected. Overheating of the luminaires (fixtures) will interrupt electrical power to the lamps. The lamps generally will light again when the luminaire (fixture) cools.

Article 411 provides specifications for lighting systems that operate at 30 volts or less. One example of such a system is low-voltage landscape lighting for gardens, walkways, decks, patios, and other building accent illumination. Lighting systems are required to be listed for the purpose. The uses not permitted are stated in *411.4*. The low-voltage secondary circuit is required to be insulated from the supply branch-circuit by an isolating transformer. Each secondary lighting circuit is not permitted to operate at more than 25 amperes. The lighting system isolating transformer is not permitted to be supplied from a branch-circuit with a rating more than 20 amperes.

Article 422 provides information and requirements for electrical appliances in any type of occupancy. Branch-circuit requirements, control of appliances, and disconnects are covered. Storage-type electric water heater wiring is covered in *422.13*. When the capacity of the electric storage-type water heater is not more than 120 gallons (450 L), the branch-circuit conductor shall be sized at not less than 1.25 times the nameplate rating of the water heater. If the water heater rating is given in watts, then a calculation must be done to determine the ampere rating of the water heater. An electric water heater with a rating of 3500 watts at 240 volts will have an ampere rating of 14.6 amperes. This is determined by dividing 3500 watts by 240 volts. The minimum branch-circuit conductor rating permitted is 1.25 times the 14.6 amperes or 18.3 amperes. The minimum circuit conductor size would be 12 AWG copper protected with a 20-ampere overcurrent device.

Article 424 on fixed electric space-heating equipment is covered including space heating cables. Branch-circuit requirements, control, disconnection, grounding, location, and wiring are covered. The first part of the article gives general wiring requirements, while the later parts of the article relate to specific types of electric heating equipment or installations.

An individual branch-circuit is permitted to supply any size space-heating load, although within an electric heating unit, the resistance loads may be subdivided. *NEC® 424.22(B)* only permits an individual resistance heating element to draw up to 48 amperes and be protected at not more than 60 amperes. If a heating branch-circuit supplies two or more fixed heating units, the branch-circuit is not permitted to have a rating of more than 30 amperes, according to *424.3(A)*. The overcurrent device protecting an electric heating circuit, as well as the conductors, is not permitted to be less than 1.25 times the load to be served. The heating units are most likely rated in watts, therefore, the wattage must be divided by the circuit operating voltage to determine the full-load current of the circuit. If a room has two 1500-watt, 240-volt resistance baseboard heating units, the total load served will be 12.5 amperes (3000 W / 240 V = 12.5 A). The overcurrent device and the conductor are required to have a rating not less than 15.6 amperes (12.5 × 1.25 = 15.6 A). The overcurrent device will be rated 20 amperes, and the conductor will be size 12 AWG copper.

Disconnecting means for electric heating equipment is covered in *Part III* of *Article 424*. Of particular importance is *424.20,* which specifies the conditions that must be met in order for a thermostat to serve as the disconnecting means for an electric heating unit. All ungrounded conductors must be opened, the thermostat must have a marked off position, and a change of temperature is not capable of energizing the circuit conductors.

Article 426 deals with the installation and use of outdoor electric de-icing and snow-melting equipment, such as heating cable embedded in concrete. Definitions of different methods of using electricity for the production of heat are covered in *426.2*. The minimum size of conductor permitted for branch-circuits for electric outdoor snow-melting and de-icing equipment is covered in *426.4* and shall be not less than 1.25 times the total load on the circuit.

Outdoor snow-melting and de-icing equipment is required by *426.28* to be provided with equipment ground-fault protection. This protection is required for the equipment, not the circuit, and the purpose is fire prevention not personnel protection. Ground-fault equipment protection (GFPE) is permitted to be provided as an integral part of the equipment, or it can be provided as part of a GFPE circuit breaker. These devices commonly are rated to detect 30 milliamperes or 50 milliamperes. The range over which they are intended to protect is between 6 and 50 milliamperes.

Article 720 deals with electrical circuit installation, either alternating current or direct current, that operates at less than 50 volts. Minimum wire size and receptacle rating are covered. Some installation types are listed that are covered elsewhere in the Code. It is important to note that even though the voltage is low, it is electrical current flow that causes heating and can result in fire. There are specific requirements for installations operating at less than 50 volts in other articles.

Article 725 covers the installation of remote-control, signaling, and power-limited circuits that are not an integral part of a device or appliance. These circuit types are divided into Class 1, Class 2, and Class 3 circuits. Line voltage control circuits for motors and other equipment are Class 1 circuits. Thermostat circuits operating at 24 volts for furnace control, and door chime and similar circuits are considered Class 2 because they have supply transformers limiting the maximum current that can flow if the wires become shorted. Wiring methods for Class 1 circuits are covered in *725.25*. The wiring methods of *Chapter 3* of the Code shall be permitted. If the conductors are of sizes 18 and 16 AWG, the insulation type is specified in *725.27(B)*. Wiring methods and materials for Class 2 and Class 3 wiring are specified in *725.52*.

Class 2, Class 3, and PLTC cables defined as abandoned in *725.2* not being terminated at equipment or tagged for future use. Accessible portions of abandoned Class 2, Class 3, and PLTC cables as stated in *725.3(B),* are required to be removed. Cables that are installed exposed, as stated in *725.8,* are required to be supported by the structure of the building in such a manner that they will not be damaged in normal use of the area in which they are installed. When installed through building materials such as bored holes in wood framing members, the same rules in *300.4(D)* must be followed as normal power circuit cables. When installed above a suspended ceiling, the rules of *300.11* must be followed, and cables are required to be supported by the structure of the building in some manner such as shown in Figure 3.12, or supported by tie wires installed specifically for the purpose of electrical wiring. Cables are not permitted to be installed above a suspended ceiling in such a manner that access to the space above is prevented by the installation of the cables according to *725.7*. Installation of Class 2 and Class 3 cables is covered in *725.56* through *725.58*. In *725.58* it is stated that Class 2 and Class 3 cables are not permitted to be attached to the exterior of conduits and raceways.

SAMPLE CALCULATIONS

Information found in *Articles 210* and *220* of the Code is used to determine the minimum number of circuits required in a dwelling or other building. The following are examples of how these calculations are used to determine the number of circuits required. The same information is used to determine the maximum number of outlets permitted on a particular circuit. An important issue is whether the loads are considered continuous loads. In many situations, it is a matter of judgment to the use of the loads on a circuit. In the case of receptacles outlets, it would depend on the particular application. If the load is considered to be continuous, then the overcurrent device protecting the circuit is not permitted to be loaded to more than 80% of its rating. The exception is when the overcurrent device and the enclosure into which it is installed are rated for operation at 100%, *210.19(A)(1)*.

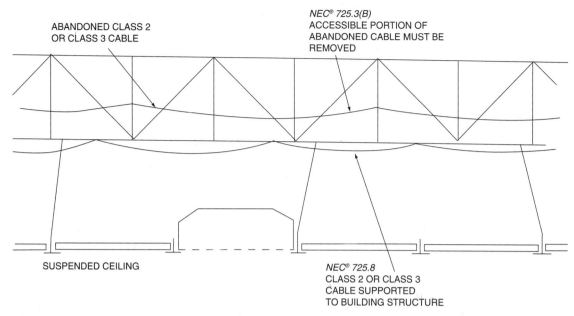

Figure 3.12 Class 2 and Class 3 cables are required to be supported by the structure of the building and all accessible portions of abandoned cable are required to be removed.

Circuit and Outlet Requirements

The minimum number of circuits for general illumination is based on the actual load when the load is known; however, the minimum permitted demand load for general illumination is specified in Code *Table 220.12*. The demand load for receptacle outlets shall be considered a load of 180 volt-amperes (VA) per strap or yoke at an outlet. In the case of a dwelling, the receptacle outlets are considered loads for general illumination, and the load is determined based on the 3 volt-amperes per square ft (33 VA/m²) from *Table 220.12* and not from the 180-volt-amperes per outlet as stated in *220.14(I)*. This information on loads has three basic purposes: (1) it is used to determine the minimum number of branch-circuits in a building or an area of a building; (2) it is used to determine the minimum number of outlets on a particular branch-circuit; and (3) it is used to determine the demand load of a building or an area of a building for the purpose of sizing a feeder or electrical service or distribution panel. This latter function of feeders and panels will be the subject of *Unit 4*.

The minimum number of circuits for general illumination for a dwelling, which includes receptacle outlets and lighting outlets, is determined by multiplying the area of the building by the unit load found in *Table 220.12*. This process is described in *210.11(A)* and *220.12*. Once the total general illumination load in volt-amperes is determined, divide by 120 volts to get the current. Assuming the circuits are made up of fixed luminaires (lighting fixtures) and receptacle outlets, the minimum number of circuits can be determined by dividing the general illumination current by the rating of the circuits. If this number of circuits is not adequate to supply the specific load, then the number of circuits must be increased. That decision cannot be made without some knowledge of the loads to be served. The following example shows how to determine the minimum number of circuits required in a dwelling for general illumination.

Example 3.1 A dwelling has 2100 square ft (195 m²) of living area. The area does not include the unfinished basement. Determine the minimum number of general illumination branch-circuits required in this dwelling assuming circuits rated 15 amperes.

Answer: First look up the minimum unit load for dwelling general illumination in *Table 220.12* and find 3 VA/ft² (33 VA/m²). Next multiply this unit load by the living area of the dwelling to get 6300 VA (6435 VA based on metric calculation). Four 15-ampere rated general illumination circuits are required for this dwelling.

$$2100 \text{ ft}^2 \times 3 \text{ VA/ft}^2 = 6300 \text{ VA} \qquad 195 \text{ m}^2 \times 33 \text{ VA/m}^2 = 6435 \text{ VA}$$

$$\frac{6300 \text{ VA}}{120 \text{ V}} = 53 \text{ amperes} \qquad \frac{6435 \text{ VA}}{120 \text{ V}} = 54 \text{ amperes}$$

$$\frac{53 \text{ A}}{15 \text{ A/circuit}} = 3.5 \text{ circuits} \qquad \frac{54 \text{ A}}{15 \text{ A/circuit}} = 3.6 \text{ circuits}$$

There is seldom justification to consider dwelling loads for general illumination to be continuous loads. There are both lighting loads and receptacle outlets on the circuits, and generally this type of load combination will not apply a heavy load to the circuit on a continuous basis.

Buildings other than dwellings have the general illumination load generally limited to fixed lighting. If the actual ampere rating of the luminaires (lighting fixtures) is known, then this load is required to be used if it is greater than the load determined using the unit load required from *Table 220.12*. The following example shows how to determine the minimum number of general illumination circuits that are required for an office building:

Example 3.2   An area of a building devoted to offices has an area of 8200 sq. ft (762 m²). Determine the minimum number of 20-ampere, 120-volt rated general illumination branch-circuits required for the office space of the building.

Answer: This is a building space listed in general illumination *Table 220.12*. The unit load for office space is 3.5 VA/ft² (39 VA/m²). Multiply the unit general illumination load by the area of the office space to get the volt-ampere for general illumination. This is a continuous load, therefore, multiply the load by 1.25. Then divide by 20 ampere per circuit to determine the minimum number of circuits required which is 15.0 (15.5 using metric factors). Even though one calculation arrives at 15 circuits and the other arrives at 16 circuits, either is considered to be in compliance according to *90.9(D)*.

$$8200 \text{ ft}^2 \times 3.5 \text{ VA/ft}^2 = 28{,}700 \text{ VA} \qquad 762 \text{ m}^2 \times 39 \text{ VA/m}^2 = 29{,}718 \text{ VA}$$

$$1.25 \times 28{,}700 \text{ VA} = 35{,}875 \text{ VA} \qquad 1.25 \times 29{,}718 \text{ VA} = 37{,}148 \text{ VA}$$

$$\frac{35{,}875 \text{ VA}}{120 \text{ V}} = 299 \text{ amperes} \qquad \frac{37{,}148 \text{ VA}}{120 \text{ V}} = 310 \text{ amperes}$$

$$\frac{299 \text{ A}}{20 \text{ A/circuit}} = 15.0 \text{ circuits} \qquad \frac{310 \text{ A}}{20 \text{ A/circuit}} = 15.5 \text{ circuits}$$

To determine the actual number of luminaires (lighting fixtures) on the circuit, these lighting loads usually are considered to be continuous loads. Continuous load means the branch-circuit is not permitted to be loaded more than 80% of the circuit rating. For a 20-ampere lighting circuit, the maximum permitted continuous load on the circuit would be 16 amperes. If the current drawn by the luminaires (lighting fixtures) is known, the number of luminaires (fixtures) on a branch-circuit can be determined. The method is shown in the following example.

Example 3.3   Electric discharge luminaires (lighting fixtures) to be installed in a building each are rated at 1.9 amperes at 120 volts. Determine the maximum number of these luminaires (fixtures) permitted to be installed on a 20-ampere branch-circuit in a commercial building.

Answer: The 20-ampere branch-circuit is only permitted to be loaded to 80% of the circuit rating, which is 16 amperes. Divide the 16 amperes by the rated current of each luminaire (fixture) to determine the maximum number of luminaires (fixtures) permitted to be installed on the circuit. It is necessary to round a fraction down to the next integer or the circuit will carry more than 16 amperes. The maximum number of luminaires (fixtures) permitted to be installed on the circuit is eight.

$$\frac{16 \text{ A/circuit}}{1.9 \text{ A/luminaire (lighting fixture)}} = 8.4 \text{ luminaires (lighting fixtures)}$$

When to consider loads to be continuous is frequently a matter of judgment. In a commercial building, for example, receptacle loads may be operated with a great amount of diversity. Therefore, the times will be infrequent when the circuit would be operated near the circuit rating, especially for three hours or longer. In this case, the receptacle circuit is not to be considered a continuous load. In another case, the receptacle circuit could be considered a continuous load.

In *Annex D* of the Code, several examples illustrate how to determine the minimum number of branch-circuits for different building types. *Example D1(a)* is an example for a single-family dwelling, and *Example D4(a)* shows how to determine the minimum number of branch-circuits for each dwelling unit of a multi-family dwelling. *Example D3* shows how to determine the minimum number of branch-circuits for a store building.

Electric Range for a Dwelling

An electric range for a dwelling seldom operates at full nameplate rating, and when it does, the load is on only for a short time. Damage to conductors requires heat-producing current over a time period. If the time period is known to be limited, the wires will not be damaged. The range circuit rating is permitted by *220.18(C)* to be based on a range demand load. The minimum rating circuit permitted for a household electric range is 40 amperes for ranges rated more than $8^3/4$ kW, as specified in *210.(A)(3)*. The range demand load is found in *Table 220.55*. Column C of *Table 220.55* gives the demand load of an electric range or ranges that have a rating of not more than 12 kW. For example, one 12-kW electric range can be taken at a demand load of 8 kW for the purpose of sizing the branch-circuit overcurrent device and minimum conductor size. The following example explains the process.

> **Example 3.4**   A 10-kW electric range in a dwelling operates from a 120/240-volt circuit. Determine the minimum ampere rating of the circuit permitted to supply this electric range.
>
> **Answer:**   The demand load for a 10-kW electric range is considered to be 8 kVA from column C of *Table 220.55*. Multiply the 8 kVA by 1000 to convert to VA and then divide by 240 volts to get the current, which is 33.3 amperes. A 35-ampere overcurrent device (*240.6*) is sufficient for the load, but the minimum permitted is 40 amperes *210.19(A)(3)*.

The minimum circuit rating for a household electric range is 40 amperes according to *210.19(A)(3)*. The circuit conductor is not permitted to have an allowable ampacity less than the rating of the range circuit, and the minimum conductor size is found in *Table 310.16*. For the previous example, where the circuit has a 40-ampere rating, the minimum copper conductor size permitted is 8 AWG. If the circuit is wired using Nonmetallic-Sheathed Cable, Type NM-B, according to *334.80*, the allowable ampacity of the conductors is found in the 60°C column of *Table 310.16*. If the circuit is wired using service-entrance cable Type SE, the installation of the cable is required to be the same as for Type NM-B, but *334.80* does not apply to Type SE cable according to *338.10(B)(4)*. This means that if the conductor insulation and terminations are rated 75°C, then the 75°C column of *Table 310.16* is permitted to be used to size the Type SE cable. If the circuit is wired as individual conductors in raceway, then the column of *Table 310.16* that is used to size the conductors depends upon the insulation rating of the conductors and the conductor terminations. Be aware of any conductor insulation minimum ratings that may be marked on an appliance terminal box.

New electric range branch-circuits are required to have the grounded-circuit conductor (neutral) separated from the equipment grounding conductor. A 4-wire circuit and a 4-wire cord are required to supply an electric range, counter-mounted cooking unit, or wall-mounted oven. If the range is cord- and plug-connected, the receptacle is required to be of the 4-wire type with the neutral terminal separate from the equipment grounding terminal. The minimum ampere rating of the receptacle is permitted to be determined in the same manner as the circuit overcurrent device, according to *210.21(B)(4)*. In the case of an existing electric range circuit, the neutral conductor is permitted by *Exception 1* of *250.142(B)* to also serve as the equipment grounding conductor. The conditions that must be satisfied for the neutral to also serve as the equipment grounding conductor are covered in *250.140*.

The minimum branch-circuit rating for a household electric range is permitted to be determined using the rules in *220.55* according to *422.10(A)*. NEC® *422.11(B)* sets the maximum branch-circuit demand load at 60 amperes. If the appliance has a demand load higher than 60 amperes, then the load of the appliance is required to be subdivided so it will not exceed 50 amperes. Column C of *Table 220.55* applies to an electric range rated not over 12 kW. *Note 1* of *Table 220.55* permits the use of Column C for ranges rated more than

12 kW by increasing the value in column C by 5% for each kW in access of 12 kW. If the range is rated 14 kW, then the value in column C is increased by 5% twice. The demand load of an electric range with a rating greater than 12 kW can be calculated according to the rule of *Note 1* of *Table 220.55* using Equation 3.1. Don't be confused by the units kW and kVA. For the purpose of these branch-circuit calculations these units are considered interchangeable.

$$\text{\textbf{Range Demand Load}} = [(\text{\textbf{Value in column C}}) \times \textbf{0.05} \times (\text{\textbf{Range kW – 12 kW}})] \\ + (\text{\textbf{Value in column C}}) \qquad\qquad \text{\textbf{Eq. 3.1}}$$

$$= [(8\text{ kW}) \times 0.05 \times (14\text{ kW} - 12\text{ kW})] + 8\text{ kVA}$$

$$= (8\text{ kW} \times 0.05 \times 2) + 8\text{ kVA}$$

$$= 0.8\text{ kVA} + 8\text{ kVA} = 8.8\text{ kVA}$$

If the rating of the electric range is a fraction, such as 14.5 kW, the number is rounded to the nearest whole number before doing the calculation. *Note 1* of *Table 220.55* requires the value in column C to be increased by 5% for every kW the range rating is greater than 12 kW, and by 5% for any major fraction of a kW. A major fraction is 0.5 or greater. This means that for a 14.5-kW electric range you must use 15 in the calculation. Do not use 14.5 kW. If an electric range has a rating of 16.4 kW, round down and use 16 kW in the calculation. The following example will illustrate how to determine the minimum range demand load when the nameplate rating of the range is a fraction.

Example 3.5   A 17.4-kW electric range is installed in a dwelling. Determine the minimum circuit rating and copper conductor size if the circuit is wired using Type NM-B Nonmetallic-Sheathed Cable.

Answer: The demand load can be determined using Equation 3.1, which is the rule in *Note 1* of *Table 220.55*. The range rating is greater than $8^3/4$ kW, therefore, look up the demand for one range rated 12 kW in column C of *Table 220.55* and find 8 kVA. The nameplate rating of the range is a fraction, so round the 17.4 kW off to 17 kW. Now use Equation 3.1 to determine the range demand load.

$$\text{Demand load} = [(8\text{ kW}) \times 0.05 \times (17\text{ kW} - 12\text{ kW})] + 8\text{ kVA}$$

$$= (8\text{ kW} \times 0.05 \times 5) + 8\text{ kVA} = 10\text{ kVA}$$

$$\text{Demand load current} = \frac{10\text{ kVA} \times 1000}{240\text{ V}} = 41.7\text{ amperes}$$

The minimum circuit rating is 45 amperes, however, it is more likely a 50-ampere overcurrent device would be used. The circuit is wired with Type NM-B Cable, therefore, the 60°C column of *Table 310.16* will be used to determine the minimum size conductor. The rating of the circuit determines the size of conductor, therefore, the minimum size permitted is 6 AWG.

An oven and cooking unit can be separate units and built into the kitchen, as shown in Figure 3.13. The range conductor is permitted to extend from the supply panel to a junction box, and then taps connect to the counter-mounted cooking unit and to the wall-mounted oven. This is permitted by *Exception 1* to *210.19(A)(3)*. The rules for determining the minimum size of conductors for a wall-mounted oven and a counter-mounted cooking unit supplied from the same branch-circuit are found in *Note 4* to *Table 220.55*. The minimum conductor size for the counter-mounted cooking unit and the wall-mounted oven is determined based on the nameplate rating of each unit. It is permitted to add together the nameplate ratings of one counter-mounted cooking unit and up to two wall-mounted ovens for a dwelling to determine the rating of the circuit. The combined nameplate ratings are treated as one range for the purpose of determining the demand load. The following example is for a counter-mounted cooking unit and a wall-mounted oven supplied by the same branch-circuit with a tap to each unit.

Example 3.6   A dwelling electric range consists of a wall-mounted oven with a nameplate rating of 6.6 kW and a counter-mounted cooking unit with a rating of 8 kW, shown in Figure 3.13. Determine the minimum rating for the range circuit, and the minimum size Type NM-B Copper Cable required for the circuit and for the tap to each cooking unit.

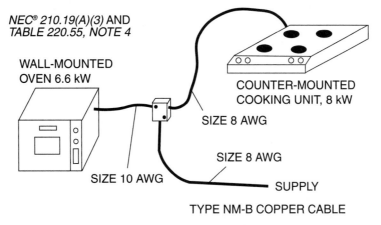

NEC® 210.19(A)(3) AND
TABLE 220.55, NOTE 4

WALL-MOUNTED
OVEN 6.6 kW

COUNTER-MOUNTED
COOKING UNIT, 8 kW

SIZE 8 AWG

SIZE 8 AWG

SIZE 10 AWG

SUPPLY

TYPE NM-B COPPER CABLE

Figure 3.13   A wall-mounted oven and a counter-mounted cooking unit are installed on a single branch-circuit.

Answer: Add the 8-kW rating of the counter-mounted cooking unit to the 6.6-kW rating of the wall-mounted oven to get a combined rating of 14.6 kW. Next determine the demand load for the range using Equation 3.1.

$$\text{Demand load} = [(8 \text{ kW}) \times 0.05 \times (15 \text{ kW} - 12 \text{ kW})] + 8 \text{ kVA}$$

$$= (8 \text{ kW} \times 0.05 \times 3) + 8 \text{ kVA} = 9.2 \text{ kVA}$$

$$\text{Demand load current} = \frac{9.2 \text{ kVA} \times 1000}{240 \text{ V}} = 38.3 \text{ amperes}$$

The minimum rating permitted for this branch-circuit serving both the counter-mounted cooking unit and the wall-mounted oven is 40 amperes. Now determine the full-load current for each cooking unit.

$$\text{Wall-mounted oven current} = \frac{6.6 \text{ kVA} \times 1000}{240 \text{ V}} = 28 \text{ amperes}$$

$$\text{Counter-mounted cooking unit current} = \frac{8 \text{ kVA} \times 1000}{240 \text{ V}} = 33 \text{ amperes}$$

Nonmetallic-Sheathed Cable is used for the circuit, therefore, the 60°C column of *Table 310.16* is used to determine the minimum conductor size. The branch-circuit conductor is size 8 AWG, the tap to the counter-mounted cooking unit is size 8 AWG, and the tap to the wall-mounted oven is 10 AWG.

A disconnecting means is required to be provided for the electric range that will disconnect all ungrounded conductors. The rule is found in *422.31*. If the branch-circuit overcurrent device is located within sight from the appliance, *422.31(B)* will permit the overcurrent device to act as the disconnecting means. The plug and receptacle is permitted to act as the disconnecting means if it is accessible or if it can be reached by removal of a range storage drawer, as stated in *422.33(B)*. *NEC® 422.16(B)(3)* permits a plug and receptacle or a plug and connector to be used to make the power connection to a wall-mounted oven or a counter-mounted cooking unit, but it does not permit these connections to serve as the disconnecting means for the cooking unit.

MAJOR CHANGES TO THE 2005 CODE

These are the changes to the 2005 *NEC®* that correspond to the Code sections studied in this unit. The following analysis explains the significance of the changes from the 2002 to the 2005 Code only, and this analysis is not intended to be used in place of the Code. Refer to the actual section of the 2005 Code for the exact wording and meaning of each section discussed. Changes are indicated in the Code with a vertical line in the margin. If material was deleted or moved to another location in the Code, the location of the deletion is indicated with a dark dot in the margin.

Article 210 **Branch Circuits**

210.4(B): This section deals with rules relative to wiring a multiwire branch-circuit. Examples of multiwire branch-circuits are shown in Figure 1.19 and Figure 1.20 in *Unit 1*. This paragraph in the previous edition of the Code applied only to multiwire branch-circuits in dwellings. If the multiwire branch-circuit supplied more than one device or piece of equipment on the same yoke, such as two receptacles, it was required to be able to disconnect power to all conductors simultaneously. A typical means of power disconnection to a multiwire branch-circuit was a 2-pole circuit breaker or two single-pole circuit breakers with handle ties. The change is that this section no longer applies just to dwellings. This rule applies to all multiwire branch-circuits in any building. It is questionable if this rule is different from *210.7(B)*. The question is whether a multiwire circuit is one circuit or multiple circuits.

210.6(D)(2): This section specifies which utilization equipment is permitted to be supplied power at higher than nominal 277 volts to ground but not more than 600 volts line-to-line. Since luminaires (lighting fixtures) were not specifically excluded from this section it was interpreted in some areas that HID luminaires (lighting fixtures) were considered utilization equipment and were permitted to be installed inside and supplied from 480-volt delta systems. This was apparently not the intent, and now it is specifically stated that luminaires (lighting fixtures) are not included in this rule that permits utilization equipment to be supplied by circuits operating at more than 277 volts to ground but not more than 600 volts line-to-line.

210.7(B): The previous edition of the Code required that when more than one receptacle on the same strap or yoke is supplied from different branch-circuits, the overcurrent device supplying the circuits was required to be of a type where all ungrounded conductors would be de-energized simultaneously. This required a multi-pole circuit breaker or single-pole circuit breakers with listed handle ties. The word receptacles was changed to devices. It is no longer required that there be only receptacles on the strap or yoke. Now this rule applies to devices such as a switch and receptacle on a single strap, or perhaps two switches on the same strap.

210.8(A)(7): All 125-volt, single-phase 15- and 20-ampere receptacles installed within 6 ft (1.8 m) of the outside edge of a dwelling laundry, utility, or wet bar sink are required to be ground-fault circuit-interrupter protected such as shown in Figure 3.14. The change is that this rule now applies to laundry and

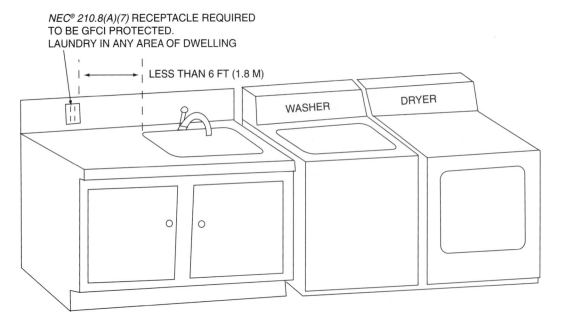

Figure 3.14 **Any receptacle in a dwelling that is located within 6 ft (1.8 m) of a utility sink or laundry sink is required to be GFCI protected.**

utility sinks in a dwelling. There is an important omission to this requirement that receptacles within 6 ft (1.8 m) of the edge of a laundry sink be GFCI protected. If the washer and gas dryer are located in the basement adjacent to the laundry sink, *Exception 2* to *210.8(A)(5)* permits a single receptacle for one appliance or a duplex receptacle for two appliances in a dedicated space to be exempt from the GFCI requirement. There is no such exception for a washer and dryer located next to the laundry sink when not in the basement.

210.8(B)(2): The previous edition of the Code required ground-fault circuit-interrupter protection for 125-volt, single-phase, 15- and 20-ampere receptacles installed in nondwelling kitchens. A definition of a kitchen was not provided, which created differences of interpretation as to where this rule would apply. Now it is specified that the rule applies to commercial and institutional kitchens. There is now also a definition of a kitchen that is an area with a sink and permanent facilities for food preparation and cooking. A counter space in a room with a sink, coffeemaker, refrigerator, and space for a microwave oven would not qualify as a kitchen since permanent space for cooking is not provided. In the definition of a dwelling space a portable microwave oven is not considered permanent provisions for cooking.

210.8(B)(4): All 125-volt, single-phase, 15- or 20-ampere receptacles located outdoors in spaces accessible to the public for nondwelling locations are required to be ground-fault circuit-interrupter protected. Figure 3.15 is an example of a commercial building where outside receptacles are required to be GFCI protected. Receptacles that are exempt would be those in restricted areas that are off limits to the public. The word public would mean nonemployees. There will most likely be a difference in interpretation as to when outside outlets are considered to be in restricted areas and when they are accessible to the public.

210.8(B)(5): A 125-volt, single-phase, 15- or 20-ampere receptacle installed outdoors as the required receptacle within 25 ft (7.5 m) of heating, air-conditioning, and refrigeration equipment is now required to be ground-fault circuit-interrupter protected. This rule now applies to all buildings where, according to the previous edition of the Code GFCI protection, previously was only required for outside rooftop receptacles installed to service this equipment.

210.8(C): A 125-volt, single-phase, 15- or 20-ampere outlet for the supply of a boat hoist at a dwelling location shall be provided with ground-fault circuit-interrupter protection.

210.12(B): Arc-fault circuit-interrupters (AFCI) continue to be required to protect circuits supplying 15- and 20-ampere, 125-volt outlets in bedrooms of dwellings. The change deals with the type of AFCI required. A new combination type AFCI is required to protect these circuits which is capable of detecting an arcing condition between a circuit conductor and ground (parallel arcing) or an arcing condition due to a loose splice, terminal or broken conductor (series arcing). A second paragraph in this section permits the presently available branch/feeder type AFCI to be used at the circuit panel until January 1, 2008. The branch/feeder type AFCI is best suited at detecting parallel arcing conditions, but not good at detecting series arcing conditions.

210.12(B) Exception: This new exception permits a receptacle-type (combination-type) AFCI device to be installed adjacent to the circuit panelboard. The receptacle-type AFCI provides arc-fault protection for

NEC® 210.8(B)(4) RECEPTACLES LOCATED OUTDOORS IN PUBLIC SPACES REQUIRED TO BE GFCI PROTECTED

Figure 3.15 **Outside receptacles on any type of building are required to be GFCI protected if in an area accessible to the public.**

lamps, equipment, and appliances plugged into the receptacle. The receptacle-type AFCI has feed-through capability so the remainder of the circuit will also be arc-fault protected. This type of AFCI is designed to be more sensitive to arc-faults, which could result in nuisance tripping. The exception required the length of circuit wire from the circuit breaker in the panelboard to the AFCI device to be not more than 6 ft (1.8 m) in length. The circuit between the panelboard and the AFCI device is also required to be run in metal raceway or metallic-sheathed cable as illustrated in Figure 3.16.

210.18: This is a new section that requires guest rooms and guest suites with permanent provisions for cooking to meet the same receptacle requirements as a dwelling, but they also must meet the branch-circuit requirements of a dwelling. Two 20-ampere, 125-volt, small appliance branch-circuits are required to be provided that serve the counter space. The bathroom receptacles are required to be on a 20-ampere, 125-volt circuit that includes no other outlets, unless there is only one bathroom in which case all outlets in the bathroom are permitted to be on the 20-ampere circuit. Receptacle spacing shall be as required in *210.52*.

210.19(A)(3) Exception 1: This exception deals with range circuit taps to a surface-mounted cooking unit or a wall-mounted oven. A new sentence was added that makes it clear that the cord used to make a flexible connection to the cooking unit is considered to be a part of the tap. The minimum cord ampacity is 20 amperes if the unit is supplied by a 50-ampere range circuit. Examples of such a branch-circuit are shown in Figure 3.2 and Figure 3.13. The connection between the permanent wiring and the cooking unit is usually made by using flexible cord with a cord connector for easy disconnection of the unit for installation and servicing.

210.52(C)(1) Exception: This new exception now requires the wall space behind a sink or a counter-mounted cooking unit to be considered in receptacle spacing requirements if the back edge of the sink or cooking unit is 12 in. (300 mm) or more out from the wall as shown in Figure 3.17. There is a similar requirement if the sink or cooking unit is installed in a corner of the kitchen counter. If the back edge of the sink or cooking unit is mounted so the distance to the corner is 18 in. (450 mm) or greater, then the wall space behind the sink or cooking unit is to be considered in the receptacle spacing require-

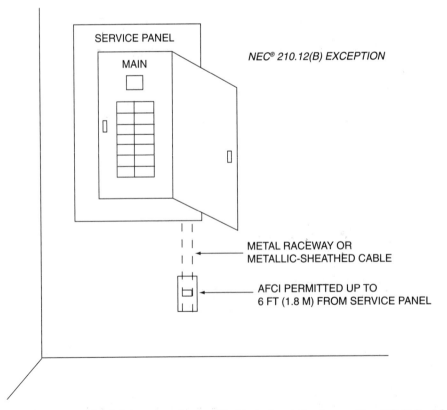

Figure 3.16 An arc-fault circuit-interrupter is permitted to be located not more than 6 ft (1.8 m) from the panelboard, and the wiring between the panelboard and the AFCI is required to be metal raceway or metal-sheathed cable.

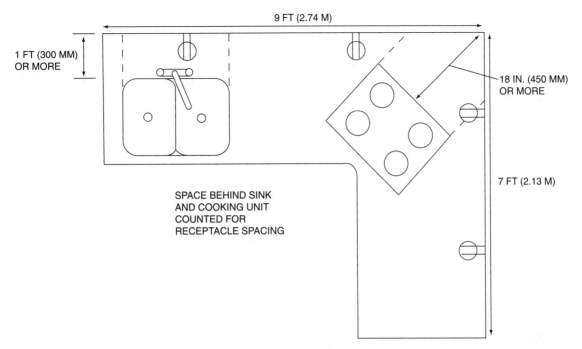

Figure 3.17   The space behind a kitchen sink or a counter-mounted cooking unit is required to be counted as a wall requiring a receptacle if the distance from the back of the unit to the wall is 12 in. (300 mm) or more and if in a corner the distance to the corner of the wall is 18 in. (450 mm) or more.

ments of *210.52(C)(1)*. If the space behind a sink or cooking unit is limited, then according to *210.52(C)(4)* the wall spaces are not considered as usable wall space.

210.52(C)(2): A new sentence was added to make it clear that when a counter-mounted cooking unit or a sink is located in an island kitchen counter, the ends of the counter are considered separated if the cooking unit or sink is mounted such that less than 12 in. (300 mm) of counter remains at the location of the sink or cooking unit. This would mean that a receptacle must be installed to serve the counter space at each end of the island if those counter spaces are at least 12 in. (300 mm) in both dimensions. This is illustrated in Figure 3.18.

210.52(D) Exception: This is a new exception to the required placement of a receptacle on the wall or partition adjacent to and within 3 ft (900 mm) of the bathroom sink. The exception permits the receptacle to

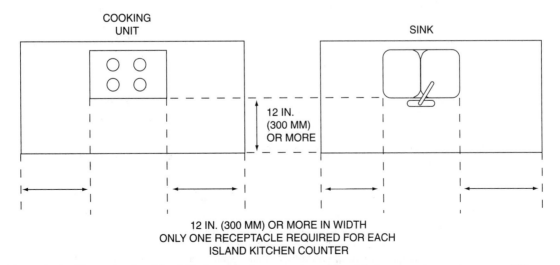

Figure 3.18   The ends of an island kitchen counter are not considered to be separate counter spaces if the area behind a sink or counter-mounted cooking unit is 12 in. (300 mm) or more in width.

be installed on the face or side of the bathroom countertop and located not more than 12 in. (300 mm) below the top of the counter space and within 3 ft (900 mm) of the edge of the sink.

210.52(E): A new second sentence was added that now requires a receptacle to be installed accessible at grade level for each dwelling unit of a multifamily dwelling that is at grade level and has an individual entrance from the dwelling unit to the outside.

210.60: This section specifies receptacle placement in guest rooms and guest suites of hotels, motels, and similar occupancies. The previous edition of the Code required receptacle placement to be the same as for a dwelling if the guest room or guest suite met the definition of a dwelling in that permanent provisions were made for cooking. The definition of a dwelling also includes permanent provisions for eating, which may not be the case with a guest room or guest suite. A specific dining area may not be provided. Now the requirement is simply that the guest room or guest suite is required to meet all of the appropriate receptacle requirements of *210.52* if there are permanent provisions for cooking.

210.63 Exception: This section requires a 125-volt, 15- or 20-ampere receptacle to be installed at an accessible location within 25 ft (7.5 m) of heating, air-conditioning, and refrigeration equipment. This new exception does not require the receptacle to be installed for the service of evaporative coolers at one- and two-family dwellings. These coolers are generally installed on nonconductive rooftops.

210.70(B): In guest rooms and guest suites of hotels and motels, every habitable room and bathroom is now required to have a wall switch-controlled lighting outlet. The term lighting outlet is interpreted as meaning a luminaire (lighting fixture) of some type permanently installed.

210.70(B) Exception 1: This is a new exception that permits a wall switch-controlled receptacle to serve as the lighting outlet in hotel and motel rooms other than the bathroom or kitchen if a kitchen is present.

210.70(B) Exception 2: Lighting outlets in a guest room or guest suite of a motel or hotel is permitted to be controlled by an occupancy sensor in addition to a wall switch. As an alternative, an occupancy sensor with a manual switch is permitted to be installed at the normal location where a wall switch would be installed.

Article 220, Parts I and II   Branch Circuit-Calculations

Article 220 was completely reorganized. A new *Figure 220.1* was added to give an overview of where information on branch-circuit calculations and feeder calculations can be found. Nearly all of the sections have different numbers than in the previous edition of the Code.

Table 220.3: This is a new table that gives the subject and location of other calculations in the Code. This is a very handy table when looking for the location of calculation methods.

220.14(K): This was *footnote b* to what was *Table 220.3(A)* and is now *Table 220.12,* which gives the minimum general illumination load to be used for calculations for some occupancies. *Footnote b* stated that a minimum of 1 VA per sq. ft was required to be included in office buildings for the receptacle load. This section makes it clear that the minimum load that is required to be used for receptacles in an office building and now banks is 180 VA for each general-use receptacle to be installed or 1 VA per sq. ft, which ever is greater. This is not a change of intent; now it is made clear in this section that the higher value is to be used for calculations.

Article 225   Outside Branch-Circuits and Feeders

225.17: This is a new requirement for a mast supporting overhead feeder conductors or overhead branch-circuit conductors. Only the feeder or branch-circuit final span conductors are permitted to be attached to the mast, which is required to be of adequate strength to support the overhead span, or to be supported by a brace or guy. The mast is the portion of the raceway that extends above the roof or beyond the point of last support. It is recommended the portion of the riser above the point of last support be solid raceway without fittings. If fittings are used in the portion of the riser above the point of last support, those fittings are required to be identified for use with a mast. This rule applied in the case of service drops in *230.28*, but it never applied to an overhead feeder or branch-circuit. What this rule means is that other conductors such as communications cables or feeders extending on to other buildings (see Figure 3.19) are no longer permitted to be attached to a mast riser except those conductors terminating at the mast riser.

225.22: Raceway installed on the exterior of a building or structure is only required to have raintight fittings when installed in wet locations. A wet location is defined in *Article 100* as unprotected and exposed to the weather. This would mean that a conduit entry on the bottom of a raintight enclosure would not be

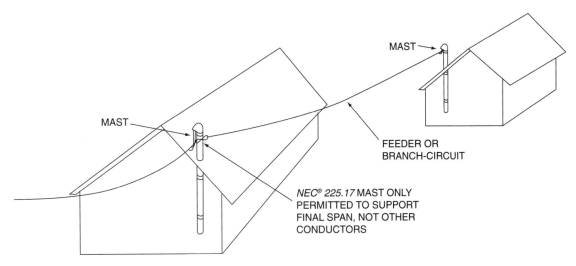

Figure 3.19   A mast riser is now only permitted to support the final span of a branch-circuit or feeder.

required to be raintight as would be the case if the entry were on the side or top of the enclosure. The previous edition of the Code required that raceway systems run on the exterior of a building or structure be raintight. That is no longer the rule if the raceway is installed in a protected location.

Article 398   **Open Wiring on Insulators**

398.15(A): Individual conductors run as open wiring on insulators are permitted to be run in continuous lengths of Flexible Nonmetallic Tubing for a distance of not greater than 15 ft (4.5 m) and secured to the surface with straps at intervals not exceeding 4½ ft. (1.4 m). This practice is only permitted where the cable is not subject to physical damage. The change is that the previous edition of the Code only required that the cable was not subject to severe physical damage.

Article 400   **Flexible Cords and Cables**

400.5: There is now a requirement that, in the case of determining allowable ampacity for flexible cords, an adjustment must be made for ambient temperatures above 30°C (86°F). The temperature correction factors to be used are to be from *Table 310.16* for the appropriate temperature column for the type of cable. Flexible cord temperature is marked on the jacket. There may be one temperature for the cord when operated in a dry area and another where the cable is operated in a wet area. The flexible cord shown in Figure 3.20 is a 90°C wire when dry and a 60°C wire when wet. The following example will illustrate how flexible-cord ampacity is adjusted for a high-ambient temperature.

Example 3.7: A SOW flexible cord with size 12 AWG copper wires supplies 3-phase equipment in a dry location where the ambient temperature when the equipment is operating is likely to be 120°F. Determine the ampacity of the flexible cord under these conditions.

Figure 3.20   Flexible cords that are installed in locations where the ambient temperature exceeds 30°C (86°F) are required to have the ampacity adjusted according to the adjustment factors provided in *Table 310.16*.

Answer: Since this is 3-phase equipment there will be three current-carrying conductors in the cable and the ampacity is found in column A of *Table 400.5* as 20 amperes. A temperature adjustment factor of 0.82 is found at the bottom of the 90°C copper column of *Table 310.16*. The ampacity of the flexible cord under these conditions is 16.4 amperes.

400.8(7): Flexible cords are not permitted to be installed in locations where they are exposed to physical damage.

400.14: Flexible cords are permitted to be protected from physical damage by running them through above ground raceway that is not more than 50 ft (15 m) in length. This is only permitted in industrial establishments where qualified personnel are present to service the installation. Ampacity adjustment is required if there are more than three current-carrying conductors in the raceway.

Article 404   Switches

404.7 Exception 2: A center pivot handle for a busway tap switch is permitted to be down when the tap is energized. The busway tap switch shown in Figure 3.21 is activated by pulling a rope, chain, or cable to "turn-on" and "turn-off" to energize and de-energize the tap. The on and off labels are required to be clearly visible from the point of operation of the tap switch.

404.8(B): This section requires a barrier between snap switches installed in the same box and supplied by different phases of a 480/277-volt system controlling electric discharge luminaires (lighting fixtures). The previous edition of the Code required permanently installed barriers between adjacent devices. Now the barrier is required to be identified and securely installed. Barrier plates are manufactured for field installation for the purpose of installation between adjacent devices. The barrier plates are not necessarily listed, and they are secure after installation, but they are not a permanent part of the box.

Article 406   Receptacles, Cord Connectors, and Attachment Plugs

406.4(D) Exceptions: This section requires a nonmetallic faceplate covering a box and receptacle to be either flush with the receptacle, or the receptacle must extend out from the faceplate. This rule pre-

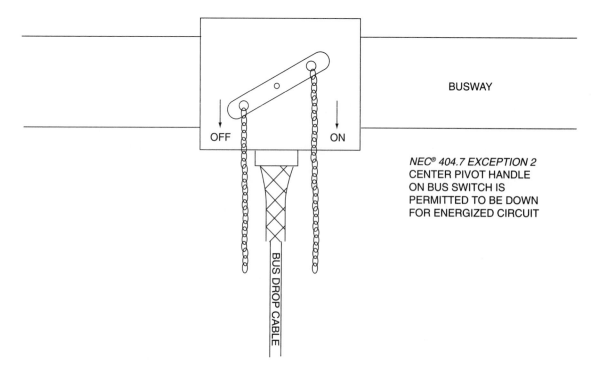

BUSWAY

NEC® 404.7 EXCEPTION 2
CENTER PIVOT HANDLE
ON BUS SWITCH IS
PERMITTED TO BE DOWN
FOR ENERGIZED CIRCUIT

OFF   ON

BUS DROP CABLE

Figure 3.21   A center pivot busway switch is permitted to have the handle down in the on position if the switch is clearly labeled from the operator's position.

vents the installation of a temper-resistant faceplate over a receptacle because the receptacle is neither flush with the faceplate nor extends out from the faceplate. This new exception permits listed
nonmetallic faceplate such as the tamper-resistant type to be installed that completely covers the
receptacle.

406.6(D): This is a new paragraph that does not permit the exposed blades of a flanged inlet to be energized
unless an energized cord connector is inserted. An example where this may be a problem is where a
flanged inlet is installed to receive a cord from a portable generator to feed into a service panel. The
flanged inlet may be supplied by a circuit breaker, which if turned-on would result in exposed energized blades. This practice of backfeeding into a service panel or any other situation where the exposed
blades may become energized is not permitted. This technique is permitted to supply power to one or
more circuits from a generator where double-pole switches are used to prevent the exposed flanged
inlet blades from becoming energized.

406.8(B)(1): The word "outdoor" was deleted so this rule now applies to receptacles installed either inside
or outside in wet locations. A 125-volt, 15- or 20-ampere receptacle installed in a wet location is
now required to be equipped with a cover that is weatherproof with or without the plug inserted.
This means that so called "in use" covers are now required for inside wet locations as well as outside wet locations. In *Article 100*, a wet location is defined as an area subject to saturation with
water or other liquids, or unprotected areas exposed to the weather. Inside areas where water is
sprayed is an example of a wet location. An example of an "in use" cover for a receptacle is shown
in Figure 3.11.

Article 410   Luminaires, Lampholders, and Lamps

410.1: The scope was expanded to include newer lighting products such as decorative lighting, holiday lighting products, and portable flexible lighting products.

410.4(D): This section establishes a zone in the area of bathtubs and showers where luuminaires (lighting
fixtures) suspended by cords, chain, or cable, track lighting, and paddle fans are not permitted. This
area is directly over the tub or shower, and within the area 8 ft (2.5 m) above and 3 ft (900 mm) extending out from the top of the bathtub rim or the shower stall threshold. The change is that any luminaire
(lighting fixture) not excluded that is installed within this area is required to be listed for damp locations or listed for wet locations if exposed to shower spray. Particularly for small bathrooms and
depending upon the placement of the sink, some common bathroom luminaires (lighting fixtures) may
not be permitted to be installed.

410.4(E): This is a new section placing specifications on the type of mercury vapor and metal halide lum-
naires (lighting fixtures) permitted to be installed for indoor sports and similar facilities where the
luminaires (lighting fixtures) may be exposed to physical damage. If the outer glass lamp envelope
is broken and the arc tube continues to operate, persons exposed to direct light from the lamp will
be exposed to short-wavelength ultraviolet radiation, which can cause eye irritation and potentially
burning of the skin. The glass envelope blocks short-wavelength ultraviolet radiation from escaping
the lamp. Mercury vapor and metal halide luminaires (lighting fixtures) subject to physical damage
and installed in playing and spectator seating areas of indoor sports and similar facilities are
required to be equipped with glass or plastic lenses. The lens will block short-wavelength ultraviolet radiation in the event the lamp envelope is broken and the lamp continues to operate. Mercury
vapor and metal halide lamps that have broken glass outer envelopes and continue to operate should
be shut off immediately and replaced. The arc tube of a metal halide lamp can rupture, showering
the area below with debris. The lens on the luminaire (lighting fixture) will contain the debris in the
event of an arc-tube rupture.

410.15(B): Now it is permitted to allow nonmetallic poles to support luminaires (lighting fixtures) as well as
serve as a raceway for the conductors to the luminaire (lighting fixture). A handhole is required to make
electrical connections within the pole. The previous edition of the Code only permitted metal poles for
this application.

410.18(B) Exception 2: This section only permits a luminaire (lighting fixture) with no exposed metallic
parts to be installed at a location where an equipment grounding conductor is not present. The new

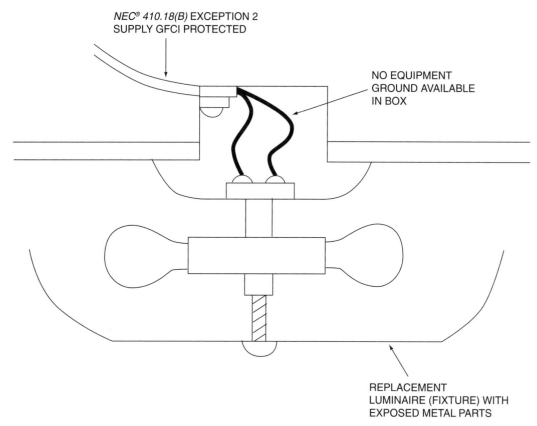

NEC® 410.18(B) EXCEPTION 2
SUPPLY GFCI PROTECTED

NO EQUIPMENT
GROUND AVAILABLE
IN BOX

REPLACEMENT
LUMINAIRE (FIXTURE) WITH
EXPOSED METAL PARTS

Figure 3.22   A replacement luminaire with exposed metal parts is permitted to be installed at an existing outlet that does not have an equipment grounding conductor if the supply conductors are ground-fault circuit-interrupter protected.

exception, as illustrated in Figure 3.22, permits luminaires (lighting fixtures) with exposed metallic parts to be installed if the luminaire (lighting fixture) is protected with a ground-fault circuit-interrupter.

410.73(E)(4): The word emergency was deleted. A fluorescent luminaire (lighting fixture) that is intended to provide egress illumination and only be energized when there is a failure of normal power is not permitted to have a ballast that is thermally protected. A loss of normal power is not necessarily an emergency situation. There was some difference of interpretation concerning this issue when the word emergency was present in the section.

410.73(F)(5): This is a new paragraph that deals with the potential problem of metal halide lamp end-of-life violent arc-tube failure. A potential fire exists if a metal halide lamp ruptures and showers flammable materials with hot lamp debris. The metal halide luminaire (lighting fixture) is required to be provided with a containment barrier that encloses the lamp except for thick-glass PAR lamps and Type O lamps. The Type O lamp has a built-in arc-tube containment feature. A closed luminaire with a plastic or glass diffuser will met this requirement. Only Type O and thick-glass PAR metal halide lamps are permitted to be installed in open luminaires (lighting fixtures) where the lamp is exposed.

410.73(G): For indoor locations other than dwellings and associated buildings, fluorescent luminaires (lighting fixtures) that use double-ended lamps (tubular type) and contain a ballast are required to be provided with an integral disconnecting means, accessible to qualified persons, to disconnect all conductors from the ballast. This requirement becomes effective January 1, 2008. There are several exceptions to this new rule. This new rule does not apply to fluorescent luminaires (lighting fixtures) installed in hazardous locations or for emergency illumination. This rule will not apply to cord- and plug-connected luminaires (lighting fixtures). The rule does not apply to industrial locations where only qualified personnel will service the luminaires (lighting fixtures). A switch in the room that opens all conductors to the luminaires (lighting fixtures) can serve in place of a disconnect in the luminaire (lighting fixture) provided the luminaires (lighting fixtures) are not supplied by multiwire circuits.

410.110: This is a new section in a new *Part XVI* on decorative lighting and associated accessories. The section requires decorative lighting and accessories used for holiday lighting to be listed products when used in accordance with *590.3(B)*, which means they are installed for a period of not more than 90 days.

Article 411   Lighting Systems Operating at 30 Volts or Less

411.4(A)(2): Lighting systems operating at 30 volts or less are generally not permitted to be run as concealed wiring. Lighting systems are available that are powered by a Class 2 power supply, in which case the wiring, if installed according to *725.52*, is now permitted to be run concealed.

411.5(C): Lighting systems operating at 30 volts or less are permitted to have bare conductors or exposed live parts located not less than 7 ft (2.1 m) above the floor. It is now made clear that this requirement applies only to inside installations. Bare conductors or exposed live parts are not permitted for outside installations.

Article 422   Appliances

422.12 Exception 2: This section required an individual branch-circuit for central heating equipment. The new exception permits central air-conditioning equipment to be on the same branch-circuit as the central heating equipment.

422.16(B)(4): Range hoods are now permitted to be cord- and plug-connected. The cord must be identified as suitable for use with a range hood. The cord must terminate at a grounded plug unless the range hood is listed as double insulated. The cord is to be not less than 18 in. (450 mm) or more than 36 in. (900 mm) in length. The receptacle must be accessible and supplied by an individual branch-circuit.

422.31(B): A permanently connected appliance rated over 300 VA or $^1/_8$ horsepower is permitted to have a disconnect such as a circuit breaker located out of sight of the appliance if the disconnect is capable of being locked in the open position. A locking device that is installed on a switch or circuit breaker is to be of a type that will remain in place even if the lock is removed. There are add-on locking devices that will only stay in place if the lock is attached. That type of locking device is not permitted to satisfy this requirement.

422.51: Cord- and plug-connected vending machines starting January 1, 2005 will be required to either have integral GFCI protection in the machine or the cord, or to be connected to a branch-circuit that is GFCI protected.

Article 424   Fixed Electric Space-Heating Equipment

424.6: This is a new section that requires fixed electric space-heating equipment to be listed and labeled such as baseboard heaters, heating cable, and duct heaters.

Article 426   Fixed Outdoor Electric De-icing and Snow-Melting Equipment

426.50(A): All disconnecting means for fixed outdoor electric de-icing and snow-melting equipment is required to be of the type that clearly indicates whether it is in the off or on position. In addition, the disconnecting means is required to be of a type that can be locked in the off position.

Article 725   Class 1, 2, and 3 Remote-Control, Signaling, and Power-Limited Circuits

725.2: **Circuit Integrity Cable (CI)** is a new definition for remote-control and signaling cable for critical circuits that is rated to survive fire conditions for a specified time.

725.3: This section specifies the installation of Class 1, Class 2, and Class 3 cables. The change is that these cables are to be installed in accordance with *300.11*, which provides specific installation directions for all types of cables including installation above suspended ceilings.

725.41(A) Exception 2: This section lists the acceptable power sources for Class 2 and Class 3 circuits. At issue is the type of wiring required for circuits. This new exception recognizes listed equipment with limited power circuits that fall within the limits of *Table 11A* or *Table 11B* are permitted to be treated as Class 2 or Class 3 power-limited circuits and wired accordingly.

725.56(F): This is a new section that prohibits Class 2 and Class 3 circuit cables to be installed in the same cable or raceway with audio-circuit conductors. The voltage and current levels of audio-circuit conductors can be sufficient to sometimes disrupt operation of the Class 2 or Class 3 circuit if adequate separation is not maintained.

725.61(A): Listed plenum signaling raceway is now permitted to be installed in other spaces used for environmental air, but not permitted to be installed in ducts or plenums. Cables permitted to be installed in the raceway are Types CL2P and CL3P as shown in Figure 3.23.

725.61(B)(1): Listed riser signaling raceway is permitted to be installed as a vertical riser in shafts between floors. Cables permitted to be run in the raceway are Types CL2R, CL3R, CL2P, and CL3P.

725.61(B)(3): Listed general-purpose signaling raceway is permitted to be installed as a riser between floors in one-family and two-family dwellings. Cables permitted to be run in the raceway are Types CL2, CL3, CL2X, and CL3X. It is also permitted to install cable Types CL2P, CL2R, CL3P, and CL3R.

725.61(E)(7): This section specifies the wiring methods and materials that can be used in other areas of a building for Class 2 and Class 3 circuits other than run in environmental air spaces or used as risers. Added to the list of acceptable cables is Power-Limited Tray Cable, Type PLTC cable, provided it meets the crush and impact requirements of Type MC cable and the cable is identified for use as wiring between a cable tray and utilization equipment. The cable shall be protected against physical damage and supported at intervals not to exceed 6 ft (1.8 m).

725.82: Nonmetallic signaling raceway is recognized for use with Class 2 and Class 3 circuits and is required to be listed as resistant to the spread of fire.

725.82(I): Plenum signaling raceway is required to be listed as being fire-resistant and low-smoke producing.

725.82(J): Riser signaling raceway is required to be listed as being fire-resistant and capable of preventing the carrying of fire from floor to floor.

725.82(K): General-purpose signaling raceway is required to be listed as being resistant to the spread of fire.

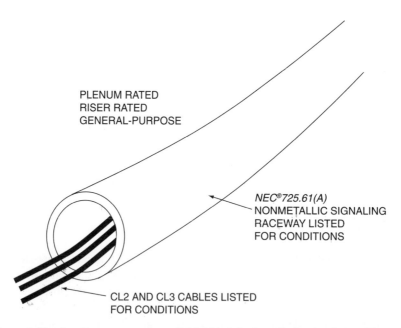

Figure 3.23 Nonmetallic signaling raceway is available listed for installation in air-handling spaces other than ducts and plenums, as a riser between floors, and for general-purpose applications. The material is flexible and available in sizes from ¹/₂ (16) to 2 (53) and the cables are required to be rated for the conditions.

WORKSHEET NO. 3—BEGINNING OUTLETS, LIGHTING, APPLIANCES, AND HEATING

Mark the single answer that most accurately completes the statement based upon the 2005 Code. Also provide, where indicated, the Code reference that gives the answer or indicates where the answer is found, as well as the Code reference where the answer is found.

1. A two-family dwelling has both living units with grade level access both from the front and the back yard. The minimum number of 125-volt, 15- or 20-ampere receptacle outlets required on the outside of this building is:
   A. not specified in the Code.
   B. one.
   C. two.
   D. three.
   E. four.

   Code reference _____

2. A living area in a dwelling has a wall that is 4 ft (1.22 m) long, then makes a 90° corner and continues for another 20 ft (6.10 m), and makes another corner and continues for 7 ft (2.13 m) as shown in Figure 3.24. The minimum number of receptacle outlets permitted to be installed on this 31 ft (9.45 m) wall section is:
   A. two.
   B. three.
   C. four.
   D. five.
   E. six.

   Code reference _____

3. A thermostat circuit in a single-family dwelling is rated as Class 2 power-limited operating at 24 volts. A listed cable less than $1/4$ in. (6 mm) in diameter permitted for this application but not permitted to be installed in a commercial building without raceway protection is Type:
   A. CL2.
   B. CL3.
   C. CL2P-CI.
   D. CL2R.
   E. CL2X.

   Code reference _____

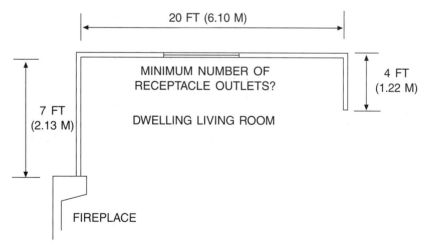

Figure 3.24 Determine the minimum number of receptacle outlets permitted to be installed on this length of unbroken wall in the living room of a dwelling.

4. Receptacle outlets are installed along the walls of living areas of a single-family dwelling where a specific load is not anticipated. These receptacle outlets are not in the kitchen, bath, or dining room. Assume also that receptacle outlets intended for an entertainment center are not included. The maximum number of receptacle outlets permitted to be installed on a 15-ampere, 125-volt branch-circuit is:

   A. not limited as long as the circuit is not overloaded.
   B. six.
   C. seven.
   D. eight.
   E. ten.

   Code reference_____

5. An overhead 120/240-volt, 3-wire set of feeder conductors supplies power from a dwelling to an outbuilding on a property considered to be strictly residential with no intended truck traffic as shown in Figure 3.25. If the conductors pass over a driveway, the minimum clearance from ground to an open conductor is to be not less than:

   A. 10 ft (3.0 m).   D. 18 ft (5.5 m).
   B. 12 ft (3.7 m).   E. 22 ft (6.7 m).
   C. 15 ft (4.5 m).

   Code reference_____

6. A single-family dwelling has three bathrooms. The minimum number of 20-ampere 125-volt branch-circuits required for the dwelling is:

   A. not specified in the Code.   D. five.
   B. three.   E. six.
   C. four.

   Code reference_____

7. A 12-kW electric range is to be installed in a single-family dwelling. The minimum rating 120/240-volt, 3-wire circuit permitted for the range is:

   A. 30 amperes.   C. 40 amperes.   E. 60 amperes.
   B. 35 amperes.   D. 50 amperes.

   Code reference_____

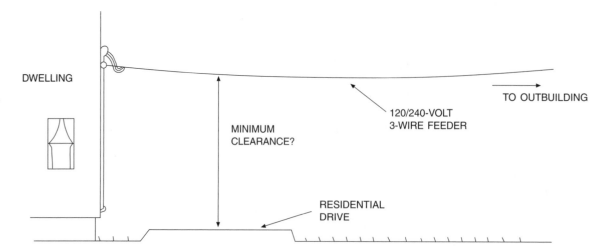

Figure 3.25 Determine the minimum clearance of overhead conductors operating at 120/240 volts and running over a residential drive.

8. A 15-ampere rated duplex receptacle on a 20-ampere, 125-volt circuit and mounted on the outside of a building and exposed to the weather is:
    A. only required to have an enclosure that is weatherproof when the receptacle is not in use.
    B. permitted to have an enclosure that is weatherproof only when the receptacle is not in use provided the intended use is only when personnel are present.
    C. not required to have a weatherproof cover if the receptacle is GFCI protected for personnel.
    D. required to have an enclosure that is weatherproof whether the receptacle is in use or not in use.
    E. not permitted to be in use during wet weather.

    Code reference _____

9. Fluorescent luminaires (lighting fixtures) in a commercial building are designed to be connected end-to-end. The luminaires (fixtures) are installed in rows in a room connected end-to-end, and supplied power for the entire row with one outlet at the first luminaire (fixture) as shown in Figure 3.26. The maximum number of 20-ampere 2-wire branch-circuits permitted to supply any one row of luminaires (fixtures) is:
    A. none, because fluorescent luminaires (lighting fixtures) are required to be individually supplied with power.
    B. one.
    C. two.
    D. three.
    E. four.

    Code reference _____

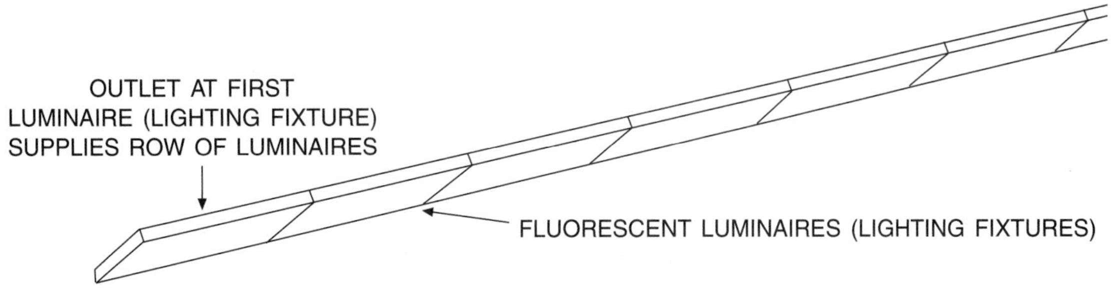

10. The allowable ampacity of a size 18 AWG copper 2-wire Type SPT-1 flexible extension cord is:
    A. 7 amperes.   C. 12 amperes.   E. 15 amperes.
    B. 10 amperes.   D. 13 amperes.

    Code reference _____

11. A receptacle outlet is located on a wall of a dwelling living room 6 in. (0.15 m) from a corner. After turning the corner, a wall section is 10 ft (3.05 m) in length. The wall then makes another corner and comes to another receptacle within 6 in. (0.15 m) as shown in Figure 3.27 on the next page. The minimum number of receptacle outlets required to be installed on the 10-ft (3.05 m) wall section is:
    A. none.   C. two.   E. four.
    B. one.   D. three.

    Code reference _____

Figure 3.26   Determine the maximum number of 2-wire circuits that are permitted to supply luminaires (lighting fixtures) mounted end-to-end in a continuous row.

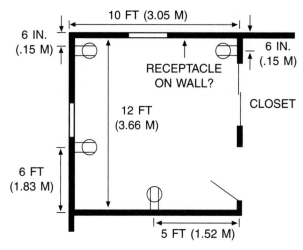

Figure 3.27   Determine the minimum number of receptacle outlets required to be installed on the 10 ft (3.05 m) section of wall in the dwelling.

12. An electric range in a single-family dwelling has a rating of 15 kW. The minimum demand load permitted to be used to determine the minimum circuit rating for the range is:
A. 8 kVA.   C. 9.2 kVA.   E. 15 kVA.
B. 8.4 kVA.   D. 12 kVA.

Code reference_____

13. A single-pole snap switch on a 15-ampere circuit in a dwelling fails and during replacement it is discovered the wire connected to the switch is aluminum size 12 AWG. If the aluminum wire is to be connected directly to the terminal screw of the replacement switch, the switch must be marked:
A. cu/al.   D. cu/al and 90° rated.
B. approved for aluminum wire.   E. CO/ALR.
C. al-only.

Code reference_____

14. A recessed incandescent luminaire (lighting fixture) is marked as thermally protected but it is not rated for direct contact with insulation (non-Type IC). Not counting the mounting points, the minimum distance required between the luminaire (fixture) and the wood joist, as shown in Figure 3.28, is:
A. $1/2$ in. (13 mm).   D. 2 in. (51 mm).
B. $3/4$ in. (19 mm).   E. 3 in. (76 mm).
C. 1 in. (25 mm).

Code reference_____

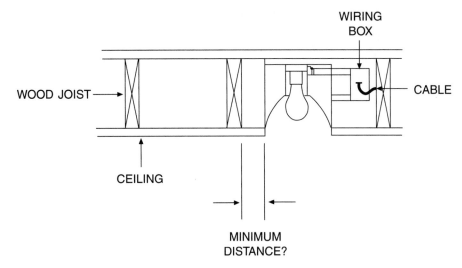

Figure 3.28 **A recessed incandescent lunminaire (lighting fixture) is marked as thermally protected but not rated for installation in direct contact with insulation. Determine the minimum spacing required between the side of the luminaire (lighting fixture) and the adjacent wood joist.**

15. A Type CL2 cable for a Class 2 circuit is run from a furnace to a thermostat location in another part of a new building under construction. The cable is required to be attached to structural components of the building at intervals not to exceed:

   A. any distance because support distances are not specified for a Class 2 circuit.
   B. 4^{1}/2 ft (1.37 m).
   C. 5 ft (1.52 m).
   D. 6 ft (1.83 m).
   E. 8 ft (2.44 m).

   Code reference _____

WORKSHEET NO. 3—ADVANCED OUTLETS, LIGHTING, APPLIANCES, AND HEATING

Mark the single answer that most accurately completes the statement based upon the 2005 Code. Also provide, where indicated, the code reference that gives the answer or indicates where the answer is found, as well as the Code reference where the answer is found.

1. A single-family dwelling has a living area of 3860 sq. ft (358.6 m²). All circuits for general illumination, in addition to those for small appliances, laundry and bathroom receptacles are rated at 20 amperes. The minimum number of general illumination branch-circuits permitted for this dwelling is:
   A. three.
   B. four.
   C. five.
   D. six.
   E. seven.

   Code reference _____

2. A 4500-watt, single-phase, 240-volt electric, 80-gallon (300 L) storage-type water heater is provided power with a copper Type NM-B Cable. The minimum size copper conductor permitted for this circuit is:
   A. 18 AWG.
   B. 16 AWG.
   C. 14 AWG.
   D. 12 AWG.
   E. 10 AWG.

   Code reference _____

3. Electric discharge luminaires (lighting fixtures) supplied by a 30-ampere branch-circuit are cord- and plug-connected to receptacles located directly above the luminaires. Each luminaire (fixture) draws 3.1 amperes. The minimum rating receptacle permitted for each luminaire (fixture) is:
   A. 5 amperes.
   B. 10 amperes.
   C. 15 amperes.
   D. 20 amperes.
   E. 30 amperes.

   Code reference _____

4. A peninsular type dwelling kitchen counter has a cupboard mounted above with a receptacle intended to serve the peninsular counter mounted to the underside of the cupboard as shown in Figure 3.29. The maximum distance from the surface of the counter to the receptacle outlet is not permitted to be more than:
   A. 18 in. (450 mm).
   B. 20 in. (500 mm).
   C. 24 in. (600 mm).
   D. 30 in. (750 mm).
   E. 36 in. (900 mm).

   Code reference _____

5. An electric range rated 16.4 kW is installed in a single-family dwelling. The minimum demand load permitted to be used to determine the rating of the circuit is:
   A. 9.6 kVA.
   B. 9.76 kVA.
   C. 10 kVA.
   D. 16 kVA.
   E. 16.4 kVA.

   Code reference _____

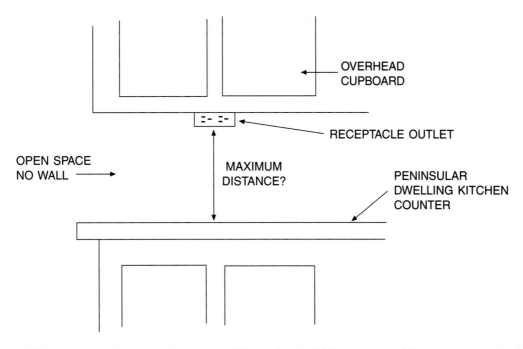

Figure 3.29   A cupboard is mounted above a swelling peninsular kitchen counter with a receptacle outlet mounted to the bottom of the cupboard. Determine the minimum distance from the counter to the receptacle.

6. A particular piece of single-phase, 240-volt portable equipment is connected to a permanent wiring system with extra hard service Type SO copper flexible cord. The minimum conductor ampacity required for the equipment is 16 amperes. The minimum size flexible cord permitted to supply this equipment is:
   A. 16 AWG.   C. 12 AWG.   E. 8 AWG.
   B. 14 AWG.   D. 10 AWG.

   Code reference_____

7. Class 2 circuit conductors installed in an elevator hoistway are *not* permitted to be run in:
   A. Rigid Metal Conduit.
   B. Rigid Nonmetallic Conduit.
   C. Liquidtight Flexible Nonmetallic Conduit.
   D. Liquidtight Flexible Metal Conduit.
   E. Electrical Metallic Tubing.

   Code reference_____

8. Three separate electric baseboard space-heating units are installed in a room of a dwelling on a single 240-volt branch-circuit. The maximum rating branch-circuit permitted for these baseboard heating units is:
   A. 15 amperes.   C. 30 amperes.   E. not limited.
   B. 20 amperes.   D. 40 amperes.

   Code reference_____

9. A dwelling has three bathrooms, and the fan and luminaires (lighting fixtures) in each bathroom are installed on the same circuit with the receptacle in the same bathroom. The minimum number of 20-ampere, 125-volt branch-circuits required for this dwelling to serve the bathrooms is:
    A. only dependent upon the load to be served.
    B. one.
    C. two.
    D. three.
    E. four.

    Code reference _____

10. The outlets in a dwelling that are required to be protected with an AFCI-type circuit breaker are:
    A. only the receptacles in the bedrooms.
    B. only the receptacles and luminaires (lighting fixtures) in the bedrooms.
    C. only lighting outlets in bedrooms.
    D. all bedroom 15- or 20-ampere, 125-volt rated outlets.
    E. all receptacle outlets in living areas of the dwelling except the bathroom, kitchen, outside, garage, and basement.

    Code reference _____

11. One building is supplied power from another building on the same property. The feeder to the second building is protected by a 200-ampere overcurrent device at the first building, and terminates immediately at a 200-ampere fusible switch upon entry to the second building as shown in Figure 3.30. This fusible switch is required to be rated as suitable for use as service equipment:
    A. only if the feeder contains a grounded circuit conductor that is also grounded at the second building and bonded to the disconnect enclosure.
    B. only if the feeder conductors are installed overhead and exposed to lightning.
    C. no matter what voltage or type of feeder supplies the second building.
    D. in all cases except where a grounded circuit conductor in the feeder is run separate from an equipment grounding conductor, and the grounded conductor is not grounded at the second building.
    E. only if the feeder has an ungrounded conductor operating at more than 250 volts-to-ground.

    Code reference _____

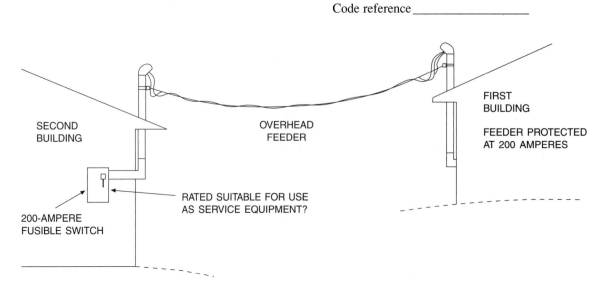

Figure 3.30 The second building receives the power from the first building. Determine the conditions under which the disconnect in the second building is required to be rated as suitable for use as service equipment.

12. A dimmer switch is *not* permitted to control:
    A. a receptacle that is intended to supply an incandescent floor or table luminaire (lighting fixture).
    B. more than one incandescent luminaire (fixture) on the same circuit in a dwelling.
    C. loads greater than 300 watts except in commercial or industrial buildings.
    D. luminaires (lighting fixtures) in a dwelling with exposed conductive parts unless protected with a GFCI for personnel.
    E. a single incandescent luminaire (fixture).

    Code reference_____

13. A complex of buildings is supplied power from a substation on the property with overhead lines to each building operating as a 4-wire grounded wye feeder at 8300/4800 volts. The conductors pass over open lands subject to vehicular traffic, pedestrian areas, and roadways suitable for truck traffic. The clearance from the roadway to the conductors, as shown in Figure 3.31, is not permitted to be less than:
    A. 15 ft (4.5 m).   C. 20 ft (6 m).   E. 30 ft (9 m).
    B. 18¹/2 ft (5.6 m).   D. 22 ft (6.7 m).

    Code reference_____

14. A 208/120-volt, 4-wire feeder supplies power from one building to a separate building on the same property. The feeder is protected by a 100-ampere overcurrent device in the first building. The property management does not have a qualified electrician on site at all times. The disconnecting means for the second building is:
    A. permitted to be located inside the first building.
    B. required to be located on the outside of the second building or inside and in either location near the point of entry of the conductors to the building.
    C. required to be located inside the building at the nearest practical point where the conductors enter the building.
    D. required to be located at the first building.
    E. permitted to be located outside the second building, not necessarily on the building, but within 50 ft (15 m) of the building.

    Code reference_____

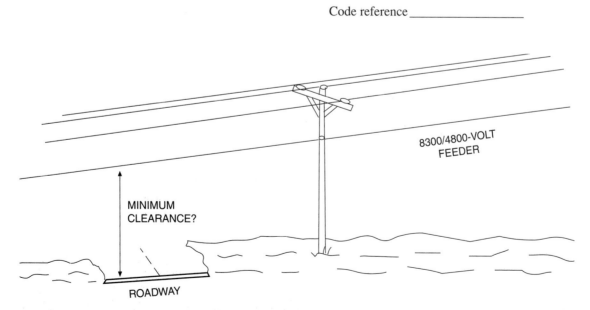

Figure 3.31 Overhead feeder conductors serving building on industrial property operate at 8300 volts between conductors and 4800 volts to ground. Determine the minimum clearance from a roadway subject to truck traffic and the overhead conductor.

15. Several Class 2 cables, with a $^1/_4$-in. (6 mm) diameter, are installed in an existing building with a suspended ceiling, and it is ruled that attaching the cables to the structural components above the ceiling, will be difficult. Assuming the cables are to be installed such that they do not prevent the removal of any ceiling panel, the maximum number of cables run across the top of any ceiling tile is:

A. zero.

B. six.

C. one.

D. two.

E. three.

Code reference _____

UNIT 4

Services and Feeder Calculations

OBJECTIVES

After completion of this unit, the student should be able to:

- determine the minimum permitted service entrance conductor size.
- determine the demand load to be included in a multifamily service calculation for more than two electric ranges in the building.
- determine the minimum permitted ampere rating for a single-family dwelling using the methods of *Article 220, Part III* or the optional calculation method.
- determine the maximum unbalance load for a single-family dwelling for the purpose of sizing the neutral conductor.
- determine the minimum ampere demand load for a small commercial building.
- determine the minimum permitted service entrance demand load for a multifamily dwelling.
- determine the minimum ampere rating of the service entrance for an existing dwelling where there has been an addition to the structure.
- determine the minimum demand load in amperes for a farm building where the loads operating with diversity and without diversity are known.
- look up the installation requirements for a service entrance from the Code.
- determine the minimum permitted ampere rating of central distribution equipment for a group of farm buildings, where the demand loads at each building are known.
- answer wiring installation questions relating to *Articles 215, 220, Parts III, IV, and V, 230,* and *Annex D, Examples.*
- state at least five significant changes that occurred from the 2002 to the 2005 Code from the previously stated articles.

CODE DISCUSSION

The installation of service equipment and calculations for the determination of the minimum rating of service equipment and feeders are covered in this unit. Grounding and bonding of service equipment are discussed in *Unit 5.*

Article 215 provides the minimum requirements for feeder conductor size and overcurrent protection. Feeders are main conductors that are ultimately subdivided into smaller circuits. The rules that are in *Article 220, Parts III, IV,* and *V* are used to determine the minimum load to be supplied by a set of feeder conductors. The total load must be subdivided into continuous and noncontinuous loads. Continuous loads, as defined in *Article 100,* are those that generally operate for three hours or longer. Lighting is usually considered to be a continuous load. Some appliances and equipment may also be considered continuous loads. Electric motors are frequently operated continuously but there is a separate rule for determining electric motor loads. Feeder load calculations are frequently figured in volt-amperes. A 150-watt incandescent lamp operating at a nominal 120 volts would be taken as a 150-volt-ampere load. A 3.5-kW electric water heater operating at 240 volts would be taken as a 3500-volt-ampere load. For some equipment, such as electric motors, the load in volt-amperes or watts is not provided. It is necessary to determine the load in volt-amperes for an electric motor by looking up the full-load current of the motor in *Table 430.248* for single-phase motors or *Table 430.250* for 3-phase motors and multiplying by the nominal voltage to get the volt-amperes. Equations 4.1 and 4.2 are used to determine load in volt-amperes for motors and similar equipment. Example 4.1 shows how to find the load for a 3-phase motor to be included in a feeder calculation.

Single-phase motors:

> **Motor Load = (Amperes from *Table 430.248*) $\times$ Nominal Voltage**   **Eq. 4.1**

Three-phase motors:

> **Motor Load = 1.73 $\times$ (Amperes from *Table 430.250*) $\times$ Nominal Voltage**   **Eq. 4.2**

Example 4.1   A 208Y/120-volt feeder in a commercial building supplies lighting and receptacle loads as well as several 3-phase electric motors, one of which is rated 10 horsepower. Determine the load in volt-amperes for the 10-horsepower, 208-volt, 3-phase electric motor to be used in the feeder calculation.

Answer: This is a 3-phase motor, therefore, Equation 4.2 can be used to calculate the motor load. Look up the full-load current of the motor in *Table 430.250* and find 30.8 amperes. The load for the motor is 11,083 VA.

$$\text{Motor Load}_{(10\ hp,\ 3\text{-phase},\ 208\ V)} = 1.73 \times 30.8\ A \times 208\ V = 11,083\ VA$$

When determining the minimum load to be supplied by a feeder, it is often convenient to add up the loads in the following three categories: continuous load, noncontinuous load, and motor load. If there is more than one motor supplied by the feeder, the rule of *430.24* will apply, and the largest motor load is multiplied by 1.25, and the other smaller motor loads are taken at their calculated value.

NEC® 215.2(A)(1) sets the minimum feeder conductor ampacity at 1.25 times the continuous load plus the noncontinuous load. Add motor loads by taking the largest motor load times 1.25 and the remaining motor loads at 100%. This is done without any consideration of adjustment factors for more than three current-carrying conductors in a raceway or cable or any temperature correction factors. The same method is used to determine the minimum ampere rating required for service conductors in *230.42(A)*. Examples of applying correction and adjustment factors are worked out in *Unit 1, Unit 2,* and *Example 03(a)* in the Code.

NEC® 215.3 specifies the same method for determining the minimum rating of overcurrent protection for the feeder as was used for determining the minimum ampacity of the feeder conductors. As illustrated in Figure 4.1, multiply the continuous load by 1.25 and add the noncontinuous load. Actually, figuring the min-

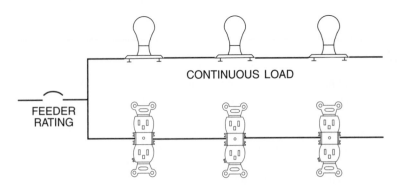

CONTINUOUS LOAD

FEEDER
RATING

ASSUMED TO BE NONCONTINUOUS LOAD IN THIS CASE

NEC® 215.3 FEEDER RATING SHALL NOT BE LESS THAN THE SUM OF:
1. NONCONTINUOUS LOAD AT 100%
2. 1.25 TIMES THE CONTINUOUS LOAD

Figure 4.1   The feeder rating is not permitted to be less than the noncontinuous load plus 1.25 times the continuous load.

imum rating of feeder overcurrent protection should be the first step in the process. Refer to *240.6(A)* for standard ratings of overcurrent devices once the minimum feeder load has been determined. The minimum size of conductor not only depends upon the load to be served but also upon the rating of the feeder overcurrent protection. *NEC® 240.4* requires a conductor to have an allowable ampacity not less than the rating of the overcurrent device. *NEC® 240.4(B)* applies in the case where the allowable ampacity of the conductor does not match a standard rating of overcurrent device. Provided the overcurrent device rating does not exceed 800 amperes, the next higher standard rating is permitted to protect a conductor. The following example will illustrate the application of *240.4(B)*.

Example 4.2   The calculated load for a feeder according to *215.2(A)(1)* and *215.3* is 260 amperes. The feeder will be protected with a 300-ampere circuit breaker. If the feeder conductors are copper with 75°C insulation and terminations and no adjustment or corrections factors apply, determine the minimum size conductor permitted for this feeder.

Answer: According to *215.2(A)(1)*, the conductor is required to have an allowable ampacity not less than 260 amperes. From *Table 310.16*, this is a size 300 kcmil conductor rated at 285 amperes. *NEC® 240.4(B)* permits this conductor to be protected at 300 amperes since it has an ampacity higher than the calculated load. The minimum size is 300 kcmil.

Solidly grounded wye electrical systems operating at more than 150 volts-to-ground from the phase conductors but not over 600 volts between conductors, are required to be provided with protection from damage due to ground faults if the disconnecting means is rated 1000 or more amperes. The arcing that can occur during a ground fault of feeders and services that have high ampere ratings can cause extensive damage and possible injury to personnel. For these services and feeders, a ground-fault sensor is required for feeders, *215.10*, and for services, *230.95*. There are two types of equipment ground-fault sensors. A current transformer, similar to a clamp-around ammeter, is installed after the main circuit breaker with the three phase conductors and neutral passing through the current transformer. This type is shown in Figure 4.2. The equipment grounding conductor does not pass through the sensor. The setting of the sensor is not permitted to be higher than 1200 amperes. Lower settings are desirable, although nuisance tripping may become a problem. With the second type of ground-fault protection, the sensor is installed around the main bonding jumper as shown in Figure 4.2. Fault current returning to the source must flow on the main bonding jumper. Care must

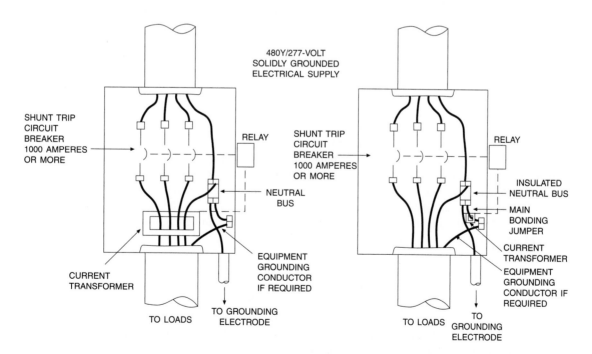

Figure 4.2   Equipment ground-fault protection is permitted to be installed around all phase conductors and the neutral, or it can be installed to sense current flow on the main bonding jumper at the service disconnect.

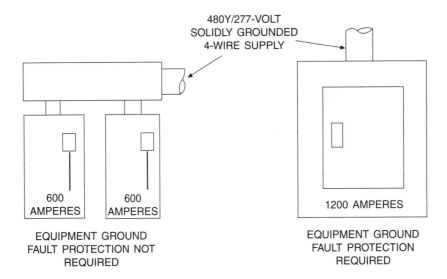

Figure 4.3   Equipment ground-fault protection is not required for a 480Y/277-volt solidly grounded electrical system if all disconnecting means are rated less than 1000 amperes.

be taken to make sure there are no parallel return paths to the source with this type of sensor. For more about grounding and grounding terminology, refer to *Unit 5*.

Equipment ground-fault protection can sometimes be avoided by using multiple disconnects each rated less than 1000 amperes as shown in Figure 4.3. The disconnect rating is the highest rating of fuse that can be installed in the disconnect, not necessarily the rating of the fuse installed. In the case of a circuit breaker used as a disconnect, the maximum rating is the maximum continuous current setting of the circuit breaker.

Article 220, Part III, specifies a basic method of calculating the load for a feeder or service. This load is used to size the conductors and the rating of overcurrent protection. It is important to note these calculations only provide for the known load. Provisions for future load growth is a matter of judgement and is not covered in the Code. This discussion will concentrate on load calculations for single-family dwellings, multifamily dwellings, a small commercial building, and farm buildings. *Part IV* provides optional methods for making some load calculations.

NEC® 220.10 requires the feeder or service calculation to include all appropriate loads described in *Part II* of *Article 220. NEC® 220.12* specifies the minimum load that is required to be included for general illumination for any type of facility listed in *Table 220.12*. This is a minimum load that is required to be included. If the actual lighting installed represents a larger load, then the larger load is used in the calculation. The feeder or service conductors must be capable of supplying the minimum load even if a smaller lighting load in installed. It is also required to provide an adequate number of circuits in the panelboard to supply the minimum listed lighting load even if a smaller lighting load is actually installed. In the case of dwellings according to *220.14(J)*, the receptacles, except for small appliance, laundry, and dedicated circuits, are included as outlets for general illumination, which means no additional load is required to be included for these receptacles. The load for general illumination is determined by multiplying the area of the building times the unit load from *Table 220.12* for the type of building. In the case of a single-family dwelling, the area to be included is the total area of the dwelling not including open porches, garages, and unused or unfinished spaces not practical for future use as living space in the dwelling. A finished portion of a basement would be included in the total floor area of the dwelling to which the unit lighting load would apply. The following example illustrates the calculation of general illumination load for a single-family dwelling.

Example 4.3   A single-family dwelling has two floors of living space with a total area of 2600 sq. ft (241.6 m^2). The basement is not finished and is not used as living space in the dwelling. Determine the minimum general illumination load required to be included in the service calculation for the dwelling.

Answer: Look up the unit load of 3 VA/ft^2 (33 VA/m^2) for a dwelling unit in *Table 220.12*. Next, multiply the living area of the dwelling by the unit general illumination load to get the minimum load of 7800 VA (7973 VA for the metric calculation).

General illumination load = 2600 ft^2 × 3 VA/ft^2 = 7800 VA

(= 241.6 m^2 × 33 VA/m^2 = 7973 VA)

In the case of receptacle loads, it may be necessary to include a specific load if the receptacle is for specific equipment. The load for receptacles on 15- and 20-ampere, 125-volt general circuits generally depends upon the actual number of receptacles installed. *NEC® 220.14(I)* specifies that general use receptacles are to be included in the feeder or service calculation as 180 volt-amperes for each receptacle strap or yoke. A duplex receptacle represents a load of 180 VA. This rule does not apply to dwellings, *220.14(J)*. *NEC® 220.14(K)* requires that for an office building, a minimum receptacle load of 1 VA/ft^2 (11 VA/m^2) is required to be included for general-use receptacles. Generally the actual receptacle load will be higher than this value and the higher load is required to be included in the calculation. *NEC® 220.44* recognizes that it is not likely that all receptacle outlets will be utilized at the 180-volt-ampere level at the same time. A 50% demand according to *Table 220.44* is permitted to be applied to receptacle load in excess of 10,000 VA. The following example will illustrate how the general use receptacle load is determined:

Example 4.4 An office building has a total floor area of 1800 sq. ft (167.3 m^2). Also, 72 general-use receptacles on 20-ampere, 125-volt circuits are installed. Determine the minimum receptacle load required to be included in the service load calculation.

Answer: First, determine the minimum load that is required to be included for general use receptacles by multiplying the area of the office by 1 VA/ft^2 (11 VA/m^2) to get 1800 VA (1840 VA). Next, determine the minimum load that is required to be included in the calculation for each receptacle installed by multiplying the number of receptacles by 180 VA/receptacle to get 12,960 VA. The larger load is included, but this load is in excess of 10,000 VA. *NEC® 220.44* permits the excess over 10,000 VA to be taken at half the value which will result in a total demand for receptacle load in this building of 11,480 VA [(12,960 − 10,000) × 0.50] + 10,000 = 11,480 VA.

NEC® 220.18(C) permits an electric range load for a dwelling to be figured using the demand factors of *220.55*. If the individual living units in a multifamily dwelling have electric ranges, *Table 220.55,* according to *Note 1,* can be used to determine the minimum load that must be included in the total building service or feeder load. If the electric ranges are not less than 8^3/4 kW or more than 12 kW in rating, the demand load for all the ranges can be looked up in column C of *Table 220.55*. The following example will illustrate the method of determining the total service demand load for electric ranges in a multifamily dwelling.

Example 4.5 A multifamily dwelling has 18 living units each with a 12-kW electric range. Determine the demand load for the ranges that is required to be included in the total service load calculation.

Answer: The ranges are not rated more than 12 kW, therefore, the demand load of 33 kVA can be looked up in column C of *Table 220.55*.

When the electric ranges in a multifamily dwelling have different ratings, it will be necessary to determine the average range rating if some of the electric ranges are rated more than 12 kW. *Note 2* of *Table 220.55* describes the method used to determine the average range size for the multifamily dwelling. All electric ranges rated grater than 8^3/4 kW but not greater than 12 kW are to be taken as though they were rated at 12 kW when determining the average. The following example will illustrate how the average range rating is determined:

Example 4.6 A multifamily dwelling has 18 living units—six with 10-kW electric ranges, six with 12-kW electric ranges, and six with 16-kW electric ranges. Determine the average rating range for the 18 living units.

Answer: *Note 2* of *Table 220.55* describes the process of determining the average range rating. All ranges 12 kW or less in rating are to be taken as though all were rated at 12 kW for the purpose of determining the average range size. The average range size is 13.3 kW.

$$12 \text{ ranges} \times 12 \text{ kW} = 144 \text{ kW}$$
$$\underline{6 \text{ ranges} \times 16 \text{ kW} = 96 \text{ kW}}$$
$$18 \text{ ranges} \qquad\qquad 240 \text{ kW}$$

$$\text{Average range rating} = \frac{240 \text{ kW}}{18 \text{ ranges}} = 13.3 \text{ kW}$$

Note 1 of *Table 220.55* describes how to determine the demand load for electric ranges in a multifamily dwelling when the average range rating is greater than 12 kW. The value in column C is increased by 0.05 for each kW the average rating exceeds 12 kW. As was described in the single electric range example of *Unit 3*, round the average range rating to the nearest whole number before figuring the range demand load. Equation 4.3 can be used to determine the range demand load to be included in the service calculation for the entire building. The following example will illustrate the process.

Range Demand Load = **[(Average range rating – 12) $\times$ 0.05 $\times$**
(Value from column C of Table 220.55)] +
(Value from column C of Table 220.55)   **Eq. 4.3**

Example 4.7   A multifamily dwelling has 18 living units with electric ranges of different ratings and the average range rating is 13.3 kW. Determine the minimum demand load for the electric ranges permitted to be included in the service calculation for the building.

Answer: *Note 1* of *Table 220.55* describes the process of determining the total minimum range demand load, which is summarized in Equation 4.3. Look up the demand load for 18 ranges rated at 12 kW from column C of *Table 220.55*. Round 13.3 kW off to 13 kW. Now put these values into Equation 4.2 and find the minimum range load for the building to be 34.65 kVA.

Range Demand Load = [(13 – 12) $\times$ 0.05 $\times$ 33 kVA] + 33 kVA = 34.65 kVA

NEC® 220.42 provides demand factors that are permitted to be applied to lighting loads for several types of buildings. In particular, the demand factors can be applied to the general illumination load for dwellings when calculating the service or feeder demand load using the method of *Article 220, Part III*. *NEC® 220.52* requires 1500 volt-amperes to be included in the load for each of the two required small appliance branch-circuits and the one required laundry branch-circuit. *NEC® 220.52* states that the small appliance and laundry branch-circuit loads are permitted to be included in the load to which the demand factors of *Table 220.42* apply. The following example will illustrate how the demand factors of *Table 220.42* are used:

Example 4.8   The demand load for a single-family dwelling service is calculated using the method of *Article 220, Part III*. The dwelling has a living area of 2400 sq. ft (223 m²). Determine the minimum load permitted to be included in the service calculation for general illumination, small appliance circuits, and the laundry circuit.

Answer: The general illumination, small appliance, and laundry circuit loads before applying the demand factors total 11,700 VA.

General illumination	2400 ft² $\times$ 3 VA/ft²	= 7,200 VA	(7359 VA)
Small appliance	(223 m² $\times$ 33 VA/m²)	3,000 VA	
Laundry		1,500 VA	
		11,700 VA	(11,859 VA)

Look up the demand factors for a dwelling in *Table 220.42*. The first 3000 VA of this load must be taken at 100% and the balance is taken at 35% to arrive at a minimum permitted load of 6045 VA (6101 VA).

[(11,700 VA – 3000 VA) $\times$ 0.35] + 3000 VA = 6045 VA

NEC® 220.53 permits a 75% demand factor to be applied to fastened in place appliance loads in a dwelling when there are four or more appliances. This demand factor can be applied when the method of *Article 220, Part III* is used for the service demand load calculation. This demand factor does not apply to the electric range, clothes dryer, space heat, or air-conditioner. Typical loads in a dwelling to which this factor does apply is the electric water heater, garbage disposer, dishwasher, and built-in microwave oven. *NEC® 220.60* permits the smaller of two loads to be disregarded in the calculation if it is not likely the loads will operate at the same time. An example would be electric heating and air-conditioning. There may be cases where they would operate at the same time, but generally the smaller of the two loads is permitted to be disregarded.

NEC® 220.61 explains the rules for determining the maximum unbalanced load for a service, which is the maximum load that could be expected on the neutral conductor. In the case of a 120/240-volt 3-wire service or feeder, even though the maximum unbalance load may be a substantial amount of current, the load on the neutral conductor is generally only a few amperes. This section is used to determine the maximum unbalance load whether the ungrounded conductors and the rating of the service is determined using the method of *Article 220, Part III,* or one of the optional methods in *Part IV.* To determine the maximum unbalanced load, identify all of the loads that operate at the lower voltage and would use the neutral. This would be 120-volt loads for a single-phase, 120/240-volt, 3-wire system and a 3-phase 208Y/120-volt, 4-wire system or a single-phase, or a 3-wire system derived from a 208Y/120-volt 3-phase system. Equipment and circuits operating at 277 volts would be considered in the determination of the maximum unbalanced load for a 3-phase 480Y/277-volt, 4-wire system.

It is not uncommon in commercial, industrial, or farm buildings for the load to operate mostly at the higher voltage and not utilize the neutral conductor. The load in the building requiring a neutral conductor may be very small. A single-phase, 120/240-volt 3-wire service to a building may supply several 240-volt single-phase motors. Lighting and receptacles are needed only for servicing the equipment in the building. In this case, the unbalanced load will be very small compared to the rating of the service disconnecting means. *NEC® 230.42(C)* sets a lower limit to the size of the neutral regardless. The neutral conductor is not permitted to be smaller than the size specified in *250.24(B)(1),* which specifies the size to be found in *Table 250.66.* The following example illustrates the method of finding the minimum neutral conductor size for a building with a small unbalanced load:

Example 4.9   A building is supplied with a 400-ampere, 3-phase, 208Y/120-volt, 4-wire service with the ungrounded conductors size 500 kcmil copper. The maximum calculated unbalanced load for this building is 20 amperes. Determine the minimum size copper neutral conductor for the service.

Answer: Look up the minimum size neutral conductor in *Table 250.66.* The neutral is not permitted to be smaller than the minimum size grounding electrode conductor for the service. The minimum neutral size is 1/0 AWG copper.

Article 220, Part V, covers the methods of determining the demand loads for farm buildings. These methods use amperes for the calculations, where the methods of the previous parts of *Article 220* use volt-amperes. It is important to keep in mind that amperes at 120 volts are different than amperes at 240 volts for the purpose of making a service calculation for a farm building. It is suggested that all loads be converted to a 240-volt basis. For example, assume there are lighting circuits that draw a total of 26 amperes at 120 volts. When making the service calculation of *220.102,* consider this lighting load as 13 amperes at 240 volts. Conversion of a farm building 120-volt demand load to a 240-volt basis is illustrated in Figure 4.4.

NEC® 220.102
FARM BUILDING DEMAND LOAD

LOAD	120 VOLTS	240 VOLTS
RECEPTACLES	22 AMPERES	(11 AMPERES)
LIGHTING (TIMES 1.25)	26 AMPERES	(13 AMPERES)
MOTOR	3.6 AMPERES	(1.8 AMPERES)
TOTAL		(25.8 AMPERES)

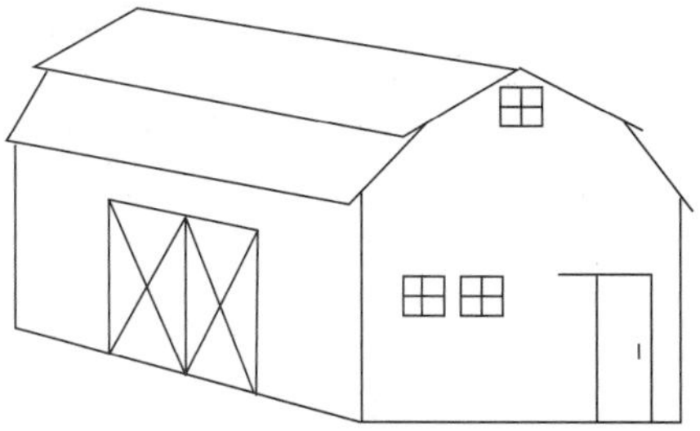

Figure 4.4   Farm building feeder demand loads are determined as current in amperes on a 240-volt basis.

Table 220.102 refers to loads that operate simultaneously and other loads. When calculating the demand load for a farm building, make a list of loads that will operate at the same time. For a milking barn, add up all equipment and lighting that will operate when cows are being milked. This will be the load that operates simultaneously. Make a second list of equipment, receptacles, and lighting that will normally operate at other times. This is equipment that is considered other load. Finally, identify any equipment that cannot operate when other equipment is operating. That equipment can be deleted from the calculation because of *220.60*. An example would be the milk tank washer in a milk house. The washer certainly will not be in operation when there is milk in the tank and the cooling compressor is operating. A control system may prevent some equipment from operating at certain times when other equipment is operating. *NEC® 220.102* requires that loads that operates simultaneously be taken at 100% in the calculation. Remember to multiply the largest motor current by 1.25 in the calculation.

A group of farm buildings, often including the dwelling, is generally supplied power from a central electrical distribution point. This may be a service at one building or it may be a service or disconnect, called a site isolation switch, at a central location. A common practice for a farm is to provide a meter pole, as shown in Figure 4.5, which shall have a minimum ampere rating determined by using *220.103*. Requirements for equipment at the central distribution point is covered in *Article 547*. Overhead feeders or underground feeders will run from the central distribution point to the various buildings.

Article 220, Part IV, describes alternative methods of making the service and feeder calculations instead of the method of *Article 220, Part III. NEC® 220.82* is an optional calculation method for a single-family dwelling. *NEC® 220.83* deals with the special case of an addition to an existing dwelling. This method can be used to determine if an addition to an existing dwelling will require an increase in the size of the service entrance. *NEC® 220.84* is the optional method for determining the demand load for a multifamily dwelling. This method cannot be used unless each living unit is equipped with electric cooking and either air-conditioning or electric heat or both.

Article 230 covers the requirements for the rating and installation of services. A service is defined as the conductors and equipment that deliver electrical energy from the utility to a wiring system. The Code does not specifically define the components that make up a service, but *Article 230* does set some specific requirements and provide rules for sizing service components. Figure 4.6 is shown as two services—one with the overcurrent protection as an integral part of the disconnecting means (circuit breaker) and one with the overcurrent protection located adjacent to the disconnecting means (fusible switch). The service conductors are protected from overcurrent on the load end of the conductors. For other electrical installations, the conductors are protected from overcurrent at their supply end. The service conductors are connected

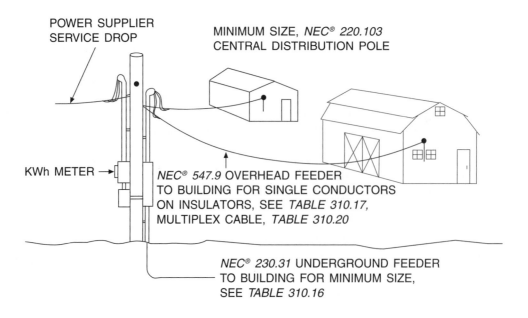

Figure 4.5   A center distribution pole provides a location for metering and a distribution point for conductors supplying power to farm buildings.

COMPONENTS OF A TYPICAL SERVICE

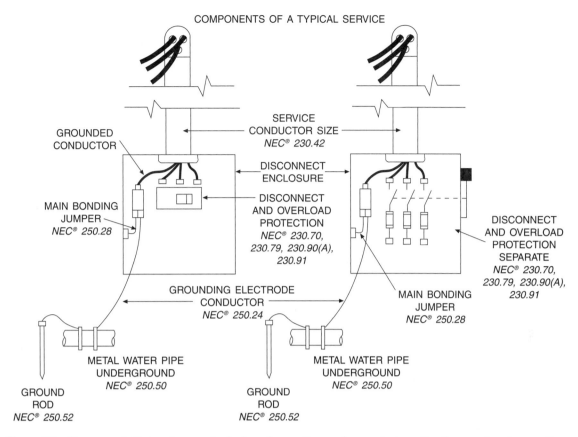

Figure 4.6   Components that make up a typical service are the service conductors, the disconnect enclosure, the disconnecting means for the ungrounded conductor, overload protection for the ungrounded conductors, a grounding electrode, a grounding electrode conductor, and if the system has a grounded conductor, a main bonding jumper.

directly to a utility system and the current that will flow during a ground fault or a short circuit is limited only by the impedance of the utility system. *NEC® 230.66* requires that equipment used as a part of a service be rated as suitable for use as service equipment. This designation will be marked on the enclosure label. In addition, *110.9* requires the equipment to be capable of interrupting whatever level of fault current that may be available from the utility supply to the service equipment.

The components that make up a typical service are the service conductors, the disconnecting means, the service conductor overcurrent protection, the grounding electrode and the grounding electrode conductor, and the main bonding jumper. These parts are pointed out in Figure 4.6. The function of the disconnecting means is to disconnect all conductors in a building from the service-entrance conductors. Each ungrounded service conductor is required to be provided with overload protection. Each service is required to have the grounded conductor connected to a grounding electrode. In the case of an ungrounded system, the metal disconnect enclosure is required to be connected to a grounding electrode. The load to be supplied by the service is determined using the methods described in *Article 220*. The disconnecting means is required to have a rating not less than the calculated demand load. The service conductors are required to have an allowable ampacity not less than the demand load. If the disconnect rating and overcurrent device rating are larger than required, then the service conductor size must be increased until *230.90(A)* is satisfied.

If a service is supplied with only one overcurrent device, then *230.90(A)* requires the conductor to have an allowable ampacity not less than the rating of the overcurrent device. The conductor may have an allowable ampacity less than the rating of the overcurrent device, according to *240.4(B)*. Example 4.2 illustrates the sizing of service conductors. *NEC® 230.71(A)* permits a set of service conductors to terminate in up to six individual disconnecting means. These disconnects can be grouped together or they can be six switches or circuit breakers in a single enclosure. *NEC® 408.36(B) Exception* permits up to six switches or circuit breakers in a power panelboard to serve as the service disconnecting means. When there are more than two disconnecting means for a service, *230.90(A) Exception 3* permits the individual overcurrent devices to add

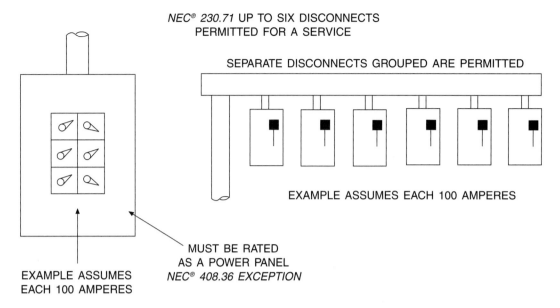

NEC® 230.71 UP TO SIX DISCONNECTS
PERMITTED FOR A SERVICE

SEPARATE DISCONNECTS GROUPED ARE PERMITTED

EXAMPLE ASSUMES EACH 100 AMPERES

MUST BE RATED
AS A POWER PANEL
NEC® 408.36 EXCEPTION

EXAMPLE ASSUMES
EACH 100 AMPERES

FOR EXAMPLE ASSUME DEMAND LOAD IS 410 AMPERES

Figure 4.7   A service is permitted to consist of up to six disconnecting means grouped in one location, or located in a single panelboard if it qualifies as a power panelboard.

up to more than the allowable ampacity of the service conductors. This is illustrated in Figure 4.7. *NEC® 230.42(A)* requires the service conductors to have an allowable ampacity not less than the calculated demand load for the service.

> **Example 4.10**   A service has a calculated demand load of 410 amperes. The service conductors terminate at six disconnecting means arranged in one location that add up to a combined rating of 600 amperes. Determine the minimum size copper conductors for this service.

> **Answer:**   When there are more than two disconnecting means for a service, it is permitted to size the conductors not less than the calculated demand load. *Exception 3* of *230.90(A)* permits the overcurrent devices to total more than the ampacity of the conductor. *NEC® 230.42* only requires the conductors to be capable of supplying the load. Look up a conductor in the 75° C column of *Table 310.16* and get size 600 kcmil.

In the case of a service for a single-family dwelling supplied with a 120/240-volt, single-phase, 3-wire service, it is permitted to size the conductors according to *Table 310.15(B)(6)*. This requirement is found in *Exception 5* of *230.90(A)*. This rule applies in the case of 120/240-volt, single-phase, 3-wire, single-family dwellings only. The rule is permitted to be applied in the case of service conductors to individual living units of two-family dwellings and multifamily dwellings. Rather than looking up the minimum conductor size in *Table 310.16*, the minimum size is found in *Table 310.15(B)(6)*. This rule applies only to the ungrounded service conductors. In the case of the neutral conductor, once the maximum unbalanced load is determined, the minimum neutral conductor size is found in *Table 310.16*.

Article 230 provides rules and specifications for the installation of services. For example, clearances of service-drop conductors and open-service conductors are found in *230.24*. A typical service supplied by an overhead service drop is illustrated in Figure 4.8. The general rule is that one set of overhead service-drop or underground-service lateral conductors supplies only one set of service-entrance conductors. Exceptions to this rule are found in *230.40*. Wiring methods permitted for service-entrance conductors are found in *Part IV*. The rules for the location of the service disconnecting means is found in *230.70*. The service disconnect is permitted to be located inside the building but as near as practical to the point where the conductors enter the building. The service disconnect is permitted to be located on the outside of the building and it is not required to be located near the point where the conductors enter the building. The disconnect is even permitted to be located outside and off the actual building as long as it is readily accessible. *Part VIII* provides rules for services with conductors operating at more than 600 volts.

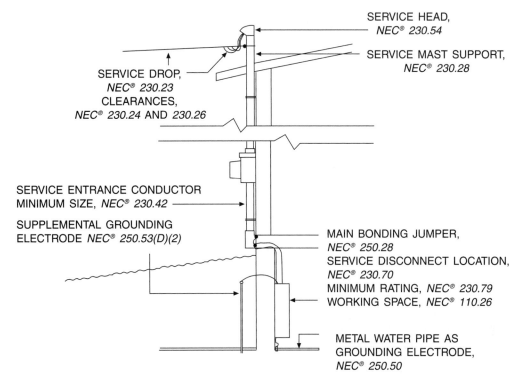

SERVICE HEAD,
NEC® 230.54

SERVICE MAST SUPPORT,
NEC® 230.28

SERVICE DROP,
NEC® 230.23
CLEARANCES,
NEC® 230.24 AND *230.26*

SERVICE ENTRANCE CONDUCTOR
MINIMUM SIZE, *NEC® 230.42*

SUPPLEMENTAL GROUNDING
ELECTRODE *NEC® 250.53(D)(2)*

MAIN BONDING JUMPER,
NEC® 250.28
SERVICE DISCONNECT LOCATION,
NEC® 230.70
MINIMUM RATING, *NEC® 230.79*
WORKING SPACE, *NEC® 110.26*

METAL WATER PIPE AS
GROUNDING ELECTRODE,
NEC® 250.50

Figure 4.8   Information needed for sizing and installing components of a service entrance is found in *Article 230* and *Article 250*.

A disconnecting means is required to be installed near the point where supply conductors enter a building or structure. The point where the conductors are considered to enter the building is important. Service conductors considered to be outside the building are covered in *230.6*. In the case of outside branch-circuits and feeders, *225.32* makes reference to *230.6* for the determination of whether the conductors are considered to be inside the building or structure. Conductors run under a building or structure and under a minimum of 2 in. (50 mm) of concrete are considered to be outside of the building. Since a concrete floor is generally about 4 in. thick (100 mm), service or feeder conductors run in raceway under a concrete floor are not considered to be in the building until the conductors emerge up though the floor. As illustrated in Figure 4.9,

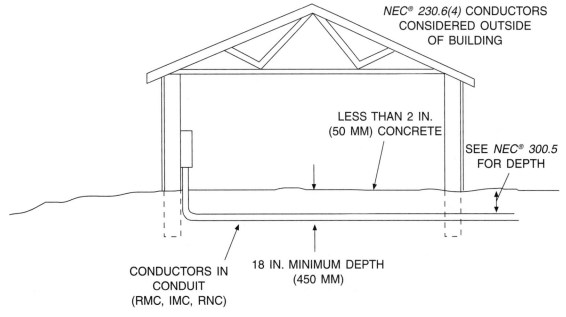

NEC® 230.6(4) CONDUCTORS
CONSIDERED OUTSIDE
OF BUILDING

LESS THAN 2 IN.
(50 MM) CONCRETE

SEE *NEC® 300.5*
FOR DEPTH

CONDUCTORS IN
CONDUIT
(RMC, IMC, RNC)

18 IN. MINIMUM DEPTH
(450 MM)

Figure 4.9   Conductors installed under a building with a minimum earth cover of 18 in. (450 mm) are considered to be outside the building.

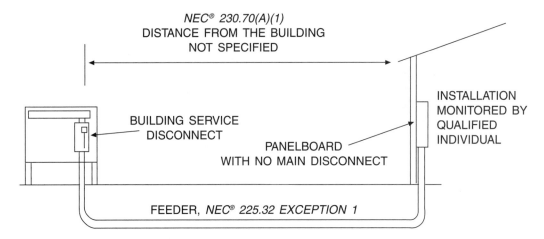

Figure 4.10   The service disconnecting means for a building is permitted to be located outside of the building.

service or feeder conductors run in Rigid Metal Conduit, Intermediate Metallic Conduit, Rigid Nonmetallic Conduit, or Liquidtight Flexible Conduit listed for direct burial are considered outside the building if buried to a depth of not less than 18 in. (450 mm). In this case, a concrete cover is not required. This rule may be useful for a dirt floor farm building or commercial building. Service or feeder conductors entering a basement where it is necessary to pass under a portion of a building that only has a crawl space is permitted by this rule. The depth of burial of the conduit below the surface of the earth is required to be not less than 18 in. (450 mm). Conductors run in conduit that is encased in at least 2 in. (50 mm) of concrete or brick are also considered to be outside the building even though this encasement may be run through a portion of the building. Vaults constructed to the specifications of *Part III* of *Article 450* are also considered to be outside the building or structure.

The service disconnecting means is permitted to be located inside the building or structure. If it is located inside the building or structure, it is required to be located as close as practical to the point where the conductors enter the building. The service disconnecting means is permitted to be located outside the building, in which case it is not required to be located near the point where the conductors enter the building or structure. It is even permitted to be located off the building as shown in Figure 4.10, and no distance from the building is specified. Whether the service disconnecting means is located off the building, or at some point on the outside of the building but away form the point of entry of the conductors to the building, the conductors on the load side of the service disconnecting means are considered feeder conductors. According to *225.32*, a disconnecting means is required to be located near the point of entry of the feeder conductors to the building, either inside or outside. A disconnecting means at the point of entry of the feeder conductors to the building can be omitted if the conditions of *Exception 1* of *225.32* are met and the installation is monitored by a qualified individual.

Annex D provides detailed examples to show the intent of the Code in determining the size of services and feeders and how to determine the minimum number of branch-circuits required for some installations. The notes at the beginning of *Annex D* state that the nominal voltages of 120, 208, and 240 shall be used to make calculations. They also state that when making calculations, fractions of amperes may be rounded down when the fraction is less than 0.5 amperes and rounded up when the fraction is 0.5 amperes and larger. This is also stated in *220.5(B)*. The notes also state that when a calculation for ranges results in a fraction of a kilowatt or kilovolt-ampere, it is permitted to round down if the fraction is less than 0.5. Table 4.1 is a summary of the type of calculations found in the examples.

Several examples of service and feeder calculations are included in *Annex D* of the Code. These examples are helpful in understanding the methods of *Article 220*. There are two variations of *Example D1*, which is of a single-family dwelling using the method of *Article 220, Part III*. *Example D2* is the optional calculation method of *220.82* for a single-family dwelling. There are three different variations of this example.

Example D3 is the service calculation for a commercial store building. This is an important example because it illustrates the use of the 1.25 factor for continuous loads. It also illustrates which loads are taken as continuous loads. Receptacle outlets in a commercial building are not always considered continuous loads.

Table 4.1.   Summary of the type of calculations found in the *Examples* of *Annex D.*

Example number	Number of branch-circuits	Service calculations NEC® 220, Part III method	Optional method	Neutral calculation	Range demand load
D1(a)	Yes	Yes	—	Yes	Yes
D1(b)	—	—	—	Yes	—
D2(a)	—	—	Yes	Yes	Yes
D2(b)	—	—	Yes	Yes	Yes
D2(c)	—	—	Yes	—	—
D3	Yes	Yes	—	—	—
D3(a)	—	Yes	—	Yes	—
D4(a)	Yes	Yes	—	Yes	Yes
D4(b)	Yes	—	Yes	Yes	Yes
D5(a)	Yes	Yes	—	Yes	Yes
D5(b)	Yes	Yes	—	Yes	—
D6	—	—	—	—	Yes

Example D3 can be confusing. The service entrance main overcurrent device is required to have a rating not less than the noncontinuous load plus 1.25 times the continuous load. This turns out to be 135 amperes for the example. Looking at *240.6(A)* for the next standard overcurrent device rating higher than 135 amperes results in a minimum service entrance rating of 150 amperes. Keep in mind that the Code specifies the minimum permitted, and not necessarily the recommended size for a particular application. The service-entrance conductors are sized based on the same current. The allowable ampacity values listed in the tables of *Article 310* are continuous currents. Therefore, the current used to size the conductors is the sum of the noncontinuous loads and 125% of the continuous loads. For *Example D3,* the current used to size the conductors is 135 amperes. The 75°C column of *Table 310.16* is used to select the minimum permitted ser-vice-entrance conductor size, which is 1/0 AWG, rated at 150 amperes.

Example 3D(a) is a load calculation for a 3-phase feeder supplying an industrial building. This example shows how a motor load is included in the calculation. The building 3-phase load in volt-amperes is converted to amperes for selection of the minimum rating of the feeder and selection of the minimum conductor size. The feeder ampacity is required to be adjusted for a high ambient temperature and eight current-carrying wires in the raceway. This is a good example of how to apply the rules in *215.2(A)(1)* to determine minimum feeder conductor size.

Example D4 is for a multifamily dwelling served with a single-phase electrical system. The service is sized for the entire building, and a calculation of the feeder size for the individual living units is included. This calculation is performed using the method of *Article 220, Part III* and the optional method of *220.84. Example D5* is a service calculation for a multifamily dwelling served with a 208/120-volt, 3-phase electrical system. *Example D6* illustrates how to determine the demand load for multiple electric ranges in a dwelling for the purpose of determining the size of service-entrance conductors or feeder conductors. Methods are described for determination of the load on the service neutral in several of the examples. There are also several calculations to determine the minimum number of lighting and receptacle branch-circuits.

SAMPLE CALCULATIONS

Examples of calculations of service entrances are included in this section to supplement the examples of *Annex D.* The examples in the Code should be studied with these service calculations.

Example of a Commercial Service Calculation

A store has an area of 3850 sq. ft (357.8 m²). There are 70 receptacle outlets and 1400 watts of outside lighting, which may operate during store hours. There is also an outside electric sign. The store also contains three refrigerated coolers with 240-volt, $^1/_2$-horsepower, single-phase motors and a walk-in cooler with a 3-horsepower, 240-volt, single-phase motor. The central furnace has a $^3/_4$-horsepower, 240-volt, single-phase motor on the blower.

1. Determine the noncontinuous, continuous, and motor load for the store to be used to size the ungrounded service-entrance conductors and main overcurrent protection.

Before proceeding with the calculation of the demand load for this store building, read *215.2(A)(1)* and *230.42(A)*. The overcurrent device rating protecting the service-entrance conductors shall have an ampacity not less than 100% of the noncontinuous load, plus 125% of the continuous load. It is required that the feeder conductors be sized to carry 125% of the continuous load. The demand load used for determining the minimum size ungrounded conductors for the service is 125% of the continuous load, plus 100% of noncontinuous loads.

Noncontinuous loads:
    Receptacles *(NEC® 220.14(I)* and *220.44)*
        70 × 180 VA = 12,600 VA

First 10 kVA at 100%	=	10,000 VA
Remainder over 10 kVA at 50%	=	1,300 VA

Total noncontinuous load   11,300 VA

Continuous loads:
    General lighting load *(NEC® 220.12)*

3850 ft² × 3 VA/ft²	=	11,550 VA
(357.8 m² × 33 VA/m²)	=	(11,807 VA)

    Outside sign circuit *(NEC® 220.14(F)* and *600.5(A))*

Not specified, so use minimum	=	1,200 VA

    Outside lighting

1,400 VA	=	1,400 VA

Total continuous load   14,150 VA
(metric total)   (14,407 VA)

Motor loads *(NEC® 220.14(C)* and *430.24)*

3 hp	240 V × 17 A × 1.25	=	5,100 VA
$^1/_2$ hp	240 V × 4.9 A × 3 motors	=	3,528 VA
$^3/_4$ hp	240 V × 6.9 A	=	1,656 VA

Total motor load   10,284 VA

2. Determine the minimum rating permitted for the overcurrent device protecting the service-entrance conductors, assuming one circuit breaker or set of fuses is protecting the ungrounded conductors.

According to *215.3*, the minimum permitted rating of the single overcurrent device protecting the service-entrance conductors shall be not less than the noncontinuous load plus 125% of the continuous load.

Noncontinuous load	11,300 VA	
Continuous load, 14,150 VA × 1.25 =	17,688 VA	(18,009 VA)
Motor load	10,284 VA	
Total	39,272 VA	(39,593 VA)

$$\frac{39{,}272 \text{ VA}}{240 \text{ V}} = 164 \text{ A} \qquad\qquad (165 \text{ A})$$

The next higher standard rating of overcurrent device listed in *240.6(A)* is 175 amperes, but from a practical standpoint, the size of main overcurrent device in a panelboard used as service equipment for this building would be 200 amperes.

3. Determine the minimum permitted size of ungrounded conductors for the service entrance to this store building, assuming that the conductors are copper with 75°C insulation and terminations and a single service-entrance main overcurrent device rated at 200 amperes.

The demand load for determining the minimum size service-entrance conductors is the same used for determining the rating of the service, *215.2(A)(1)*. The minimum size copper conductor, with 75°C insulation and terminations permitted for the calculated demand load of 164 amperes (165 A), is size 2/0 AWG as determined using *Table 310.16* and *240.4(B)*. But the maximum overcurrent device permitted for this conductor is 175 amperes. Therefore, the minimum permitted size of service-entrance conductors when the main overcurrent device is 200 amperes is size 3/0 AWG copper with 75°C insulation.

If according to *230.90, Exception 3* there were two to six overcurrent devices rather than only one main overcurrent device, then the service-entrance conductor would only have to be sized to carry the load of 164 amperes. In that case, the minimum permitted copper service-entrance conductor size with 75°C insulation and terminations would then be size 2/0 AWG.

4. Determine the minimum permitted size of neutral service conductor for the store, assuming that the conductor is copper with type THWN insulation.

The minimum size of neutral conductor permitted for this service is determined by the rules of *220.61*. The maximum unbalanced load is 50% of all line-to-neutral load in the case of a 3-wire, single-phase and 34% of line-to-neutral load in the case of a 4-wire wye, 3-phase system.

Total noncontinuous load		11,300 VA
Total continuous load (metric total)	14,150 VA × 1.25 = (14,407 VA)	17,688 VA (18,009 VA)
Total		28,988 VA (29,309 VA)

$$\frac{28,988 \text{ VA}}{240 \text{ V}} = 121 \text{ A} \qquad (122 \text{ A})$$

The minimum size neutral conductor permitted is 1 AWG, with 75°C insulation and terminations. It may be necessary to check to see if this is smaller than the minimum permitted neutral conductor size, according to *230.42(C)*.

5. Establish a grounding electrode and determine the minimum copper grounding electrode conductor size permitted. See *Unit 5* for more information.

Select an available electrode from *250.52*. If a metal underground water-pipe is used, then supplement it with one additional electrode *(250.53(D)(2))*. The minimum size grounding electrode conductor is found in *Table 250.66*. If the only electrode available is a rod, pipe, or plate, then the minimum size required is size 6 AWG copper. In this example of a 200-ampere service, the minimum size bare copper grounding electrode conductor is 4 AWG.

6. Determine the minimum number of 20-ampere, 120-volt general lighting circuits for the inside of the store. Continuous load is considered because the rating of the circuit is the size of overcurrent protection. The overcurrent device is not permitted to be loaded continuously at more than 80% of the overcurrent device rating. The minimum number of general illumination branch-circuits is determined based on the actual lighting load even if it is less than the calculated load based on *220.12*. The actual general illumination load was not stated in the problem, so the calculated value will be used to determine the minimum number of lighting circuits.

11,550 × 1.25 = 14,438 VA
(11,807 × 1.25 = 14,759 VA)

$$\frac{14,438 \text{ VA}}{120 \text{ V}} = 120 \text{ A} \qquad\qquad \frac{14,759 \text{ VA}}{120 \text{ V}} = 123 \text{ A}$$

$$\frac{120 \text{ A}}{20 \text{ A / circuit}} = 6 \text{ circuits} \qquad\qquad \frac{123 \text{ A}}{20 \text{ A / circuit}} = 6.14 \text{ circuits}$$

7. Determine the minimum number of 20-ampere, 120-volt receptacle circuits for the store. Some inspectors may judge a receptacle circuit to be a continuous load; however, this issue is not clear in the Code. It depends on the use of the circuit. For this example, we will consider the receptacle load to be a noncontinuous load. A minimum of five circuits is required.

$$\frac{11{,}300 \text{ VA}}{120 \text{ V}} = 94 \text{ amperes}$$

$$\frac{94 \text{ A}}{20 \text{ A / circuit}} = 4.7 \text{ circuits}$$

Example of a Single-Family Dwelling Demand Load

A single-family dwelling has a living area of 1800 sq. ft (167.3 m²), and the dwelling is to contain the following appliances at the time of construction:

Electric range	240 V	14 kW
Microwave oven (built-in)	12 A	120 V
Electric water heater	240 V	6 kW
Dishwasher	120 V	1.8 kW
Clothes dryer	240 V	5 kW
Water pump	8 A	240 V
Food waste disposer	7.2 A	120 V
Baseboard electric heat (8 total units)	240 V	15 kW
Air-conditioner	3 at 8 A	240 V

1. Determine the service demand load using the method of *Article 220, Part III.*

General lighting load *(220.12)*		
1800 ft² × 3 VA/ft² =	5,400 VA	
(167.3 m² x 33 VA/m² = 5521 VA)	(5,521 VA)	
Small-appliance circuits *(220.52(A))*		
2 × 1500 VA =	3,000 VA	
Laundry circuit *(220.52(B))*		
1 × 1500 VA =	1,500 VA	
Subtotal *(220.42, 220.52)*	9,900 VA	
	(10,021 VA)	
First 3000 VA at 100%		3,000 VA
Remainder at 35%		2,415 VA
		(2,457 VA)
Electric range *(Table 220.55)*		8,800 VA
Electric space heating *(220.51)*		15,000 VA
Air conditioning *(220.60)*		
3 × 8 A = 24 A		
24 A × 240 V = 5760 VA		0 VA
Clothes dryer *(220.54)*		
5 kW at 100%		5,000 VA
Other appliances *(220.53)*		
Microwave oven (built-in)		
12 A × 120 V =	1,440 VA	
Electric water heater =	6,000 VA	
Dishwasher	1,800 VA	
Water pump *(430.24)*		
8 A × 240 V × 1.25 =	2,400 VA	
Food waste disposer		
7.2 A × 120 V =	864 VA	
Subtotal	12,504 VA	
12,504 VA × 0.75 =		9,378 VA
Total demand load		43,593 VA
Service load		(43,635 VA)

$$\frac{43{,}593 \text{ VA}}{240 \text{ V}} = 182 \text{ amperes}$$

2. Determine the minimum size neutral conductor for the service entrance. The rule for sizing the neutral is found in *220.61*.

General lighting, small-appliance, and laundry load		5,415 VA
		(5,457 VA)
Electric range	8800 VA × 0.7 =	6,160 VA
Electric clothes dryer	5000 VA × 0.7 =	3,500 VA
Other electric appliances		

Microwave oven (built-in)	1,440 VA	
Dishwasher	1,800 VA	
Food waste disposer		
7.2 A × 120 V =	864 VA	

	4,104 VA
Total 120-volt load (unbalanced load)	19,179 VA
Unbalanced load	(19,221 VA)

$$\frac{19,179 \text{ VA}}{240 \text{ V}} = 80 \text{ amperes}$$

Optional Dwelling Service Calculation

The optional method of determining demand load for a single-family dwelling usually, *220.82*, results in a smaller value than the previous method in *220, Part III*. The minimum size service wires, however, is 100 amperes.

Electric space heater 15,000 × 0.4 =		6,000 VA
(Each room separately controlled)		
Air conditioner *(NEC® 220.82(C))*		OMIT
General lighting load		
1800 ft² × 3 VA/ft² =	5,400 VA	
(167.3 m² × 33 VA/m²=)	(5,521 VA)	
Small appliance circuits		
2 × 1500 VA =	3,000 VA	
Laundry circuit		
1 × 1500 VA =	1,500 VA	
Electric range	14,000 VA	
Clothes dryer	5,000 VA	
Microwave oven (built-in)	1,440 VA	
Electric water heater	6,000 VA	
Dishwasher	1,800 VA	
Water pump	2,400 VA	
Garbage disposer	864 VA	
Subtotal	41,404 VA	
	(41,525 VA)	

Apply the demand factors of *NEC® 220.82*:		
First 10 kVA of all other load at 100%		10,000 VA
Remainder of other load at 40%		
31,404 VA × 0.4 =		12,562 VA
(31,525 VA × 0.4 =)		(12,610 VA)
Total load		28,562 VA
Service load		(28,610 VA)

$$\frac{28,562 \text{ VA}}{240 \text{ V}} = 119 \text{ amperes}$$

Note: Earlier, the demand load was determined to be 182 amperes. The minimum neutral size is determined using *220.61,* as shown earlier.

Farm Building Demand Load

A hog-farrowing barn contains outlets for twenty heat lamps at 250 watts each, six 2-lamp, 40-watt fluorescent strips with a load of 1.6 amperes at 120 volts, and three electric fans operating at 240 volts: 1/6 horsepower, 2.2 amperes; $^1/4$ horsepower, 2.9 amperes; and $^1/2$ horsepower, 4.9 amperes. In addition, there are eight general-purpose, 120-volt receptacle outlets.

1. Determine the minimum service demand load for the building. All loads are considered to operate simultaneously except the general-purpose receptacle outlets. The heat lamps are considered a continuous load. The 120-volt heat lamp, receptacles, and lighting loads are divided by 240 volts rather than 120 volts, because all amperes for the service calculation must be on a 240-volt basis.

Loads operating simultaneously

Heat lamps	$\dfrac{20 \times 250\text{ W} \times 1.25}{240\text{ V}} = 26.0\text{ A}$	
Lights	$\dfrac{1.6\text{ A} \times 120\text{ V} \times 1.25}{240\text{ V}} =$	6.0 A
Fans	$4.9\text{ A} \times 1.25 =$	6.1 A
	$2.9\text{ A} =$	2.9 A
	$2.2\text{ A} =$	2.2 A
		43.2 A

Other loads:

Receptacles	$\dfrac{8 \times 180\text{ VA}}{240\text{ V}} =$	6.0 A

Loads operating simultaneously 100%	43.2 A
Other loads, but not	6.0 A
less than first 60 amperes of all loads at 100%	
Total load	49.2 A

MAJOR CHANGES TO THE 2005 CODE

These are the changes to the 2005 *NEC*® that correspond to the Code sections studied in this unit. The following analysis explains the significance of the changes from the 2002 to the 2005 Code only, and this analysis is not intended to be used in place of the Code. Refer to the actual section of the 2005 Code for the exact wording and meaning of each section discussed. Changes are indicated in the Code with a vertical line in the margin. If material was deleted or moved to another location in the Code, the location of the deletion is indicated with a dark dot in the margin.

Article 215   **Feeders**

215.2(A)(1): A new sentence was added to the last paragraph in this section dealing with the minimum permitted size grounded feeder conductor. Feeder conductor size, including the grounded conductor, is required to be not less than the load to be served as calculated according to the rules in *Article 220.* There are cases where the unbalanced load as calculated according to *220.61* only requires a small size grounded conductor (usually the neutral) compared to the size of the ungrounded conductors. There has never been a minimum size grounded conductor requirement for feeders. Now the grounded-circuit conductor minimum size is to be not smaller than the size of equipment grounding conductor required for the feeder as determined by *Table 250.122.* Figure 4.11 is an example of a feeder where the unbalanced load is small and this rule will apply.

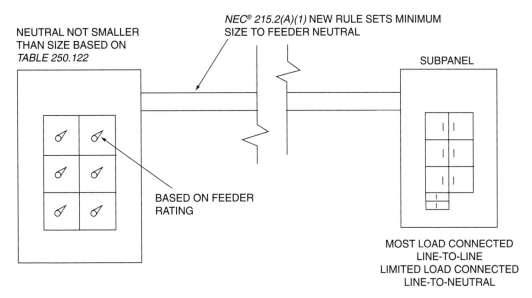

NEUTRAL NOT SMALLER
THAN SIZE BASED ON
TABLE 250.122

NEC® 215.2(A)(1) NEW RULE SETS MINIMUM
SIZE TO FEEDER NEUTRAL

SUBPANEL

BASED ON FEEDER
RATING

MOST LOAD CONNECTED
LINE-TO-LINE
LIMITED LOAD CONNECTED
LINE-TO-NEUTRAL

Figure 4.11   A feeder-grounded conductor (neutral) shall be sized not less than the calculated load, and not smaller than the equipment grounding conductor required for the feeder based on *Table 250.122*.

Example 4.11: A 200-ampere feeder has a calculated unbalanced load of only 30 amperes. Determine the minimum size copper grounded-circuit conductor permitted for the feeder.

Answer: Based upon the calculated unbalance load of 30 amperes, *Table 310.16* requires only a size 10 AWG grounded conductor. Based upon *Table 250.122*, the actual minimum size copper grounded conductor is 6 AWG.

In the case of a feeder consisting of parallel sets of conductors run in separate raceways or cables, the rule in *250.122(F)* does not apply. In the case of parallel sets of conductors run in separate raceways or cables, the minimum grounded conductor size permitted in each raceway or cable is the cross-sectional area of the minimum size grounded conductor determined using *Table 250.122* divided by the number of parallel sets, but not smaller than size 1/0 AWG.

Example 4.12: A 600-ampere feeder is run as two parallel sets of size 300 kcmil, THWN copper conductors in separate Rigid Metal Conduit. The calculated demand load for the feeder is 500 amperes, and the calculated unbalanced load is only 40 amperes. Determine the minimum size copper grounded conductor permitted for each set of conductors.

Answer: According to *310.4*, the minimum size grounded conductor for each parallel set is 1/0 AWG. Based upon *Table 250.122*, the minimum size grounded conductor for the feeder is 1 AWG copper, which according to *Table 8* in *Chapter 9* has a cross-sectional area of 83,690 cmil (42.41 sq. mm). Dividing the cross-sectional area of the 1 AWG wire by two gives 41,845 cmil (21.2 sq. mm). From *Table 8*, a wire with a cross-sectional area of 41, 845 cmil is a size 3 AWG. The minimum, however, is size 1/0 AWG.

215.2(A)(2): This subsection from the previous edition of the Code was deleted and the remainder of the section was renumbered. The portion of the section that was deleted required a minimum feeder rating of 30 amperes when the feeder supplied multiple branch-circuits. Now there is no minimum conductor size and feeder rating except as required to supply the load. Since the feeder conductors are required to have overcurrent protection at the supply end in accordance with *240.4*, a danger is not considered to be created when the feeder is subdivided into several branch-circuits. For example, a feeder consisting of size 14 copper wires and protected at 15 amperes is permitted to supply a panelboard with several circuit breakers acting as disconnecting means for small individual loads. The previous edition of the Code would have required a 30-ampere circuit with size 10 AWG copper wire even though the loads might only have required a 15-ampere feeder and size 14 AWG wire.

215.2(B): There is a new minimum size requirement for the grounded-circuit conductor operating at over 600 volts. If the calculation of unbalanced load only requires a small size grounded conductor, the size is not permitted to be smaller than the size of the equipment grounding conductor for the feeder. If the feeder conductors are run as parallel sets in separate raceway or cable, the rule of *250.122(F)* for an equipment grounding conductor minimum size does not apply to the minimum size of grounded-circuit conductor for the feeder.

215.8: This section from the previous edition of the Code was deleted because the requirement is in *110.15*. The section deleted specified the marking of the phase with the higher voltage to ground of a 4-wire, 3-phase, 240/120-volt delta electrical system.

215.12: This is a new section that specifies the method of identifying the grounded conductor, equipment grounding conductors, and ungrounded conductors of different electrical systems that serve the same building or structure. The inclusion of this new section and *210.5* for branch-circuit is redundant since *310.12* already applies to all types of electrical conductors. There is a change in all of these sections in that the color-coding method or other identification method to distinguish between the different electrical systems is now required to be permanently posted at each feeder panelboard or other distribution equipment such as a disconnect switch.

Article 220, Parts III, IV, and V   **Feeder and Service Calculations**

There was some reorganization of the first portions of the article and complete renumbering of the sections in the latter part of this article. The changes of significance in this article are as follows:

Part III: The general method used to calculate the load on a feeder or service conductors was *Part II* in the previous edition of the Code and is now *Part III*. All of the sections were given new numbers, but there are essentially no significant changes.

220.61: This was *220.22* in the previous edition of the Code and explains how to calculate the minimum required load on a neutral conductor. There are no changes in the section, but it now consists of three main subsections, which make the rule easier to follow.

220.82(C)(2) and (4): This was *220.30(C) (2)* and *(4)* in the previous edition of the Code. This is the optional feeder or service conductor calculation for a single-family dwelling with a heat pump and supplemental electric heat. These two paragraphs were revised to make the rule clear on how to calculate the demand load when a heat pump is used as the main heating unit in the dwelling. The previous edition of the Code required the heat pump to be included in the calculation at 100% and the supplemental electric heat to be included in the calculation at 100%. Now the heat pump is figured at 100% and the supplemental heat figured at 65%. This change is summarized in Figure 4.12. This can make a significant difference in the size of service for a dwelling. This change will result in a change in the heat pump calculation in *Example D2(c)* in *Annex D*. The load required to be added to the load calculation for heating in the previous edition of the Code was 20.76 kVA. With this change the heating load in *Example D2(c)* is now 15.51 kVA. The following example will illustrate the change in calculation:

Example 4.13: Assume a single-family dwelling is supplied by a 120/240-volt, 3-wire, single-phase service and is heated with a heat pump rated 30 amperes at 240 volts, and 20 kW of supplemental electric heat at 240 volts. The heat pump and the supplemental electric heating can operate at the same time. Determine the demand load for this heating system required to be included in the service conductor calculation.

Answer: The heat pump is figured into the calculation at 7200 VA (30 A × 240 V = 7200 VA). The supplemental electric heat according to *220.82(C)(4)* is permitted to be included in the calculation at 65%, which is 13.0 kVA (20 kVA × 0.65 = 13.0 kVA). The total heating load that is required to be included is 20.2 kVA (7.2k VA + 13.0 kVA = 20.2 kVA).

Table 220.102: This was *Table 220.40* in the previous edition of the Code, and the section deals with the method of calculating the load for a farm building. In *Table 220.40* of the previous edition of the Code, the term load "without diversity" was used in the table to indicate those loads in a farm building that normally operate at the same time when the building is in use. Now *Table 220.102* uses the expression

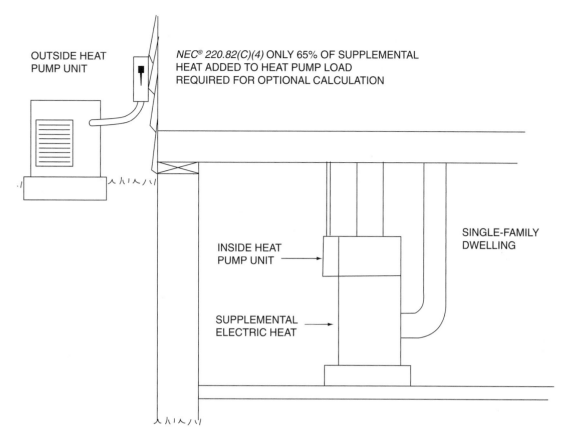

Figure 4.12   When a single-family dwelling service load calculation is performed using the optional method of *NEC® 220.82* **and there is a heat pump with supplemental electric heating, the supplemental heating can be taken at a demand factor of 65%.**

"loads expected to operate simultaneously." Other than a change in terminology to help improve understanding, there is no change in the method used to calculate the load for a farm building.

Article 230   Services

230.2(A)(6): The basic rule of this section is for only one service to a building. The subsections then state the situations where other services are permitted to be installed. A situation that did not seem to be specifically covered is where a building or structure is supplied with independent separate sets of service conductors from different sources for the purpose of creating a redundant system to enhance reliability of electrical supply to the facility. Such systems have been designed by engineers and installed by special permission even though the Code does not seem to allow for such independent redundant systems. A new item *(6)* was added to permit a service to consist of multiple sources for the purpose of enhanced reliability. If there is a failure of one set of service conductors or the distribution line supplying the service, within a short period of time the facility can be restored to full service through the redundant service.

230.40 Exception 1: This section requires that a service drop or service lateral supply only one set of service conductors. This means that several services are not permitted to be connected to the service drop or lateral. There are several exceptions to this rule. The meaning of *Exception 1* was confusing. It now states that each service as permitted by *230.2* is only permitted to have one set of service conductors. There is no actual change of intent with this rewording of *Exception 1*.

230.44: Cable trays are permitted to support service-entrance conductors, and now there is a rule that only permits the cable tray to contain service conductors. There is an exception that permits conductors other than service conductors to be installed in the cable tray provided there is a solid fixed barrier of

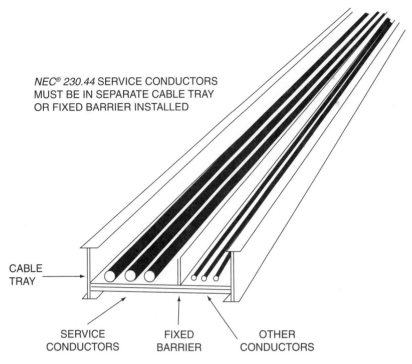

NEC® 230.44 SERVICE CONDUCTORS
MUST BE IN SEPARATE CABLE TRAY
OR FIXED BARRIER INSTALLED

CABLE
TRAY

SERVICE   FIXED   OTHER
CONDUCTORS   BARRIER   CONDUCTORS

Figure 4.13   When service conductors are run in cable tray there shall be no other conductors in the cable tray, or a fixed barrier compatible with the cable tray material shall be installed to separate the service conductors from the other conductors.

material compatible with the cable tray material between the service conductors and other conductors in the same cable tray. This is illustrated in Figure 4.13.

230.71(A): A service is permitted to consist of up to six separate disconnecting means either located in the same enclosure or located in a group. A disconnecting means for a transient voltage surge suppressor (TVSS) is now permitted to be installed in addition to the six service disconnects.

230.72(B): The previous edition of the Code omitted emergency systems from the list of service disconnects in addition to the six disconnects permitted and required to be located remote from the service disconnects. This is not a change of intent. It was always required that the emergency system disconnecting means be located remote from the service disconnecting means.

230.82(3): A meter disconnect switch is permitted to be installed on the supply side of the service disconnecting means for a service operating at not more than 600 volts. Now there is an additional requirement that the meter disconnect switch have a short-circuit rating not less than the short-circuit current available at the service.

230.82(8): A transient voltage surge suppressor (TVSS) is now permitted to be tapped ahead of the service disconnecting means provided it is installed on the load side of separate service equipment. What this means is that a TVSS is permitted to be tapped to the service conductors, but it is required to have a disconnecting means, overcurrent protection, a main bonding jumper, and a grounding electrode conductor. The disconnecting means is also required to be suitable for use as service equipment.

230.95: Equipment ground-fault protection is required for all 3-phase, solidly grounded, wye-connected services with disconnects rated 1000 amperes or more and operating at more than 150 volts to ground, but not more than 600 volts phase-to-phase. The intent did not change, but there were some issues with the term solidly grounded. Solidly grounded means that the grounded conductor of the service is connected directly to ground without installing any resistor or other impedance device. What this means is that this rule does not apply in the case of an impedance grounded system as described in *250.36*. A high-impedance grounded system is illustrated in Figure 5.18 of *Unit 5*.

Annex D   Examples

Example D2(c): The change has to do with the method of determining the heating load. This is due to a change in *220.82(D)(4)*. The heat pump is included in the calculation at 100%, but now the supplemental electric heat load in included in the calculation at 65% rather than 100%, which may result in a smaller size service required when a heat pump in installed in a single-family dwelling.

Example D3(a): This is an example of an industrial feeder calculation where two outbuildings are supplied 3-phase, 480/277-volt, 4-wire power from a main building. It is necessary to determine the continuous load and noncontinuous load in the outbuildings and to determine the minimum permitted overcurrent device rating for the feeders. The two feeders are run in the same conduit for a portion of the run, therefore, it is necessary to adjust the conductor ampacity for more than three current-carrying conductors in the raceway. The conduit is run such that the ambient temperature is 35°C, which requires an ampacity adjustment for high ambient temperature. This is an excellent example that demonstrates how to determine whether the neutral conductors are to be counted as current-carrying conductors, and how to determine the minimum permitted wire size when ampacity adjustments are required. This example should help clarify the rules of *210.19(A)(1)* and *215.2(A)(1)* with respect to determining wire size when ampacity adjustments are required. The last part of the example explains how to determine the maximum unbalanced load needed to determine the minimum neutral size for the feeder. That portion of the example does fall short in that it does not apply the same ampacity adjustment factors that were applied to the ungrounded conductors. There is also likely to be some debate as to whether assuming that all of the lighting load will be placed between only one of the phase conductors and the neutral is a reasonable assumption. It must also be pointed out that if the neutral load is small, the minimum size neutral conductor is based upon the minimum required size equipment grounding conductor for the feeder, which if one were required it would be size 6 AWG copper, as required by *215.2(A)(1)*.

WORKSHEET NO. 4—BEGINNING SERVICES AND FEEDER CALCULATIONS

Mark the single answer that most accurately completes the statement based upon the 2005 Code. Also provide, where indicated, the Code reference that gives the answer, or indicates where the answer is found, as well as the Code reference where the answer is found.

1. Service entrance cable Type SE style U is installed on the outside of a building for a service entrance as shown in Figure 4.14. The weather head is fastened to the side of the building and the cable is supported within 12 in. (300 mm) of the weather head. The maximum distance permitted between supports for the cable is:
   A. 18 in. (450 mm).   D. 3 ft (900 mm).
   B. 24 in. (60 mm).   E. 4¹/₂ ft (1.4 m).
   C. 30 in. (750 mm).

   Code reference _____

2. The calculated demand load for the ungrounded conductors of a single-family living unit in an apartment building is 68 amperes. The rating of the disconnecting means and feeder from the meter location to the service panel in the living unit is not permitted to be less than:
   A. 70 amperes.   C. 90 amperes.   E. 110 amperes.
   B. 80 amperes.   D. 100 amperes.

   Code reference _____

3. A 5-horsepower, single-phase, 240-volt electric motor when included in a service calculation represents a load of not less than: (not necessarily the largest motor)
   A. 4245 VA.   C. 5000 VA.   E. 8800 VA.
   B. 4820 VA.   D. 6720 VA.

   Code reference _____

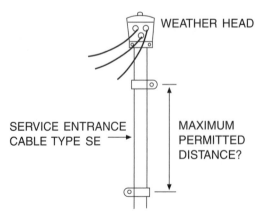

Figure 4.14   Determine the maximum permitted spacing between supports for Type SE Cable used as service conductors.

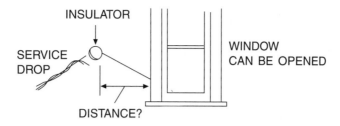

Figure 4.15 Determine the minimum distance permitted from the edge of a window that can be opened and open conductors of a service.

4. A service-drop terminates at an insulator attached to the side of a building next to a window that can be opened as shown in Figure 4.15. The minimum distance from the edge of the window to the open conductors is not permitted to be less than:

A. 3 in. (75 mm). D. 3 ft (900 mm).
B. 12 in. (300 mm). E. 6 ft (1.8 m).
C. 24 in. (600 mm).

Code reference _____

5. A 120/240-volt, 3-wire, single-phase feeder supplies a small panelboard with four 2-wire branch-circuits where the calculated load is only 15 amperes on each ungrounded conductor. The minimum conductor ampacity permitted for this feeder is:

A. 15 amperes. C. 30 amperes. E. 50 amperes.
B. 20 amperes. D. 40 amperes.

Code reference _____

6. A single-family dwelling is served with single-phase, 3-wire, 120/240-volt service using aluminum conductors with 75°C insulation and terminations. The demand load for the dwelling is 149 amperes, and the main circuit breaker is rated 200 amperes as shown in Figure 4.16. The minimum size ungrounded conductors permitted for this service is:

A. 1/0 AWG. C. 3/0 AWG. E. 250 kcmil.
B. 2/0 AWG. D. 4/0 AWG.

Code reference _____

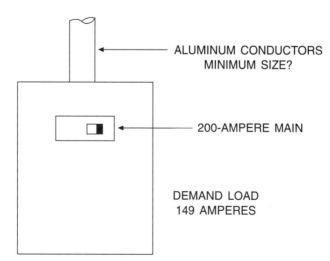

Figure 4.16 Determine the minimum size aluminum conductors for a service with a demand load of 149 amperes, and terminating at a 200-ampere main circuit breaker.

7. A wiring method installed, suitable for the conditions but not permitted for use as a part of a service entrance installation, is:
    A. insulated conductors run in Rigid Nonmetallic Conduit.
    B. insulated conductors run in Electrical Nonmetallic Tubing.
    C. Metal Clad Cable, Type MC.
    D. insulated conductors in Liquidtight Flexible Nonmetallic Conduit.
    E. Type AC Armored Cable.

    Code reference _____

8. In the case where a service entrance receives the source of power from an overhead service drop, the connection of service-entrance conductors to the service drop:
    A. has no specific requirements other than a tight connection.
    B. is required to be made above the level of the service head.
    C. shall be located below the level of the service head.
    D. is required to be formed into a drip loop and the actual connection is then permitted to be located above or below the service head.
    E. is only permitted to be made with an irreversible listed crimp-type connector.

    Code reference _____

9. A single-family dwelling has four 20-ampere, 125-volt small-appliance branch-circuits installed to serve the receptacle outlets in the kitchen and dining room. When calculating the service demand load for the dwelling, the minimum load that is permitted to be included for the small-appliance branch-circuits in this dwelling is:
    A. 1500 VA.   C. 4500 VA.   E. 7500 VA.
    B. 3000 VA.   D. 6000 VA.

    Code reference _____

10. Service conductors are not considered outside a building when:
    A. installed on the outside surface of the building as Type SE Cable.
    B. run within an interior wall encased within 2 in. (50 mm) of concrete.
    C. installed in Rigid Nonmetallic Conduit under a building with a minimum of 2-in. (50 mm) of concrete above the conduit.
    D. installed beneath a building in Rigid Nonmetallic Conduit with at least 18-in. (450 mm) of earth cover.
    E. installed within Rigid Metal Conduit under a building with a minimum 6-in. (150 mm) earth cover.

    Code reference _____

11. A dwelling 3-wire, 120/240-volt service has a trade size 2 (53) Rigid Metal Conduit mast extending up through the roof overhang, as shown in Figure 4.17, with a 4-in. by 12-in. (100 mm by 300 mm) roof slope. The roof overhang is 20 in. If the service drop terminates at an insulator mounted on the mast riser, the minimum clearance permitted from the roof to the conductors is:
    A. 18 in. (450 mm).   D. 3 ft (900 mm).
    B. 24 in. (600 mm).   E. 4 ft (1.2 m).
    C. 30 in. (750 mm).

    Code reference _____

12. A building on a farm has a calculated load that is expected to operate simultaneously of 92 amperes and other load of 56 amperes. These load amperes are on a 240-volt basis. The minimum demand load permitted for this building to be used to size the conductors and rating of service equipment is:
    A. 106 amperes.   C. 136 amperes.   E. 171 amperes.
    B. 120 amperes.   D. 148 amperes.

    Code reference _____

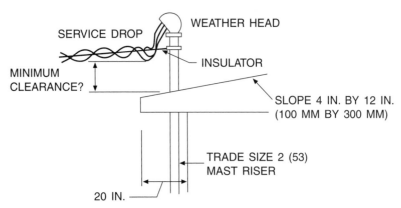

Figure 4.17 Determine the minimum distance from the roof to the open conductors when a mast riser extending up through a roof overhang, where the roof slope is 4 in. by 12 in. (100 mm by 300 mm), and the overhang is not more than 4 ft (1.2 m).

13. The calculated demand load for a commercial building served with 120/240-volt, single-phase power is 360 amperes. A single main 500-ampere overcurrent device is installed as the disconnect for the service as shown in Figure 4.18. The minimum size copper service-entrance conductors permitted for this service is:

A. 500 kcmil.   C. 700 kcmil.   E. 800 kcmil.
B. 600 kcmil.   D. 750 kcmil.

Code reference_____

14. A multifamily dwelling consists of eight individual living units each with a 12-kW electric range. The method of *Article 220, Part III,* is used for the load calculation for the service conductors for the multifamily dwelling. The minimum load permitted to be included in the calculation for the electric ranges is:

A. 23 kVA.   C. 76.8 kVA.   E. 120 kVA.
B. 64 kVA.   D. 96 kVA.

Code reference_____

15. A total length of 45 ft (13.7 m) of track lighting is installed in a commercial building. For the purpose of calculating the load on the feeder supplying the track lighting, the minimum load that is required to be included is:

A. 3450 VA.   C. 4219 VA.   E. 4800 VA.
B. 3840 VA.   D. 4313 VA.

Code reference_____

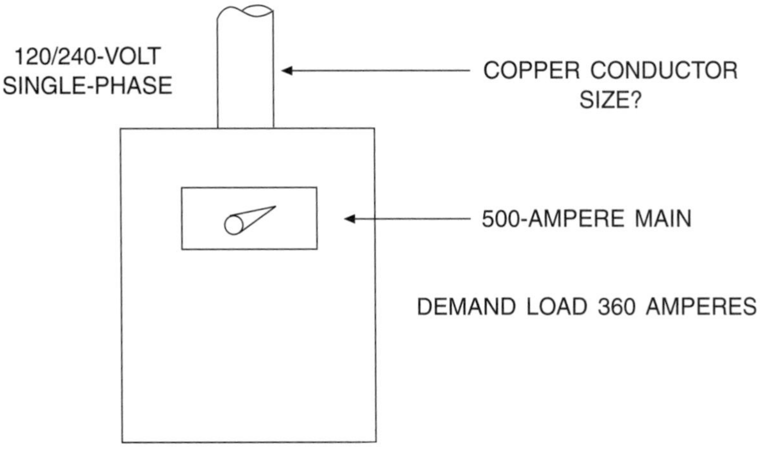

**120/240-VOLT
SINGLE-PHASE**

COPPER CONDUCTOR
SIZE?

500-AMPERE MAIN

DEMAND LOAD 360 AMPERES

Figure 4.18 Determine the minimum size copper conductors for a service where the single main overcurrent device is rated 500 amperes and the demand load is 360 amperes.

WORKSHEET NO. 4—ADVANCED SERVICES AND FEEDER CALCULATIONS

Mark the single answer that most accurately completes the statement based upon the 2005 Code. Also provide, where indicated, the code reference that gives the answer or indicates where the answer is found, as well as the Code reference where the answer is found.

1. Equipment ground-fault protection is required to be installed on the service of a commercial building served with a 3-phase solidly grounded:
   A. 208Y/120-volt system greater than 1000 amperes.
   B. 480Y/277-volt system of any size.
   C. 240/120-volt delta 4-wire system over 1000 amperes.
   D. 480-volt corner-grounded delta 3-wire system 1000 amperes and larger.
   E. 480Y/277-volt system 1000 amperes and larger.

   Code reference _____

2. A commercial building has a 208Y/120-volt, 3-phase demand load of 130 amperes. However, the electrician installs a service with a 200-ampere main overcurrent device. The service conductors installed are Type THWN aluminum as shown in Figure 4.19. The minimum size ungrounded conductors permitted for this service is:
   A. 2/0 AWG.   C. 4/0 AWG.   E. 300 kcmil.
   B. 3/0 AWG.   D. 250 kcmil.

   Code reference _____

3. A 208-volt, 3-phase, 5-horsepower electric motor when included in the service calculation for a commercial building is included at: (not necessarily the largest motor)
   A. 5000 VA.   C. 6406 VA.   E. 11,083 VA.
   B. 6009 VA.   D. 8245 VA.

   Code reference _____

4. The calculated demand load for a commercial building served with 120/240-volt single-phase power is 360 amperes. Rather than installing a single main overcurrent device and disconnect for the service, a 200-ampere panelboard and three 100-ampere

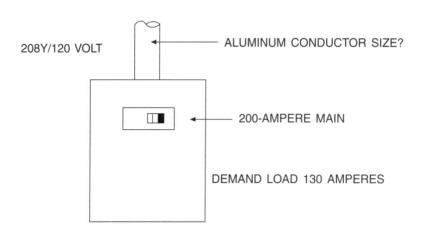

Figure 4.19   Determine the minimum size aluminum conductors for a 208Y/120-volt 3-phase service with a single 200-ampere main overcurrent device and a demand load of 130 amperes.

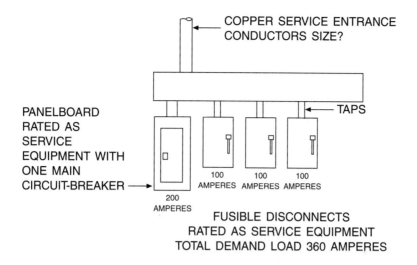

COPPER SERVICE ENTRANCE
CONDUCTORS SIZE?

PANELBOARD
RATED AS
SERVICE
EQUIPMENT WITH
ONE MAIN
CIRCUIT-BREAKER →

TAPS

100
AMPERES

100
AMPERES

100
AMPERES

200
AMPERES

FUSIBLE DISCONNECTS
RATED AS SERVICE EQUIPMENT
TOTAL DEMAND LOAD 360 AMPERES

Figure 4.20   Determine the minimum rating copper service conductors when there are four disconnecting means adding up to 500 amperes and the demand load for the service is 360 amperes.

disconnect switches are installed in a group, as shown in Figure 4.20, and tapped from the service conductors entering the auxiliary gutter. The panelboard and disconnect switches are all rated as suitable for use as service equipment. The minimum size copper service-entrance conductors permitted for this service is:

A.   500 kcmil.   C.   700 kcmil.   E.   800 kcmil.

B.   600 kcmil.   D.   750 kcmil.

Code reference_____

5.   The disconnecting means for a service to a building is:
   A.   required to be located only inside the building near the point of entry of the service conductors.
   B.   permitted to be located on the outside of the building provided it is not more than 10 ft (3 m) from the point where the conductors enter the building.
   C.   permitted to be located on the outside of the building as long as it is located nearest the point where the conductors enter the building.
   D.   required to be located on the outside of the building, or inside, and in either location must be nearest the point of entry of the service conductors.
   E.   permitted to be located outdoors, and not necessarily on the building.

Code reference_____

6.   A commercial office building supplied with a 208Y/120-volt, 3-phase electrical system has a calculated demand load of 127,380 VA. If the ungrounded service conductors terminate at a single main fixed rating circuit breaker, the minimum standard rating permitted for this service is:
   A.   350 amperes.   C.   500 amperes.   E.   700 amperes.
   B.   400 amperes.   D.   600 amperes.

Code reference_____

7.   A single-family dwelling with a total living area of 2100 ft² (195.2 m²) is served with a single-phase, 120/240-volt electrical system. Appliances in the dwelling are a 12-kW electric range and a 5-kW electric clothes dryer operating at 120/240-volts, a 3.5-kW, 240-volt electric water heater, a 1.2-kW, 120-volt dishwasher, an 800-watt, 120-volt built-in microwave, a ¹/₂ horsepower, 120-volt garbage disposer, and a cen-

tral air conditioner with a nameplate rated-load current of 17 amperes at 240 volts. The minimum service demand load for the dwelling using the method of *Article 220, Part III* is:

A. 99 amperes.　　　　C. 123 amperes.　　　　E. 137 amperes.
B. 116 amperes.　　　　D. 133 amperes.

Code reference _____

8. For the single-family dwelling described in the previous question, the ampere rating of the minimum size neutral permitted for the service is: (maximum unbalanced load)

A. 60 amperes.　　　　C. 85 amperes.　　　　E. 116 amperes.
B. 75 amperes.　　　　D. 99 amperes.

Code reference _____

9. A commercial building is supplied with a 120/240-volt, single-phase service with 200 amperes main overcurrent protection and size 3/0 AWG copper ungrounded service conductors as shown in Figure 4.21. The load in the building is primarily 240-volt motors, and the maximum unbalanced load is only 24 amperes. Based upon the use of the building, additional 120-volt loads are not likely. The minimum size copper neutral conductor permitted for this service is:

A. 10 AWG.　　　　C. 6 AWG.　　　　E. 1/0 AWG.
B. 8 AWG.　　　　D. 4 AWG.

Code reference _____

10. A single-family dwelling with a total living area of 2100 ft² (195.2 m²) is served with a single-phase, 120/240-volt electrical system. Appliances in the dwelling are a 12-kW electric range and a 5-kW electric clothes dryer operating at 120/240 volts, a 3.5-kW, 240-volt electric water heater, a 1.2-kW, 120-volt dishwasher, an 800-watt, 120-volt built-in microwave, a 1/2-horsepower, 120-volt garbage disposer, and a central air conditioner with a nameplate rated-load current of 17 amperes at 240 volts. The minimum service demand load for the dwelling using the optional method of *220.82* is:

A. 74 amperes.　　　　C. 89 amperes.　　　　E. 99 amperes.
B. 79 amperes.　　　　D. 93 amperes.

Code reference _____

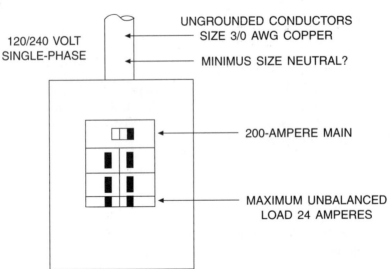

Figure 4.21 Determine the minimum size copper neutral for a single-phase 120/240-volt service with a 200–ampere main, size 3/0 AWG copper ungrounded conductors and a maximum calculated unbalanced load of 24 amperes.

11. A group of farm buildings is supplied from a center distribution point where the metering equipment and the main disconnecting means is located. The calculated demand loads for the individual buildings, shown in Figure 4.22, are listed.

dwelling	74 amperes
barn	66 amperes
shop	36 amperes
cattle barn	24 amperes
storage building	20 amperes

The minimum rating of the ungrounded conductors and equipment at the center distribution point is:

A. 119 amperes. C. 176 amperes. E. 220 amperes.
B. 169 amperes. D. 193 amperes.

Code reference _____

12. An office building is supplied with 208Y/120-volt, 3-phase electrical service. The office space in the building is 3800 sq. ft (353.2 m²). In addition to general lighting, other continuous loads in the building are 3.8 kW of 120-volt outside lighting, and an electric sign figured at the minimum load. The noncontinuous load consists of 68 receptacle outlets. In addition, there are several motors for the air conditioner and air circulation systems that are to be included into the service calculation at 8546 VA. Based upon these loads, the minimum demand load permitted for this office building used to determine the minimum service conductor size is:

A. 35,842 VA. C. 39,086 VA. E. 43,661 VA.
B. 37,966 VA. D. 42,541 VA.

Code reference _____

13. A multifamily dwelling is supplied by 120/240-volt, single-phase power and has 24 individual living units each with an electric range. Eight living units have 10-kW electric ranges, another eight have 12-kW electric ranges, and the remaining eight have 15-kW electric ranges. When the service load is determined using the method

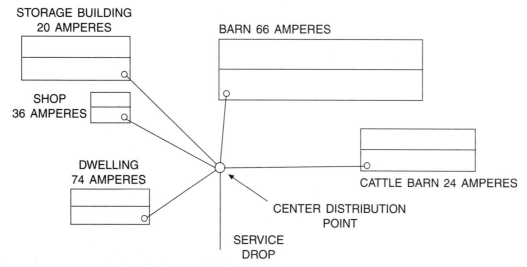

Figure 4.22 Determine the minimum rating of the center distribution conductors and equipment serving farm building with demand loads of 74 amperes for the dwelling, 66 amperes for the barn, 36 amperes for the shop, 24 amperes for the cattle barn, and 20 amperes for the storage building.

of *Article 220, Part III,* the minimum demand load permitted to be included in the service calculation for the building is:

A. 38.20 kVA.   C. 40.95 kVA.   E. 42.90 kVA.
B. 39.00 kVA.   D. 42.32 kVA.

Code reference_____

14. A multifamily dwelling consists of 12 single-family living units, each with a 1.2-kVA dishwasher. When the service load is determined using the method of *Article 220, Part III,* the minimum permitted demand load to be included in the service calculation for the building for the dishwashers is:

A. 10.8 kVA.   C. 12.6 kVA.   E. 18.0 kVA.
B. 11.0 kVA.   D. 14.4 kVA.

Code reference_____

15. A multifamily dwelling is supplied 120/240-volt, single-phase power, and consists of eight individual dwelling units with a living area of 800 sq. ft (74.3 m²). In each living unit is a 10-kW electric range, a 2.5-kW electric water heater, a 1.2-kW dishwasher, 2 kW of electric space heat, and an air conditioner with a nameplate rated-load current of 8 amperes at 240 volts. Laundry facilities are provided in a common area of the building. The total house load to be included in the service demand load calculation is 16,250 VA and includes common area lighting, outside lighting, and laundry facilities. The minimum demand load, using the optional method of *220.84,* permitted for sizing the service conductors and service equipment is:

A. 324 amperes.   C. 370 amperes.   E. 465 amperes.
B. 358 amperes.   D. 420 amperes.

Code reference_____

UNIT 5

Grounding and Bonding

OBJECTIVES

After completion of this unit, the student should be able to:

- explain the purpose of equipment grounding.
- explain the purpose of electrical system grounding, and define bonding.
- diagram a single-phase, dual-voltage, 3-wire electrical system and label the voltages between the wires.
- diagram a 3-phase wye and a 3-phase delta electrical system and label the voltages between the wires.
- state when a single-phase and a 3-phase electrical system are required to be grounded if the voltages between the wires are known.
- show which wire of an electrical system is required to be grounded if the electrical system is one required to be grounded.
- name at least five materials considered by the Code as acceptable as a means of grounding electrical equipment not in a location with specific requirements.
- determine the minimum size equipment grounding conductor permitted for a branch-circuit and feeder if the rating of overcurrent protection is known.
- state the minimum requirement for a grounding electrode for a service entrance to a building.
- determine the minimum size of grounding electrode conductor permitted for a particular service entrance.
- specify the type required and the minimum size permitted of bonding conductor for a swimming pool.
- answer wiring installation questions relating to *Articles 250, 280, 285*, and *680*.
- state at least five significant changes that occurred from the 2002 to the 2005 Code for *Articles 250, 280, 285*, and *680*.

CODE DISCUSSION

The emphasis of this unit is to discuss the purpose of equipment grounding and of system grounding. Equipment grounding is easily understood if the electrical installer has a clear understanding of the purpose of equipment grounding, an understanding of the circuit involved, and an understanding of the requirements of a good equipment grounding conductor. This unit deals with grounding and bonding of electrical systems in general. There are several locations or types of materials with special grounding requirements. These cases will be discussed in later units. For example, requirements for cable tray to be used as an equipment grounding conductor will be covered in *Unit 12*, grounding of electrical equipment in agricultural buildings will be covered in *Unit 14*, hazardous locations in *Unit 9*, and health care facilities in *Unit 10*. Only the specific grounding and bonding requirements different from *Article 250* are covered in other articles of the Code.

Article 250 deals with the grounding of an electrical system and the grounding of equipment. The requirements of this article apply to all wiring installations, unless either specifically different or additional grounding requirements are covered elsewhere in the Code. *NEC® 250.4* references the other locations in

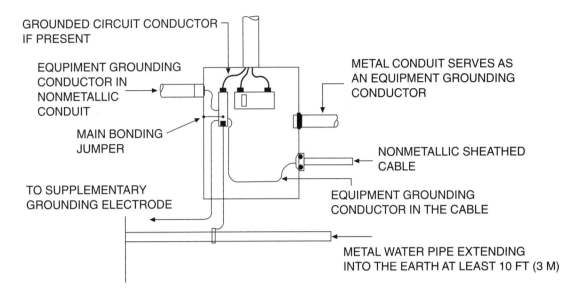

GROUNDED CIRCUIT CONDUCTOR
IF PRESENT

EQUIPMENT GROUNDING
CONDUCTOR IN
NONMETALLIC
CONDUIT

METAL CONDUIT SERVES AS
AN EQUIPMENT GROUNDING
CONDUCTOR

MAIN BONDING
JUMPER

NONMETALLIC SHEATHED
CABLE

TO SUPPLEMENTARY
GROUNDING ELECTRODE

EQUIPMENT GROUNDING
CONDUCTOR IN THE CABLE

METAL WATER PIPE EXTENDING
INTO THE EARTH AT LEAST 10 FT (3 M)

Figure 5.1   System grounding and equipment grounding are required to be considered in all electrical systems.

the Code where there are different or additional grounding requirements for specific materials or locations. System grounding and equipment grounding are illustrated in Figure 5.1. The purpose of electrical system grounding and equipment grounding is explained in *250.4*. *Part IV* of *Article 250* requires that exposed metallic parts of equipment likely to become energized are required to be grounded. Methods of equipment grounding are explained in *Part VII* of *Article 250* and acceptable equipment grounding conductors are described in *Part VI* of *Article 250*, in particular in *250.118*.

The metal enclosure of the service disconnect or a grounding bus at that location is the collecting point for all feeder and branch-circuit equipment grounding conductors. The basic standard for equipment grounding is explained in *250.2, 250.4(A)(5)*, and *250.4(B)(4)*. An electrically continuous and permanent path is required to be established to allow enough current to flow during a fault to operate overcurrent devices such as circuit breakers and fuses. To accomplish this task, all equipment likely to become energized must be permanently grounded back to a common point. If a ground-fault occurs, current tries to return to the grounded terminal of the electrical supply of a grounded electrical system. That point is the grounded terminal of a transformer, generator, or other power source. In the case of a grounded electrical supply to a building, the grounded service conductor entering the building is bonded or connected to the metal enclosure of the service disconnect or equipment grounding bus. This **main bonding jumper** is the connecting link between the equipment grounding conductors in the building and the grounded terminal of the electrical supply to the building. The current path during a ground fault in a building supplied by a grounded electrical system is illustrated in Figure 5.2.

Some electrical systems are not grounded. A 3-phase, 3-wire, 480-volt electrical system may not be grounded. *NEC® 250.4(B)(1)* and *250.24(D)* require a grounding electrode to be established for an ungrounded electrical system and connected to the service disconnect enclosure. All equipment served by the ungrounded electrical system is grounded the same as if the system was grounded. The only difference is that there is no grounded circuit conductor with the service and there will be no main bonding jumper.

Grounding an electrical system to the earth has a different purpose than equipment grounding. *NEC® 250.4(A)(1)* states that the purpose is to stabilize voltages to ground during normal operation and to limit voltages due to lightning, line surges, and unintentional contact with high voltage conductors. There are cases where grounding is required and some where grounding is optional. These rules are discussed and illustrated in detail later in this unit. Which conductor is required to be grounded is described in *250.26*.

Part III of *Article 250* describes the types of grounding electrodes that are acceptable. All of the grounding electrodes described that are available at a building are required to be bonded together and used as a grounding electrode for the building electrical service. In many cases, there are no grounding electrodes available and one must be established. *NEC® 250.52(A)(1)* requires that a metal underground water pipe that is in contact with the earth for a minimum of 10 ft (3 m) be used as a grounding electrode for the service.

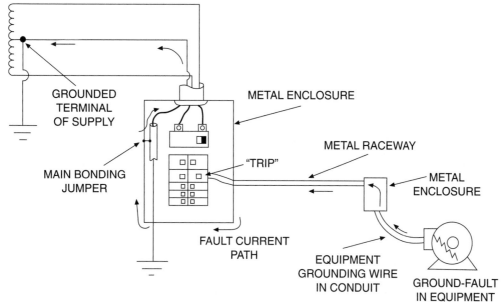

Figure 5.2   Ground-fault current returns to the grounded terminal of a grounded electrical system, and must be provided a low-resistance metallic path from all equipment back to the metal enclosure or grounding bus of the service disconnect.

The connection to the water pipe is required to be made within the first 5 ft (1.5 m) of the point where the water pipe enters the building. That connection is frequently permitted to be made a greater distance from the point of entry for many commercial and industrial buildings. *NEC®250.53(D)(2)* requires a supplemental grounding electrode be established when using the metal water pipe as a grounding electrode. That would mean that if the metal water pipe was the only grounding electrode, then an additional grounding electrode would be required to be installed. Generally it is easiest to add a ground rod as the supplemental grounding electrode. When a ground rod is installed, it is required to have a resistance-to-earth of not more than 25 ohms. If that is not the case, then an additional electrode is needed. Generally it is easiest to add a second ground rod which is required by *250.56* to be located a minimum of 6 ft (1.8 m) from the first ground rod. This type of installation is shown in Figure 5.3.

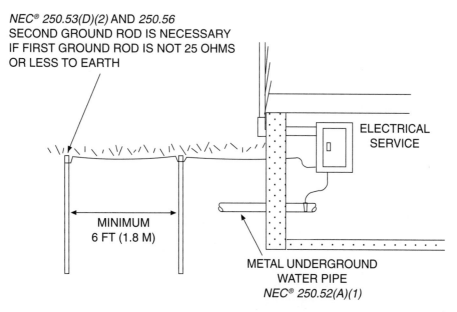

Figure 5.3   When a grounding electrode is installed as a supplementary grounding electrode for a water pipe electrode, if the ground rod does not have a resistance of 25 ohms or less to earth, it is required to be supplemented with one additional electrode.

The metal frame of the building is also a common grounding electrode for many commercial and industrial buildings. The stipulation is that the metal building frame must be effectively grounded. Simply bolting the steel frame to concrete pilings that extend above ground is not sufficient. The concrete is generally too dry to be a satisfactory conductor. Acceptable means of making earth connections for the metal building frame are listed in *250.52(A)(2)*. If the mounting bolts are bonded to the reinforcing steel in the piling and the reinforcing steel extends down to a minimum of 3 ft (900 mm) below grade level, the steel frame is generally considered to be effectively grounded. *NEC® 250.52(A)(3)* permits bare or zinc galvanized steel reinforcing bars placed near the bottom of a building footing to serve as a grounding electrode if the length is not less than 20 ft (6 m). Steel tie wires used for the purpose of connecting sections of reinforcing steel is considered to be an adequate bonding means when measuring the 20 ft (6 m) minimum distance. One continuous length is not required.

The minimum size of grounding electrode conductor is specified in *250.66*. The size of the largest ungrounded conductor is used as the basis for determining the minimum grounding electrode conductor size from *Table 250.66*. Grounding electrodes generally have a resistance-to-earth ranging from several ohms to hundreds of ohms depending upon the soil conditions. *Table 250.66*, therefore, sets size 3/0 AWG copper as the maximum size grounding electrode conductor required in any situation. If the grounding electrode is a ground rod, the maximum grounding electrode conductor size required is size 6 AWG copper, *250.66(A)*. In the case of a concrete-encased electrode such as a reinforcing steel in a footing, the maximum required is size 4 AWG copper, *250.66(B)*.

A service is unique in that the conductors are protected from overcurrent at the load end of the conductors. For this reason, it is important that any metal service enclosure or raceway be electrically connected in such a way that enough fault-current may flow to open the utility primary overcurrent device. As a result, extra ordinary bonding is required for service equipment and raceways. Figure 5.4 is a diagram of a typical service panel with Code references where details on bonding and grounding are found. *NEC® 250.28(D)*

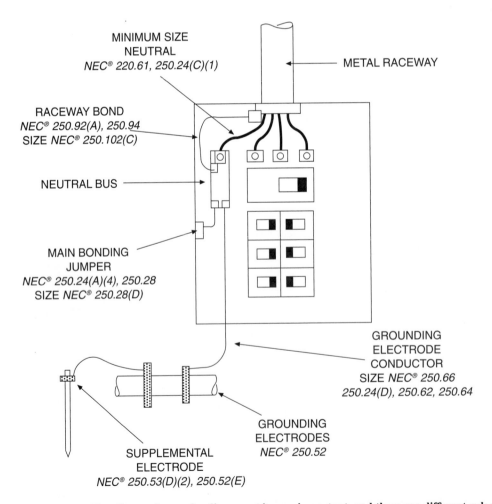

Figure 5.4   Grounding and bonding at the service disconnect is very important, and there are different rules for sizing the different components of the grounding system at the service.

specifies the minimum size of the main bonding jumper at not less than the minimum size of the grounding electrode conductor as determined from *Table 250.66*. If the ungrounded conductor size is larger than the maximum given in *Table 250.66*, then the minimum main bonding jumper size is determined by multiplying the cross-sectional area of the ungrounded conductor by 0.125. *NEC® 250.102(C)* specifies the minimum size bonding jumper for a metal service raceway. The rule is the same as for the main bonding jumper except it is based upon the cross-sectional area of the largest ungrounded conductor in the service raceway. When the service conductors are run in parallel in separate raceways, the bonding jumper for the metal raceways will be smaller than the main bonding jumper.

  NEC® 250.32 deals with supplying power to one building from another building. Power can be extended to the second building either overhead or underground. Rules for installation of the conductors are found in *Article 225*. Much of the confusion centers around grounding and whether an equipment grounding conductor separate from the neutral is required. The rules are summarized in Figure 5.5, which illustrates

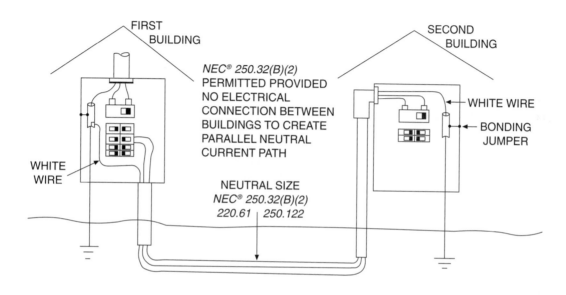

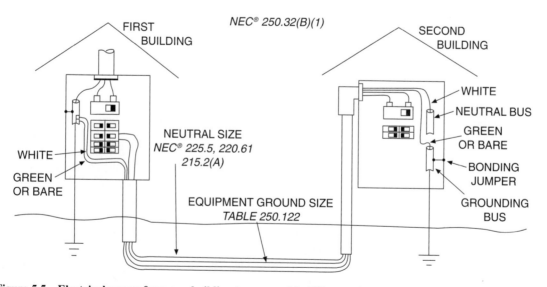

Figure 5.5   Electrical power from one building to a second building on the same property is permitted to be by means of a feeder with the grounded conductor acting both as the neutral and the equipment grounding conductor as in the top diagram, or with a separate neutral and equipment grounding conductor as in the bottom diagram.

a single-phase, 120/240-volt, 3-wire electrical service to the property. The rules would be the same for a 3-phase, 4-wire electrical system with one additional ungrounded conductor. The first building is considered to be the service to the property. *NEC® 250.32(B)(2)* permits the supply to the second building to consist of two ungrounded conductors and a neutral conductor, with the neutral conductor grounded and bonded to the service enclosure at the second building. The minimum neutral conductor size is determined from the appropriate allowable ampacity table in *310.15* depending upon the calculation of unbalanced load according to *220.61*. The neutral conductor is not permitted to be sized smaller than the size given in *Table 250.122* based upon the rating of the feeder overcurrent device.

The neutral is permitted to serve also as the equipment grounding conductor provided there is no metal connection between the two buildings that could act as a parallel path for neutral current. If the property is served with a metal underground water-piping system, also used as the grounding electrode for both buildings, neutral current would flow on the metal water pipe between the two buildings as well as on the neutral conductor. If this was the case, then it is not permitted to use the neutral conductor as the equipment grounding conductor and ground the neutral at the second building. The method of *250.32(B)(1)* is then required.

For the method of *250.32(B)(1)*, a neutral conductor and a separate equipment grounding conductor are run to the second building, also illustrated in Figure 5.5. A separate neutral bus and equipment grounding bus are installed in the panel in the second building. The neutral conductor is *not* bonded at the second building and it is *not* connected to a grounding electrode at the second building. Because the neutral and equipment grounds are separate in the second building, neutral current is forced to flow only on the neutral conductor. The minimum size of the neutral conductor is determined from the appropriate allowable ampacity table in *310.15* depending upon the calculation of unbalanced load according to *220.61*. This is a feeder; therefore, the minimum size equipment grounding conductor is found in *Table 250.122* using the rating of the feeder overcurrent device.

The rules for distributing power to the various buildings on a farm are slightly different than described in *250.32* due to several special considerations. Power on a farm is frequently supplied to buildings from a central distribution point in the yard where only a disconnect is required. The feeders to the various buildings may not have individual overcurrent protection. The conductors are required to be installed according to the rules in *Article 225*, while minimum neutral and grounding conductor size is specified in *547.9*.

NEC® 250.30 covers the rules with respect to grounding of a separately derived alternating current electrical system. A transformer installed within a building is a common example of a separately derived system. This subject is discussed in detail in *Unit 8*.

Article 280 covers specifications for and the installation of a surge arrester installed on an electrical system. These requirements do not apply to a surge arrester associated with a primary electrical distribution system of an electrical power supplier. The *National Electrical Safety Code®* covers these installations.

High voltages can be induced into outdoor electric lines during lightning storms. These voltages produce surges of high voltage that travel along the lines. These high-voltage surges can enter buildings and equipment and cause damage. Surge arresters are used to limit these voltages to a safe level by providing a path to ground to dissipate the energy. A surge arrester is connected between an ungrounded conductor and the earth. The connection to earth is frequently made by connecting the surge arrester to the neutral conductor.

Here is how a surge arrester works. When the voltage on the ungrounded conductor is normal, the surge arrester has a high resistance and acts like an open switch preventing current flow through the surge arrester to the neutral conductor. A surge voltage or impulse voltage is a sudden short duration large rise in voltage, usually positive or negative, but not both. Lightning striking near a power line is a frequent cause of surge or impulse voltages, which travel in both directions along a wire from the point of origin. When a surge or impulse reaches a surge arrester, the device must act like a closed switch to the voltage, higher than normal, allowing the voltage to dissipate by passing current through the surge arrester to earth. Once the voltage on the ungrounded conductor returns to normal, the surge arrester must once again act like a high resistance, preventing current from flowing to ground at normal system voltage. Voltage surges or impulses are frequently only a few microseconds in duration, therefore, the surge arrester must be fast acting. It must also be capable of passing the energy of the surge so it will survive to be ready to repeat the process. A typical device that can provide this function for power systems in buildings is a low-voltage metal-oxide varistor (MOV). The material has a high resistance at normal operating voltages, but switches to a low resistance when the voltage across the MOV rises above a critical value. The MOV conducts current until the voltage drops below the critical value and the resistance once again goes very high to prevent current flow.

A surge arrester is installed between the ungrounded conductors and the grounded conductor or earth usually near the point where service or feeder conductors enter the building. Surge arresters installed on electrical systems operating at not more than 1000 volts are required to be listed *280.4(A)*. They are permitted to be installed indoors or outdoors. Figure 5.6 shows a metal-oxide varistor-type surge arrester (MOV)

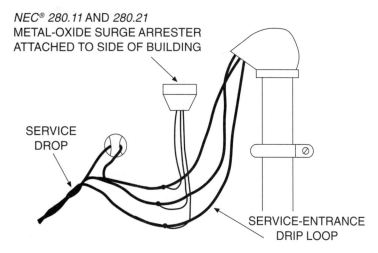

NEC® 280.11 AND 280.21
METAL-OXIDE SURGE ARRESTER
ATTACHED TO SIDE OF BUILDING

SERVICE
DROP

SERVICE-ENTRANCE
DRIP LOOP

Figure 5.6   A surge arrester, which provides lightning protection, is permitted to be installed outside the building at a service-entrance drip loop.

installed at the servic-entrance drip loop of a building. For an ungrounded electrical system, the surge arrester is connected to the grounding electrode conductor for the service disconnecting means. If the surge arrester grounding conductor is run in metal raceway, *280.25* requires the grounding conductor to be bonded to the raceway at the point where it enters and exits the raceway.

Article 285 covers the installation of transient voltage surge suppressors (TVSS) on electrical systems. This article does not apply to portable plug-in TVSS devices. Transient voltage surge suppressors are only permitted to be installed on grounded electrical systems operating at under 600 volts. A TVSS is not permitted to be installed on an ungrounded electrical system, because under fault conditions within the building, the TVSS can be exposed to an excessive overvoltage. They are permitted to be installed on any circuit within a building but they are always required to be installed on the load side of the first overcurrent device for the conductors entering the building. A TVSS is permitted to be installed either inside or outside a building but always on the load side of the first overcurrent device. Figure 5.7 shows a typical TVSS installation in a building with a device at the main service and one at a down-line panelboard. A voltage surge or impulse originating on a branch-circuit would travel back to the panelboard and would be conducted to ground at the panelboard by the TVSS, thus preventing the voltage impulse from feeding to other circuits supplied from that panelboard.

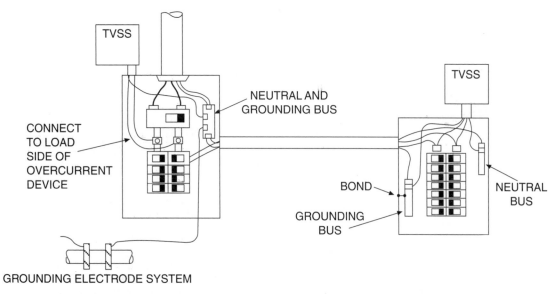

TVSS

TVSS

NEUTRAL AND
GROUNDING BUS

CONNECT
TO LOAD
SIDE OF
OVERCURRENT
DEVICE

BOND

NEUTRAL
BUS

GROUNDING
BUS

GROUNDING ELECTRODE SYSTEM

Figure 5.7   Transient voltage surge suppressors (TVSS) are required to be installed on the load side of the first overcurrent device of the building, and they help protect sensitive electrical equipment from voltage surges and impulses.

The basic function of a transient voltage surge suppressor and a surge arrester are similar, except a TVSS begins to become conductive at a voltage only slightly higher than the ungrounded conductor operating voltage. Like a surge arrester, the TVSS is generally connected between the ungrounded conductor and the neutral or equipment grounding conductor. However, a transient voltage surge suppressor is permitted to be installed between ungrounded conductors. The device becomes conductive when a surge voltage or an impulse voltage develops on the ungrounded conductors. That surge or impulse voltage is not necessarily due to lightning. It may be due to operation of switches and other equipment within or near the building. The TVSS is generally installed to protect sensitive electronic equipment from over voltages. Because the TVSS goes into conduction mode and passes surge current much more easily than a surge arrester, it must be installed on the load side of an overcurrent device to provide it with backup protection.

NEC® 285.6 requires a transient voltage surge suppressor to be marked with a fault-current rating. The available fault current at the point of installation is not permitted to be in excess of the fault-current rating of the TVSS. A transient voltage surge suppressor can be a one-port device that is tapped to the ungrounded conductors, the grounded conductor, and the equipment grounding conductor. With a one-port device, the load current does not pass through the TVSS. With the two-port TVSS, there is an input port and an output port. The load current does pass through the device and the TVSS must have a short-circuit current rating.

Article 680 covers the wiring of equipment associated with swimming pools, fountains, and similar equipment. Requirements are placed on the wiring installed in the area of pools, fountains, and similar installations. The types of installations covered are swimming, wading, therapeutic and decorative pools; fountains; hot tubs; spas; and hydromassage bathtubs. The terms **pool** and **fountain** are defined in *680.2*. Definitions important to the installation of equipment in these areas are covered in *680.2*.

Differences in voltage between two places around the pool, fountain, or similar installation may be of a sufficient level to create a hazard. To help prevent hazardous conditions from developing, metal parts of the swimming pool are required by *680.26* to be bonded together with a solid copper wire not smaller than size 8 AWG. Metal raceway, metal-sheathed cable, metal piping, and other metal parts fixed in place within 5 ft (1.5 m) horizontally of the inside edge of the pool shall also be bonded together. All metal parts shall be bonded to an equipotential bonding grid. This is illustrated in Figure 5.8. The specifications for an equipotential bonding grid are not specifically stated in the Code in *680.26*, which states that structural reinforcing steel of a concrete pool or the wall of a bolted or welded metal pool is permitted to serve as the required equipotential bonding grid.

NEC® 680.26 PERMANENT SWIMMING POOL
BONDING TO FORM AN EQIPOTENTIAL BONDING GRID

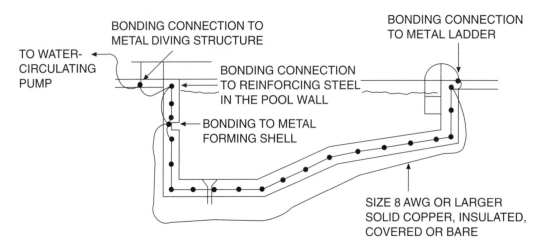

Figure 5.8   Metal parts of a permanent swimming pool are required to be bonded together to reduce the likelihood of voltage gradients in the pool area.

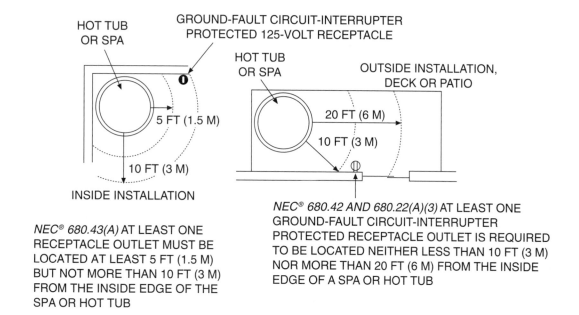

HOT TUB
OR SPA

GROUND-FAULT CIRCUIT-INTERRUPTER
PROTECTED 125-VOLT RECEPTACLE

HOT TUB
OR SPA

OUTSIDE INSTALLATION,
DECK OR PATIO

5 FT (1.5 M)

20 FT (6 M)

10 FT (3 M)

10 FT (3 M)

INSIDE INSTALLATION

NEC® 680.43(A) AT LEAST ONE
RECEPTACLE OUTLET MUST BE
LOCATED AT LEAST 5 FT (1.5 M)
BUT NOT MORE THAN 10 FT (3 M)
FROM THE INSIDE EDGE OF THE
SPA OR HOT TUB

NEC® 680.42 AND 680.22(A)(3) AT LEAST ONE
GROUND-FAULT CIRCUIT-INTERRUPTER
PROTECTED RECEPTACLE OUTLET IS REQUIRED
TO BE LOCATED NEITHER LESS THAN 10 FT (3 M)
NOR MORE THAN 20 FT (6 M) FROM THE INSIDE
EDGE OF A SPA OR HOT TUB

Figure 5.9   At least one ground-fault circuit-interrupter protected 125-volt receptacle outlet for an inside instal-lation is required to be installed at least 5 ft (1.5 m), but not more than 10 ft (3 m) from the inside edge of an inside spa or hot tub, and between 10 ft (3 m) and 20 ft (6 m) for an outside spa or hot tub.

At least one 125-volt receptacle outlet, not for the water-circulating pump, is required to be installed not more than 20 ft (6 m) and not closer than 10 ft (3 m) from the inside edge of a dwelling permanent swimming pool as stated in *680.22(A)(3)*. *NEC® 680.22(A)(6)* covers the situation in which a receptacle out-let beyond a barrier or wall is within 20 ft (6 m) of the edge of a permanent swimming pool. All 125-volt receptacle outlets within 20 ft (6 m) of the inside edge of the pool shall be protected by a GFCI. These same rules apply in the case of a spa or hot tub installed on the outside of a building, as stated in *680.42* and illus-trated in Figure 5.9. In the case of a spa or hot tub installed on the inside of a building, at least one 125-volt rated receptacle outlet, which is required to be GFCI protected, is required to be installed at a distance of not more than 10 ft (3 m) from the inside edge of the spa or hot tub, and not closer than 5 ft (1.5 m).

GROUNDING AND BONDING FUNDAMENTALS

Grounding and bonding is one of the most important parts of an electrical wiring system, and at the same time it is one subject of the Code that seems to be confusing. It is essential to understand the purpose of grounding and bonding, and then it will be easier to make sure the installation has met the requirements of the Code. Grounding and bonding is the key element to the backup safety system for an electrical circuit or electrical equipment. If the grounding system does not function properly, then other safety devices such as circuit breakers and fuses may not function when a problem develops. Grounding and bonding must be installed with as much care as any other part of the electrical system. One Code section in particular dis-cusses important information necessary to understand system and equipment grounding and bonding. Read *250.4* to get an explanation of the purpose of grounding.

Equipment Grounding

The equipment grounding conductor is permitted to be a metal raceway or a metal box, cabinet, or frame of equipment. These and other permitted means of providing an equipment grounding conductor are covered in *250.118*. The equipment grounding conductor is permitted to be copper or aluminum wire. There is a restriction in the case of aluminum wire used for grounding. Aluminum as an equipment grounding wire that is not insulated or covered is not permitted to be installed where it is in direct contact with masonry or the earth, or where subject to corrosive conditions. When aluminum wire is exposed and used outside for grounding, it is not permitted to be installed within 18 in. (450 mm) of the earth. These restrictions on aluminum and copper-clad aluminum conductors are found in *250.120* and *250.64.*

Equipment grounding wires are permitted to be solid or stranded, insulated or bare. If the wire is insulated, it is required to have insulation that is green or green with yellow stripes. These requirements are found in *250.119,* and for flexible cords, *400.23.* If the equipment grounding wire is size 4 AWG or larger, it is permitted to be identified as an equipment grounding wire at the time of installation at every location where it is accessible. Acceptable means of identifying an equipment grounding conductor of size 4 AWG or larger that is not bare, green, or green with yellow stripes are: (1) to strip the insulation from the entire exposed wire, leaving the wire bare, or (2) to cover the entire exposed wire with green tape or green paint.

All fittings, conduit, splices, and any other connections in the equipment grounding conductor shall be made up tight with proper tools. Flexible Metal Conduit is not considered to be as effective an equipment grounding conductor as Rigid Metal Conduit, Intermediate Metal Conduit (IMC), or Electrical Metallic Tubing (EMT). Therefore, limitations are placed on the use of Flexible Metal Conduit and Liquidtight Flexible Metal Conduit for use as an equipment grounding conductor. These restrictions are given in *250.118.* If these requirements are not satisfied, then an equipment grounding wire shall be provided to bond from the enclosure at one end of the Flexible Metal Conduit to the enclosure at the other end of the Flexible Metal Conduit.

There are several methods of installing and terminating an equipment bonding jumper from one end to the other of flexible conduit. Installation of equipment bonding jumpers is covered in *250.102.* The equipment bonding or grounding jumper is permitted to be run on the inside of a raceway, or it is permitted to be run on the outside. If run on the outside, the equipment bonding jumper is not permitted to be more than 6 ft (1.8 m) long. The bonding jumper shall be routed with the raceway or enclosure. Special fittings are manufactured for terminating the bonding wire on the outside of the Flexible Metal Conduit and Liquidtight Flexible Metal Conduit. Various methods are acceptable for terminating an equipment bonding jumper or an equipment grounding wire at an enclosure.

The methods permitted to terminate an equipment grounding conductor or an equipment bonding jumper to an enclosure are covered in *250.8.* Connecting an equipment grounding conductor to a metal box or enclosure is required to be done with a device listed for the purpose or by means of a grounding screw used for no other purpose. Sheet metal screws are not permitted to be used to connect equipment grounding conductors to boxes and enclosures. When an equipment grounding conductor is terminated at a lug in an enclosure, the screw or bolt used to connect the lug to the enclosure shall be used for no other purpose. Connections or fittings that depend on solder are not permitted. This does not prevent the use of solder at a connection where the connection is made mechanically secure before soldering. *NEC® 250.148* requires that the removal of a device at a box or enclosure shall not interrupt the continuity of the equipment grounding conductor.

Sizing Equipment Grounding Conductors

An equipment grounding wire size is based on the size of the overcurrent device protecting the circuit. *Table 250.122* in the Code lists the minimum wire size permitted for various overcurrent device ratings. In the case of electrical cable, the manufacturer has installed the correct size of equipment grounding wire based on the maximum size overcurrent device rating permitted to be used for that cable. When raceway is permitted as an equipment grounding conductor, cross-sectional area of the metal is adequate for the largest size wires permitted in the raceway. Usually, the only time when an equipment grounding wire must be sized is in the case of nonmetallic conduit or for bonding jumpers. The following examples illustrate how to determine the minimum permitted size of equipment grounding conductor for a branch-circuit or a feeder.

Example 5.1   Determine the minimum size copper equipment grounding conductor for a 30-ampere circuit consisting of Type THWN copper wires size 10 AWG installed in Rigid Nonmetallic Conduit.

Answer: From *Table 250.122,* the minimum size copper equipment grounding conductor permitted is 10 AWG.

Example 5.2   A 7-ft (2.13-m) length of Liquidtight Flexible Metal Conduit extends from a fusible disconnect switch to an electric furnace. The wires inside the flexible conduit are Type THHN copper size 4 AWG with 70-ampere overcurrent protection. Determine the minimum size copper equipment grounding conductor to bond around the Liquidtight Flexible Metal Conduit. There is an equipment grounding lug in the furnace for terminating the equipment grounding conductor.

Answer: A 70-ampere overcurrent device is not listed in *Table 250.122*; therefore, the next larger overcurrent device rating shall be used. For this example, use the equipment grounding wire size required for a 100-ampere overcurrent device. The minimum size copper equipment grounding wire size for this 70-ampere circuit is size 8 AWG.

Parallel Equipment Grounding Conductors

In *310.4,* conductors are permitted to be run in parallel. In general, the minimum conductor size permitted to be installed in parallel is 1/0 AWG. This minimum size applies to ungrounded conductors and to grounded conductors. There is no mention of equipment grounding conductors, bonding conductors, or grounding electrode conductors. *NEC® 250.122(F)* covers the situation where conductors are run in parallel. When the conductors are run in parallel sets within more than one raceway where an equipment grounding conductor is required, the equipment grounding conductors are required to be run in parallel. Each of the separate equipment grounding conductors required is determined using *Table 250.122* and based on the size of the circuit overcurrent device rating. Example 5.3 illustrates how the size of parallel equipment grounding conductors is determined.

Example 5.3   A feeder protected with 500-ampere time-delay fuses is run with two parallel sets of conductors in separate Rigid Nonmetallic Conduits as shown in Figure 5.10. Determine the minimum size of copper equipment grounding conductor required in each conduit run of the feeder.

Answer: A separate equipment grounding conductor is required to be run with each set of conductors. Each of the equipment grounding conductors is required to be sized according to *Table 250.122*. The minimum size copper equipment grounding conductor required for a 500-ampere overcurrent device is 2 AWG.

Equipment Grounding Conductor Size Adjustments

Equipment grounding conductors are sized in the Code to be capable of conducting sufficient current during fault conditions to open the circuit or feeder overcurrent device. The longer the circuit conductors, the higher the resistance of the equipment grounding conductor, and the more difficult it becomes to conduct sufficient current to open the overcurrent device. If a circuit or feeder is long enough to require an increase in ungrounded wire size to limit voltage drop, then a corresponding adjustment in equipment grounding

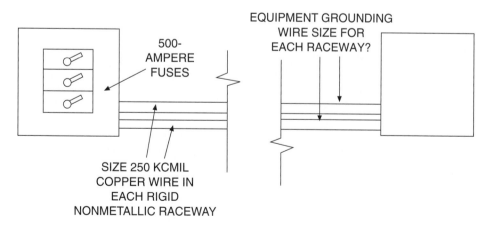

Figure 5.10   When conductors run in parallel in separate raceways require an equipment grounding conductor run with the circuit conductors, the equipment grounding conductor in each raceway is sized based upon the rating of the branch-circuit or feeder.

conductor size seems appropriate. *NEC® 250.122(B)* requires an adjustment in equipment grounding conductor cross-sectional area in proportion to the cross-sectional area increase of the ungrounded conductors. Equation 5.1 can be used to determine the adjusted equipment grounding conductor size.

Adjusted equipment grounding conductor size =
$$\textbf{Area of minimum grounding wire size} \times \frac{\textbf{Adjusted ungrounded wire area}}{\textbf{Minimum ungrounded wire area}} \qquad \textbf{Eq. 5.1}$$

To determine the adjusted equipment grounding conductor size, start by finding the minimum size equipment grounding conductor for the circuit from *Table 250.122*. Next, look up the minimum ungrounded conductor size required for the branch-circuit or feeder. Assuming no adjustment factors apply, look up the minimum size conductor from *Table 310.16* in most cases. A larger size ungrounded conductor is actually being used to limit the voltage drop of the circuit. From *Table 8* in *Chapter 9*, look up the cross-sectional area of all of these conductors and put them into Equation 5.1 to determine the adjusted equipment grounding conductor size. Use *Table 8* again to convert the equipment grounding conductor area into an AWG size. The following example will illustrate how this process works:

Example 5.4   A 3-phase, 480-volt feeder is protected by a 100-ampere overcurrent device. The ungrounded conductors of the feeder are size 1/0 AWG copper to limit the voltage drop on the feeder. It will be assumed conductor insulation and terminations are 75°C rated. If the feeder is run in Rigid Nonmetallic Conduit, determine the minimum size copper equipment grounding conductor permitted.

Answer: First look up the minimum copper equipment grounding conductor size permitted for this feeder in *Table 250.122*. Next, look up the minimum ungrounded conductor size permitted for the feeder in *Table 310.16*, assuming voltage drop is not an issue. The minimum is size 3 AWG. Now look up the cross-sectional area of the conductors in *Table 8*. Then calculate the adjusted equipment grounding conductor area using Equation 5.1. The result will be 33,133 cmil (16.79 mm^2), which is a size 4 AWG from *Table 8*.

size 8 AWG	16,510 cmil	8.37 mm^2
size 3 AWG	52,620 cmil	26.67 mm^2
size 1/0 AWG	105,600 cmil	53.49 mm^2

$$\text{Adjusted eqpt. grnd. conductor size} = 16,510 \text{ cmil} \times \frac{105,600 \text{ cmil}}{52,620 \text{ cmil}} = 33,133 \text{ cmil}$$

$$\text{Adjusted eqpt. grnd. conductor size} = 8.37 \text{ mm}^2 \times \frac{53.49 \text{ mm}^2}{26.67 \text{ mm}^2} = 16.79 \text{ mm}^2$$

Electrical Shock

Electrical shock is caused by the flow of electrical current through the body or a part of the body. Current passing through the body can cause burns, muscle reaction, and injury to body organs. If the current travels through the head or central part of the body, severe injury or even death may occur. Dry skin has a high resistance to the flow of electricity. With under 15 volts, enough current generally will not flow through the body to be harmful to a human. This is not necessarily true for a human in wet conditions or for an animal standing in a moist environment. The moist hoof or foot of a farm animal offers low resistance to the flow of electrical current. The main factors involved in electrical shock to an animal or human are:

- voltage
- duration of the electrical shock
- condition of skin or body surface
- surface area contact with the source
- path the current takes

Voltage and condition of the skin usually determine the amount of current that flows. A normal person usually will just begin to feel the sensation of shock if $1/1000$ ampere or 1 milliampere of 60 hertz alternating current flows through the body. Shock may become painful to humans with a continuous flow of 8 milliamperes or more. Muscular contraction (cannot let go) may occur in humans with a continuous current flow of 15 milliamperes or more. Greater flows of current through the body are usually very serious. An effective equipment grounding system is important for the protection of people and animals with the damp and wet conditions that exist in many residential, commercial, and farm locations.

Electrical System Grounding

A building wiring system usually must have one conductor grounded to the earth. This grounding helps limit voltages caused by lightning surges. Grounding prevents high voltages from occurring on the wiring if a primary line should accidentally contact a secondary wire. Grounding limits the maximum voltage to ground from the hot wires. For most residential, commercial, and farm wiring systems, the maximum voltage to ground is not more than 125 volts. Several decisions must be made in the grounding of an electrical system to the earth. Three major decisions are which wire is permitted to be grounded, the type of grounding electrode, and determining the minimum permitted size of grounding electrode conductor.

Which Conductor of the Electrical System to Ground

The service-entrance wire required to be grounded is described in *250.20(B)* for an alternating current system operating between 50 volts and 1000 volts. The grounded-circuit conductor shall have white or gray insulation. Identification of the grounded-circuit conductor is covered in *200.6*. If the wire is larger than size 6 AWG, the wire is permitted to be labeled with white or gray tape or paint where the wire is visible. The following electrical systems are required to be grounded:

- Single-phase, 2-wire, 120 volts
- Single-phase, 3-wire, 120/240 volts
- Three-phase, 4-wire, 208/120 volts, wye
- Three-phase, 4-wire, 240/120 volts, delta
- Three-phase, 4-wire, 480/277 volts, wye, when supplying phase to neutral loads
- Three-phase, 4-wire, 600/347 volts, wye, when supplying phase to neutral loads

Grounding Electrode

The grounding electrode is the means by which electrical contact is made with the earth. The type of electrode permitted to be used for grounding an electrical system is given in *250.52*. The following grounding electrodes are covered in *250.52* and shall be used if available:

- Metal underground water pipe in direct contact with the earth for at least 10 ft (3 m). A metal underground water pipe shall be supplemented by at least one additional electrode.
- The metal frame of the building, where the metal frame is effectively grounded
- A bare size 4 AWG or larger copper wire or reinforcing steel at least $1/2$ in. (13 mm) in diameter at least 20 ft (6 m) long and encased in concrete in contact with the earth, with at least 2 in. of concrete around all sides of the wire or steel
- A bare copper wire circling the building at a depth of at least $2^1/2$ ft (750 mm), size 2 AWG or larger, and at least 20 ft (6 m) long
- A rod or pipe electrode driven to a depth of at least 8 ft (2.5 m) into the soil. If rock bottom is encountered, it shall be permitted to drive the pipe or rod at an angle of not less than 45° from the horizontal. Or it may be laid in a trench at least $2^1/2$ ft (750 mm) deep. The acceptable rods and pipes shall have the following minimum diameters: (1) trade size $3/4$ (21) galvanized steel pipe, (2) $5/8$-in. (16 mm) iron or steel rod, and (3) $1/2$-in. (13 mm) copper-coated rod
- A plate electrode shall be used in areas where soil conditions prevent the use of a pipe or rod electrode. The plate shall make contact with at least 2 sq ft (0.186 m²) of soil.
- Other metal underground structures or equipment such as a metal well casing

Rod, pipe, and plate electrodes sometimes do not make good low-resistance contact with the earth. They must be placed in areas not subject to damage but where the soil is most likely to be moist. They should not be installed in areas protected from the weather, such as under roof overhangs. The Code specifies in *250.56* that a single rod, pipe, or plate electrode shall have a resistance to ground not exceeding 25 ohms. If the resistance to earth is greater than 25 ohms, then one additional rod, pipe, plate, or other electrode shall be installed. Rod, pipe, or plate electrodes shall be at least 6 ft (1.8 m) apart.

Lightning-protection system grounding electrodes are not permitted to be used as one of the required electrodes for the electrical system, *250.60*. However, electrodes for the electrical and lightning systems shall be electrically connected together, as specified in *250.106*.

Grounding and Bonding for Service with Parallel Conductors

Several components of the grounding and bonding system must be sized when a service entrance consists of two or more parallel sets of service-entrance conductors. The grounding and bonding conductors may all be different sizes as illustrated in Figure 5.11. The minimum permitted size of grounding electrode conductor is found in *250.66*. When there are two or more parallel sets of service-entrance conductors, the minimum size of grounding electrode conductor is determined based on the sum of the cross-sectional areas of the ungrounded conductors for any one particular leg or phase. This information is found in *Note 1* to *Table 250.66*. With some 3-phase electrical systems, such as a 4-wire delta system, the phase conductors may not all be the same size. In this case, the phase that yields the largest sum of the cross-sectional areas of the conductors is used for the grounding electrode conductor determination. Note from *Table 250.66* that the largest size of grounding electrode conductor required to be installed for the service of Figure 5.11 is size 3/0 AWG copper or 250-kcmil aluminum. Some grounding electrodes, such as a ground rod, have a limited ability to dissipate ground fault current. Therefore, *250.66(A)* permits the grounding electrode conductor to be smaller in some cases than would normally be required by *Table 250.66*.

Bonding of metal enclosures and raceways is required for service equipment. What must be bonded is covered in *250.92(A)*, the method of bonding is covered in *250.92(B)*, and the minimum permitted size of bonding jumper is specified in *250.102*. A bonding means may be provided with the service equipment, or a copper or aluminum wire may be used as the bonding means. The main bonding jumper connects the service equipment enclosure to the grounding electrode conductor and to the grounded service conductor if the supply system has a grounded conductor. In the case of Figure 5.11, the main bonding jumper connects the grounding bus to the service equipment enclosure. The minimum permitted size of this main bonding jumper is determined from *250.28(D)*. This section is somewhat confusing in the case of parallel service-entrance

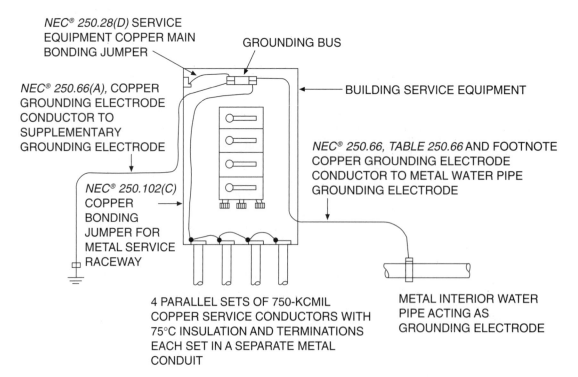

Figure 5.11   The minimum permitted size of the grounding electrode conductor and main bonding jumper are determined based on the sum of the cross-sectional areas of the conductors of any one phase.

conductors. First, the main bonding jumper is not permitted to be sized smaller than the size specified in *Table 250.66* for the service-entrance conductors where the cross-sectional area to use is the sum of the cross-sectional areas of all conductors of one phase. For example, if there are two sets of 500-kcmil conductors in parallel for the service entrance, the cross-sectional area to be used is 1000 kcmil. Secondly, when the cross-sectional area of the service-entrance conductors for one phase is determined to be larger than 1100-kcmil copper or 1750-kcmil aluminum, then the minimum permitted size of conductor for the main bonding jumper is 12.5% (0.125 times the cross-sectional area) of the cross-sectional area of the conductors. The main bonding jumper in this latter case will be larger than the grounding electrode conductor.

If the service-entrance conductors are run in metal raceway, then the metal raceway must be bonded to the service equipment enclosure or to the service grounding bus. *NEC® 250.102(C)* explains how to determine the minimum size of service raceway bonding jumper. The minimum size is based on the size of the largest conductors in the raceway. If more than one set of parallel conductors are in a single raceway for each phase, then the cross-sectional area is taken as the sum of the cross-sectional areas of the individual conductors for that phase. Example 5.5 illustrates sizing of grounding electrode conductors and bonding jumpers for a service entrance with parallel sets of service-entrance conductors.

Example 5.5   The conductors for a service entrance consist of four parallel sets of 750-kcmil copper conductors with 75°C insulation and terminations. Each set of conductors is run in a separate Rigid Metal Conduit as shown in Figure 5.11. All bonding and grounding electrode conductors are copper. The metal water pipe is used as a grounding electrode, and it is supplemented with a driven ground rod. A separate copper grounding electrode conductor is run from the grounding bus in the service equipment to each grounding electrode. The grounding bus is bonded to the service equipment enclosure with a copper main bonding jumper. A single copper conductor is run from the service equipment grounding bus to a bonding bushing on each service entrance conduit. Determine the minimum permitted size of the following:

a. the copper grounding electrode conductor to the water pipe
b. the copper grounding electrode conductor to the ground rod
c. the copper main bonding jumper from the grounding bus to the service equipment enclosure
d. the copper bonding jumper from the grounding bus to each service conductor conduit

Answer: a. The minimum permitted size of the copper grounding electrode conductor to the water pipe is determined from *250.66, Table 250.66,* and the footnote to the table. First, determine the total cross-sectional area of the conductors of one phase by adding the cross-sectional areas of all conductors of that phase to get 3000 kcmil. This is larger than 1100 kcmil; therefore, the minimum permitted size is 3/0 AWG.

$$750 \text{ kcmil} \times 4 = 3000 \text{ kcmil}$$

b. *NEC® 250.66(A)* does not require the grounding electrode conductor to a made electrode such as a ground rod to be larger than size 6 AWG.

c. *NEC® 250.28(D)* requires the main bonding jumper to not be smaller than specified in *Table 250.66* or not less than 0.125 times the cross-sectional area of the conductors of one phase, whichever is larger. In this case, it is 0.125 times 3000 kcmil that gives 375 kcmil. If one of the standard wire sizes from the Code is used, the minimum wire size can be determined from *Table 310.16* or *Table 8* of *Chapter 9.* The minimum standard wire size would be 400 kcmil.

$$3000 \text{ kcmil} \times 0.125 = 375 \text{ kcmil}$$

d. The minimum permitted size of bonding jumper for the service-entrance conduit is based on the phase conductor cross-sectional area contained in the conduit according to *250.102(C)*. In this case, the phase conductor size is 750 kcmil, and the minimum permitted raceway bonding jumper size from *Table 250.66* is size 2/0 AWG copper.

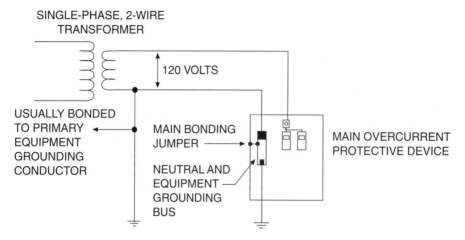

Figure 5.12   A 120-volt, 2-wire, single-phase electrical system.

ELECTRICAL SYSTEM TYPES AND VOLTAGES

Selecting the type of electrical supply system for a commercial, farm, or industrial application involves careful consideration of several factors, some of which are (1) the size of the electrical service required, (2) the voltages required, (3) the anticipated kilowatt-hour usage, (4) the electrical energy rates available, and (5) the size and number of electrical motors. An electric power supplier customer service representative is usually available to help evaluate these points when making a choice for the type of electrical system for a particular application.

Single-Phase Electrical Systems

Some buildings are served by a 2-wire, 120-volt electrical system, as shown in Figure 5.12. These systems are permitted as long as the building needs are limited to two circuits. This type of system is permitted to be installed in an outbuilding on a property where there is little electrical usage in the building. A 2-wire, 120-volt electrical system can be derived from a separate transformer, or it can be derived from a 120/240-volt, 3-wire, single-phase system or from a 208Y/120-volt, 4-wire, 3-phase electrical system.

The most common electrical system for dwellings is the 120/240-volt, 3-wire, single-phase system shown in Figure 5.13. The wire originating at the transformer center tap is grounded. The voltage from the grounded neutral wire to each ungrounded wire is 120 volts. A load powered at 240 volts, such as a motor, is supplied by the two ungrounded wires. The power supplier grounds the common conductor at the transformer, and the neutral conductor is required by *250.24* to be grounded again, usually at the service disconnecting means.

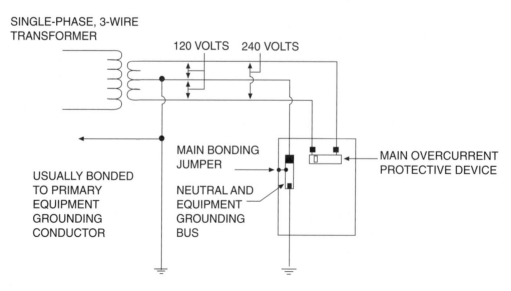

Figure 5.13   A 120/240-volt, 3-wire, single-phase electrical system.

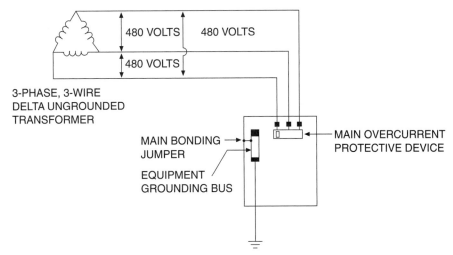

Figure 5.14   A 3-phase, 3-wire, ungrounded delta electrical system operating at 480 volts (sometimes 240 volts) is required to have the service disconnect enclosure grounded.

It is important to balance the 120-volt load so that half is connected between the neutral and each ungrounded wire. If the 120-volt loads are perfectly balanced, the current flowing on the neutral will be zero. Loads operating at 240 volts do not use the neutral; therefore, they do not place current on the neutral.

Three-Phase Electrical Systems

Three-phase electrical systems are of different types and voltages to fit different needs. There are three common types of delta electrical systems. One type of delta 3-phase electrical system is the 3-wire ungrounded electrical system. *NEC® 250.20* permits the 240-volt and the 480-volt 3-phase delta 3-wire systems to be operated without one of the conductors grounded. A 480-volt ungrounded 3-wire delta electrical system is illustrated in Figure 5.14.

It is important to remember that equipment is required to be grounded as specified in *250.110* and *250.112*. *NEC® 250.130(B)* states that the equipment grounding conductors shall be bonded to the grounding electrode conductor of the ungrounded system. This can be confusing. The system is considered ungrounded because one circuit conductor is not grounded. But a grounding electrode is required for the service.

A 3-phase, 3-wire electrical system operating either at 240 volts between phases or 480 volts between phases is permitted to be operated with one of the phase conductors grounded. This is frequently called a corner grounded delta 3-phase electrical system. A 3-wire, corner grounded, 3-phase electrical system is shown in Figure 5.15. *NEC® 240.22* states that an overcurrent device is not permitted to be installed in a

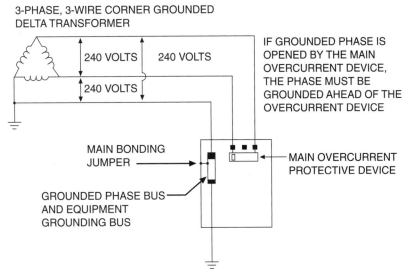

Figure 5.15   A 3-phase, 3-wire, delta electrical system, operating at either 240 volts or at 480 volts, is permitted to have one phase conductor grounded.

grounded conductor unless all ungrounded conductors are opened when the overcurrent device operates. An overcurrent device is permitted to be installed in the grounded phase conductor when it serves as overload protection for an electric motor, as permitted in *430.36*.

The 4-wire, 240/120-volt delta 3-phase electrical system provides single-phase and 3-phase service at 240 volts and single-phase at 120 volts. This system is shown in Figure 5.16. There is a single-phase neutral conductor with this system. The voltage from two of the ungrounded phase conductors to the neutral is 120 volts. Voltage from one of the phase conductors to the neutral is 208 volts. This phase conductor is called the phase with the higher voltage to ground in *110.15, 230.56*, and *408.3(E)*. In *408.3(E)*, the phase conductor with the higher voltage to ground is required to be the B-phase, which is required to be the center conductor in a disconnect or panelboard. There is an exception in the case of some metering equipment. The phase conductor with the higher voltage to ground is not intended to be combined with the neutral because the voltage is 208 instead of 120. Single-phase and 3-phase loads are permitted to be served from the same panel, but this must be done with care to keep the loads balanced on the phase conductors and the neutral conductor.

Any 3-phase delta electrical system can be operated with only two transformers. This is called an open delta system. If the electrical system of Figure 5.16 was operated as an open delta electrical system, the transformer between the B-phase and the A-phase or the transformer between the B-phase and the C-phase would be missing. The voltages delivered by this open delta electrical system should be the same as for the full delta 3-phase system. There may be a slight voltage variation between phase conductors.

Three-phase power may be supplied using a 4-wire wye electrical system. Figure 5.17 shows a 208/120-volt, 3-phase wye electrical system. A 480/277-volt, 4-wire wye, 3-phase system is also available. The voltage from any one of the ungrounded phase conductors to the neutral is the same with this system.

It is possible to install the neutral-to-phase circuits to balance the neutral-to-phase loads on the three transformers. This is frequently done using multiwire branch-circuits with one neutral common to three circuit conductors. Motors normally designed for 240-volt operation must not be connected to a 208/120-volt wye system unless marked on the nameplate as suitable for operation at 208 volts. Motors rated at 200 to 240 volts are available. With this system, equipment must be specified with 208-volt motors. The 3-phase, 4-wire, 480/277-volt wye electrical system is used commonly for industrial applications. Motors are powered using 3-phase, 480-volt circuits, and single-phase electric discharge lighting circuits are operated at 277 volts.

High-Impedance Grounding

High-impedance (high-resistance) grounding is permitted for some industrial applications at 480 volts and above and where qualified maintenance staff are always on duty to immediately investigate ground faults. This type of electrical system is carefully engineered for each application. It provides an alternative to the ungrounded 480- and 600-volt industrial electrical system. The typical application is for industrial processes where orderly shutdown is necessary. If a ground-fault occurs somewhere on the system, it may

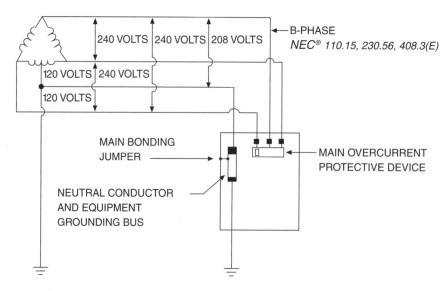

Figure 5.16   A 240/120-volt, 4-wire, 3-phase delta electrical system provides single-phase power at 120/240 volts and 3-phase power at 240 volts. One phase has a higher voltage to ground than the other two phases, and that phase must be identified with an orange marking.

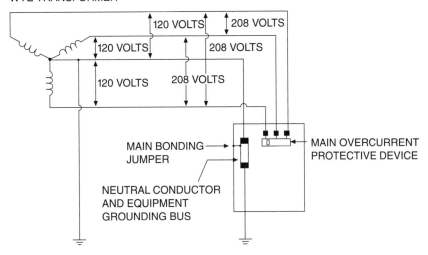

Figure 5.17   A 208/120-volt, 4-wire, 3-phase wye electrical system, common for commercial buildings, provides single-phase power at 120 and 208 volts and three-phase power at 208 volts. The wye electrical system can also operate at 480/277 volts.

not be desirable to allow that circuit or feeder to immediately shut down. On an ungrounded or high-resistance grounded system, a signal indicates that a ground fault has occurred on a particular phase, and the maintenance personnel immediately find the fault and assess the seriousness of the condition. To make repairs, usually a controlled shutdown can be accomplished. Sometimes repairs can be made without shutting down the system. The resistor installed in series with the equipment grounding conductor limits the value of fault current to a level that an overcurrent device will not open, but the fault condition will be indicated. A high-impedance grounded electrical system is shown in Figure 5.18.

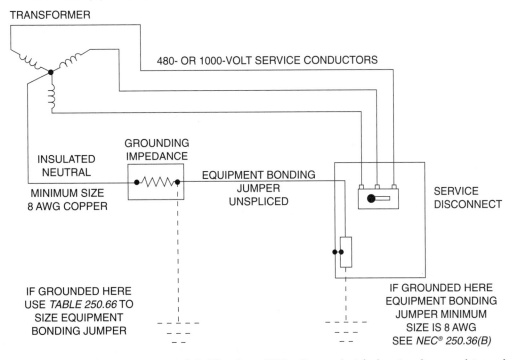

Figure 5.18   A high-impedance grounded 480-volt or 1000-volt wye electrical system has a resistance installed between the grounding electrode conductor connection and the neutral point of the supply transformer that limits the ground-fault current to a level that will sound a trouble alarm, but will not cause an overcurrent device to open.

With this type of system, the neutral conductors and the equipment grounding conductors are separated, and a resistor is connected between the neutral bus and the equipment grounding bus. Specifications for installing a high-impedance grounded electrical system are found in *250.36*.

MAJOR CHANGES TO THE 2005 CODE

These are the changes to the 2005 *NEC®* that correspond to the Code sections studied in this unit. The following analysis explains the significance of the changes from the 2002 to the 2005 Code only, and this analysis is not intended to be used in place of the Code. Refer to the actual section of the 2005 Code for the exact wording and meaning of each section discussed. Changes are indicated in the Code with a vertical line in the margin. If material was deleted or moved to another location in the Code, the location of the deletion is indicated with a dark dot in the margin.

Article 250 **Grounding and Bonding**

The title of the article was expanded to include bonding. There seems to be a debate as to whether much of what is covered in this article is grounding or bonding of metal parts. Whatever it is called, the point is that a path is required to be established from all exposed metal parts likely to become energized back to the grounding point at the service disconnecting means.

250.21: Ungrounded ac systems operating at 120 volts or more are now required to have ground detection. An example of an ungrounded ac electrical system with ground detection is shown in Figure 5.19. An ungrounded electrical system can be created by a separately derived system within the premises and many ground detectors may be required.

250.28: The term system bonding jumper was added to this section. The main bonding jumper is understood to be at the service to the building or structure. When a separately derived system is created, such as at a transformer, a bonding jumper is required between the grounded conductor and the equipment grounding terminal or bar. At a separately derived system, this bonding connection is called the system bonding jumper. Since this main bonding jumper was considered located at the service, there was no clear rule for sizing the system bonding jumper. Now the rules for sizing the main bonding jumper and the system bonding jumper are stated in this section.

250.30(A)(4)(a): It is permitted to ground multiple separately derived systems to a common grounding electrode conductor run throughout a building or structure. Each separately derived system is then bonded to the common grounding electrode conductor. There is now one minimum size required for a com-

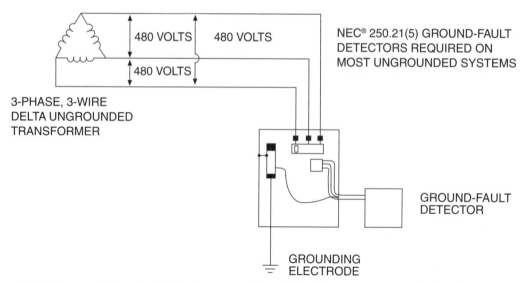

Figure 5.19 Ungrounded ac electrical systems operating at 120 volts or more are required to be equipped with a ground-fault detection system.

mon grounding electrode conductor used for two or more separately derived systems. The minimum size is 3/0 AWG copper or 250 kcmil aluminum. For a large number of installations, this is an increase in size as compared with the rule in the previous edition of the Code. For installations with low kVA ratings of separately derived systems, it is more practical to ground each separately rather than run a common grounding electrode conductor. The following example will illustrate how such a common grounding electrode system is sized:

Example 5.6: Four single-phase transformers rated 3 kVA are installed in one area of a building and connected to a 480-volt supply to provide a 20-ampere, 120-volt circuit at each location. If all four transformers are grounded to a common copper grounding electrode conductor, determine the minimum size grounding electrode conductor permitted.

Answer: The common grounding electrode conductor is required to be not smaller than size 3/0 AWG copper according to *250.30(A)(4)(a)*. The grounding electrode conductor from each transformer to the common grounding electrode conductor is required in this case to be not smaller than size 8 AWG copper according to *250.30(A)(3)* using *Table 250.66* and knowing the transformer output conductors are size 12 AWG copper. The previous edition of the Code would have permitted a size 8 AWG copper common grounding electrode conductor for these transformers.

250.50: This section specifies the grounding electrode required for a service to a building or structure. The previous edition of the Code required that all of the electrodes described in *250.50(A)* that were available were required to be bonded together to form a grounding electrode system. The words "if available" were deleted, and now all of the types of grounding electrodes described in *250.50(A)* are required to be used. The real issue is the concrete-encased electrode such as reinforcing steel in a foundation. If steel reinforcing is installed, it is now required to be used as a grounding electrode. This means the steel reinforcing in new construction, if installed, is required to be made available so it can be used as a grounding electrode. This is illustrated in Figure 5.20. There is an exception for existing buildings.

250.50 Exception: This new exception only applies to existing buildings and structures. If it is not possible to get at reinforcing steel without penetrating the concrete, then it is not required to use the reinforcing steel as a grounding electrode.

250.52(A)(2): This subsection now specifies when the metal frame of a building is considered to be a grounding electrode. The previous edition of the Code required the building metal frame to be used as a grounding electrode when effectively grounded. Now the specific methods of effectively grounding the metal building frame are specified. At least 10 ft (3 m) of the metal extends into the earth or is

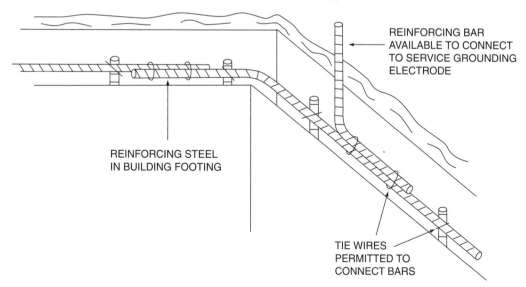

NEC® 250.50 IF INSTALLED REINFORCING STEEL IN FOOTING SHALL BE MADE AVAILABLE TO BE USED AS GROUNDING ELECTRODE

REINFORCING BAR AVAILABLE TO CONNECT TO SERVICE GROUNDING ELECTRODE

REINFORCING STEEL IN BUILDING FOOTING

TIE WIRES PERMITTED TO CONNECT BARS

Figure 5.20   Reinforcing steel that is installed in building and structure footings and foundations is required to be installed in such a manner that a connection can be made to a grounding electrode conductor from the service.

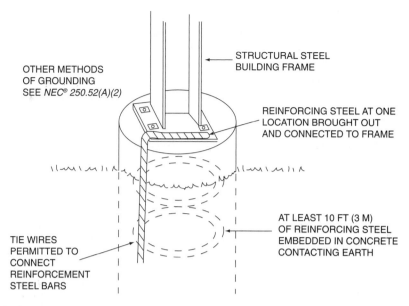

OTHER METHODS
OF GROUNDING
SEE *NEC*® 250.52(A)(2)

STRUCTURAL STEEL
BUILDING FRAME

REINFORCING STEEL AT ONE
LOCATION BROUGHT OUT
AND CONNECTED TO FRAME

AT LEAST 10 FT (3 M)
OF REINFORCING STEEL
EMBEDDED IN CONCRETE
CONTACTING EARTH

TIE WIRES
PERMITTED TO
CONNECT
REINFORCEMENT
STEEL BARS

Figure 5.21   One method of effectively grounding the metal frame of a building is to make a connection to footing or foundation reinforcing steel that is a minimum of 10 ft (3 m) in length and embedded in concrete that is in contact with the earth.

encased in concrete that is in contact with the earth. The structural metal frame is bonded to the water pipe, or a ground ring, or reinforcing steel in concrete that is in contact with the earth. An example is shown in Figure 5.21. The metal building frame can be grounded by bonding to a ground rod, or a pipe, or a metal plate. Other methods of grounding a metal building frame can also be used if approved.

250.52(A)(7): An underground metal well casing that is not effectively bonded to a metal underground water-piping system is now considered another local metal underground system acceptable to be used as a grounding electrode. A metal well casing is not considered to be a metal underground water-piping system unless it is effectively bonded to the metal water pipe. Now a metal well casing can be used as the grounding electrode for a service, and a supplemental grounding electrode is not required since it is not considered to be a metal underground water-piping system. If a metal well casing is installed and connected with nonmetallic water piping, the metal well casing is permitted to serve as the only grounding electrode as illustrated in Figure 5.22.

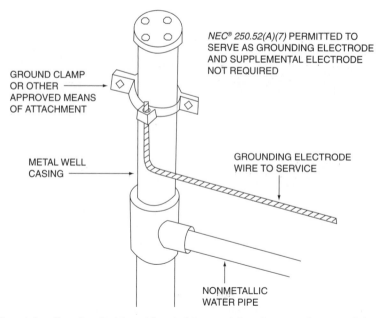

NEC® 250.52(A)(7) PERMITTED TO
SERVE AS GROUNDING ELECTRODE
AND SUPPLEMENTAL ELECTRODE
NOT REQUIRED

GROUND CLAMP
OR OTHER
APPROVED MEANS
OF ATTACHMENT

METAL WELL
CASING

GROUNDING ELECTRODE
WIRE TO SERVICE

NONMETALLIC
WATER PIPE

Figure 5.22   A metal well casing that is not bonded to a metal underground water-piping system is considered an independent grounding electrode and is permitted to be used as the grounding electrode for a building.

250.64(C)(3): This section requires that the grounding electrode conductor be run as a continuous length from the grounding point at the service disconnecting means to the grounding electrode. This is a new paragraph that permits a copper or aluminum busbar with cross-section dimensions of not less than 1/4 in. by 2 in. (6 mm by 50 mm) to be securely fastened in an accessible location as a common connecting point between grounding electrodes and grounding electrode conductors. Connections to the busbar are to be made using listed connectors or by exothermic welding. An example where this technique may be applicable is when there are several disconnecting means for a service and several grounding electrodes such as illustrated in Figure 5.23. Since an aluminum busbar is permitted for this purpose, a new paragraph *(4)* was added that requires aluminum busbars to be installed in accordance with *250.64(A)*, which does not permit bare aluminum to be attached to a masonry surface or if installed outside aluminum conductors are not to be terminated within 18 in. (450 mm) of the earth.

250.64(E): When a grounding electrode conductor is run through an enclosure or conduit for protection from physical damage, the grounding electrode conductor is required to be bonded to the enclosure or conduit at both ends. The issue is that steel (a ferrous metal) is a magnetic material and when a grounding electrode conductor carries lightning current, the steel envelope around the conductor creates an impedance that reduces the ability of the grounding electrode conductor to carry current if the steel is not bonded to the wire at both ends. The previous edition of the Code required bonding to the grounding electrode conductor for all metal enclosures and conduits. Now this section makes it clear that bonding is only required when the enclosure for the grounding electrode conductor is ferrous metal such as steel.

250.68(A) Exception No. 2: This section requires connections to grounding electrodes to be accessible. This new exception does not require the connection to a fireproofed steel building frame to be accessible.

250.104(D)(3): When a common grounding conductor for multiple separately derived systems is run through a building or structure and there is a metal water pipe or exposed metal building frame in the area served by any of the separately derived systems, the common grounding conductor is required to be bonded to the metal water-piping system and metal building frame. According to the exception, it is not necessary to provide a bond at each separately derived system if the bonding is made to the common grounding conductor.

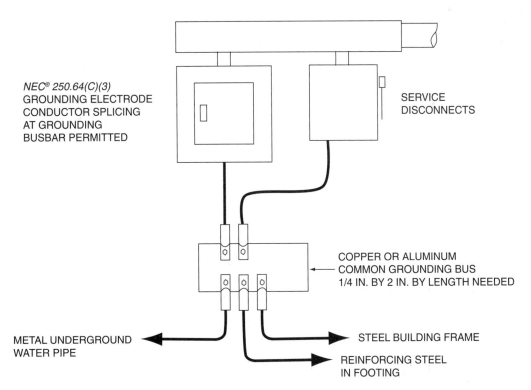

NEC® 250.64(C)(3)
GROUNDING ELECTRODE
CONDUCTOR SPLICING
AT GROUNDING
BUSBAR PERMITTED

SERVICE
DISCONNECTS

COPPER OR ALUMINUM
COMMON GROUNDING BUS
1/4 IN. BY 2 IN. BY LENGTH NEEDED

METAL UNDERGROUND
WATER PIPE

STEEL BUILDING FRAME

REINFORCING STEEL
IN FOOTING

Figure 5.23 **A copper or aluminum bus with a cross-sectional dimensions not less than ¹/₄ in. (6 mm) by 2 in. (50 mm) is permitted to act as a collecting point between service disconnecting means grounding electrode conductors and grounding electrode conductors to various grounding electrodes.**

250.118(5): This paragraph of the previous edition of the Code was deleted. This section lists acceptable materials permitted to serve as equipment grounding conductors. The deleted paragraph stated that Flexible Metal Conduit was permitted as a grounding means if both the conduit and fittings were listed for grounding. Flexible Metal Conduit is not presently listed for grounding in lengths greater than 6 ft (900 mm). The deletion of this paragraph makes it clear that under no circumstances is Flexible Metal Conduit in lengths greater than 6 ft (900 mm) is permitted to serve as an equipment grounding conductor.

250.118(5)(d) and (6)(e): If Flexible Metal Conduit or Liquidtight Flexible Metal Conduit is installed in lengths not exceeding 6 ft (900 mm) but flexibility is required, it is necessary to install an equipment bonding jumper across the flexible conduit section. The issue was the definition of flexibility since these materials are usually only installed because flexibility is desired for some reason. These sections now make it clear that flexibility requiring a bonding jumper is flexibility that is necessary after installation such as where the connected equipment actually moves.

250.119: A new last sentence was added to this section that specifies how equipment grounding conductors are to be identified. Now there is a rule that only permits equipment grounding conductors to have insulation or covering that is green or green with yellow stripes.

250.119(A) Exception: This section permits conductors larger than size 6 AWG to be identified as an equipment grounding conductor at the time of installation at each end and at every point where the conductors are accessible. This new exception does not require the equipment grounding conductor to be identified at conduit bodies where there are no splices and connections, provided there are no unused hubs where another conductor could enter in the future and thus require splicing.

250.122(G): This is a new section that specifies the minimum size equipment grounding conductor required for a tap to a feeder. Since the equipment grounding conductor of the tap must be capable of opening the feeder overcurrent device in the case of a short circuit or ground fault, the minimum size is based on the rating of the feeder overcurrent device using *Table 250.122*. The equipment grounding conductor is not required to be larger than the ungrounded tap conductors.

Example 5.7: A tap is made to a 3-phase, 480-volt feeder consisting of 600-kcmil copper conductors protected by a 400-ampere circuit breaker. The tap is 25 ft (7.5 m) in length and ends at a disconnect containing 100-ampere time-delay fuses. The minimum size ungrounded tap conductor permitted is 1/0 AWG copper. If these tap conductors are run in Rigid Nonmetallic Conduit, determine the minimum size equipment grounding conductor required for this tap.

Answer: Look up the minimum size equipment grounding conductor based upon *Table 250.122* using the 400 ampere feeder circuit breaker rating, but the size is not required to be larger than 1/0 AWG in this case. The minimum in this case is a size 3 AWG copper equipment grounding conductor.

250.146(A): When a metal box is surface mounted, direct metal-to-metal contact between a device yoke and the box is permitted to serve as the equipment grounding connection between the box and the device. This means the equipment grounding conductor, connected to the metal box, is not required to be connected to the grounding screw of the device. Devices that are not of the self-grounding type usually have a nonmetallic washer on each mounting screw to hold it in place. Sometimes this nonmetallic washer, if not removed, can prevent metal-to-metal contact between the box and the device yoke. Now it is required that at least one of the nonmetallic mounting-screw washers be removed to ensure contact between the device yoke and the metal box.

250.184(B): This section provides basic minimum guidelines for the design and installation of customer-owned solidly grounded high-voltage distribution systems. The previous edition of the Code only addressed multigrounded distribution systems, but did not prohibit single-point grounded distribution systems. This new subsection (B) provides basic installation requirements for single-point grounded distribution systems operating at 1000 volts and higher. A multigrounded distribution system will likely result in some earth current flow and may cause significant current flow on metal piping and structures depending upon the particular circumstances. To prevent objectionable current flow, some customers and system designers utilize a single-point grounded distribution system. With the single-point grounded distribution system, an equipment grounding conductor is run to all load locations. The separate insulated neutral is only required to be run to locations where loads are connected line-to-neutral.

A grounding electrode is provided at the source of the single-point grounded distribution system and bonded to the neutral and equipment grounding conductors at that point. At all other points along the system the neutral shall be insulated and isolated from the earth. The equipment grounding conductor is permitted to be grounded to the earth at locations along the system.

Article 280 Surge Arresters

280.4(A)(3): This is a new requirement for surge arresters installed on electrical systems operating at under 1000 volts. The surge arrester is required to be marked with a short-circuit current rating, which is required to be higher than the available fault current at the installation location on the system.

280.4(A)(4): This is a new rule that requires surge arresters to be specifically listed for the purpose when installed on an ungrounded electrical system, an impedance grounded system, or a corner grounded system.

Article 285 Transient Voltage Surge Suppressors

285.3(2): The previous edition of the Code did not permit a transient voltage surge suppressor (TVSS) to be installed on an ungrounded electrical system. There are transient voltage surge suppressors listed for installation on ungrounded electrical systems. This paragraph now permits a TVSS listed for the specific purpose to be installed on an ungrounded electrical system, an impedance grounded electrical system, or a corner grounded electrical system.

285.21(A)(1): A transient voltage surge suppressor (TVSS) is permitted to be installed as a tap ahead of the service disconnecting means as permitted in *230.82(5)*. When this installation technique is used, a disconnecting means, overcurrent protection, and grounding is required to be provided for the TVSS that satisfies the requirements of a service.

Article 680 Swimming Pools, Fountains, and Similar Installations

680.21(A)(1): Wiring supplying motors associated with a permanent swimming pool is required to contain an equipment grounding conductor not smaller than size 12 AWG insulated copper. The change is that the equipment grounding wire is now required to be insulated.

680.23(A)(2): A transformer used to supply an underwater luminaire (lighting fixture) at a permanent swimming pool is now required to be listed as a swimming pool and spa transformer.

680.25(B)(2): This subsection deals with the case where permanent swimming pool equipment is supplied from a panelboard in a separate building or structure that receives power by means of a feeder from another building. The previous edition of the Code permitted the grounded conductor (neutral) to serve as the equipment grounding conductor if installed in accordance with *250.32* as illustrated in the top diagram of Figure 5.5. This practice is no longer permitted. There must be an equipment grounding conductor in the feeder separate from the grounded conductor, and it must be installed to meet the requirements of *250.32(B)(1)*, which is illustrated in the bottom diagram of Figure 5.5.

Permanent swimming pool equipment is permitted to be installed from an existing panelboard in a separate building where the existing feeder from the service equipment is run in flexible metal conduit or an approved cable assembly with an equipment grounding conductor. The neutral and equipment grounding conductors in the feeder and panelboard supplied are required to be electrically separated. In this case, the last sentence of *680.25(B)* permits the equipment grounding conductor to be uninsulated.

A feeder run to a separate building that supplies permanent swimming pool equipment is now required to have the neutral and equipment ground separated as described in *250.32(B)(1)* and shown in the bottom diagram of Figure 5.5. All of the conductors in the feeder, including the equipment grounding conductor, are required to be insulated. The wiring methods for the feeder to the separate building or structure are described in *680.25(A)* and do not include cables or overhead messenger-supported wiring for new installations.

680.26: The title of this section was changed from bonding to equipotential bonding. The term equipotential plane is defined in *547.2,* which is the same concept of attempting to keep step and touch potentials below the level of human perception. An equipotential plane in a milking barn is shown in Figure 14.6.

680.26(B)(1): This subsection specifies which pool equipment and material is required to be bonded together. The last sentence requires an alternate means to be installed if reinforcing steel is coated with a non-conductive material and is not bonded to the pool equipotential grid. The change is that the deck area around the pool is now included. If the reinforcing steel in the deck is coated with a noncondutive material and is not bonded, an alternate means is now required to be installed in the deck to create an equipotential grid.

680.26(C): A solid size 8 AWG copper wire is not necessarily required to bond pool equipment to form an equipotential grid. Brass Rigid Metal Conduit, brass Intermediate Metal Conduit, and other identified corrosion-resistant metal conduit connected to metal equipment requiring bonding is now permitted to serve as the bonding conductor.

680.26(C): The equipotential bonding grid required to be installed at a permanent swimming pool is now required to extend under walkways around the pool extending out from the inside edge of the pool a distance of not less than 3 ft (1 m).

680.26(C): The equipotential bonding grid is no longer permitted to consist only of a size 8 AWG bare, insulated, or covered copper wire used to connect metal equipment. Brass Rigid Metal Conduit and brass Intermediate Metal Conduit by itself also are no longer permitted to satisfy the requirement of establishing an equipotential bonding grid.

680.26(C)(3): There are now three methods permitted for establishing an equipotential bonding grid for a permanent swimming pool. The two methods permitted in the past and still permitted is the reinforcing steel in the pool floor, walls, and surrounding deck, and a bolted or welded metal pool. A new method of establishing an equipotential bonding grid consists of size 8 AWG bare solid copper wire forming 1 ft (300 mm) squares on the bottom, sides, and deck extending out under the walkway a distance not less than 3 ft (1 m) from the inside edge of the pool. An example is shown in Figure 5.24. This method of establishing an equipotential grid is not necessary if the method of 680.26(C)(1) is used to establish the grid.

680.32: All 125-volt, 15- and 20-ampere receptacles located within 20 ft (6 m) of the inside wall of a storable swimming pool is now required to be ground-fault circuit-interrupter protected.

680.34: This is a new section that does not permit any receptacle to be installed less than 10 ft (3 m) from the inside edge of a storable swimming pool.

680.52(B)(2)(b): Underwater enclosures in fountains are now permitted to be supported by stainless steel conduit.

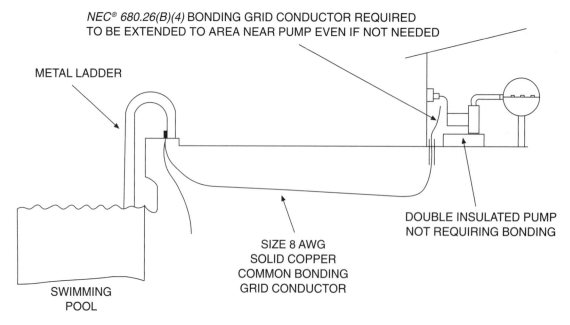

Figure 5.24 Determine the maximum length of Flexible Metal Conduit permitted to be installed between a motor controller and a motor terminal housing and supported only by the end connectors.

WORKSHEET NO. 5—BEGINNING GROUNDING AND BONDING

Mark the single answer that most accurately completes the statement based upon the 2005 Code. Also provide, where indicated, the Code reference that gives the answer or indicates where the answer is found, as well as the Code reference where the answer is found.

1. A single-family dwelling has a service with a 100-ampere main circuit breaker and size 2 AWG aluminum ungrounded conductors. A grounding electrode for this service is a ground rod driven so the top is below the surface of the earth and the clamp is listed as suitable for direct burial. As shown in Figure 5.25, the grounding electrode conductor is attached to the outside of the dwelling and through the earth to the ground rod. If the copper grounding electrode conductor is in a location not subject to physical damage, the minimum size permitted for this service is:
    A. 8 AWG.   C. 4 AWG.   E. 2 AWG.
    B. 6 AWG.   D. 3 AWG.

                                    Code reference_____

2. A commercial service entrance with a 200-ampere main circuit breaker has size 3/0 AWG copper ungrounded service conductors. The service is grounded to a metal underground water pipe and a ground rod as the supplemental electrode. The minimum size copper grounding electrode conductor permitted to the metal water pipe is:
    A. 10 AWG.   C. 6 AWG.   E. 2 AWG.
    B. 8 AWG.   D. 4 AWG.

                                    Code reference_____

3. A building is provided with a single-phase, 120/240-volt service with a 200-ampere main circuit breaker and size 4/0 AWG aluminum service conductors. The water pipe entering the building is nonmetallic and the grounding electrode is a $^5/_8$ in. (16

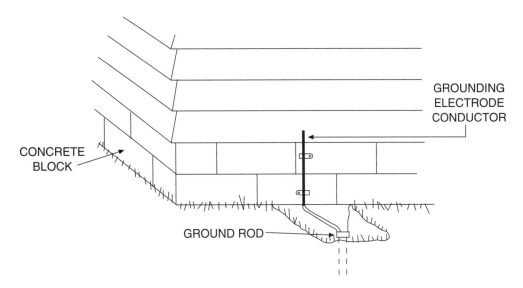

Figure 5.25 Determine the minimum size copper grounding electrode conductor run unprotected on the outside surface of a dwelling for a 100-ampere service entrance.

mm) diameter galvanized steel rod driven 8 ft (2.5 m) into the earth. The minimum size copper grounding electrode conductor permitted for this service is:

A. 8 AWG.   C. 4 AWG.   E. 2 AWG.
B. 6 AWG.   D. 3 AWG.

Code reference _____

4. A building is supplied water by a metal underground pipe and the metal frame of the building is considered to be effectively grounded.
    A. Both the metal water pipe and the metal building structure are required to be bonded together and used as the grounding electrode system for the service.
    B. Even if both are used as the grounding electrode, a ground rod is still required to augment the metal water pipe.
    C. A ground rod is permitted to be used in place of either the metal water pipe or the metal building structure.
    D. The metal water pipe or the metal building frame, but not both, are required to serve as the grounding electrode.
    E. Only the metal building frame is required to be used as the grounding electrode.

Code reference _____

5. Flexible Metal Conduit trade size ³/4 (21) extends from the controller to a motor terminal housing to allow for ease of installation and maintenance and is not installed for flexibility as shown in Figure 5.26. The end fittings provide the only support for the Flexible Metal Conduit and they are listed as suitable for equipment grounding but not the Flexible Metal Conduit. The circuit to the motor is protected by 20-ampere rated time-delay fuses. The maximum length of Flexible Metal Conduit permitted for this installation is:

A. not limited.   C. 18 in. (450 mm).   E. 6 ft (1.8 m).
B. 12 in. (300 mm).   D. 3 ft (900 mm).

Code reference _____

6. An 8 ft (2.5 m) ground rod is installed at a service to augment a metal underground water pipe as the grounding electrode. The ground rod has a resistance of more than 25 ohms, therefore, it will be supplemented by a second 8 ft (2.5 m) ground rod, with the two rods bonded together. These two ground rods shall be spaced a distance from

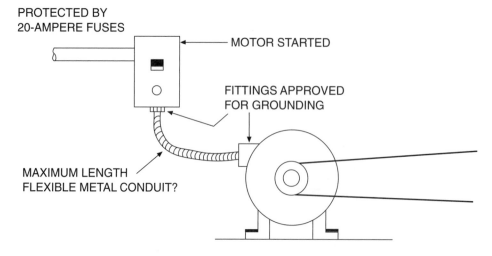

PROTECTED BY
20-AMPERE FUSES

MOTOR STARTED

FITTINGS APPROVED
FOR GROUNDING

MAXIMUM LENGTH
FLEXIBLE METAL CONDUIT?

Figure 5.26   Determine the maximum length of Flexible Metal Conduit permitted to be installed between a motor controller and a motor terminal housing and supported only by the end connectors.

each other of not less than:

A. 6 in. (150 mm).   C. 4 ft (1.2 m).   E. 10 ft (3 m).
B. 2 ft (600 mm).   D. 6 ft (1.8 m).

Code reference _____

7. An equipment grounding means *not* permitted is:
   A. Liquidtight Flexible Metal Conduit not smaller than trade size 1½ (41).
   B. the metal sheath of Type MC Cable.
   C. Electrical Metallic Tubing.
   D. a bare aluminum wire.
   E. the armor of Type AC cable.

Code reference _____

8. A motor control center in a building is fed under the floor in Rigid Nonmetallic Conduit from a disconnect at the main service. The feeder is protected with 200-ampere time-delay fuses, and the ungrounded conductors are size 3/0 AWG copper with THHN insulation as shown in Figure 5.27. The minimum size copper equipment grounding conductor permitted to be run in this conduit for this feeder is:
   A. 8 AWG.   C. 4 AWG.   E. 2 AWG.
   B. 6 AWG.   D. 3 AWG.

Code reference _____

9. A conductor that is not bare or does not have green covering or insulation, or green covering or insulation with yellow stripes, is permitted to be re-identified as an equipment grounding conductor. Acceptable methods of re-identifying the conductor are by stripping the covering or insulation from the entire exposed length, coloring the exposed insulation green, or marking the exposed insulation with green tape or green adhesive labels. This re-identification is only permitted for conductors sizes:
   A. smaller than 4 AWG.
   B. larger than 8 AWG copper and 4 AWG aluminum.
   C. larger than 12 AWG copper and 10 AWG aluminum.
   D. smaller than 8 AWG copper and 6 AWG aluminum.
   E. larger than 6 AWG.

Code reference _____

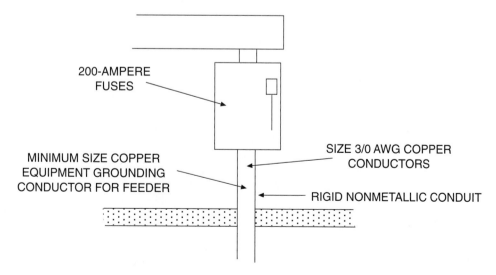

Figure 5.27 Determine the minimum size copper equipment grounding conductor permitted for a 200-ampere rated feeder run in Rigid Nonmetallic Conduit.

10. An ac electrical system not required to be grounded is a:
    A. 120/240-volt, 3-wire, single-phase system.
    B. 208Y/120-volt, 4-wire, 3-phase system.
    C. 480Y/277-volt, 4-wire, 3-phase system serving 277-volt lighting loads.
    D. 240/120-volt, 4-wire, 3-phase delta system.
    E. 240-volt, 3-wire, 3-phase delta system.

Code reference_____

11. At least one receptacle on a general purpose 125-volt, 15- or 20-ampere branch-circuit is required to be installed at a dwelling not more than 20 ft (6 m) from the inside edge of a permanent swimming pool as shown in Figure 5.28. That receptacle, where space is not restricted, is not permitted to be located closer to the inside edge of the swimming pool than:
    A. 3 ft (750 mm).   C. 5 ft (1.5 m).   E. 12 ft (3.7 m).
    B. 4 ft (1.2 m).   D. 10 ft (3 m).

Code reference_____

12. An equipotential bonding grid for a permanent swimming pool is formed by connecting together with a bonding conductor all metal parts associated with the pool including the reinforcing steel in the pool floor, walls, and deck. These include metallic parts of the pool structure, forming shells and mounting brackets of luminaires (lighting fixtures), metal ladders and other metal fixtures attached to the pool, metal equipment associated with the water circulating system, pool covers, and similar metal parts. The conductor used for this bonding is:
    A. required to be solid copper not smaller than size 8 AWG.
    B. permitted to be solid aluminum if insulated and not smaller than size 8 AWG.
    C. required to be insulated copper, stranded, and not smaller than size 8 AWG.
    D. permitted to be bare, insulated, or covered copper if not smaller than size 6 AWG.
    E. required to be copper not smaller than size 2 AWG.

Code reference_____

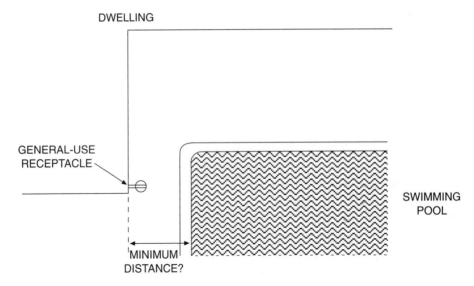

Figure 5.28 Determine the minimum distance a general-use receptacle installed on the side of a dwelling is permitted to be from the inside edge of a permanent swimming pool.

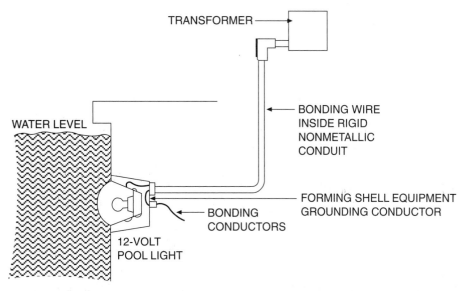

TRANSFORMER

BONDING WIRE
INSIDE RIGID
NONMETALLIC
CONDUIT

WATER LEVEL

FORMING SHELL EQUIPMENT
GROUNDING CONDUCTOR

BONDING
CONDUCTORS

12-VOLT
POOL LIGHT

Figure 5.29   The conductors from a transformer enclosure to the forming shell of a wet-niche luminaire (lighting fixture) in a permanent swimming pool are run in Rigid Nonmetallic Conduit.

13. Rigid Nonmetallic Conduit is used to connect the forming shell of a wet-niche, low-voltage swimming pool light to a transformer enclosure as shown in Figure 5.29. This is not a listed low-voltage luminaire (lighting fixture) not requiring grounding. The minimum size copper forming shell equipment grounding conductor permitted to be run inside the conduit to the forming shell is 8 AWG:
   A. insulated solid only.
   B. insulated stranded only.
   C. insulated solid or stranded.
   D. bare, insulated, covered, and solid or stranded.
   E. bare solid or stranded.

   Code reference_____

14. A surge arrester is:
   A. only permitted to be installed outside of a building.
   B. only permitted to be installed inside of a building.
   C. permitted to be installed inside or outside of a building.
   D. only permitted to be installed by a utility worker with training in high-voltage installations.
   E. installed between an ungrounded conductor of the electrical system and a lightning rod.

   Code reference_____

15. A Nonmetallic-Sheathed Cable, Type NM-B, enters a metallic device box containing a duplex receptacle that has a device to maintain continuity between the mounting screw and the metal device yoke. There is also a grounding screw in the box. The equipment grounding conductor in the supply cable is:
   A. required to be terminated to the grounding screw on the duplex receptacle.
   B. required to be terminated both to the grounding screw on the duplex receptacle and the grounding screw of the box.
   C. permitted to be terminated to either the grounding screw of the duplex receptacle or the grounding screw of the box.
   D. not required to be terminated in the box if grounded metal objects or masonry surfaces are located not closer than 6 ft (1.8 m) from the box.
   E. required to be terminated to the grounding screw of the box.

   Code reference_____

WORKSHEET NO. 5—ADVANCED GROUNDING AND BONDING

Mark the single answer that most accurately completes the statement based upon the 2005 Code. Also provide, where indicated, the code reference that gives the answer or indicates where the answer is found, as well as the Code reference where the answer is found.

1.   A 480Y/277-volt service entrance with a 400-ampere main circuit breaker and size 500-kcmil copper ungrounded service conductors is shown in Figure 5.30. A metal underground water pipe entering the building is used as a grounding electrode for the service. The minimum size of a copper grounding electrode conductor permitted to be run to the water pipe is:

A.   3 AWG.   C.   1 AWG.   E.   2/0 AWG.
B.   2 AWG.   D.   1/0 AWG.

Code reference _____

2.   A commercial building service is 208Y/120-volt, 3-phase, 4-wire wye with a 400-ampere rated main circuit breaker, and size 500-kcmil copper ungrounded conductors. Galvanized steel, $1/2$ in. (13 mm) diameter reinforcing bars, a minimum of 20 ft (6 m) long, are installed in the bottom of the foundation which has direct contact with earth. The minimum size copper grounding electrode conductor permitted to be run to the reinforcing bars is size:

A.   4 AWG.   C.   1 AWG.   E.   2/0 AWG.
B.   2 AWG.   D.   1/0 AWG.

Code reference _____

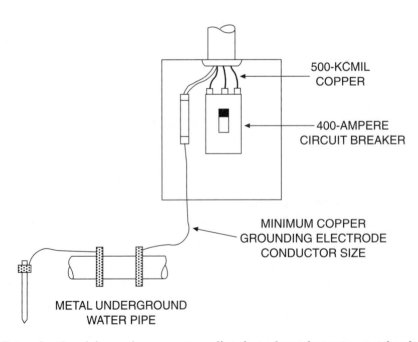

Figure 5.30   Determine the minimum size copper grounding electrode conductor to a metal underground water pipe for a service consisting of a 400-ampere main circuit breaker supplied with size 500-kcmil copper conductors.

3. An aluminum conductor is permitted to be used as a grounding electrode conductor:
   A. as a bare conductor in dry locations and shall be insulated when in direct contact with masonry or the earth.
   B. the same as a copper conductor if it is copper-clad aluminum.
   C. as an insulated conductor run on a masonry surface but not terminated within 18 in. (450 mm) of the earth when run outside a building.
   D. except within 18 in. (450 mm) of the earth either inside or outside a building.
   E. except when in direct contact with masonry or outside within 18 in. (450 mm) of the earth.

   Code reference _____

4. Two parallel sets of size 350-kcmil copper conductors are run to a panelboard from the service equipment through two Rigid Nonmetallic Conduits with one set of the parallel conductors in each conduit as shown in Figure 5.31. The feeder to the panelboard is protected with 600-ampere fuses located in the service panel. A copper equipment grounding conductor is routed in each conduit with each set of parallel conductors. The minimum size equipment grounding conductor permitted to be run in each conduit is:

   A. 4 AWG.   C. 1 AWG.   E. 2/0 AWG.
   B. 2 AWG.   D. 1/0 AWG.

   Code reference _____

5. Two separate circuits are run in the same nonmetallic conduit. One circuit consists of size 8 AWG copper insulated conductors protected at 50 amperes, and the other circuit run with size 3 copper conductors protected at 100 amperes. One copper equipment grounding conductor is run in this conduit to serve both circuits. The minimum size equipment grounding conductor permitted is:

   A. 10 AWG.   C. 6 AWG.   E. 3 AWG.
   B. 8 AWG.   D. 4 AWG.

   Code reference _____

6. A 480-volt delta service entrance has two parallel sets of size 900-kcmil copper Type THWN ungrounded service conductors with each set in a separate service conduit as shown in Figure 5.32. The metal water line is used as a grounding electrode and is

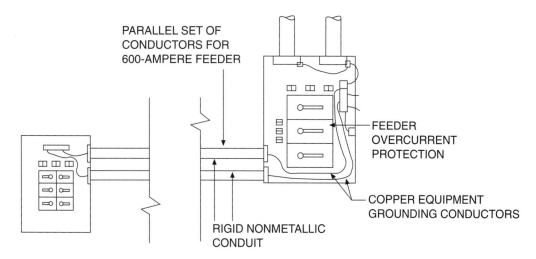

Figure 5.31   Determine the minimum size copper equipment grounding conductor permitted to be run in each conduit where a feeder protected with 600-ampere fuses is run as two parallel sets of conductors in Rigid Nonmetallic Conduit.

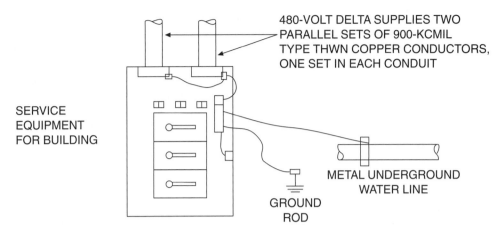

Figure 5.32   Determine the minimum size grounding electrode conductor to a metal water pipe for a service entrance consisting of two parallel sets of copper size 900-kcmil conductors.

supplemented with a driven ground rod. The copper grounding electrode conductor to the ground rods is size 6 AWG. The copper main bonding jumper to the service enclosure size 250 kcmil, and the copper bonding jumper to the metal service conduits is size 2/0 AWG. The minimum permitted size of copper grounding electrode conductor to the metal water pipe is:

A.   1/0 AWG.   C.   3/0 AWG.   E.   250 kcmil.

B.   2/0 AWG.   D.   4/0 AWG.

Code reference_____

7.   Two panelboards rated as suitable for use as service equipment make up a service for a building as shown in Figure 5.33. One panelboard has a 200-ampere main circuit breaker and the other has a 100-ampere main circuit breaker. The ungrounded service conductors are size 350-kcmil copper. The grounded circuit conductor is run to each service enclosure. The grounding electrode conductor is run from the 200-

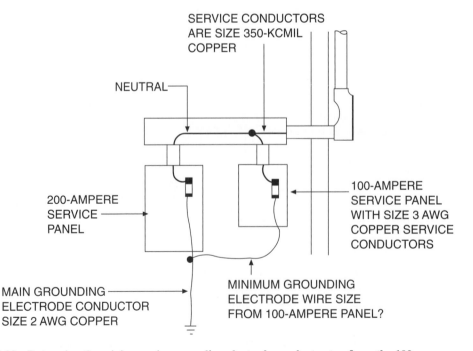

Figure 5.33   Determine the minimum size grounding electrode conductor tap from the 100-ampere panelboard grounded conductor termination point to the main grounding electrode conductor.

ampere panelboard to the grounding electrode. A grounding electrode conductor tap is run from the grounding bus in the 100-ampere panelboard to the grounding electrode conductor. The minimum permitted size grounding electrode conductor tap for the 100-ampere panelboard is:

A. 8 AWG.   C. 4 AWG.   E. 2 AWG.
B. 6 AWG.   D. 3 AWG.

Code reference _____

8. The copper Type THWN conductors for a 480-volt, 3-phase feeder protected by 150-ampere fuses is increased from a size 1/0 AWG to a size 3/0 AWG to compensate for voltage drop. If the feeder is run in Rigid Nonmetallic Conduit, the minimum size copper equipment grounding conductor permitted to be run for this feeder is:

A. 8 AWG.   C. 4 AWG.   E. 2 AWG.
B. 6 AWG.   D. 3 AWG.

Code reference _____

9. One building on the same property is supplied 120/240-volt, single-phase power from another building as shown in Figure 5.34. There is no metallic water pipe or other metal equipment connecting the buildings. The second building is supplied from a 3-wire feeder with two ungrounded conductors and a neutral. The neutral conductor is:

A. required to be bonded to the disconnect enclosure in the second building and connected to a grounding electrode.
B. not permitted to be bonded to the disconnect enclosure or connected to a grounding electrode in the second building.
C. only permitted to be connected to a grounding electrode at the supply end of the feeder in the first building.
D. only permitted to be connected to a grounding electrode at the second building.
E. not permitted to be connected to a grounding electrode at either building.

Code reference _____

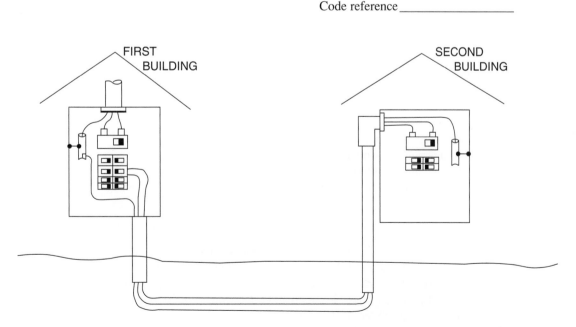

Figure 5.34 Electrical power in one building is supplied from another building on the same property with a 120/240-volt, 3-wire, single-phase feeder consisting of two ungrounded conductors and a neutral conductor.

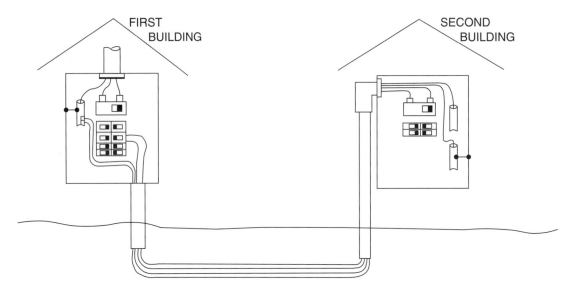

Figure 5.35   Electrical power in one building is supplied from another building on the same property with a 120/240-volt, 3-wire, single-phase feeder consisting of two ungrounded conductors, a neutral conductor, and a separate equipment grounding conductor.

10. One building on the same property is supplied 120/240-volt single-phase power from another building. The second building is supplied from a 4-wire feeder with two ungrounded conductors, a neutral conductor, and an equipment grounding conductor as shown in Figure 5.35. The neutral conductor at the second building is:
   A. not permitted to be connected to a grounding electrode at either building.
   B. required to be bonded to the equipment grounding conductor at the load end in the second building.
   C. required to be bonded to the disconnect enclosure in the second building and connected to a grounding electrode.
   D. only permitted to be connected to a grounding electrode at the second building.
   E. only permitted to be connected to a grounding electrode and bonded to the disconnect enclosure at the supply end of the feeder in the first building.

   Code reference _____

11. A 15- or 20-ampere, 125-volt receptacle used for the reduction of electrical noise where the grounding terminal is isolated from the receptacle mounting yoke is required to be identified by:
   A. a green grounding symbol.
   B. the words "isolated ground."
   C. being completely orange in color.
   D. having an orange triangle on the face of the receptacle.
   E. having a triangle and a grounding symbol on the face of the receptacle.

   Code reference _____

12. A duplex receptacle is installed in a commercial building that is wired with Electrical Metallic Tubing and metal device boxes. The receptacle is an isolated ground type with the grounding terminal insulated from the metal mounting yoke as shown in Figure 5.36. The receptacle is identified on the front with an orange triangle. The receptacle grounding terminal is required to be:
   A. bonded to the metal box.
   B. grounded with an insulated grounding wire run all the way back to the grounding bus at the service panel.

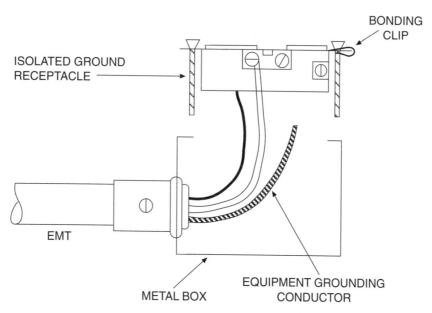

ISOLATED GROUND
RECEPTACLE

BONDING
CLIP

EMT

METAL BOX

EQUIPMENT GROUNDING
CONDUCTOR

Figure 5.36 An insulated ground receptacle marked with an orange triangle on the front is installed in a metal surface mounted device box supplied by conductors run in Electrical Metallic Tubing.

C. grounded with a bare grounding wire run all the way back to the grounding bus at the service panel.
D. left open with no conductor connected to the grounding terminal.
E. bonded with a short jumper to the neutral terminal on the receptacle.

Code reference_____

13. A hot tub installed outdoors at a single-family dwelling is required to have at least one receptacle on a general purpose 125-volt, 15- or 20-ampere branch-circuit installed not closer to the hot tub than 10 ft (3 m) nor a distance of more than:
A. 15 ft (4.5 m).   C. 25 ft (7.5 m).   E. 50 ft (15 m).
B. 20 ft (6 m).   D. 30 ft (9 m).

Code reference_____

14. A building is supplied electrical power from another building by means of a 208Y/120-volt, 4-wire overhead feeder where the grounded-circuit conductor acts as the equipment grounding conductor for the building as permitted in *250.32(B)(2)*. A transient voltage surge suppressor (TVSS) installed at the electrical supply to the building is:
A. required to be installed on the supply side of the disconnecting means to the building.
B. permitted to be installed on the supply side of the disconnecting means to the building.
C. required to be installed outside the building and preferably at the drip loop.
D. only permitted to be installed on individual circuit requiring protection from transient surges.
E. required to be installed on the load side of the first overcurrent device in the building.

Code reference_____

15. A transformer supplies a 120/240-volt, single-phase, 200-ampere panelboard from a 480-volt, 3-phase supply. The secondary ungrounded conductors from the transformer to the panelboard are size 3/0 AWG copper. This separately derived system installation is shown in Figure 5.37. This transformer is near the service disconnect to the building and it is grounded to the metal underground water pipe. The minimum size copper grounding electrode conductor permitted for this transformer installation is:

A. 10 AWG.

C. 6 AWG.

E. 2 AWG.

B. 8 AWG.

D. 4 AWG.

Code reference_____

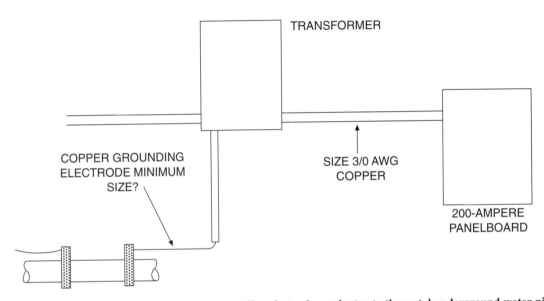

Figure 5.37 Determine the minimum size grounding electrode conductor to the metal underground water pipe for a transformer supplying a panelboard with a 200-ampere main circuit breaker with size 3/0 AWG copper conductors.

UNIT 6

Overcurrent Protection

OBJECTIVES

After completion of this unit, the student should be able to:

- name two types of overcurrent conditions.
- explain two ways in which conductors and equipment are damaged during overcurrent conditions.
- name two types of electrical fault conditions.
- explain interrupting rating as related to electrical equipment.
- name the two types of overcurrent conditions from which an overcurrent device protects.
- determine the voltage drop along a length of conductor.
- determine the minimum size wire that will limit voltage drop to a specific value for a given length of run and load.
- answer wiring installation questions relating to *Articles 240, 408, 550, 551,* and *552.*
- state at least five significant changes that occurred from the 2002 to the 2005 Code for *Articles 240, 408, 550, 551,* and *552.*

CODE DISCUSSION

The main emphasis of this unit is overcurrent protection. Several other articles are covered in which overcurrent protection is of particular importance. These additional topics are switchboards, panelboards, mobile homes, recreational vehicles, and park trailers.

Article 240 is the subject of overcurrent protection. The FPN to *240.1* describes the purpose of overcurrent protection. Rules on overcurrent protection for specific equipment are placed in the articles covering that specific equipment. Those articles and subjects are listed in *Table 240.3.* Some of those other rules will be discussed in other units of this text. The Code lists standard ratings of fuses and fixed-trip circuit breakers in *240.6(A).* In this and other articles of the Code, reference is sometimes made to rounding up, or rounding down to the next standard rating of overcurrent device. Fuses or circuit breakers are not necessarily available at all of these standard ratings. Manufacturers also have overcurrent protective devices available in other rated sizes. Common ratings of time-delay fuses from 1/10 to 30 amperes are listed in Table 7.2 in Unit 7.

NEC® 240.20(A) is fundamentally important in that a fuse, overload trip unit, or a circuit breaker is required to be installed in series with each ungrounded conductor. *NEC® 240.21* requires that the location of the overcurrent protective device be at the point where the conductors receive their supply. The cases where this is not true are for service conductors, sometimes conductors from generators, and for a special type of conductor called a tap. The definition of a tap is found in *240.2,* and an example of a tap is shown in Figure 6.1. The rules for sizing and installing feeder and transformer tap conductors are found in *240.21.* The tap rules for services are found in *230.46,* for motors circuits tap rules are found in *430.28* and *430.53.* A tap is permitted to be made to a branch-circuit to serve specific loads, but not to serve general use receptacles. This rule is in *210.19(A),* which also give rules for taps from range circuits to serve a counter-mounted cooking unit or a wall-mounted oven.

NEC® 240.4 requires that conductors be protected from overcurrent in accordance with the conductor ampacity as determined in *310.15.* This was discussed in several of the earlier units of this text. There are several specific situations where this rule does not apply, and those situations are listed in *Table 240.4(G).* According to *240.4(B),* if the overcurrent device protecting the conductors is not rated more than 800 amperes,

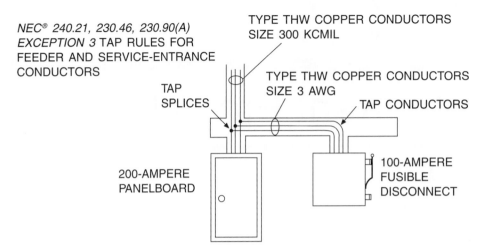

NEC® 240.21, 230.46, 230.90(A)
EXCEPTION 3 TAP RULES FOR
FEEDER AND SERVICE-ENTRANCE
CONDUCTORS

TYPE THW COPPER CONDUCTORS
SIZE 300 KCMIL

TYPE THW COPPER CONDUCTORS
SIZE 3 AWG

TAP
SPLICES

TAP CONDUCTORS

200-AMPERE
PANELBOARD

100-AMPERE
FUSIBLE
DISCONNECT

Figure 6.1   A conductor of a smaller size is permitted to be tapped from a conductor that is larger provided the tap conductor ends at an overcurrent device sized not larger than the allowable ampacity of the tap conductor.

and the allowable ampacity of the conductor does not correspond with a standard rating of overcurrent device, then it is permitted to protect that set of conductors with the next higher rating of standard overcurrent device. Assume for example a set of feeder conductors adequate to supply the intended load is size 300-kcmil copper. According to *Table 310.16*, this conductor is good for 285 amperes if there are no adjustments or corrections. It is permitted to protect this conductor with a 300-ampere-rated set of fuses or a circuit breaker. This rule is not permitted to be used in the case of branch-circuits serving multiple receptacles intended for plug-connected portable loads. According to *240.4(C)*, a conductor is required to have an allowable ampacity not less than the rating of an overcurrent device rated more than 800 amperes.

NEC® 240.4(D) is a special rule that applies to conductor sizes 10 AWG copper and smaller. This rule sets a maximum rating of overcurrent device that is permitted to protect circuits using sizes 14, 12, and 10 AWG copper wires, and sizes 12 and 10 AWG aluminum wires. There are some special cases like motor circuits where this rule does not apply.

Part V of *Article 240* deals with screw shell or plug fuses, which are rated 125 volts and up to 30 amperes. These fuses are permitted to be used to protect a circuit or equipment operating at 120 volts or 208 volts line-to-line where the voltage from either conductor is not more than 150 volts. This is permitted by *240.50(A)(2)* because two fuses are working together in series for a line-to-line short circuit, and they are rated to handle up to 150 volts on a ground fault. They are of the Edison-base type, which is the same as a standard medium-base incandescent lamp. They are also available as Type S noninterchangeable fuses where an adapter is inserted into the standard Edison-base fuse holder. The internal threads of the adapter will only fit the threads on a specific ampere range. The Code recognizes Type S fuses with adapters in the range of up to 15 amperes, 16 to 20 amperes, and 21 to 30 amperes. The adapter is required to be nonremovable. Figure 6.2 shows a Type S fuse and adapter being inserted into an Edison-base fuse screw shell fuse holder.

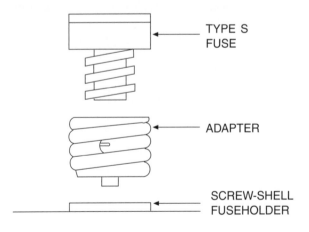

TYPE S
FUSE

ADAPTER

SCREW-SHELL
FUSEHOLDER

Figure 6.2   A Type S screw shell fuse with an adapter prevents replacing a fuse with one with a much higher rating.

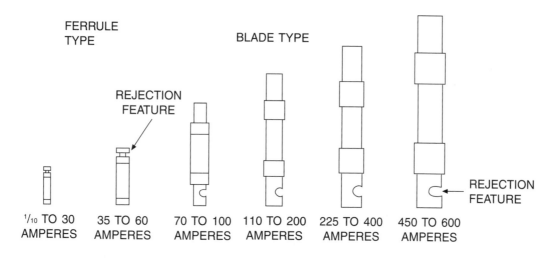

250-VOLT CARTRIDGE FUSE SIZES

Figure 6.3   Cartridge fuses are grouped into ranges of ratings that prevent replacement by a fuse with a rating from a higher range.

NEC® 240.52 requires that when fuse holders of the Edison base are installed, they are to be fitted with a Type S fuse adapter. This prevents 20- or 30-ampere fuses from being used on a circuit that requires not larger than a 15-ampere fuse. A 30-ampere fuse will not fit into an adapter that is rated for 20-ampere fuses.

Part VI of *Article 240* deals with cartridge fuses. Low-voltage cartridge fuses are rated either at 250 volts, 300 volts, or 600 volts. Fuses rated at 300 volts are considered to be Class G, and they are commonly used inside equipment. Fuses rated at 250 volts and at 600 volts are generally used for circuit protection. They are physically different in size so they cannot be interchanged. Fuses are grouped into sizes to limit the ratings that can be interchanged, and to be capable of handling the intended current as well as interrupting the current during an overload or fault. Relative sizes of cartridge fuses and the ratings are shown in Figure 6.3 for ratings up to 600 amperes. The physical size and rating of disconnect enclosure will change as the physical size of the fuses changes.

Fuses are made as one-time blow, or as time-delay. Figure 6.4 shows the internal arrangement of a time-delay fuse. There is a time-delay section and one or two short-circuit sections. There are several different methods of obtaining a time delay. With the type shown in Figure 6.4, solder holds contacts together. An overload that persists long enough will melt the solder and the spring will pull the contacts apart. The short-circuit section may consist of one conducting ribbon or several in parallel. Frequently there is one or

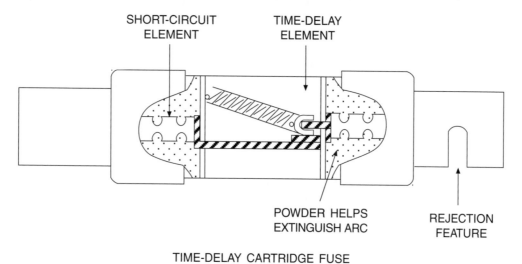

TIME-DELAY CARTRIDGE FUSE

Figure 6.4   A time-delay cartridge fuse has a time-delay element and one or more short-circuit elements.

more narrow points on the metal ribbon that will burn apart when a high current flow occurs. It usually takes more than six times the fuse rating to heat the short-circuit elements enough to burn them apart. There are frequently multiple narrow sections on the short-circuit element in series to increase the gap quickly during a short circuit to stop the arcing. On some fuses, the short-circuit elements are surrounded by a powder that also helps stop the arc quickly. Overcurrent devices that are capable of opening very quickly have the ability to limit the amount of current that is allowed to flow during a short circuit. These overcurrent devices have a high interrupting rating, and are often considered to be current limiting. *NEC® 240.60(C)* requires cartridge fuses to have a minimum interrupting rating of 10,000 amperes, and many are rated as high as 200,000 amperes. Circuit breakers are required to have an interrupting rating of 5000 amperes; however, most are rated at least 10,000 amperes. Other typical circuit-breaker interrupting ratings are 22,000, 42,000, and 65,000 amperes.

Circuit breakers are sometimes used as switches for lighting in a building. Electric discharge lighting such as fluorescent luminaires (lighting fixtures) or high-intensity discharge luminaires (lighting fixtures) have a ballast that contains inductors and capacitors. These components can store energy that gets released when the circuit is opened. To deal with this extra energy that may be present, a circuit breaker that is used frequently as a switch to interrupt current flow needs to be rated to handle the extra energy that must be interrupted. *NEC® 240.83(D)* requires circuit breakers that are intended to be used as switches for fluorescent luminaires (lighting fixtures) to be marked SWD, and if used as a switch for high-intensity discharge luminaires (lighting fixtures) to be marked HID. A circuit breaker marked HID is permitted to be used as a switch for both fluorescent and high-intensity discharge luminaires (lighting fixtures).

NEC® 210.12 requires receptacles in dwelling bedrooms on 15- or 20-ampere, 125-volt-rated branch-circuit to be protected by an arc-fault interrupter. Arcing faults can develop at outlets, in appliance cords, or in the circuit wiring that persist for long periods of time at levels too low for a fuse or circuit breaker to detect. These arcing faults over time can build up enough heat to start a fire. Arcing faults have an intermittent current-flow pattern similar to the current shown in Figure 6.5. The arc-fault interrupter is a circuit breaker that contains an electronic device to identify the current flow pattern of an arcing fault and trip off power to the circuit. Arc-fault interrupters are larger than a standard circuit breaker, but generally will fit in the space of a single-pole circuit breaker. They have a factory installed white insulated wire that is to be connected to the neutral bus. The circuit ungrounded and neutral wire are connected directly to the arc-fault interrupter as shown in Figure 6.6. A combination-type arc-fault circuit-interrupter (AFCI) is built into a receptacle similar to that of a GFCI receptacle. This type of AFCI is permitted to be installed up to 6 ft (1.8 m) from the service panel. The combination type is more sensitive to what is known as a series-arcing condition and provides a higher level of protection than the circuit-breaker-type AFCI.

NEC® 240.85 lists important voltage ratings of circuit breakers that are necessary to understand in order to avoid potentially dangerous installations. The voltage rating of the circuit breaker may be marked on the side of the circuit breaker, or it may be printed on a small label on the end next to the wire terminal. For low-voltage molded-case circuit breakers, voltages are listed as either a single voltage such as 240 Vac or two voltages such as 120/240 Vac. A circuit breaker may be called upon to clear a line-to-line short circuit, or a line-to-ground fault. A circuit breaker with a single voltage, such as 240 Vac, means the circuit breaker is able to withstand a line-to-line short circuit at 240 volts nominal. The circuit breaker is also able to withstand a line-to-ground fault of up to 240 volts nominal. This type of circuit breaker would be required on a 240-volt, 3-phase, 3-wire ungrounded electrical system. If one phase becomes faulted to ground, the other two phases will be operating at 240 volts-to-ground.

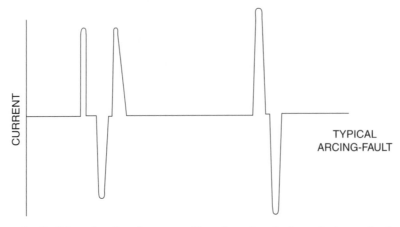

Figure 6.5 **An arcing fault in a circuit such as caused by a loose terminal or a fastener piercing a cable is characterized by an intermittent current flow.**

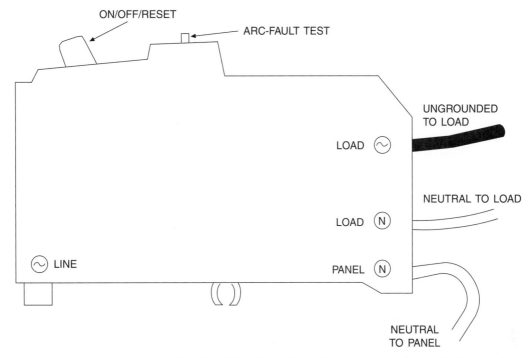

Figure 6.6 An arc-fault interrupter is a circuit breaker with an internal electronic circuit that can detect an arcing-fault current pattern and disconnect power to the circuit.

When a circuit breaker lists two voltages, the higher is the voltage line-to-line. A 120/240-volt rated circuit breaker is able to withstand a line-to-line short circuit at up to 240 volts nominal. The lower voltage is the maximum permitted line-to-ground voltage the circuit breaker can handle during a ground fault. Table 6.1 lists the different types of electrical systems and the typical circuit breaker voltage ratings available that

Table 6.1 Circuit breaker voltage ratings permitted for various circuits and different types of low voltage electrical systems.

Electrical System	1-pole circuit breaker	2-pole circuit breaker	3-pole circuit breaker
1-phase, 120/240 Vac 3-wire	120/240 Vac	120/240 Vac or 240 Vac	
1-phase, 120/208 Vac 3-wire	120/240 Vac	120/240 Vac or 240 Vac	
3-phase, 208Y/120 Vac 4-wire		120/240 Vac or 240 Vac	240 Vac
3-phase, 480Y/277 Vac 4-wire		480Y/277 Vac or 480 Vac	480 Vac
3-phase, 600Y/347 Vac 4-wire		600 Y/347 Vac or 600 Vac	600 Vac
3-phase, 240 Vac 3-wire ungrounded		240 Vac	240 Vac
3-phase, 480 Vac 3-wire, ungrounded		480 Vac	480 Vac
3-phase, 600 Vac 3-wire, ungrounded		600 Vac	600 Vac
3-phase, 240 Vac 3-wire, corner grounded		240 Vac	240 Vac
3-phase, 480 Vac 3-wire, corner grounded		480 Vac	480 Vac
3-phase, 240/120 Vac 4-wire, delta	120/240 Vac	120/240 Vac or 240 Vac	240 Vac
(using B-phase)	(Do not use)	240 Vac	
3-phase, 480Y/277 Vac 3-wire, high impedance grounded	(Do not use)	480 Vac	480 Vac

can be used for various circuits. Two-pole circuit breakers with a rating of 120/240 volts are not permitted to be installed in a 240/120-volt, 3-phase, 4-wire electrical system for a single-phase 240-volt load except between the A and C phases. The voltage from the B phase to ground is approximately 208 volts, which exceeds the line-to-ground voltage rating of the circuit breaker. A two-pole circuit breaker used in this panel for a 240-volt load connected to the B phase is required to have a single voltage rating of 240 volts.

Part VIII of Article 240 gives some special rules for what is called a supervised industrial installation. NEC® 240.92 gives some special rules for outside taps from feeders and taps from transformers that are different than for normal tap installations. The definition of a supervised industrial installation is found in 240.2. There must be at least one service operating at more than 150 volts-to-ground and over 300 volts between conductors. And, the minimum load is required to be 2500 kVA, which at 480 volts would be more than 3000 amperes. The installation is also required to have a qualified electrical maintenance staff on duty. A key difference in the tap rule for a set of conductors from the secondary of a transformer is that conductors are permitted to be up to 100 ft (30 m) in length, and are permitted to terminate in up to six overcurrent devices grouped together with a combined rating not exceeding the ampacity of the tap conductor. Taps that are run outside buildings are permitted to be of any length and terminate in up to six overcurrent devices grouped together with a combined rating not exceeding the ampacity of the tap conductors.

Article 408 covers the use, installation, and protection of switchboards and panelboards. A switchboard is defined in Article 100, and it is a large panel, frame, or assembly of panels in which are mounted switches, buses, overcurrent devices, and other protective equipment. Measurement instruments are also frequently included. This equipment is generally accessible from both the front and from the rear. A panelboard is defined in Article 100, and it is an assembly of electrical buses, overcurrent devices, and sometimes switches. It is used for the control of light, heat, and power circuits. The parts are usually installed in a cabinet accessible only from the front.

Panelboards are classified as power panels or as lighting and appliance branch-circuit panelboards. This classification is not necessarily how the panelboard is constructed, but rather how it is used. The same basic panelboard may be classified as a power panel in one installation and as a lighting and appliance branch-circuit panelboard in another installation. NEC® 408.34 defines a lighting and appliance branch-circuit as one that has a connection to the neutral and is rated 30 amperes or less. Several circuits are illustrated in Figure 6.7 with indication when they are considered lighting and appliance branch-circuits. The key is that the circuit must have a connection to the neutral, and the circuit must be rated not larger than 30 amperes. A 30-ampere circuit with two ungrounded conductors and a neutral to a clothes dryer is a lighting and appliance branch-circuit. A 40-ampere circuit to an electric range with two ungrounded conductors and a neutral is not a lighting and appliance branch-circuit. It is necessary to define the circuit in order to define the use of the panelboard.

NEC® 408.34(A) defines a lighting and appliance branch-circuit panelboard as one that has more than 10% of the overcurrent devices protecting lighting and appliance branch-circuits. NEC® 408.35 requires that

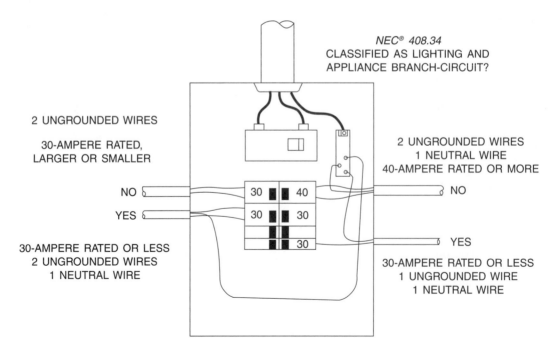

Figure 6.7   A lighting and appliance branch-circuit is one that is rated at not over 30 amperes and uses the neutral as a circuit conductor.

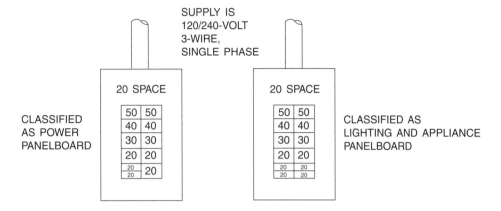

NEC® 408.34(A) LIGHTING AND APPLIANCE BRANCH-
CIRCUIT PANELBOARD HAS MORE THAN 10% OF CIRCUITS
30 AMPERES OR LESS USING A NEUTRAL CONDUCTOR

ASSUME NEUTRAL IS NOT USED WITH 20- AND 30-AMPERE,
2-POLE CIRCUIT BREAKERS

Figure 6.8 **A panelboard is classified as a lighting and appliance branch-circuit panelboard if more than 10% of the circuits are rated 30 amperes or less and the circuits have a neutral conductor.**

for the purpose of counting overcurrent devices, a 2-pole device is considered two devices and a 3-pole device is considered three devices. If a panelboard is not classified as lighting and appliance, it is a power panelboard. An example of both types of panelboards is illustrated in Figure 6.8. For the 20-space panelboard on the left, the circuit spaces rated 30 amperes or less with a connection to the neutral is two, which is 10% of the total spaces. The panelboard on the right has four spaces rated 30 amperes or less with a connection to the neutral, which is 20%.

NEC® 408.36(A) requires that a lighting and appliance branch-circuit panelboard be protected on the supply side by not more than two overcurrent devices with a combined rating not greater than that of the panelboard. Individual protection at the panelboard is not required if the panelboard is supplied by a feeder that has overcurrent protection rated not greater than the rating of the panelboard. Individual overcurrent protection for a power panel is not required unless there is a neutral in the panelboard, and more than 10% of the overcurrent devices are rated 30 amperes or less. This would mean that the power panelboard in Figure 6.8 would either require a main in the panelboard, or located somewhere on the supply side of the panelboard such as protecting the feeder. That overcurrent device is not permitted to have a rating greater than the rating of the power panelboard. If the neutral is not present in the panelboard, or if the there are 10% or fewer overcurrent devices rated 30 amperes or less, then the panelboard is not required to be protected on the supply side with an individual overcurrent device. *NEC® 408.30* does require the panelboard to have a rating not less than the load to be served.

NEC® 408.36(B) Exception makes a special case when a power panelboard is installed as service equipment. A panelboard that qualifies as a power panelboard and has not more than six disconnects, is permitted to be installed as service equipment without individual overcurrent protection.

Only one neutral conductor is permitted to be connected to a terminal of the neutral bus in a panelboard unless the terminal is specifically listed for more than one wire. This rule is in *408.41* and is illustrated in Figure 6.9. There is usually a label located on the inside of a panelboard enclosure that states the maximum number of equipment grounding wires of various sizes that are permitted to be installed at a single terminal.

Article 550 begins with some definitions in *550.2* that are important for understanding the intent of the provisions of this article. *Part II* applied to wiring within the mobile home or manufactured home, the electrical disconnecting means, and the distribution equipment for the mobile home. It is important to understand that a mobile home is movable; therefore, the service drop or service lateral would logically connect to service equipment that is not mobile. The service equipment or disconnecting means for a mobile home is required by *550.32(A)* to be located adjacent to the mobile home, and not on or within the mobile home. This is illustrated in Figure 6.10. *NEC® 550.18* gives a method to determine the minimum permitted rating of the electrical power supply to the mobile home. The minimum permitted size of power supply cord or feeder to the mobile home is given in *550.10. Part III* covers services for mobile homes and

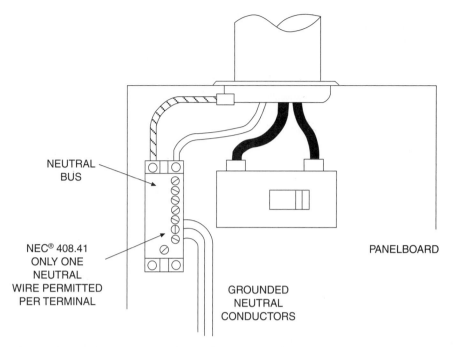

NEUTRAL
BUS

NEC® 408.41
ONLY ONE
NEUTRAL
WIRE PERMITTED
PER TERMINAL

GROUNDED
NEUTRAL
CONDUCTORS

PANELBOARD

Figure 6.9   Only a single neutral conductor is permitted to be attached to a terminal in a panelboard.

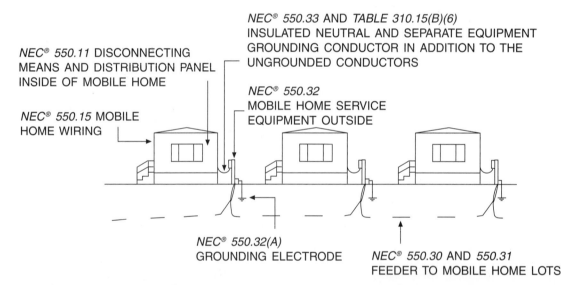

NEC® 550.11 DISCONNECTING
MEANS AND DISTRIBUTION PANEL
INSIDE OF MOBILE HOME

NEC® 550.15 MOBILE
HOME WIRING

NEC® 550.33 AND TABLE 310.15(B)(6)
INSULATED NEUTRAL AND SEPARATE EQUIPMENT
GROUNDING CONDUCTOR IN ADDITION TO THE
UNGROUNDED CONDUCTORS

NEC® 550.32
MOBILE HOME SERVICE
EQUIPMENT OUTSIDE

NEC® 550.32(A)
GROUNDING ELECTRODE

NEC® 550.30 AND 550.31
FEEDER TO MOBILE HOME LOTS

Figure 6.10   Service equipment for a mobile home is required to be located adjacent to the mobile home and not on or within the mobile home.

feeders in a mobile home park. The service for a mobile home in a mobile home park is permitted to be remote from the mobile home, but a disconnecting means for the mobile home is required to be installed according to the rules of *550.32(A)*.

The minimum rating of service equipment for a mobile home is 100 amperes. The mobile home is permitted to be supplied power with permanent wiring or with a cord. Cords rated 50 amperes are most common, although in some cases, a 40-ampere cord is permitted. After installation, there is to be a minimum length of 20 ft (6 m) of cord from the point of attachment to the mobile home to the end of the plug. *NEC® 550.11(A)* requires a single main disconnect in the mobile home near the point of entrance of the service conductors. This is generally in the distribution panel. The rating of a circuit breaker acting as the disconnect is 50 amperes for a 50-ampere supply cord, and 40 amperes for a 40-ampere supply cord. Whether the supply conductors are by means of permanent wiring or a supply cord, the neutral and equipment ground are separated in the distribution panel. The neutral bus is insulated from the enclosure, and the equipment grounding bus is required to be bonded to the enclosure. The supply conductors from the service equipment located adjacent to the mobile home to the distribution panel in the mobile home is required to have an insulated green equipment grounding conductor and an insulated white neutral conductor. These rules are summarized in Figure 6.11 along with the minimum circuit requirements for a mobile home.

Article 551 contains definitions in *551.2* that are important for understanding wiring requirements for recreational vehicles and recreational vehicle parks. Low voltage for recreational vehicle applications is defined in *551.2* as 24 volts or less either ac or dc. Low-voltage wire types used in recreational vehicles are specified by the Society of Automotive Engineers (SAE). It is important that a separation be maintained between conductors carrying direct current and those carrying alternating current. *Part III* of *Article 551* deals with the installation of wiring intended to operate from a combination of power sources. For example, the circuits may be supplied from a battery, or from a 120-volt alternating current source. *Part IV* applies to the installation of power sources, such as an engine-driven generator or batteries. When there are multiple power sources, such as a generator and a power supply cord, a transfer switch shall be installed to prevent interconnection of the power sources if such a transfer device is not supplied as part of the generator.

Part V of *Article 551* covers the installation of wiring to be supplied from a nominal 120-volt or 120/240-volt electrical system. *NEC® 551.41* provides the requirements for the spacing and type of receptacle outlets. The receptacle outlet next to a lavatory in the bath is required to be ground-fault circuit-interrupter protected. Also, all receptacle outlets installed to serve a counter space that are within 6 ft (1.8 m) of a lavatory or sink are required to be ground-fault circuit-interrupter protected. This provision primarily applies to kitchen counters. *NEC® 551.41(C)(2)* does not apply to receptacles located within 6 ft (1.8 m) of

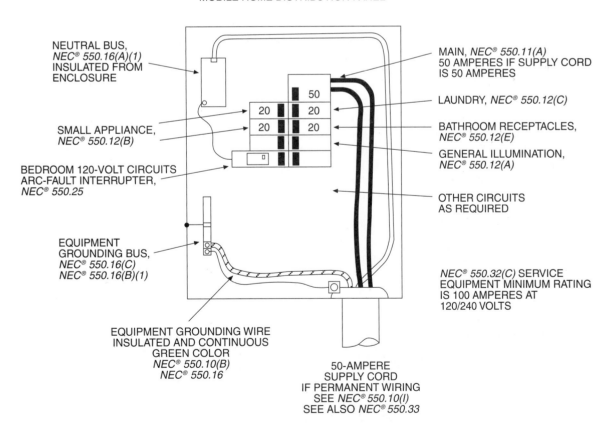

Figure 6.11   A distribution panel in a mobile home is supplied with a feeder consisting of all insulated conductors, two ungrounded, one neutral, and one equipment grounding conductor where the neutral terminal block is not bonded to the panel enclosure.

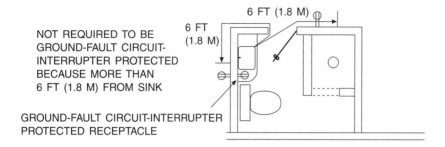

NOT REQUIRED TO BE GROUND-FAULT
CIRCUIT-INTERRUPTER PROTECTED
BECAUSE NOT INTENDED TO SERVE
THE COUNTERTOP

6 FT (1.8 M)

6 FT
(1.8 M)

NOT REQUIRED TO BE
GROUND-FAULT CIRCUIT-
INTERRUPTER PROTECTED
BECAUSE MORE THAN
6 FT (1.8 M) FROM SINK

GROUND-FAULT CIRCUIT-INTERRUPTER
PROTECTED RECEPTACLE

NEC® 551.41(C)(2) RECEPTACLES WITHIN 6 FT (1.8 M) OF THE
BATHROOM SINK INTENDED TO SERVE COUNTERTOPS ARE
REQUIRED TO BE GROUND-FAULT CIRCUIT-INTERRUPTER PROTECTED

Figure 6.12   Receptacle outlets installed within 6 ft (1.8 m) of the lavatory in the bathroom of a recreational vehicle are not required to be ground-fault circuit-interrupter protected unless intended to serve the counter tops.

a bathroom lavatory and not intended to serve a counter space as shown in Figure 6.12. *NEC® 551.42* covers the minimum branch-circuit requirements for a recreational vehicle. The type and installation of the distribution panelboard including working clearances are covered in *551.45*. The means of connecting the recreational vehicle to an external power supply are covered in *551.46*. *NEC® 551.47* describes the wiring methods permitted to be used in a recreational vehicle for 120-volt or 120/240-volt electrical systems. The methods of grounding equipment are covered in *551.55*.

Part VII provides requirements for determining the minimum size and installation of equipment at each recreational vehicle site and the feeders supplying the sites. Recreational vehicle parks are only permitted to be supplied power by a single-phase, 120/240-volt, 3-wire electrical system according to *551.72*. Of the recreational vehicle sites with electrical power, every one is required to provide a 20-ampere, 125-volt receptacle, according to *551.71*. A minimum of 70% are required to be equipped with a 30-ampere, 125-volt receptacle, and a minimum of 20% of the sites are required to be equipped with a 120/240-volt, 50-ampere receptacle. Other receptacle configurations are permitted to be provided. All 15- or 20-ampere, 125-volt recreational vehicle site receptacles are required to be ground-fault circuit-interrupter protected. The recreational vehicle site supply equipment is required to contain a disconnect for the power and one or more receptacles as required for the park. The location where the recreational vehicle can be parked is called the stand. The receptacle and disconnect are to be located at the side of the stand with the dimensions described in *551.77*. The equipment is to be located at least 2 ft (600 mm) above ground, but not more than $6^{1}/_{2}$ ft (2 m). The location of the power supply pedestal for a back-in recreational vehicle site is on the driver side of the vehicle as shown in Figure 6.13. For a drive-through recreational vehicle site, the power supply pedestal is to be located from the centerline of the stand as shown in Figure 6.14.

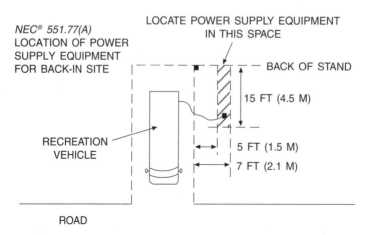

NEC® 551.77(A)
LOCATION OF POWER
SUPPLY EQUIPMENT
FOR BACK-IN SITE

LOCATE POWER SUPPLY EQUIPMENT
IN THIS SPACE

BACK OF STAND

15 FT (4.5 M)

RECREATION
VEHICLE

5 FT (1.5 M)

7 FT (2.1 M)

ROAD

Figure 6.13   The power supply pedestal for a back-in recreational vehicle site is located on the driver's side of the vehicle and distances are measured from the back corner of the stand.

NEC® 551.77(A) LOCATION OF POWER SUPPLY
EQUIPMENT FOR DRIVE-THROUGH SITE

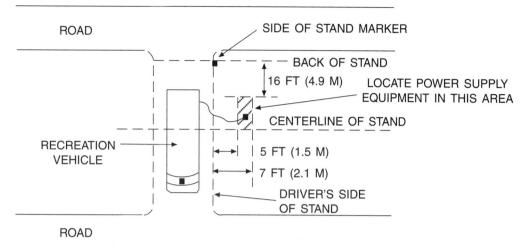

Figure 6.14 **The power supply pedestal for a pull-through recreational vehicle site is located on the driver's side of the vehicle, and distances are measured from the centerline of the stand.**

Article 552 specifies the rules for wiring on park trailers. A park trailer is defined in *552.2* as a unit that is built on a single chassis which is mounted on wheels, and the unit gross area does not exceed 400 sq. ft (37 m²) when set up for occupancy. These units are ones that are intended for seasonal use, and they are not intended to have commercial uses or to be occupied as a permanent dwelling. A park trailer has differences that are not addressed in *Article 550* on mobile homes or in *Article 551* on recreational vehicles. This article does cover the wiring of lighting required to transport the trailer on public roads and highways. Electrical wiring operating from an alternating current source of 120 volts or 120/240 volts is specified in *Part IV.* The minimum permitted size of distribution panelboard and power supply cord or feeder for the park trailer can be determined by using the method of *Article 220, Part III,* or the size determination can be made by using *552.47.*

OVERCURRENT PROTECTION FUNDAMENTALS

The purpose of overcurrent protection is to open a circuit if the temperature in the wiring or equipment rises to an excessive or dangerous level. This is discussed in the FPN to *240.1.*

Every circuit shall be protected by two levels of overcurrent protection: overload and short circuits or ground faults. Usually one overcurrent device provides both levels of protection, but in the case of a motor circuit, this protection is often provided at two locations. In a typical motor branch-circuit, short-circuit and ground-fault protection is provided by the fuse or circuit breaker, while overload protection is provided by the thermal unit in the motor starter.

Overloads

An overload is a gradual rise in current level above the current rating of the wires in the circuit or in the equipment. Heat is produced as the current flows through the wires. The two important variables for conductor heating are current level and time, in addition to resistance. Equation 1.16 discussed in *Unit 1* may be used to determine the heat produced as electrical current flows in a conductor for a given length of time.

$$\text{Heat} = \text{Current}^2 \times \text{Resistance} \times \text{Time}$$

Heat is in joules
Current is in amperes
Resistance is in ohms
Time is in seconds

Faults

A high level of current generally flows during a fault. Common faults are short circuits between wires and faults from an ungrounded wire to ground. This condition is called a **ground fault.** If arcing occurs, then the fault is also referred to as an **arcing fault,** such as the case when a wire touches the conduit causing an **arcing ground-fault.** If two conductors with voltage between them are accidentally connected together, arcing generally will not occur. If the conductors are connected together so that arcing does not occur, this is termed a **bolted fault.**

Symmetrical and Asymmetrical Current

Understanding overcurrent protection requires an understanding of the behavior of electrical current during a ground fault or a short circuit. The current that flows generally is many times higher than the rating of the overcurrent device. When a graph of the current is made over time, it draws out the shape of a sine wave as shown in Figure 6.15. If the current wave is centered about the zero axis, then it is called symmetrical current. If a direct current is added to the voltage supply, then the current wave will be offset either positively or negatively as shown in Figure 6.15. During a fault, reactive loads supplied by the electrical system such as motors, can cause a dc offset. This is called asymmetrical current. The effect of this offset is that the peak current can be considerably higher than if the offset is not present. During an actual fault in the circuit, there is often a dc offset, but it generally only lasts for a few cycles. The fault current may start out asymmetrical and quickly become symmetrical as shown in Figure 6.15. Even though the offset lasts only for a very short time, the peak current will occur in just one quarter cycle or about 0.004 seconds (4 ms). Another important number is the rms current. For a steady current flow, this is 0.707 times the peak current. The rms current is represented by the dashed line in Figure 6.15. During an actual fault, the rms current starts out high and quickly decreases to a steady value.

How Damage Occurs

When a fault occurs, the amount of current that flows depends on the impedance (resistance) of the fault—the impedance of the wires, transformer, and electrical system feeding the fault. The circuit voltage between the faulted wires is also important in determining the amount of fault current that will flow. The current flow during a fault is often in the thousands of amperes.

High current flow during a fault will cause strong magnetic forces that can move wires and deform electrical parts if the parts are not adequately secured and braced to withstand these forces. This mechanical force is proportional to the square of the peak current (I_{peak}^2). Arcing will occur when parts with a voltage between them touch together. This may be a secondary effect after the magnetic forces deform the parts. Heating will occur in conductors and components of the electrical system. Contacts may be welded together from heat produced by the rms current flow. Heating during a fault is proportional to the square of the rms current times time ($I_{rms}^2 \times t$).

If an electrical conductor is overloaded with current for a long enough period of time, the conductor will get hot. The conductor insulation will begin to sustain damage when the temperature rating of the conductor is exceeded, such as 75°C (167°F). Short circuits only last for a brief time, but the heating of the conductor is con-

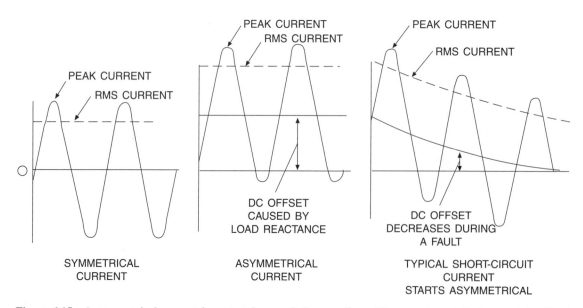

Figure 6.15 A symmetrical current is centered around the zero line while an asymmetrical current is off-set either positively or negatively from the zero line resulting in a higher than normal peak and rms current.

trolled by the length of time and the square of the rms current. Testing by the Insulated Cable Engineers Association[1] has revealed that a size 8 AWG copper conductor with thermoplastic insulation can withstand up to 6800 amperes for one cycle (0.017 sec.) without damage. For one full second (60 cycles) the same conductor can withstand only 850 amperes without damage. A size 3/0 AWG copper conductor with thermoplastic insulation can withstand up to approximately 68,000 amperes for one cycle and 8800 amperes for one second. The goal of an overcurrent device is to cut off current in time to prevent damage to conductors and equipment.

Interrupting Ratings of Fuses and Circuit Breakers

NEC® 110.9 requires that equipment intended to interrupt current at fault levels must have an interrupting rating sufficient for the circuit voltage and the current that is available at the supply terminals of the equipment. Circuit breakers, fuses, knife switches, and contactors are examples of components that are required to interrupt circuits at fault-current levels. An interrupting rating is the highest current a device is intended to interrupt at rated voltage under standard test conditions. *NEC® 110.10* requires overcurrent devices to be capable of opening the circuit during a fault so as to prevent extensive damage to electrical components of the circuit. The component of the electrical system that is most frequently exposed to the highest available current during a short circuit or a fault is the service equipment. This is why in *230.66* the service equipment is required to be marked as suitable for use as service equipment. The withstand rating of equipment is the maximum current equipment can safely handle at the system voltage.

The responsibility of the electrician is to know when excessive short-circuit currents can occur. Standard circuit breakers are only required to withstand 5000 amperes of short-circuit current. However, most are rated at 10,000 amperes as stamped on the circuit breaker. Fuses are required to have an interrupting rating of at least 10,000 amperes. Fuses are commonly available having interrupting ratings up to 200,000 amperes. Some fuses have an interrupting rating as high as 300,000 amperes.

Fuses that have an interrupting rating higher than the minimum 10,000 amperes are manufactured with a rejection feature as shown in Figure 6.3 and Figure 6.4. A pin can be installed into the fuse holder that will accept only fuses with the rejection feature. This prevents fuses that only have a 10,000-ampere interrupting rating from being installed in rejection fuse holders where short-circuit currents higher than 10,000 amperes are available.

Current Limiting

The amount of short-circuit current that will flow if there is a short circuit or a ground fault near service equipment may be higher than the interrupting rating of the overcurrent device at the service equipment. In this case the overcurrent device can be replaced with one that has an interrupting rating greater than the available short-circuit current. The dashed sine wave in Figure 6.16 shows the level of current that would

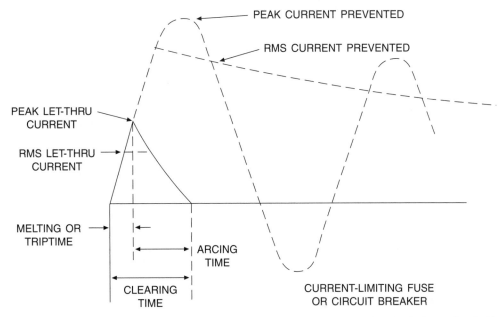

Figure 6.16 An overcurrent device that is current limiting is able to open the circuit within the first half-cycle of fault-current flow, which limits the peak and rms current that are let-thru the device.

---

[1]Insulated Cable Engineers Association, P.O. Box 1568, Carrollton, GA 30112.

flow if the overcurrent device was incapable of limiting the flow. If a circuit-breaker panelboard is installed adjacent to this main overcurrent device then all of the circuit breakers must have an interrupting rating not less than this current. Many fuses, and some circuit breakers open the circuit so rapidly the peak current of the first half cycle never gets through to load side equipment. This type of overcurrent device is called current limiting. The solid line in Figure 6.16 represents the short-circuit current during a fault. For a fuse, the current increases until the short-circuit element melts. Then there is a short period of arcing within the fuse as the current flow is terminated. The peak let-thru current of the fuse is much less than the original peak current. And the rms let-thru current is much less than the previous rms current.

Series Rating

In the case where a fusible switch acts as the service disconnect and it is followed immediately by a circuit-breaker panelboard, the two overcurrent devices may operate at the same time to open when there is a short circuit or a ground fault. If the circuit breaker immediately follows the fuse as shown in Figure 6.17, and they both act at the same time to clear a fault, they are considered to be a series-rated system. The same would be true for a circuit-breaker panelboard with a main circuit breaker acting as the service disconnect. The main circuit breaker immediately followed by a branch-circuit or feeder circuit breakers are considered to be a series-rated system. If both overcurrent devices have an interrupting rating not less than the available short-circuit current, then this is considered to be a fully-rated system. Each overcurrent device is capable of opening the circuit during a ground fault or short circuit without the help of the other circuit breaker.

With a series-rated system, the first circuit breaker or fuse must have an interrupting rating not less than the available short-circuit current. The circuit breaker that immediately follows is permitted to have an interrupting rating less than the available short-circuit current if the two overcurrent devices acting at the same time have a listed series interrupting rating greater than the available short-circuit current. Example interrupting ratings are shown in Figure 6.17. In each case, the first overcurrent device has an interrupting rating greater than the available short-circuit current. The circuit breakers in the panelboard have an interrupting rating less than the available short-circuit current. Note in both cases in Figure 6.17, the circuit breakers in the panelboard acting in series with the first overcurrent device each have a series-interrupting rating that is greater than the available short-circuit current. These are acceptable installations provided the equipment is labeled as a series-rated system, *240.86(A)* and *110.22*. Suggested wording for the series rated-label is given in *110.22*. Specific replacement parts required are listed as manufacturer and model number of fuses and circuit breakers.

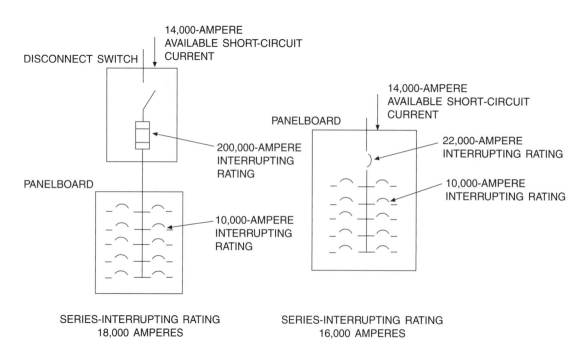

Figure 6.17 **Two overcurrent devices in series such as a fuse and circuit breaker or two circuit breakers opening at the same time can act to limit the let-thru current during a short circuit or ground fault to protect load side equipment from excessive current. This arrangement is called a series-rated system.**

Feeder Taps

A conductor is permitted to be tapped from a feeder where the feeder overcurrent device has a higher rating than the allowable ampacity of the tap conductor. This is permitted where the tap conductor terminates at an overcurrent device with a rating not higher than the allowable ampacity of the tap conductor. This practice, illustrated in Figure 6.18, is permitted by *240.4(E)*. The overload protection is provided by the fuse or circuit breaker at the tap termination. The short-circuit and ground-fault protection is provided by the fuses or circuit breaker protecting the feeder. There are several rules for determining the size of tap conductor with respect to the rating of the feeder overcurrent protection. These rules are found in *240.21* and are based upon the distance from the tap point to the tap overcurrent device and the degree of hazard due to location. A transformer is permitted to be installed along a tap, but that case will be discussed in *Unit 8*. The secondary conductors leaving a transformer are also considered to be tap conductors and those cases will be discussed in *Unit 8*.

The rules for a 10 ft (3 m) tap are covered in *240.21(B)(1)*. The distance from the tap point to the tap overcurrent device is not permitted to be more than 10 ft (3 m). The tap conductor is required to be enclosed in raceway. The tap conductor is required to have an allowable ampacity not less than the load to be served and not less than the rating of the fuse or circuit breaker at its termination. And finally, the tap conductor must have an allowable ampacity not less than 10% of the rating of the overcurrent device protecting the feeder. These rules are illustrated in Example 6.1.

Example 6.1   A feeder protected by a 200-ampere circuit breaker consists of size 4/0 AWG copper Type THW conductors run in Rigid Metal Conduit with several pull boxes located along the length of the feeder. The feeder is tapped at one of the pull boxes to supply an electrical resistance heater with a full-load current of 37 amperes. The distance from the tap point to a 50 ampere rated overcurrent device is just under 10 ft (3 m) as shown in Figure 6.18. The copper Type THWN tap conductors are run in EMT with all terminations rated 75°C. Determine the minimum size tap conductor.

Answer: For this situation there are two points to be considered. First, meet the conditions of *240.21(B)(1)(1)b*. The tap conductor must have an allowable ampacity not less than the overcurrent device at its termination. The overcurrent device has a 50-ampere rating. Therefore, find a copper conductor size that has an allowable ampacity of not less than 50 amperes in the 75°C column of *Table 310.16*. The minimum size permitted is 8 AWG. The other condition is that of *240.21(B)(1)(4)*. The rating of the overcurrent device protecting the feeder is not permitted to have a rating higher than 10 times the allowable ampacity of the tap conductor, which in this case is 500 amperes (50 A × 10 = 500 A). Another approach is to divide the feeder overcurrent device rating by 10 to determine the minimum allowable tap conductor ampacity which is 20 amperes (200 A / 10 = 20 A). The size 8 AWG copper conductor required to supply the load is adequate to also meet this rule. The minimum size tap conductor for this installation is 8 AWG copper.

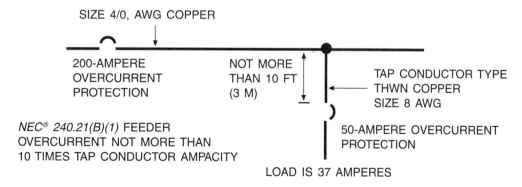

Figure 6.18   A tap is a conductor of lower ampacity connected to a feeder conductor and ending at a single overcurrent device that has a rating not greater than the ampacity of the tap conductor. If the distance from the tap point to the overcurrent device is not more than 10 ft (3 m), the tap conductor is permitted to have an ampacity as low as 10% of the feeder overcurrent device rating.

The tap is permitted to have a total length of 25 ft (7.5 m) from the tap point to the overcurrent device in cases where it is not practical to limit the length of the tap to 10 ft (3 m). The rules for a 25 ft (7.5 m) tap are found in *240.21(B)(2)*. The tap conductor is required to be suitably protected or run in raceway. The tap conductor is required to terminate at a single set of fuses or a circuit breaker with a rating not higher than the allowable ampacity of the tap conductor. And finally, the allowable ampacity of tap conductor is not permitted to be less than one-third the rating of the feeder overcurrent device. Determining the size of a tap conductor for a load using the 25 ft (7.5 m) tap rule is demonstrated by Example 6.2.

Example 6.2   A feeder protected by a set of 400-ampere fuses consists of size 500-kcmil copper Type THWN conductors run in Electrical Metallic Tubing with several pull boxes located along the length of the feeder. The feeder is tapped at one of the pull boxes to supply a load that draws 46 amperes, with the tap terminating at a switch containing 60-ampere fuses. The distance from the tap point to the 60-ampere fuses is just under 25 ft (7.5 m) as shown in Figure 6.19. The tap conductors are copper, Type THWN, all terminations are rated 75°C, and the tap conductors are suitably protected from physical damage. Determine the minimum size tap conductor.

Answer: For this situation there are two points to be considered. First, meet the conditions of *240.21(B)(2)(2)*. The tap conductor must have an allowable ampacity not less than the overcurrent device at its termination. The fuses have a rating of 60 ampere. Therefore, find a copper conductor size that has an allowable ampacity of not less than 60 amperes in the 75°C column of *Table 310.16*. The minimum size copper tap conductor permitted is 6 AWG. The other condition is that of *240.21(B)(2)(1)*. The rating of the overcurrent device protecting the feeder is not permitted to be less than one-third the rating of the overcurrent device protecting the feeder which would be 133 amperes (400 A / 3 = 133A). The tap conductor is required to have an allowable ampacity of not less than 133 amperes even if the load served is much smaller. This rule requires a size 1/0 AWG copper conductor from the tap point to the disconnect switch. The conductors on the load side of the 60-ampere fuses is permitted to be size 6 AWG. The minimum tap conductor size permitted for this installation is 1/0 AWG copper.

In the case of a manufacturing building with a room that is not less than 35 ft (11 m) above the floor at the walls, it would be impractical to tap a feeder run along the underside of the roof. It would not be possible to get to an overcurrent device accessible from the floor within the 25 ft (7.5 m) limit. *NEC® 240.21(B)(4)* permits the tap to be up to 100 ft (30 m) in length in the high ceiling area of a manufacturing building where the tap is actually made not less than 30 ft (9 m) above the floor. The tap is not required to be directly below the tap point. The single overcurrent device at the tap termination is permitted to be located anywhere within 25 ft (7.5 m) measured horizontally of a point directly below the tap to the feeder as illus-

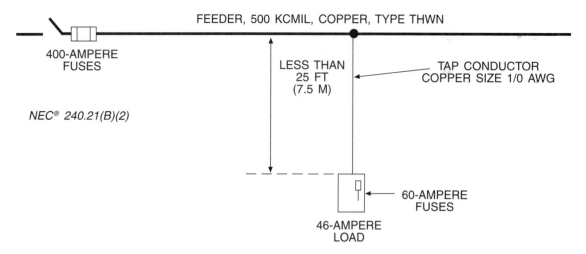

Figure 6.19   **If the distance from the tap point to the overcurrent device is not more than 25 ft (7.5 m) and more than 10 ft (3 m), the minimum ampacity of the tap conductor is not permitted to be less than the single overcurrent device at its termination or one-third the rating of the feeder overcurrent device, whichever is higher.**

trated in Figure 6.20. This conductor is required to be not smaller than size 6 AWG copper or size 4 AWG aluminum and they are not permitted to penetrate any walls, floors, or ceilings. The remaining rules are the same as for 25 ft (7.5 m) feeder taps. Example 6.3 is a case in point of a feeder supplying two panelboards in a high ceiling manufacturing building.

Example 6.3   A feeder protected by a 400-ampere circuit breaker consists of size 600-kcmil copper Type THWN conductors run in Electrical Metallic Tubing across the ceiling of a manufacturing building to supply two panelboards each with a 200-ampere main overcurrent device. The feeder is run 40 ft above the floor with the tap points also 40 ft above the floor. The tap conductors are run in Rigid Metal Conduit down a structural support to the panelboard mounted with the top 6 ft (1.83 m) above the floor. The tap conductors are copper, Type THWN, and all terminations are rated 75°C. Determine the minimum size tap conductor.

Answer: For this situation there are two points to be considered. First, meet the conditions of *240.21(B)(4)(4)*. The tap conductor must have an allowable ampacity not less than the overcurrent device at its termination. The overcurrent device has a 200-ampere rating, therefore, find a copper conductor size that has an allowable ampacity of not less than 200 amperes in the 75°C column of *Table 310.16*. The minimum size permitted is 3/0 AWG. The other condition is that of *240.21(B)(4)(3)*. The tap conductor ampacity is not permitted to be less than one-third the rating of the feeder overcurrent device. According to this rule, the tap conductor is required to have an ampacity not less than 133 amperes (400 A / 3 = 133 A). The load requires size 3/0 AWG. Therefore, the ampacity required by this rule is easily exceeded. The minimum size tap conductor for this installation is 3/0 AWG copper with a total length of approximately 34 ft (10.36 m).

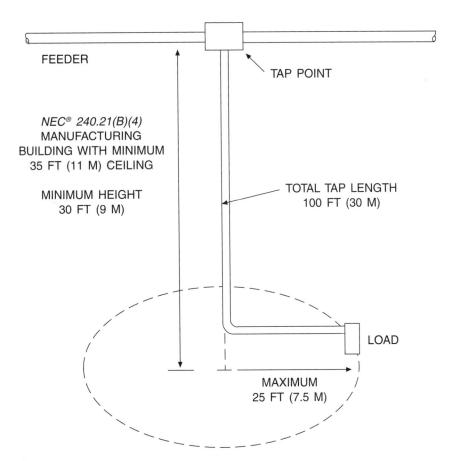

Figure 6.20   **In a manufacturing building with a ceiling not less than 35 ft above the floor at the walls, a tap to a feeder made at least 30 ft above the floor is permitted to have a total length of not more than 100 ft and terminate at a single overcurrent device rated not greater than the ampacity of the conductor. The tap conductor ampacity is also not permitted to be less than one-third the rating of the feeder overcurrent device.**

Tap conductors, either run overhead or underground, are frequently outdoors except at the point of entry to a building. *NEC® 240.21(B)(5)* permits a feeder conductor that is run outside a building to be tapped, outside the building, to supply another load provided the feeder has an adequate ampacity, as determined by *215.2,* to supply all loads. The tap conductor is permitted to be of any length provided it is run completely outside except at the point of entry to the building supplied. The tap conductor is required to be sized to the load served, but it is not required to be limited by the rating of the feeder overcurrent device. Example 6.4 will show an application of the outside tap rule.

Example 6.4   A group of several buildings are supplied power from a central distribution point consisting of a 400-ampere fusible switch rated as suitable for use as service equipment, and located outdoors near the utility transformer pole. A tap box is installed at this location and the buildings are supplied using underground aluminum Type USE single-conductor direct burial cable. A tap box is installed beside the service disconnect and no overcurrent protection other than the 400 amperes fuses in the service disconnect protects the taps to the buildings. One of the taps is 150 ft in length and terminates at a panelboard with a 100-ampere main circuit breaker. The tap is illustrated in Figure 6.21. All terminations are 75°C rated. Determine the minimum size aluminum tap conductor.

Answer: There is only one condition that must be met to size the conductor—provided this qualifies as an outside tap. The allowable ampacity of the conductor is not permitted to be less than the rating of the 100-ampere overcurrent device at its termination. Look up the minimum conductor size in the 75°C column of *Table 310.16*. The minimum tap conductor size for this situation is 1 AWG aluminum, Type USE.

VOLTAGE DROP

The minimum size conductor for a branch-circuit or feeder may not be adequate if the circuit or feeder length causes an excessive voltage drop. Excessive voltage drop results in inefficient operation of electrical equipment, sometimes malfunctioning of control systems, and premature motor failure. Voltage drop is caused by the current experiencing the resistance of the circuit. The more current, the more voltage drop caused by the circuit conductors. This is simple Ohm's law such as described in Equation 1.4 in *Unit 1*. The solution is to reduce the resistance of the conductors which is described by Equation 1.2 in *Unit 1*. An actual circuit is two wires for single-phase and three wires for 3-phase. Therefore, Equation 1.4 must be modified to account for all circuit conductors. Equation 6.1 can be used to determine voltage drop in a single-phase circuit. Equation 6.2 accounts for the current flowing on all three conductors and gives the voltage drop for a 3-phase circuit.

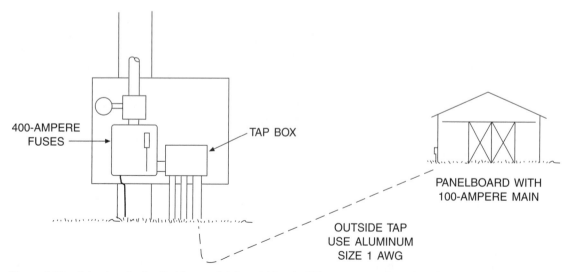

Figure 6.21   A tap to a feeder that is completely outside a building except at the point of termination to the building is permitted to be of any length and is only required to have an ampacity not less than the single overcurrent device at its termination.

$$\textbf{Voltage = Current} \times \textbf{Resistance} \qquad\qquad\qquad \textbf{Eq. 1.4}$$

$$\textbf{Resistance of Wire} = \frac{\textbf{Resistivity (K)} \times \textbf{Length (L)}}{\textbf{Area (A)}} \qquad\qquad \textbf{Eq. 1.2}$$

Single-phase circuit

$$\textbf{Voltage Drop = 2} \times \textbf{Current} \times \textbf{Resistance} \qquad\qquad \textbf{Eq. 6.1}$$

Three-phase circuit

$$\textbf{Voltage Drop = 1.73} \times \textbf{Current} \times \textbf{Resistance} \qquad\qquad \textbf{Eq. 6.2}$$

The Code does not have a requirement on the amount of voltage drop permitted on branch-circuits and feeders in general. It is suggested, however, that a maximum of 5% voltage drop be permitted on a feeder plus branch-circuit to the most distant point of power use. This is in the form of a FPN in *215.2* and *210.19(A) FPN 4*. Many tables, charts, and formulas are available to size wires that limit voltage drop to within a specified percentage. Voltage drop is the extent to which the line voltage at the source end of a branch-circuit or feeder is reduced as the load current flows along the conductors. The way to measure voltage drop is to operate the load, then measure the voltage at the source end of the branch-circuit or feeder and at the load end. Equation 6.3 can be used to determine the percent voltage drop when the line-to-line voltage is measured at the source and load ends of a conductor carrying current. Example 6.5 shows how percentage voltage drop can be determined for an actual circuit.

$$\textbf{Percent Voltage Drop} = \frac{\textbf{Voltage}_{\textbf{Source}} - \textbf{Voltage}_{\textbf{Load}}}{\textbf{Voltage}_{\textbf{Source}}} \times \textbf{100} \qquad\qquad \textbf{Eq. 6.3}$$

Example 6.5   A single-phase, 5-horsepower, 240-volt electric motor draws 28 amperes and is supplied by size 8 AWG copper conductors that have a one-way length from the panelboard to the motor of 360 ft (109.7 m). During the normal operation within the building and the motor running, there was 240 volts at the panelboard and 225.3 volts at the motor. Determine the percent voltage drop for this circuit.

Answer: Use Equation 6.3 to determine the percentage voltage drop. The 240 volts is at the source and the 225.3 volts is at the load. For this circuit, the voltage drop is 6.1%.

$$\text{Percent Voltage Drop} = \frac{240 \text{ V} - 225.3 \text{ V}}{240 \text{ V}} \times 100 = 6.1\%$$

Calculating the expected voltage drop for a circuit involves looking up the resistance of the conductor in *Table 8, Chapter 9* of the Code for the proper wire size, and making an adjustment for the actual length and operating temperature. A typical operating temperature for branch-circuits and feeders can be assumed to be about 50°C. The values of resistance in *Table 8* are at 75°C. Another way to determine expected voltage drop is to substitute Equation 1.2 into Equation 6.1 or 6.2 and come up with a new equation where all you need to do is remember the typical value of resistivity for the conductor. The result is the following two equations for voltage drop, where Equation 6.4 is for single-phase voltage drop and Equation 6.5 is for 3-phase voltage drop. Values of resistivity **K** for copper and aluminum conductors in both English and metric units are given in Table 1.2 in *Unit 1*. Suggested values for **K** are at approximately 50°C and are listed below

Single-phase

$$\textbf{Voltage Drop} = \frac{\textbf{2} \times \textbf{K} \times \textbf{Current} \times \textbf{One-way length}}{\textbf{Wire cross-sectional area}} \qquad\qquad \textbf{Eq. 6.4}$$

Three-phase

$$\text{Voltage Drop} = \frac{1.73 \times \text{K} \times \text{Current} \times \text{One-way length}}{\text{Wire cross-sectional area}} \qquad \text{Eq. 6.5}$$

English Units: K = 12 for copper and 19 for aluminum
One-way length is in ft
Wire cross-sectional area is in circular mils (cmil)

Metric Units: K = 0.02 for copper and 0.032 for aluminum
One-way length is in meters (m)
Wire cross-sectional area is in square millimeters (mm²)

Example 6.6   A building with a 120/240-volt single-phase, 100-ampere panelboard is supplied with a size 3 AWG overhead aluminum triplex cable, which is 180 ft (54.86 m) in length. If the typical load in the building is evenly balanced with 80 amperes, on each ungrounded conductor, determine if the voltage drop on the overhead conductor is in excess of 2% assuming 240 volts at the supply end of the triplex cable.

Answer: First multiply 240 volts by 0.02 to get 4.8 volts as a 2% voltage drop. Next, use Equation 6.4 to determine the expected voltage drop. It will be assumed the conductor temperature will be approximately 50°C. Use 19 (0.032 metric) as the value of **K** in Equation 6.4. It will be necessary to look up the cross-sectional area of a size 3 AWG conductor in *Table 8, Chapter 9* in the Code and find it to be 52,620 cmil (26.7 mm²). The expected voltage drop is 10.4 volts, which is more than 2%.

$$\text{Voltage Drop} = \frac{2 \times 19 \times 80A \times 180 \text{ ft}}{52,620 \text{ cmil}} = 10.4 \text{ volts}$$

$$\text{Voltage Drop} = \frac{2 \times 0.032 \times 80 \text{ A} \times 54.86 \text{ m}}{26.7 \text{ mm}^2} = 10.5 \text{ volts}$$

If voltage drop was expected to be excessive, then it would be desirable to have an easy way to select a wire size that would limit the voltage drop to the desired percentage. That can easily be accomplished by a slight rearrangement of Equations 6.4 and 6.5. The following Equations 6.6 and 6.7 can be used to determine the wire size required to limit the voltage drop. The result of the calculation will be the cross-sectional area of the wire. The actual wire size is found in *Table 8, Chapter 9* in the Code. The value for voltage drop in the equation is actually in decimal. For example, if the desired voltage drop is 3%, insert 0.03 into the equation.

Single-phase

$$\text{Wire cross-sectional area} = \frac{2 \times \text{K} \times \text{Current} \times \text{One-way length}}{\% \text{ voltage drop} \times \text{Voltage}_{\text{Supply}}} \qquad \text{Eq. 6.6}$$

Three-phase

$$\text{Wire cross-sectional area} = \frac{1.73 \times \text{K} \times \text{Current} \times \text{One-way length}}{\% \text{ voltage drop} \times \text{Voltage}_{\text{Supply}}} \qquad \text{Eq. 6.7}$$

Example 6.7   A building with a 120/240-volt single-phase, 100-ampere panelboard is supplied with an overhead aluminum triplex cable which is 180 ft (54.86 m) in length. If the typical load in the building is evenly balanced with 80 amperes on each ungrounded conductor, determine the minimum size conductor required to limit the voltage drop on the feeder to 2%.

Answer: Use Equation 6.6 to find the cross-sectional area of the conductor required to limit the voltage drop to 2%. Assume the conductor operating temperature is approximately 50°C and the value of **K** for the equation will be 19 (0.032 metric). The cross-sectional area of conductor needed is 114,000 cmil (58.5 mm²). Now look up the conductor size in *Table 8, Chapter 9* of the Code and find size 2/0 AWG.

$$\text{Wire cross-sectional area} = \frac{2 \times 19 \times 80 \text{ A} \times 180 \text{ ft}}{0.02 \times 240 \text{ V}} = 114{,}000 \text{ cmil}$$

$$\text{Wire cross-sectional area} = \frac{2 \times 0.032 \times 80 \text{ A} \times 54.86 \text{ m}}{0.02 \times 240 \text{ V}} = 58.5 \text{ mm}^2$$

For wire sizes larger than AWG number 4/0, skin effect and inductive reactance frequently become significant and must be considered. *Table 9, Chapter 9* of the Code contains impedance values of wires in different types of raceway where the power factor is not unity. If a motor load is being supplied by a branch-circuit or feeder, the voltage drop can be estimated using the impedance values from *Table 9*. If the power factor of the circuit or feeder is approximately 0.85, then impedance values are given in *Table 9*.

MAJOR CHANGES TO THE 2005 CODE

These are the changes to the *2005 NEC®* that correspond to the Code sections studied in this unit. The following analysis explains the significance of the changes from the 2002 to the 2005 Code only and this analysis is not intended to be used in place of the Code. Refer to the actual section of the 2005 Code for the exact wording and meaning of each section discussed. Changes are indicated in the Code with a vertical line in the margin. If material was deleted or moved to another location in the Code, the location of the deletion is indicated with a dark dot in the margin.

Article 240 **Overcurrent Protection**

240.5(1) and (3): Minimum wire sizes with respect to the circuit rating were deleted for appliance cords, portable lamp cords, and commercially available extension cord sets. The products are required to be listed and indicate the maximum circuit rating for the particular cord.

240.20(B): This section was revised to make it clear that a circuit breaker is required to open all ungrounded conductors of a circuit both manually and automatically. For example, it is not permitted to use two single-pole circuit breakers with a handle tie for a 240-volt circuit as illustrated in Figure 6.22. The

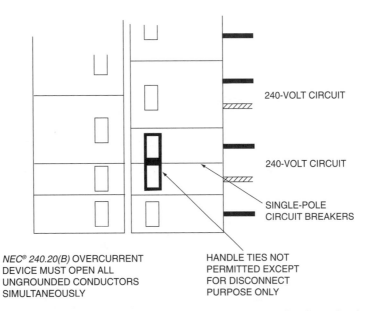

240-VOLT CIRCUIT

240-VOLT CIRCUIT

SINGLE-POLE
CIRCUIT BREAKERS

NEC® 240.20(B) OVERCURRENT
DEVICE MUST OPEN ALL
UNGROUNDED CONDUCTORS
SIMULTANEOUSLY

HANDLE TIES NOT
PERMITTED EXCEPT
FOR DISCONNECT
PURPOSE ONLY

Figure 6.22   A circuit breaker that is installed to provide overcurrent protection for a circuit is required to be capable of automatically opening all ungrounded conductors of the circuit, which cannot be accomplished with handle ties on single-pole circuit breakers.

change is the addition of the word "automatically." This means multipole circuit breakers are required to be common-trip. Common-trip means when one pole opens, the other poles will also open. This section lists three cases where individual circuit breakers with or without handle ties are permitted to be used where all ungrounded poles of the circuit are not required to open simultaneously. An example is a multiwire branch-circuit with two or three ungrounded conductors and a neutral. The handle ties will act as a disconnecting means to open all ungrounded poles simultaneously, but will not necessarily ensure that automatic trip caused by an overcurrent condition will open all ungrounded poles. Another change in this section is that identified handle ties are required not simply approved. The handle tie must be manufactured for the purpose and identified as acting as a handle tie for the circuit breaker.

240.21(B): A new last sentence was added that actually is not a change. The new sentence simply states for a tap to a feeder, the provisions of *240.4(B)* do not apply. This means the ampacity of the tap conductor is required to be equal to or greater than the rating of the single circuit breaker or single set of fuses at the termination of the tap conductor. This was already a requirement in each of the subsections of *240.21(B)*.

240.21(B)(2)(3): Tap conductors are required to be protected from physical damage or be enclosed in approved raceway. The key here is the new word "approved." It is recognized that all raceways do not necessarily provide adequate protection from physical damage. The raceway used to protect a tap conductor from physical damage is required to be approved. This same provisions appears in several other parts of this section on tap conductors.

240.21(C)(2)(1)c.: This is a new requirement for transformers that sets a minimum size for secondary tap conductors that are not more than 10 ft (3 m) in length. The secondary tap conductors are not permitted to have an ampacity less than 10% of the transformer primary overcurrent device rating times the primary voltage over the secondary voltage. This new rule is illustrated in Figure 6.23 and the following example:

$$\text{Minimum Tap Wire Ampacity} = \frac{\text{Primary Overcurrent Device Rating}}{10} \times \frac{\text{Primary Voltage}}{\text{Secondary Voltage}}$$

Example 6.8: A 3-phase, 112½-kVA transformer with 480 volt primary has wires protected by 150 ampere fuses. There are several taps from the 208Y/120-volt secondary each of which is not over 10 ft (3 m) in length. One tap ends at a 30-ampere circuit breaker to supply a 5-horsepower, 3-phase, 208-volt motor. Determine the minimum size copper tap conductor between the transformer and the 30-ampere circuit breaker if the insulation and terminations are rated 75°C.

Answer: The tap conductor is required to have an ampacity not less than the rating of the overcurrent device at the end of the tap which is 30 amperes. From *Table 310.16* using the 75°C column, the minimum is size 10 AWG. Next check the minimum tap conductor ampacity permitted for this transformer regardless of the load to be served and find 35 amperes. Checking *Table 310.16* shows that a size 10 AWG is just large enough to also meet this requirement.

$$\text{Minimum Tap Wire Ampacity} = \frac{150 \text{ A}}{10} \times \frac{480 \text{ V}}{208 \text{ V}} = 35 \text{ A}$$

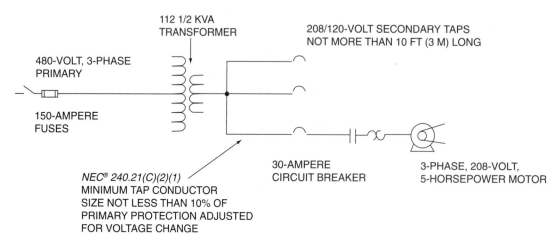

Figure 6.23   A secondary tap to a transformer that is not more than 10 ft (3 m) in length is required to have an ampacity not less than the overcurrent device it supplies and not less than 10% of the primary overcurrent device adjusted for voltage change.

240.24(A): A new requirement was added that establishes a maximum height for the center of the operating handle of a switch or circuit breaker. The maximum height is 6 ft 7 in. (2 m). This is illustrated in Figure 6.24.

240.60(D): A new section was added the prohibits use of Class H renewable link fuses for new installations. This type of fuse is only permitted to be installed as a replacement fuse for existing installations where there is no evidence of over-fusing or tampering. Fuse links of various ampere ratings can be placed into the fuse cartridge, which makes over-fusing easy to accomplish. Since these fuses are only rated as Class H they only have a 10,000-ampere interrupting rating.

240.86: This section places requirements on the installation of overcurrent devices that have a series rating. This means two overcurrent devices acting in series have the ability to limit fault current to a level that is less than the short-circuit current rating of the load-side overcurrent device. These installations are required to be labeled as series-rated systems. Many overcurrent devices are tested for series rating and listing of those devices is available. This section now permits a series-rated system to be designed by a registered professional engineer where the components have not been tested. When the series-rated system has been designed by an engineer, documentation of the design and engineer seal are required to be on file at the location of the installation and available to maintenance, installation, and inspection personnel.

Article 408   Switchboards and Panelboards

408.4: It is now required to put sufficient details on the panelboard or switchboard directory to identify the specific location of outlets on each circuit. An example is shown in Figure 6.25.

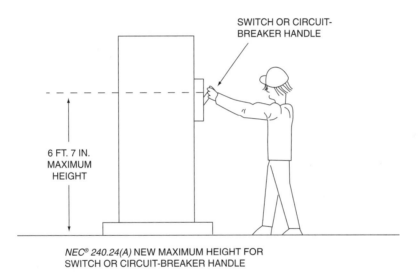

SWITCH OR CIRCUIT-BREAKER HANDLE

6 FT. 7 IN. MAXIMUM HEIGHT

NEC® 240.24(A) NEW MAXIMUM HEIGHT FOR SWITCH OR CIRCUIT-BREAKER HANDLE

Figure 6.24   The handle of a switch or circuit breaker when in the up position is not permitted to be more than 6 ft 7 in. (2 m) above the floor or platform from which it will be operated.

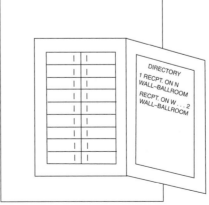

NEC® 408.4 SUFFICIENT DETAIL REQUIRED ON DIRECTORY OF PANELBOARD AND SWITCHBOARD TO IDENTIFY OUTLETS ON EVERY CIRCUIT

DIRECTORY
1 RECPT. ON N WALL–BALLROOM
RECPT. ON W . . . 2 WALL–BALLROOM

Figure 6.25   The directory on panelboards and switchboards is required to provide sufficient detail to locate all outlets or equipment supplied by each circuit.

408.7: A new section was added that requires unused openings for circuit breakers and switches to be closed with a device listed for the purpose. The requirement in *110.12(A)* deals with closing unused openings for cable and raceway entries not openings in panel covers for circuit breakers and switches. An example is shown in Figure 6.26.

Article 550   Mobile Homes, Manufactured Homes, and Mobile Home Parks

550.12(B) Exception 3: This is the section that requires all general-use receptacles in the kitchen, dining, and similar areas of a mobile home to be supplied by a 20-ampere branch-circuit. This new exception permits a dedicated circuit for refrigeration equipment to be rated at 15 amperes.

550.12(C): This is a new requirement that, except for equipment associated with the laundry in the mobile home, the 20-ampere laundry circuit shall have no other outlets. It is not permitted to put lights or other general-use receptacles on the laundry circuit.

550.13(F)(1): A receptacle is not permitted to be installed in a bathtub or shower space. The change is that the previous edition of the Code did not permit a receptacle to be located within 30 in. (750 mm) of a bathtub or shower space.

550.13(G): This section lists several locations within a mobile home where it was stated receptacles were not permitted to be located. Now the section reads "not required," which means receptacles are permitted to be installed. An example is a wall space behind a bar-type counter other than the kitchen counter.

550.25(B): There was no change in the arc-fault circuit-interrupter requirement for mobile homes from the previous edition of the Code. All outlets on 15- or 20-ampere rated 125-volt circuits in mobile home bedrooms are required to be arc-fault circuit-interrupter protected. This is now different than for all other dwelling units where the arc-fault circuit-interrupter is required to be of the combination type starting in January, 2008.

Article 551   Recreational Vehicles and Recreational Vehicle Parks

Part II: This part dealing with the installation of low-voltage systems in recreational vehicles was deleted from the Code. The installation of low-voltage systems for recreational vehicles will no longer be the responsibility of the Code.

551.32: Engine generator sets installed as sources of power in recreational vehicles are now required to be listed for use in recreational vehicles.

551.40(A): Equipment in a recreational vehicle that is to be connected line-to-line such as air-conditioning equipment or an electric heater is now required to have a dual rating of 208 to 240 volts. This requirement will solve a voltage incompatibility problem when the recreational vehicle park electrical system line-to-line voltage does not correspond with the voltage rating of line-to-line equipment.

551.46(E) Exception 3: A recreational vehicle that is also adapted for transporting livestock such as horses is now permitted to have the electrical point of entrance on either side or the front of the vehicle.

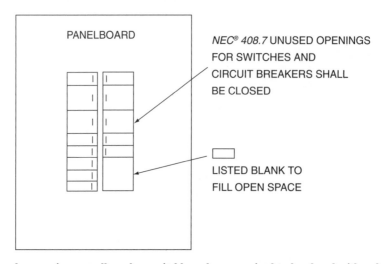

Figure 6.26   Unused spaces in a panelboard or switchboard are required to be closed with a device identified for the purpose.

Typical recreational vehicles have the power entrance point on the driver's side of the vehicle. Vehicles intended for the transport of livestock must be able to accommodate the loading and unloading of livestock, and it is necessary to locate the power supply cord and point of entry on the side away from normal livestock traffic. For a recreational vehicle that is also used to transport livestock, the power supply is permitted to be located on the passenger side or front of the vehicle as shown in Figure 6.27. Exhibition areas that have electrical power available for recreational vehicles are usually structured to allow for power entrance to a recreational vehicle from either side of the vehicle.

551.47(P): Hard-usage flexible cords in lengths not to exceed 24 in. (600 mm) are now permitted to be used as a permanent interior wiring method in a recreational vehicle to supply power between the main body of the recreational vehicle and an expandable unit of the recreational vehicle.

551.71: Recreational vehicle parks with electrical power are now required to supply 20% of the sites with 50 ampere, 125/250-volt electrical service. The previous edition of the Code only required 5% of the sites to provide 50-amperes at 125/250 volts.

551.71 FPN: There is a new fine print note pointing out that seasonal recreational vehicle parks may have need for a high percentage of 50-ampere, 125/250-volt sites. The number of recreational vehicles with high power requirements has increased over the past years. At some locations, it may be desirable to provide 50-ampere, 125/250-volt power at all of the sites.

551.73(A): Where a single power pedestal provides power to two recreational vehicle sites, the load required to be used at such a site is determined based upon the two receptacles at the site with the highest ampere rating. For example, if the pedestal has a 20-ampere and a 30-ampere receptacle at 125 volts and a 50-ampere receptacle at 125/250 volts, the load at that site would be based upon the 50-ampere and the 30-ampere receptacles, which would be a load of 13,200 VA (9600 VA + 3600 VA). The previous edition of the Code did not specify a means of making a load calculation at a pedestal that served two sites.

Article 552 **Park Trailers**

552.41(C): This section specifies the receptacle outlets required for a park trailer that are supplied 120-volt power or 120/240-volt power. Receptacles rated 15- or 20-ampere, 125-volts installed to serve a kitchen counter or located within 6 ft (1.8 m) of any lavatory or sink are required to be ground-fault circuit-interrupter protected. The previous edition of the Code required GFCI protection for 15- and 20-ampere, 125-volt receptacles located in a bathroom or within 6 ft (1.8 m) of a lavatory or sink. The real change here is that receptacles located more than 6 ft (1.8 m) from a sink, but which serve a kitchen counter, are now required to be GFCI protected.

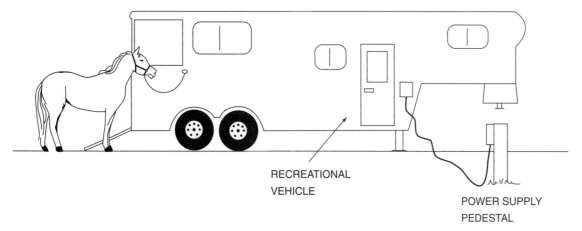

NEC® 551.46(E) EXCEPTION 3 PERMITS THE POWER SUPPLY TO BE LOCATED ON THE PASSENGER SIDE OR FRONT OF A RECREATIONAL VEHICLE THAT IS ALSO USED TO TRANSPORT LIVESTOCK

RECREATIONAL VEHICLE

POWER SUPPLY PEDESTAL

Figure 6.27   A recreational vehicle that is also used to transport livestock is permitted to have the power supply located on the passenger side or front of the vehicle.

WORKSHEET NO. 6—BEGINNING OVERCURRENT PROTECTION

Mark the single answer that most accurately completes the statement based upon the 2005 Code. Also provide, where indicated, the Code reference that gives the answer or indicates where the answer is found, as well as the Code reference where the answer is found.

1. A standard rating of an overcurrent device as recognized by the Code is:
    A. 55 amperes.
    B. 130 amperes.
    C. 700 amperes.
    D. 750 amperes.
    E. 1500 amperes.

    Code reference _____

2. A molded-case circuit breaker permitted to be used as a switch for high-pressure sodium luminaires (lighting fixtures) in a gymnasium is required to be labeled:
    A. with circuit rating.
    B. SWD.
    C. suitable as switch for luminaires (lighting fixtures).
    D. HPS loads.
    E. HID.

    Code reference _____

3. The screw shell fuse with a hexangular window as shown in Figure 6.28 has a rating of not over:
    A. 10 amperes.
    B. 15 amperes.
    C. 20 amperes.
    D. 25 amperes.
    E. 30 amperes.

    Code reference _____

4. A size 14 AWG copper conductor serves less than 12 ampere of fixed lighting load in a commercial building. The conductor insulation and terminations are 75°C rated and the allowable ampacity of the conductors from *Table 310.16* is 20 amperes. The maximum rating overcurrent protection permitted for the circuit is:
    A. 12 amperes.
    B. 15 amperes.
    C. 17.5 amperes.
    D. 20 amperes.
    E. 25 amperes.

    Code reference _____

Figure 6.28   What is the maximum rating of a plug fuse with a hexangular window?

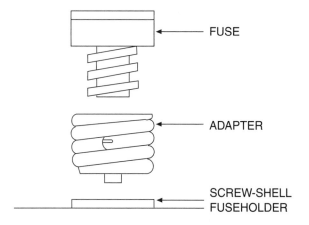

FUSE

ADAPTER

SCREW-SHELL
FUSEHOLDER

Figure 6.29   What kind of screw shell fuse uses and adapter in the screw shell fuseholder?

5. When an electrician installs a device that accepts screw shell fuses, they are to be of the type shown in Figure 6.29 that have an adapter that will only accept a fuse up to a particular maximum rating. This type of fuse in the Code is called:
    A. a Type S fuse.
    B. an Edison-base fuse.
    C. a current-limiting fuse.
    D. a class K fuse.
    E. an instantaneous fuse.

    Code reference_____

6. When terminating neutral conductors and equipment grounding conductors to the neutral terminal bus in a service panelboard:
    A. only a single neutral conductor is permitted to be connected to a terminal.
    B. only one neutral and one equipment grounding conductor of the same circuit are permitted to be connected to a terminal.
    C. any number of neutral and equipment grounding conductors are permitted to be connected to a terminal if they will fit the space available.
    D. not more than two neutral conductors are permitted to be connected to a terminal.
    E. there is no limit to the number of conductors for any terminal provided a minimum torque is applied to the terminal.

    Code reference_____

7. A feeder consisting of size 350 kcmil copper conductors is protected with 300-ampere fuses and is run through a commercial building in Rigid Metal Conduit. The feeder is tapped to supply a 100-ampere panelboard with size 3 AWG copper tap conductors run in Rigid Metal Conduit. All conductor insulation and terminations are rated 75°C shown in Figure 6.30. The maximum distance permitted from the tap point on the feeder to the 100-ampere rated overcurrent device in the panelboard is:
    A. 10 ft (3 m).   D. 50 ft (15 m).
    B. 25 ft (7.5 m).   E. 100 ft (30 m).
    C. 30 ft (9 m).

    Code reference_____

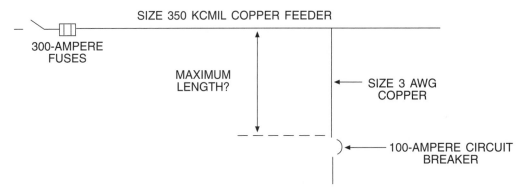

Figure 6.30   Based upon the feeder overcurrent protection and sizes of conductors, what is the maximum distance permitted from the tap point to the 100-ampere circuit breaker?

8. The maximum number of single-pole circuit breakers that are permitted to be installed in a lighting and appliance branch-circuit panelboard is:

A. 24.   C. 36.   E. not limited.

B. 30.   D. 42.

Code reference _____

9. A panelboard marked as suitable for use as service equipment is used as the service for a building supplied by a 3-phase, 240/120-volt delta, 4-wire electrical system where the midpoint of one phase is grounded and one of the three phase conductors is 208 volts to ground. The phase with the higher voltage to ground is required to be placed in the center ungrounded lug of the panel and the conductor is required to be identified with tape, paint, or other suitable marking that is the color:

A. red.   C. yellow.   E. orange.

B. blue.   D. brown.

Code reference _____

10. A mobile home served with 120/240-volt 3-wire service that has a calculated load greater than 50 amperes is required to be supplied power using a permanent wiring method from the adjacent power supply pole to the mobile home. The power supply feeder is required to consist of:

A. three insulated conductors and an equipment ground permitted to be bare.

B. only three insulated conductors if the mobile home panel is grounded to the earth.

C. four insulated and color-coded conductors—one of which is an equipment grounding conductor.

D. two insulated and color-coded ungrounded conductors with the neutral permitted to be bare.

E. three insulated conductors and a bare equipment grounding conductor with the insulated conductors permitted to be identified with colored tape.

Code reference _____

11. A luminaire (lighting fixture) mounted in the ceiling of a mobile home bedroom is required to be:

A. arc-fault circuit-interrupter protected.

B. installed on a circuit with other than receptacles in the same room.

C. ground-fault circuit-interrupter protected.

D. thermally protected.

E. only an electric discharge type luminaire (lighting fixture).

Code reference _____

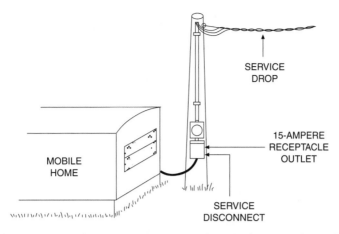

Figure 6.31 What are the requirements on the receptacle outlet in the service equipment for a mobile home?

12. A 15- or 20-ampere, 125-volt rated receptacle located in service equipment for a mobile home, as shown in Figure 6.31, is required to be:
    A. rated as suitable for use as service equipment.
    B. mounted at a height of not less than 2 ft (600 mm) above grade.
    C. arc-fault circuit-interrupter protected.
    D. of the grounding type and grounded with an insulated copper equipment grounding conductor.
    E. ground-fault circuit-interrupter protected for personnel.
    Code reference_____

13. A mobile home receives power with a 50-ampere rated power cord. After installation, the minimum length of the power cord permitted from the face of the attachment plug cap to the point where the cord enters the mobile home, as illustrated in Figure 6.32, is:
    A. 6 ft (1.8 m).   D. 30 ft (9 m).
    B. 12 ft (3.7 m).   E. 36¹/2 ft (11 m).
    C. 20 ft (6 m).
    Code reference_____

14. The recreational vehicle site electrical supply pedestal is:
    A. permitted to have a disconnect for all ungrounded conductors at a remote panelboard.
    B. required to have the disconnect for all ungrounded conductors located at a remote panelboard.

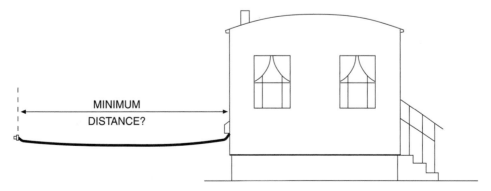

Figure 6.32 What is the minimum length of power cord for a mobile home after installation?

C. required to have a disconnect capable of being locked in the open position.

D. required to be a switch or circuit breaker located at the pedestal.

E. required to have the disconnect and receptacles located not less than 4 ft (1.2 m) above grade.

Code reference_____

15. A park trailer has a circuit prewired for a future air conditioner with the wires terminating at a junction box with a blank cover and the ends of the conductors capped and taped. This circuit is:

A. a violation of *Article 552*.

B. permitted to be used for a purpose other than an air conditioner in the future.

C. only permitted to supply a load which is 50% of the circuit rating.

D. required to have a label at the junction box stating the conductors are for future connection of an air conditioner.

E. is only permitted for use with an air conditioner operating at not more than 24 volts.

Code reference_____

WORKSHEET NO. 6—ADVANCED OVERCURRENT PROTECTION

Mark the single answer that most accurately completes the statement based upon the 2005 Code. Also provide, where indicated, the code reference that gives the answer or indicates where the answer is found, as well as the Code reference where the answer is found.

1. If a feeder conductor in a manufacturing building area with a ceiling that is at least 35 ft (11 m) high at the walls is tapped, with the tap point at least 30 ft (9 m) above the floor, the total length of the tap is permitted to be greater than 25 ft (7.5 m) in length. The tap conductor is to end at a single overcurrent device with a rating not greater than the ampacity of the tap conductor and the feeder overcurrent device is not permitted to have a rating greater than three times the ampacity of the tap conductor. In this case, the tap conductor is permitted to have a total length not to exceed:
   A. whatever length is needed to reach the load.
   B. 30 ft (9 m).
   C. 50 ft (15 m).
   D. 75 ft (22.9 m).
   E. 100 ft (30 m).

   Code reference _____

2. A cartridge fuse that does not have an interrupting rating marked on the exterior of the fuse has an interrupting rating of:
   A. 5000 amperes.   D. 100,000 amperes.
   B. 10,000 amperes.   E. 200,000 amperes.
   C. 50,000 amperes.

   Code reference _____

3. A 2-pole circuit breaker for a single-phase 240-volt load is installed in a panelboard supplied by a 3-phase, 240/120-volt, 4-wire delta electrical system with the midpoint of one phase grounded and one phase operating with a higher voltage to ground than the other two phases, as shown in Figure 6.33. A circuit breaker installed in a panelboard and connected to the phase with the higher voltage to ground:
   A. shall be marked 240 volts.
   B. shall be marked 120/240 volts.
   C. shall be two single-pole circuit breakers connected with handle ties.
   D. is permitted to be two single-pole circuit breakers connected with handle ties.
   E. is permitted to be marked 120/240 volts.

   Code reference _____

4. A molded-case circuit breaker permitted to be used as a switch for fluorescent luminaires (lighting fixtures) in a commercial building can be labeled:
   A. switch duty.
   B. FLU.
   C. HID.
   D. HPS loads.
   E. suitable as switch for luminaires.

   Code reference _____

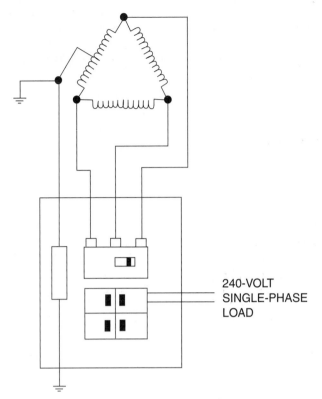

Figure 6.33   What are the requirements on the two-pole circuit breaker in this case connected to the system B phase?

5. Electrical equipment intended to disconnect power at fault levels shall, for the nominal circuit voltage and current that is available, have a sufficient:
   - A. continuous current rating.
   - B. demand factor.
   - C. interrupting rating.
   - D. overload capacity.
   - E. circuit impedance.

   Code reference _____

6. A 300-ampere service entrance consists of a panelboard with a single main overcurrent device rated at 200 amperes and a 100-ampere fused disconnect switch. The panelboard and the disconnect switch are both rated as suitable for use as service equipment and they are tapped from a size 500 kcmil, Type THWN aluminum ser-vice-entrance conductors as shown in Figure 6.34. All conductor terminations are rated 75°C. If the service tap conductors to the 100-ampere disconnect are Type THWN aluminum, the minimum size permitted is:
   - A. 3 AWG.
   - B. 1 AWG.
   - C. 1/0 AWG.
   - D. 4/0 AWG.
   - E. 500 kcmil.

   Code reference _____

7. A feeder consisting of size 3/0 AWG copper conductors supplies two 100-ampere rated panelboards for lighting and receptacle loads. This installation is permitted:
   - A. provided the panelboards have 100-ampere main overcurrent protection if the feeder is protected at 200 amperes.
   - B. provided the feeder overcurrent device is not rated more than 200 amperes and the panelboards are main lug only.
   - C. only if both panelboards have 100-ampere main overcurrent protection as well as the feeder having 100-ampere overcurrent protection.

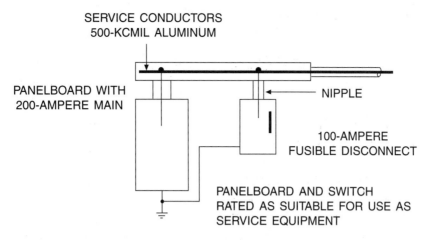

Figure 6.34 Determine the minimum size tap conductor to the 100-ampere service panel.

    D. with 100-ampere main overcurrent protection at the panelboards and no overcurrent protection on the feeder if it is not more than 25 ft (7.5 m) in length.

    E. if there is 200-ampere overcurrent protection at the source of the feeder and also at the load end ahead of the two panelboards.

Code reference_____

8. A 3-phase 208/120-volt electrical system supplies a building and the service equipment consists of an 18-space circuit-breaker panelboard. The building load calculated according to *Article 220* is 130 amperes, but the panelboard is rated 200 amperes and the three ungrounded service-entrance conductors are size 4/0 AWG copper Type THHN. A neutral conductor enters the panel and it is size 1/0 AWG copper Type THHN. There are six 3-pole circuit breakers, one rated 100 amperes, three rated 50 amperes, one rated 30 amperes, and one rated 20 amperes, as shown in Figure 6.35. The neutral conductor is not run as a part of the 20- or 30-ampere, 3-phase circuits. This service equipment:

    A. is classified as a power panel and is permitted without a main overcurrent device.

    B. is in violation because it requires a single main overcurrent device.

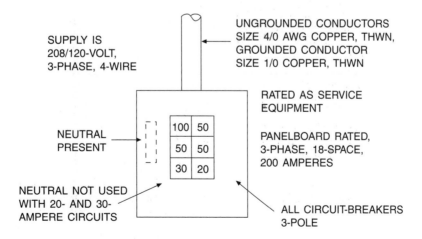

Figure 6.35 The neutral is present in the panelboard used as service equipment, but it is not used with the 20- or 30-ampere circuits.

C. is classified as a lighting and appliance branch-circuit panelboard.

D. would be considered a power panelboard if the neutral was not present in the panel.

E. would not be in violation if it had not more than two main overcurrent devices.

Code reference _____

9. A tap is made to supply a circuit-breaker panelboard from a 400-ampere, 208/120-volt, 3-phase, 4-wire feeder. The feeder ungrounded conductors are 600-kcmil copper with Type THWN insulation. The distance from the tap point on the feeder to the 125-ampere main circuit breaker in the panelboard is 25 ft (7.5 m). The tap conductor is copper, Type THWN, and all conductor terminations are rated at 75°C, as shown in Figure 6.36. The minimum size tap conductor permitted is

A. 3 AWG.   C. 1 AWG.   E. 2/0 AWG.
B. 2 AWG.   D. 1/0 AWG.

Code reference _____

10. A 208-volt, 3-phase feeder in a building, protected by 200-ampere fuses, is run in Electrical Metallic Tubing and has a total length of 275 ft (83.82 m). The expected maximum load on the feeder is 160 amperes and the power factor is close to 1.0. The operating temperature of the feeder is approximately 50°C. The minimum size copper conductors permitted that will also limit the voltage drop on the feeder to not more than 2% with the maximum expected load is:

A. 2/0 AWG.   C. 4/0 AWG.   E. 300 kcmil.
B. 3/0 AWG.   D. 250 kcmil.

Code reference _____

11. The service equipment or a disconnecting means listed as suitable for use as service equipment is required to be installed adjacent to a mobile home and within sight of the mobile home and located from the exterior wall of the mobile home not more than:

A. 6 ft (1.8 m).   C. 15 ft (4.5 m).   E. 50 ft (15 m).
B. 10 ft (3 m).   D. 30 ft (9 m).

Code reference _____

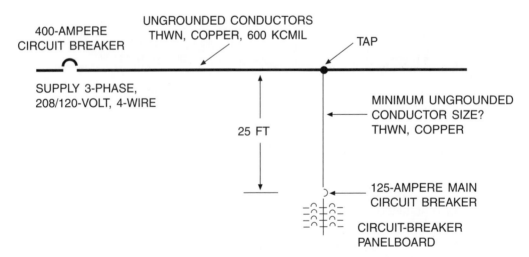

Figure 6.36   Determine the minimum size tap conductor connected to this feeder protected at 400 amperes.

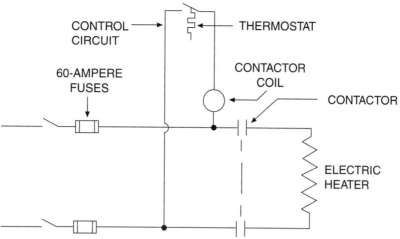

Figure 6.37   Determine the minimum size Class 1 control conductors run to the thermostat when they are copper and only protected by the branch circuit 60-ampere fuses.

12. An electric heater circuit is protected with 60-ampere fuses and the heater is controlled with a contactor. A line voltage thermostat operates the coil of the contactor, as shown in Figure 6.37. All wires are copper, run in conduit, and have 75°C rated insulation and terminations. The Class 1 control circuit wires are protected from overcurrent only by the heater circuit fuses. The minimum size Class 1 control circuit conductors permitted is:
    A. 16 AWG.   C. 12 AWG.   E. 8 AWG.
    B. 14 AWG.   D. 10 AWG.

Code reference_____

13. A recreational vehicle park has 52 recreational vehicle sites with electrical power. The minimum number of sites required to be equipped with 20-ampere, 125-volt receptacles is:
    A. 3 sites.   C. 37 sites.   E. 52 sites.
    B. 15 sites.   D. 49 sites.

Code reference_____

14. A power supply pedestal is to be installed at a back-in recreational vehicle site where a marker is placed at the back left corner of the stand. The power pedestal is permitted to be located in the space from 5 ft (1.5 m) to 7 ft (2.1 m) from the left side of the stand as shown in Figure 6.38 on the next page. The pedestal is not permitted to be located from the back of the stand toward the road a distance of more than:
    A. 9 ft (2.7 m).   D. 30 ft (9 m).
    B. 12 ft (3.7 m).   E. 35 ft (10.6 m).
    C. 15 ft (4.5 m).

Code reference_____

15. A park trailer with not more than five 120-volt circuits that is required to be supplied power has a rating of:
    A. 120 volts, 30 amperes, 2 wire.
    B. 120/240 volts, 50 amperes, 3 wire.
    C. 120/240 volts, 30 amperes, 3 wire.
    D. 120 volts 50 amperes, 2 wire.
    E. 120/240 volts, 20 amperes, 2 wire.

Code reference_____

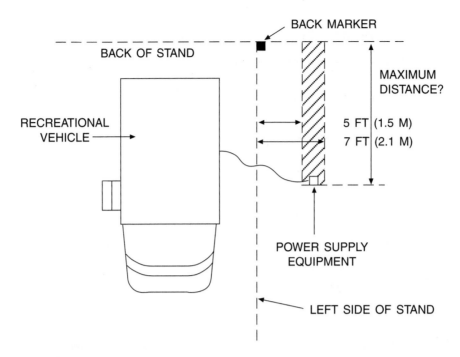

Figure 6.38   What is the maximum distance the power supply equipment is permitted to be installed from the back of the stand for a back-in recreational vehicle site?

UNIT 7

Motor-Circuit Wiring

OBJECTIVES

After completion of this unit, the student should be able to:

- read and explain the use of the information on a motor nameplate.
- determine the minimum size conductor permitted for a motor branch-circuit.
- determine the maximum rating of the overcurrent device for the motor branch-circuit short-circuit and ground-fault protection.
- select the type and rating of disconnect required for a motor circuit.
- select the type and rating of the controller for a motor.
- determine the maximum permitted rating of motor overload protection.
- determine the minimum size conductor permitted for a feeder supplying several motors.
- explain when overcurrent protection for the motor-control circuit is needed in addition to the motor branch-circuit overcurrent device.
- explain the meaning of NEMA enclosure type for motor controllers.
- diagram a control circuit for a magnetic motor starter.
- answer wiring installation questions relating to *Articles 409, 430, 440, 455, 460, 470, 675, 685,* and *Example D8* in *Annex D.*
- state three significant changes that occurred from the 2002 to the 2005 Code for *Articles 409, 430, 440, 455, 460, 470, 675, 685,* or *Example D8* in *Annex D.*

CODE DISCUSSION

Sizing and installation of components of an electric motor circuit are covered in this unit. An electric motor is a device to convert electrical energy into mechanical energy or power. An electric motor is a perfect servant. It tries to power any load to which it is connected. For this reason, overcurrent protection must be provided that will prevent self-destruction of the motor and possible fire or an electrical shock hazard. The task of sizing overcurrent protection for a motor is difficult because of the current characteristics of the motor. The motor will typically draw five to six times as much current during starting as it will when running under constant full load. Figure 7.1 shows the typical motor current draw during starting, compared with the full-load running current. The real challenge for overcurrent protection is to protect for overloads during running and yet not experience opening of the overcurrent device during the starting in-rush current.

Article 409 provides specific rules for the installation of an industrial control panel. Although the term is industrial control panel, they are not limited to installation in an industrial building or structure. An industrial control panel contains control equipment such as push buttons, selector switches, timers switches, control relays, terminal blocks, and pilot lights. It also may contain circuit breakers, fuses, disconnect switches, overload relays, and controllers. The main difference between an industrial control panel and a motor control center is that the industrial control panel contains controllers to operate loads and it carries the total load of the control system and the loads. An industrial control panel is permitted to serve as service equipment, although it is only permitted to contain one main circuit breaker or set of fuses. It must be rated as suitable for use as service equipment if used for that purpose. All industrial control panels must be marked with a short-circuit current rating. The conductors supplying an industrial control panel are sized by the same rule as a feeder to a specific motor load as described in *430.24.* The maximum rating of overcurrent protection for the industrial control panel is determined by the same rule as the overcurrent protection for a feeder for a specific motor load in *430.62.*

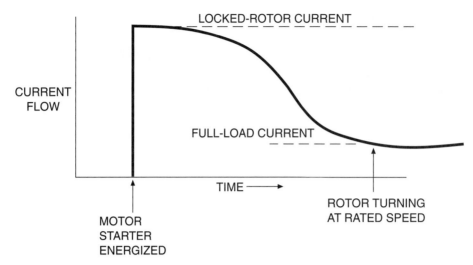

Figure 7.1   When a motor is started, there is an in-rush of current until the motor obtains running speed.

Article 430 deals with motors, motor branch-circuits and feeders, conductors and their protection, motor overload protection, motor-control circuits, motor controllers, and motor-control centers. *Figure 430.1* in the Code shows a motor circuit and can be used as an index to find the section of the article dealing with the various circuit components. *Part I* of this article provides general information about various motors and general wiring requirements for the motor circuit. There are several tables at the end of the article, which will be used frequently when sizing components for a motor circuit. *Tables 430.247* through *430.250* give the motor full-load current for various types of motors. *NEC® 430.6(A)(1)* requires that these values are used except when sizing the motor running overcurrent protection, and unless the nameplate full-load current for the motor is a higher value than given in the table. For a motor-operated appliance with both the horsepower and full-load current listed on the nameplate, the full-load current is to be used for sizing conductors and other circuit components, *430.6(A)(1) Exception 3.*

The rules for determining the minimum permitted size of single motor branch-circuit conductors are found in *430.22*. The minimum wire size for a feeder serving a specific group of motors is determined according to *430.24*. The wire size for a motor circuit is determined directly from *Table 310.16* once the ampacity has been determined in *Part II*. *NEC® 240.4(D),* which states that the overcurrent protection for a 14 AWG copper wire is not to exceed 15 amperes, 20 amperes for a 12 AWG, and 30 amperes for a 10 AWG, does not apply in the case of a motor circuit. Other references in the Code that clarify that overcurrent protection for motor branch-circuits is to be sized according to the provisions of *Article 430* are *430.1* and *240.4(G).*

Part III discusses the type of overload protection permitted for an electric motor and the sizing of the overload protection. *Part IV* explains the short-circuit and ground-fault protection for the motor branch-circuit. Frequently, a single overcurrent device does not protect for overloads, ground-faults, and short-circuits. *Part V* deals with the overcurrent protection of motor feeders. This would be a conductor supplying power to more than one motor branch-circuit. *Table 430.52* is used when determining the maximum permitted rating of motor branch-circuit short-circuit and ground-fault protection.

The electrical current to many motors is controlled by a motor-control circuit. The wiring to a start-stop station for a magnetic motor starter is a Class 1 control circuit. *Part VI* covers the wiring and over-current of motor-control circuits. *Part VII* covers controllers for motors. This part provides requirements to what is permitted to be a motor controller and the ratings of controllers. *Part VIII* deals with motor-control centers. A motor-control center is an assembly of one to several sections in which there is a bus on which motor-control units are attached. The requirements for the motor-control center are in this part of *Article 430,* including installation requirements. *Part IX* covers the type and ratings of disconnecting means for motors and controllers.

An important safety issue with motors and the machinery they power is the disconnection of electri-cal power to the controller and to the motor. Rules for disconnects are found in *430.102*. Each controller is required to have an individual disconnecting means. The disconnecting means is required to be located in

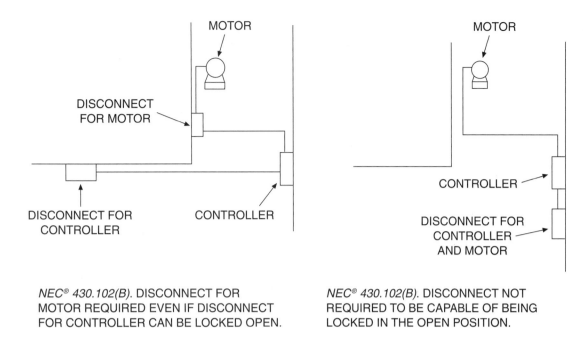

NEC® 430.102(B). DISCONNECT FOR
MOTOR REQUIRED EVEN IF DISCONNECT
FOR CONTROLLER CAN BE LOCKED OPEN.

NEC® 430.102(B). DISCONNECT NOT
REQUIRED TO BE CAPABLE OF BEING
LOCKED IN THE OPEN POSITION.

Figure 7.2   A disconnect is required to be located in sight of the controller and the motor. One disconnect is permitted if located in sight of both the controller and the motor.

sight from the controller. *NEC® 430.102(A) Exception 2* does permit a single disconnect for a group of coordinated controllers. *NEC® 430.102(B)* requires a disconnecting means to be located in sight from the motor and the driven machinery. A single disconnecting means is permitted for the controller, the motor, and the driven machinery provided it is located within sight of all three and it is not required to be of a type that can be locked in the open position. The rules requiring a disconnect are illustrated in Figure 7.2.

    NEC® 430.102(B) Exception only permits a single disconnect to serve both the controller and motor and be out of sight of the motor if it is impractical to locate the disconnect within sight of the motor. In the case of industrial buildings where there is a qualified maintenance staff on duty at all times, it is permitted to have one disconnect, capable of being locked in the open position, serve both the controller and the motor and still not be located in sight from the motor.

    Article 440 deals with the branch-circuits and motors associated with air-conditioning and refrigeration equipment. These are special cases because of frequent use of hermetically sealed motor compressors, and the use of multimotor branch-circuits. Branch-circuit selection current is covered in *440.4(C)*. The minimum permitted size of the branch-circuit conductors is found in *Article 440, Part IV*. The minimum permitted rating of the disconnecting means for the refrigerant motor compressor is determined from *Article 440, Part II*. *NEC® 440.22(C)* states that if the maximum permitted rating of branch-circuit short-circuit and ground-fault protective device is marked on the nameplate, that value shall not be exceeded. Overload protection for the motor compressor and for the branch-circuit conductors shall not exceed the value required in *Article 440, Part VI*. The installation of room air conditioners is covered in *Part VII*. When a cord-and-plug connected room air conditioner is installed on a dedicated branch-circuit, the marked ampere rating of the air-conditioner is not permitted to exceed 80% of the rating of the branch-circuit. Sometimes a cord-and plug-connected room air conditioner may be supplied by an existing general-purpose branch-circuit that supplies lights and receptacles. This is permitted provided the addition of the air-conditioner load does not overload the circuit, and provided the marked ampere rating of the air-conditioner does not exceed 50% of the rating of the branch circuit, *440.62(C)*.

    Article 455 deals with the installation of phase converters. A phase converter is an electrical device that permits a 3-phase electric motor or other 3-phase equipment to be operated from a single-phase electrical supply. The sizing of components of the circuit or feeder for a phase converter installation is difficult because of the 1.73-to-1 current ratio between the single-phase and 3-phase portions of the circuit. It theoretically takes 17.3 amperes flowing on the single-phase input conductors to a phase converter to supply 10 amperes

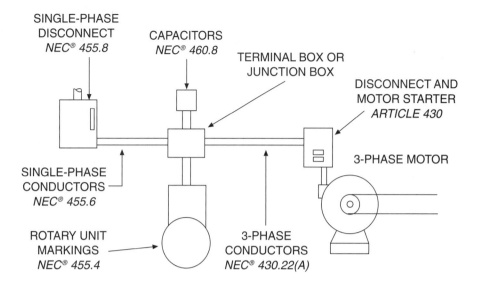

Figure 7.3   A phase converter permits a 3-phase motor or load to be operated from a single-phase electrical supply.

at 3-phase to a load connected to the phase converter. A phase-converter circuit is illustrated in Figure 7.3. An electric motor is permitted to be operated at the nameplate rated horsepower, but the motor-starting torque will be greatly reduced when operated from a phase converter. Motors powering hard starting loads should not be operated from a phase converter unless recommended by the manufacturer.

Phase converters may be a static type with no moving parts. A static phase converter usually serves only one 3-phase load, and it is sized specifically for that load. Most static phase converters consist primarily of capacitors. A transformer can be used to provide an output voltage different from the input voltage.

Rotary phase converters are capable of supplying one or more loads. If an installation consists of several 3-phase motors, then it may be possible to supply the 3-phase power from a single-phase supply with one rotary phase converter. The basic components of a rotary phase converter are a rotating unit that is actually a 3-phase motor, a capacitor bank, and if the output voltage is different than the input voltage, a transformer is included. With this type of unit, it is necessary to start the phase converter and bring the rotating unit to full speed before any 3-phase load is applied.

When sizing the conductors and overcurrent protection for a phase converter installation, it is necessary to know the "rated single-phase input full-load amperes" from the nameplate or the actual load to be served. *NEC® 455.6(A)(1)* requires that the minimum single-phase input conductor ampacity not be less than 1.25 times the rated single-phase input full-load current as marked on the nameplate. It is not uncommon for a phase converter to be oversized for the load to be served. In this case, *455.6(A)(2)* permits the single-phase input conductors to have an ampacity not less than 2.5 times the full-load current of the 3-phase load supplied by the phase converter.

Overcurrent protection is required for the single-phase input conductors and the phase converter. That overcurrent protection is located at the supply end of the single-phase input conductors. *NEC® 455.7(A)* requires that the overcurrent protection not be more than 1.25 times the rated single-phase input full-load current as marked on the phase converter nameplate. *NEC® 455.7* permits the calculated value to be rounded up to the next standard rating of overcurrent device as given in *240.6* when the calculated value does not correspond with a standard rating. *NEC® 455.7(B)* permits the maximum rating of overcurrent protection to be sized not more than 2.5 times the sum of the 3-phase loads supplied by the phase converter.

A disconnecting means is required to be located within sight of the phase converter. *NEC® 455.8* permits that disconnecting means to be a switch rated in horsepower, a circuit breaker, or a molded-case switch. If only nonmotor loads are served and a switch is used as the disconnecting means, the switch is not required to have a horsepower rating. The disconnecting means is required to have an ampere rating not less than 1.15 times the rated single-phase input full-load current as marked on the phase converter nameplate. *NEC® 455.8(C)(1)* permits a circuit breaker or a molded-case switch to be sized not less than 2.5 times the sum of the 3-phase full-load currents of the loads supplied. *NEC® 455.8(C)(2)* explains how to determine the horsepower rating of the load served by the phase converter. The disconnecting means must be capable of opening the circuit under full-load or locked-rotor conditions. The horsepower rating is selected from either *Table 430.251A* or *Table 430.251B* by using a calculated equivalent locked-rotor current. That equivalent locked-rotor current is 2.0 times the locked-rotor current of the largest motor served, plus the full-load current of all other motors served, plus the full-load current of all nonmotor loads served.

Example 7.1 A 20-kVA rotary phase converter with a rated single-phase input full-load current of 105 amperes is connected to a 240-volt single-phase electrical system to supply several 230-volt, 3-phase design B motors. The electric motors supplied are 10, 3, and 2 horsepower. The layout of the circuit is shown in Figure 7.4. A fusible switch is the disconnect for the single-phase power to the rotary phase converter. This is the switch being sized in this question. Another fusible switch on the load side of the phase converter serves as the disconnect for the feeder to the individual motor circuits. Determine the minimum horsepower rating of fusible switch on the single-phase side of the phase converter permitted to supply this specific motor load.

Answer: *NEC® 455.8(C)(2)* requires adding the locked-rotor current of the 10-horsepower, 230-volt, 3-phase design B motor (*Table 430.250B*) to the full-load currents of the 3- and 2-horsepower motors and then multiply the sum of these numbers by 2.0.

$$\text{Equivalent locked-rotor current} = 2.0 \times (162\text{ A} + 9.6\text{ A} + 6.8\text{ A}) = 357\text{ A}$$

Now find a single-phase horsepower rating from *Table 430.251A* that has a locked-rotor current equal to or greater than 357 amperes. Note that there is no single-phase switch shown in *Table 430.251A* with a horsepower rating high enough to handle the equivalent locked-rotor current of 357 amperes. Manufacturers usually have single-phase switches with horsepower ratings higher than listed in *Table 430.252A*. If a single-phase switch is not available, a 3-phase switch can be used as the phase-converter disconnecting means. In this case, one of the switch poles is not used. It is important that the 3-phase switch has a locked-rotor current rating equal to or higher than the calculated equivalent locked-rotor current for the load. In this example, the equivalent locked-rotor current was 357 amperes. Look down the 230-volt 3-phase column of *Table 430.252B* until a locked-rotor current value is found that is equal to or greater than 357 amperes. A switch rated 20-horsepower 3-phase is not adequate because it only has a locked-rotor current rating of 290 amperes. The 25-horsepower 3-phase switch has a locked-rotor current rating of 365 amperes, which is larger than 357 amperes. Therefore, a 25-horsepower 3-phase switch can be used as the disconnecting means for the phase converter.

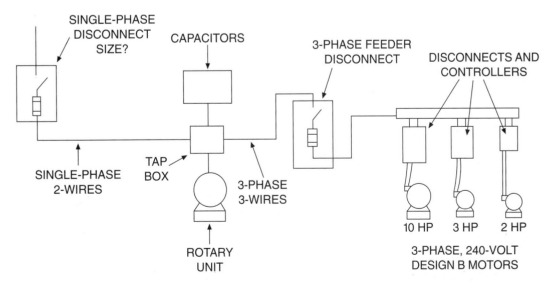

Figure 7.4 A fusible switch is frequently used as the single-phase disconnect for a rotary phase converter supplying several 3-phase motors.

Article 460 deals with the installation of capacitors wired as part of electrical circuits, and not as a component part of electrical equipment. Typical examples would be capacitors added for power factor correction or auxiliary capacitors added to motor circuits. An important aspect of this article is the discharging of the capacitor when it is de-energized. A capacitor will store a charge when power is shut off. If the charge is not removed from the capacitor, it can in some cases become a serious electrical hazard even though power has been disconnected. Capacitors are a part of the power supply in an adjustable-speed drive, and they take time to discharge when power is disconnected from the drive. Capacitors operating at 600 volts or less are required to be discharged to 50 volts in not more than one minute after power is disconnected.

NEC® 460.12 gives the requirements for marking capacitors. The rating of capacitors connected to 60-hertz alternating current systems is required to be given in amperes or kilovars (kVAR), which is reactive volt-amperes. Conductors supplying capacitors and ratings of switches for capacitors are required to have an ampacity not less than 135% of the rated capacitor current, *460.8*. If rated current is marked on the capacitors, determining the conductor size is easy. If the capacitors are rated in kVARs, then a calculation is required to get the rated current. Capacitor rated current for single-phase capacitors is determined from kVARs and the circuit voltage using Equation 7.1. For a 3-phase capacitor bank use Equation 7.2. Individual capacitors are rated in microfarads (μf) and calculating the current for a specific application is beyond the scope of this text.

Single-phase capacitor rated current:

$$\textbf{Rated capacitor current} = \frac{\textbf{kVAR} \times \textbf{1000}}{\textbf{Voltage}} \qquad \textbf{Eq. 7.1}$$

Three-phase capacitor bank rated current:

$$\textbf{Rated capacitor current} = \frac{\textbf{kVAR} \times \textbf{1000}}{\textbf{1.73} \times \textbf{Voltage}} \qquad \textbf{Eq. 7.2}$$

Article 470 covers the installation of resistors and reactors, which are sometimes used as a part of a motor or other equipment circuit. The main points of this article are the prevention of physical damage to the components and the prevention of overheating of wiring. Resistors and reactors are sometimes used in motor controllers for soft starting to reduce initial motor in-rush current.

Article 675 covers the installation of electrically driven or controlled irrigation machines. These are devices consisting of aluminum water pipe with periodic sprinkler nozzles to irrigate crop land. These devices are frequently propelled through the field with electric motor-driven wheels at regular intervals along the machine. Some irrigation machines rotate about a central pivot point, while others laterally move across the field. A typical center pivot irrigation machine is shown in Figure 7.5. The equivalent current rating of *675.7* is used to determine the size of conductors, overcurrent protection, disconnecting means, and controller for the irrigation machine. The drive motors for some irrigation machines, such as the center pivot type, operate intermittently. Therefore, a duty cycle is permitted to be applied when determining the

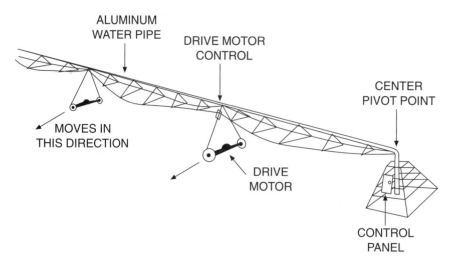

Figure 7.5   The control panel and main disconnect are located at the center pivot point of an irrigation machine with an electrical drive motor and controller located at each support tower radially out along the machine.

equivalent current rating of the machine. *NEC® 675.22* provides the value to use for the duty cycle for the center pivot machine. *NEC® 675.8(B)* requires a disconnecting means at the point where electrical power connects to the machine or within 50 ft (15 m) of that point. *NEC® 675.15* requires a grounding electrode system for lightning protection.

Article 685 applies to integrated electrical systems in industrial applications where equipment must be shut down in an orderly manner for a particular reason. In the event of a malfunction in a component machine of a continuous process, automatic shutdown of that particular machine may cause great economic loss or even endanger human life. A malfunction signal to the required on-duty maintenance personnel is permitted in place of automatic shutdown.

Annex D, Example D8 is an excellent example of an installation of three motors served by a common feeder. Each motor is on a separate branch-circuit. This example gives a clear understanding of the meaning of the various rules in *Article 430*. It is highly recommended that the person learning about motor-circuit and feeder installations work through this example. This example shows the method of determining the maximum permitted rating of the branch-circuit short-circuit and ground-fault protective device. *NEC® 430.52(C)(1) Exception 1* permits rounding up to the next standard rating of overcurrent device listed in *240.6*. The full-load current of the motor as determined from *Table 430.248* or *Table 430.250* is multiplied by the factor found in *Table 430.52*. It is permitted to round up to the next higher standard overcurrent device rating.

Two of the motors in the example are wound-rotor motors. However, for the purpose of this example, they are treated in a similar manner as other induction motors. A wound-rotor motor is an induction motor that has windings on the rotor that leave the motor through slip rings. Resistors are connected in series with the rotor windings to control the in-rush current and starting speed of the motor. *NEC® 430.23* gives the rules for sizing the branch-circuit wires and the secondary conductors from the rotor. The secondary full-load current will be marked on the motor nameplate. *NEC® 430.32(E)* permits the motor overload protection to also protect the secondary conductors.

MOTOR CIRCUITS AND CALCULATIONS

Information is provided in the Code that is necessary to size and install the components of a motor circuit. It is also necessary to determine information from the specific motor to be installed. The type of environment must be known to choose the proper enclosures for the motor and other equipment of the circuit.

Motor Nameplate and Other Information

A motor circuit is wired to fit the specifications of a specific motor. Information for determining the size or rating of the various parts is found on the motor nameplate, and the full-load current for single-phase and 3-phase motors from *Table 430.248* and *Table 430.250*. A motor nameplate is shown in Figure 7.6. The most important information needed for wiring the circuit is (1) horsepower, (2) phase, (3) voltage, (4) full-

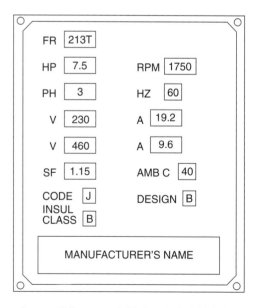

Figure 7.6 The motor nameplate contains essential information for sizing components of a motor circuit. The nameplate current may be different than the current listed for the same motor in *Article 430*.

load current, (5) temperature rise above ambient or service factor, and (6) design letter. Also, it is important to know about the physical environment, the location of the installation, the type of controller desired, and the type of load that will be powered (easy starting or hard starting load). The motor nameplate of Figure 7.6 shows ambient temperature, and not temperature rise above ambient. Ambient temperature is the maximum environmental temperature in which the motor is to be operated. If the surrounding temperature is higher than the ambient temperature marked on the nameplate, the motor will be in danger of overheating if operated at the nameplate horsepower rating.

The in-rush current of a motor is affected by the design of the motor. The National Electrical Manufacturers Association (NEMA) has established specifications and designations for motor design. Design letters are used to group motors into categories of similar operating characteristics. These characteristics include rotor slip at 100% load, locked-rotor current, and torques at various speeds. The most common type of motor in use is the design B motor. Motors are available that are designed especially as high-efficiency motors. These are designated as "energy efficient" motors. A motor intended to be a high-efficiency motor is the prime efficiency motor. Motors operating at low speeds or high torque may have full-load currents in excess of the values listed in *Table 430.148* or *Table 430.150*. In these cases, the nameplate current shall be used if it is higher than the values given in the tables.

Motor Circuit

A typical motor circuit is diagrammed in Figure 7.7. There is a chart in *430.1* that gives the location in the article where necessary information is found for sizing and wiring the circuit. The following components shall be sized or specified for a motor branch-circuit:

1. Branch-circuit disconnecting means minimum rating, *Part IX*
2. Branch-circuit short-circuit and ground-fault protection rating, *Part IV*
3. Branch-circuit conductors minimum size, *Part II*
4. Motor controller minimum size, *Part VII*
5. Motor and branch-circuit overload protection maximum rating, *Part III*
6. Motor-control circuit conductor minimum size, *Part VI*
7. Motor-control circuit overcurrent protection maximum rating, *Part VI*
8. Motor feeder conductor minimum size, *Part II*
9. Motor feeder short-circuit and ground-fault protection maximum rating, *Part V*
10. Motor-control center, *Part VIII*
11. Grounding, *Part XIII*

Methods of Controlling Motors

The motor controller directly controls the flow of electrical current to the motor. The definition is found in *Article 100*. Several methods are permitted to control a motor, as the following list indicates:

- For portable motors rated $1/3$ horsepower and less, the cord-and-plug, as stated in *430.81(B)*
- For stationary motors not over $1/8$ horsepower, normally operating continuously, the branch-circuit protective device, as stated in *430.81(A)*
- For stationary motors rated not more than 2 horsepower, a snap switch, as stated in *430.83(C)(1)*

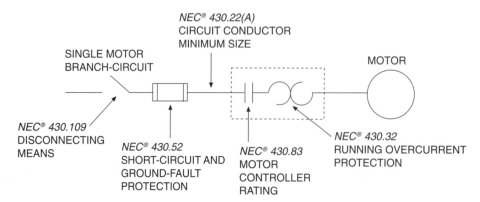

Figure 7.7   Required components of a motor branch-circuit are the disconnect, short-circuit and ground-fault protection, the controller, running overload protection, and proper conductor size.

- For a stationary motor, an inverse-time circuit breaker rated in amperes, as stated in *430.83(A)(2)*
- For a stationary motor, a fusible switch rated in horsepower, as stated in *430.90*
- A manual motor starter rated in horsepower, as stated in *430.83(A)(1)*
- A magnetic motor starter rated in horsepower, as stated in *430.83(A)(1)*

The National Electrical Manufacturers Association (NEMA) has established a size numbering system for horsepower ratings of electric motor starters. The NEMA sizes are shown in Table 7.1. A 3-pole motor starter may be used to control a single-phase motor. If only 3-phase horsepower is listed on a 3-pole motor starter to be used for a single-phase motor, divide the 3-phase horsepower rating for the desired voltage by 2 to obtain the maximum single-phase horsepower rating permitted for that motor starter. A single-phase motor with the same horsepower and voltage rating as a 3-phase motor will draw 1.73 times as much current as the 3-phase motor.

Enclosure Types

Enclosure type ratings for different environmental conditions have been established by NEMA. Enclosure types for motor starters for applications other than hazardous locations are listed in *Table 430.91*. The following is a general description of some common NEMA enclosure types and their typical applications:

NEMA 1:   General-purpose enclosure used in any location that is dry and free from dust and flying flammable materials.

NEMA 3:   Weather-resistant enclosure suitable for use outdoors. Not suitable for use in dusty locations. This type is no longer available for motor starters from some manufacturers.

NEMA 3X: Weather-resistant and corrosion-resistant enclosure suitable for use outdoors. Not suitable for use in dusty locations.

NEMA 4:   Watertight and dusttight enclosure suitable for outdoor locations and inside wet locations. Water can be sprayed directly on the enclosure without leaking inside. Suitable for most agricultural locations, provided corrosion is not a problem.

NEMA 4X: Watertight, dusttight, and corrosion-resistant enclosure suitable for outside and inside wet, dusty, and corrosive areas. Suitable for agricultural buildings.

NEMA 7:   Explosion-proof enclosure suitable for installation in Class I areas containing hazardous vapors. Required to be rated for type of hazardous vapor such as gasoline, Group D or Group IIA.

NEMA 9:   Dust-ignition-proof enclosure suitable for installation in Class II hazardous areas such as grain elevators. Required to be rated for the type of dust such as grain dust, Group G. There is a tendency for manufacturers to build one enclosure rated as NEMA 7 and 9.

Table 7.1   Motor horsepower and voltage ratings for NEMA size motor starters.

NEMA size	Single-phase		Three-phase		
	120 Volts	240 Volts	208 Volts	240 Volts	480 Volts
00	⅓	1	1½	1½	2
0	1	2	3	3	5
1	2	3	7½	7½	10
1P	3	5	—	—	—
2	3	7½	10	15	25
3	7½	15	25	30	50
4	—	25	40	50	100
5	—	50	75	100	200
6	—	—	150	200	400
7	—	—	—	300	600
8	—	—	—	450	900

Overcurrent Device Ratings

Manufacturers' ratings of time-delay fuses not larger than 30 amperes are frequently used to provide overload protection for electric motors. Table 7.2 is a listing of some generally available fuses of sizes not listed in the Code as standard ratings. Refer to *240.6* for the list of standard ratings of fuses and circuit breakers. The standard ratings of fuses less than 15 amperes recognized by the Code are 1, 3, 6, and 10 amperes.

Motor Overload Protection

Electric motors are required to be protected against overload, according to the rules in *430.32*. Overload protection is usually provided as a device responsive to motor current or as a thermal protector integral with the motor. A device responsive to motor current could be a fuse, a circuit breaker, an overload current sensor, or a thermal reset switch in the motor housing. An automatically resetting thermal switch placed in the windings will sense winding temperature directly.

The service factor or temperature rise must be known from the motor nameplate when selecting the proper size motor overload protection. These are indicators of the amount of overload a motor can withstand. If a motor has a service factor of 1.15 or greater, the manufacturer has designed extra overload capacity into the motor. In this case, the overload protection shall be permitted to be sized not greater than 125% of the **nameplate** full-load current. *NEC® 430.32(C)* permits the maximum setting of running overcurrent protection to be increased if the size determined in *430.32(A)(1)* is not sufficient to permit the motor to start.

Internal heat is damaging to motor winding insulation. A motor with a temperature rise of 40°C (104°F) or less has been designed to run relatively cool; therefore, it has greater overload capacity. The overload protection is permitted to be sized not greater than 125% of the nameplate full-load current. A service factor of less than 1.15 or a temperature rise of more than 40°C (104°F) indicates little overload capacity. The overload protection under these circumstances is permitted to be sized not greater than 115% of the nameplate full-load current.

A time-delay fuse is permitted to serve as motor overload protection. Screw-shell fuses or cartridge fuses are used for small motors. The fuse size is determined by selecting the proper multiplying factor, 1.15 or 1.25, based on the service factor or temperature rise marked on the motor nameplate. Time-delay fuse ratings 30 amperes and smaller are listed in Table 7.2.

Circuit breakers are permitted to be used as running overload protection, but they are not generally available in sizes smaller than 15 amperes. If they are used for large motors, they will usually trip on starting if they are sized small enough to provide overload protection.

Magnetic and manual motor starters have an overload relay or trip mechanism activated by a heater sensitive to the motor current. The manufacturer of the motor starter usually provides a chart inside the motor starter listing the part number for thermal overload unit. The heaters are sized according to the actual full-load current listed on the motor nameplate. Find the thermal overload unit number from the manufacturer's list corresponding to the motor nameplate full-load current. A typical manufacturer's thermal overload unit selection chart is shown in Figure 7.8. An example will help show how the thermal overload sensing unit chart is used.

> **Example 7.2**   A 3-phase, 240-volt motor has a nameplate full-load current marked as 1.5 amperes, and the service factor is 1.15. Assume the manufacturer part numbers for the overload relay sensing element are listed in Figure 7.8. Select the maximum overload relay sensing element part number permitted for this motor assuming normal starting.

Table 7.2   Typical time-delay fuse ratings available up to 30 amperes.

$\frac{1}{10}$	$\frac{15}{100}$	$\frac{2}{10}$	$\frac{4}{10}$	$\frac{1}{2}$	$\frac{6}{10}$	$\frac{8}{10}$	
1	$1\frac{1}{8}$	$1\frac{1}{4}$	$1\frac{4}{10}$	$1\frac{6}{10}$	$1\frac{8}{10}$		
2	$2\frac{1}{4}$	$2\frac{1}{2}$	$2\frac{8}{10}$	$3\frac{2}{10}$	$3\frac{1}{2}$		
4	$4\frac{1}{2}$	5	$5\frac{6}{10}$	$6\frac{1}{4}$	7	8	
9	10	12	15	$17\frac{1}{2}$	20	25	30

Motor Full-Load Current (Amperes)	Thermal Unit No.	Maximum Fuse Rating (Amperes)	Motor Full-Load Current (Amperes)	Thermal Unit No.	Maximum Fuse Rating (Amperes)
0.28–0.30	JR 0.44	0.6	2.33–2.51	JR 3.70	5
0.31–0.34	JR 0.51	0.6	2.52–2.99	JR 4.15	5.6
0.35–0.37	JR 0.57	0.6	3.00–3.42	JR 4.85	6.25
0.38–0.44	JR 0.63	0.8	3.43–3.75	JR 5.50	7
0.38–0.53	JR 0.71	1.0	3.76–3.98	JR 6.25	8
0.54–0.59	JR 0.81	1.125	3.99–4.48	JR 6.90	8
0.60–0.64	JR 0.92	1.25	4.49–4.93	JR 7.70	10
0.65–0.72	JR 1.03	1.4	4.94–5.21	JR 8.20	10
0.73–0.80	JR 1.16	1.6	5.22–5.84	JR 9.10	10
0.81–0.90	JR 1.30	1.8	5.85–6.67	JR 10.2	12
0.91–1.03	JR 1.45	2.0	6.68–7.54	JR 11.5	15
1.04–1.14	JR 1.67	2.25	7.55–8.14	JR 12.8	15
1.15–1.27	JR 1.88	2.5	8.15–8.72	JR 14.0	17.5
1.28–1.43	JR 2.10	2.8	8.73–9.66	JR 15.5	17.5
1.44–1.62	JR 2.40	3.2	9.67–10.5	JR 17.5	20
1.63–1.77	JR 2.65	3.5	10.6–11.3	JR 19.5	20
1.78–1.97	JR 3.00	4.0	11.4–12.7	JR 22	25
1.98–2.32	JR 3.30	4.0	12.8–14.1	JR 25	25

Figure 7.8   Chart for selecting thermal overload sensing units.

Answer: The manufacturer's part number charts for overload relay sensing elements are based upon a service factor of 1.15 or larger or a temperature rise of 40°C or lower. This means the sensing element has already included the 1.25 multiplier permitted by *430.32(A)(1)*. Therefore, simply look up the nameplate full-load current in the chart (Figure 7.8) and find the part number of JR 2.40.

If the service factor of the motor is less than 1.15 or the temperature rise is above 40°C, then the overload relay sensing element is not permitted to have a rating greater than 1.15 times the motor nameplate full-load current. The manufacturer chart for selecting the overload sensing element is approximately 10% too high. Before selecting the overload relay sensing element from a manufacturer chart if the service factor is less than 1.15 adjust the nameplate full-load current to a lower value by using Equation 7.3. If the service factor for the motor in example 7.2 had been 1.0, then the adjusted nameplate full-load current would have been 1.38 amperes (1.5 A × 1.15 / 1.25 = 1.38 A) and the manufacturer part number for the overload sensing element would be JR 2.10.

$$\text{Adjusted Nameplate Current}_{SF < 1.15} = \text{Nameplate Current} \times \frac{1.15}{1.25} \qquad \text{Eq. 7.3}$$

Sometimes a motor is required to start a load where the start-up time is long enough to overheat and trip the overload relay. One way to deal with this problem is to install a sensing element with a longer time rating. *NEC® 430.32(C)* does permit increasing the sensing element rating but not to exceed 140% when the service factor is 1.15 or greater, and to 130% when the service factor is less than 1.15. Once again, the nameplate current must be adjusted before selecting a manufacturer part number from a chart already based upon 125% of the motor nameplate full-load current. Equation 7.4 can be used for making the adjustment in nameplate current when the overload sensing element is being sized not to exceed 140%. In the case of the motor in Example 7.2, the adjusted nameplate current would be 1.68 amperes and the manufacturer part number would be JR 2.65.

$$\text{Adjusted Nameplate Current}_{\text{Applying } 430.32(C)} = \text{Nameplate Current} \times \frac{1.40}{1.25} \qquad \text{Eq. 7.4}$$

Remote Control Circuit Wires

A magnetic motor starter is operated with an electric solenoid coil. A control circuit is installed to operate the solenoid. One or more operating devices may be on a control circuit. The control circuit wires are permitted to be smaller than the motor-circuit wires, and they are considered to be protected by the motor branch-circuit short-circuit and ground-fault protection. There is a limit as to how high a rating is permitted for this branch-circuit protection before the control circuit is no longer considered to be adequately protected. The rules for determining the minimum control circuit wire size permitted based on the branch-circuit protection rating are given in *430.72*. When the rating of the branch-circuit protective device is too high, overcurrent protection shall be installed to protect the control circuit.

Motor-Circuit Examples

Some examples of motor-circuit component sizing will help illustrate the application of *Article 430*. Nameplate information is given for the motor in each example. The motor nameplate current, as stated in *430.6(A)(1)*, is used to determine the rating of the running overcurrent protection. The current as found in *Tables 430.247, 430.248, 430.249,* and *430.250* is generally used to determine the size of branch-circuit conductors and branch-circuit short-circuit and ground-fault protection. Example 7.3 illustrates how components of a circuit are sized for a small horsepower motor.

Example 7.3   A $^1/_4$-horsepower, 115-volt, single-phase electric motor is operated from an automatic controller, as shown in Figure 7.9. The nameplate full-load current is 5.4 amperes and the service factor is 1.15. A fusible switch acts as the disconnect and contains fuses that act as both short-circuit and ground-fault protection, as well as running overload protection. The motor is not thermally protected, and it is not powering a hard starting load. Determine (1) the minimum size copper, Type THWN branch-circuit conductor permitted assuming 75°C terminations, and (2) the maximum permitted rating of time-delay fuse to protect the motor from overloads and the branch-circuit from short circuits and ground faults.

Answer: First, look up the motor full-load current from *Table 430.248*. The value of full-load current will be 5.8 amperes. The minimum permitted branch-circuit wire size is determined according to *430.22(A)*. The ampacity of the conductor is determined by multiplying the full-load current for the motor by 1.25, then the minimum wire size is found in *Table 310.16*. The smallest wire size permitted for a branch-circuit is 14 AWG, even though the calculated value for the motor was 7.25 amperes.

$$1.25 \times 5.8 \text{ A} = 7.25 \text{ A}$$

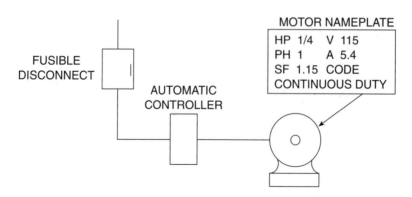

Figure 7.9   An automatically controlled, $^1/_4$-horsepower, single-phase, 115-volt electric motor branch-circuit.

A single set of fuses will serve as both branch-circuit short-circuit and ground-fault protection, as well as overload protection for the motor. Fuses can often provide both functions. The overload protection for the motor is a more restrictive requirement than short-circuit and ground-fault protection; therefore, size the fuses for the overload condition. Use the nameplate full-load current of 5.4 amperes to determine overload protection. The maximum overload rating selection is covered in *430.32(B)(1)*. It is not permitted to round up to the next standard rating fuse unless the size selected using *430.32(B)(1)* is not adequate to start the motor. When an increase in rating is required, it is not permitted to exceed 140% of the nameplate full-load current, as stated in *430.32(C)*. It is not permitted to select the 10-ampere standard rating of overcurrent device, as listed in *240.6,* because it is larger than 140%. Select a set of fuses with a rating of 6.0 or 6.25 amperes from Table 7.2. If the 6.25-ampere fuse is not adequate to start the motor, then it is permitted to choose a 7-ampere fuse.

$$1.25 \times 5.4 \text{ A} = 6.75 \text{ A}$$

$$1.40 \times 5.4 \text{ A} = 7.56 \text{ A}$$

Example 7.4   A 10-horsepower, design B, 3-phase, 460-volt electric motor is controlled by a magnetic motor starter on a branch-circuit with a fusible switch as the disconnecting means. The circuit is shown in Figure 7.10. The nameplate full-load current is 14 amperes and the service factor is 1.15. The conductors are in conduit, and the motor is not powering a difficult starting load. Determine the following for the motor circuit:

1. The minimum permitted size copper, Type THWN branch-circuit conductor with 75°C terminations
2. The minimum permitted rating of the circuit disconnect
3. The minimum NEMA size motor starter permitted
4. The maximum permitted rating of time-delay fuses to be used for branch-circuit short-circuit and ground-fault protection
5. The maximum permitted rating of motor overload device as found in the sample manufacturer chart of Figure 7.8
6. The minimum permitted size of Type THWN control circuit wire when protected from overcurrent by the branch-circuit fuses and assuming 75°C terminations

Answer: (1) Look up the motor full-load current rating of 14 amperes from *Table 430.250* in the Code. Next, determine the minimum permitted rating in amperes of the branch-circuit wire. The wire size is determined from *Table 310.16. NEC® 240.4(D),* which limits the overcurrent protection for sizes 14, 12, and 10 AWG, does not apply in the case of motor circuits. The minimum wire size permitted is 14 AWG copper.

$$1.25 \times 14 \text{ A} = 17.5 \text{ A}$$

(2) The disconnect is required to be rated in horsepower for the operating voltage of the motor. Electrical equipment typically has voltage ratings of 150, 250, and 600 volts. In the case of this motor, choose a 600-volt rated disconnect switch with a minimum 3-phase rating of 10 horsepower.

(3) The motor starter (controller) is required to have a minimum rating of 10-horsepower, 3-phase at 480 volts. Find the minimum permitted NEMA size 1 from Table 7.1.

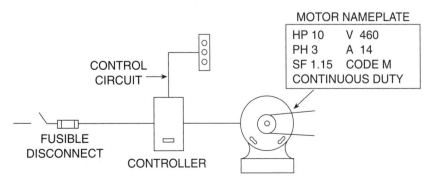

Figure 7.10   A 10-horsepower, design B, 3-phase, 460-volt squirrel-cage motor controlled with a magnetic motor starter.

(4) The maximum rating of the branch-circuit short-circuit and ground-fault protection is determined using the information from the following Code sections:

- *NEC® 430.52(C)(1)* and *Table 430.52*
- *NEC® 430.52(C)(1), Exception 1*, which permits rounding up to the next standard rating of overcurrent device, as listed in *240.6*, when the size is determined according to *Table 430.52*
- *NEC® 430.52(C)(1), Exception 2(b)*. It is permitted to increase size if high motor starting current causes the overcurrent device to open, but the overcurrent device is not permitted to have a rating in excess of 225% of motor full-load current when time-delay fuses are used.

$$1.75 \times 14 \text{ A} = 24.5 \text{ A} \quad \text{maximum fuse ampacity}$$

$$2.25 \times 14 \text{ A} = 31.5 \text{ A} \quad \text{absolute maximum ampacity}$$

It is permitted to round the 24.5 amperes up to the next standard rating of fuse, which would be 25 amperes. If this fuse rating is too small to prevent fuse opening during difficult motor starting, then it is permitted to choose higher rating fuses, but it is not permitted to exceed 31.5 amperes. In this case, it would be permitted to use a 30-ampere fuse only if the 25-ampere fuse was not of a sufficiently high rating to carry the starting current.

(5) The motor overload protection for this motor is determined by using the nameplate full-load current of 14 amperes. In this case, the nameplate current and the current from *Table 430.250* are identical. Next, check the service factor (SF) or the temperature rise on the motor nameplate. This motor has a service factor of 1.15. This means that the overload protection is permitted to be sized at a maximum of 125% of the nameplate full-load current. The manufacturer has already taken the 125% into account; therefore, use the nameplate full-load current and look up the manufacturer's number for the overload thermal unit to be installed into the motor starter. Using the nameplate full-load current of 14 amperes, the manufacturer's thermal unit number using the chart of figure 7.8 is JR 25.

(6) The branch-circuit fuses for this motor circuit are rated at 25 amperes. The control circuit is covered by *430.72(B)(2)*. The maximum permitted branch-circuit overcurrent device rating for various wire sizes is given in column C of *Table 430.72(B)*. If a size 14 AWG wire is used, the branch-circuit fuses are permitted to be rated at 45 amperes. For this circuit, the fuses are 25 amperes, and the minimum permitted wire size is 14 AWG.

Example 7.5   A 3-phase, ³/4-horsepower, design B, 230-volt motor is operated by a motor starter from a circuit protected with an inverse-time circuit breaker, as shown in Figure 7.11. The nameplate full-load current is 2.8 amperes and the service factor is 1.15. Is the 15-ampere circuit breaker permitted to act as short-circuit and ground-fault protection for the motor and controller?

Answer: The minimum branch-circuit wire size permitted is 14 AWG, as determined by multiplying the 2.8-ampere full-load current for the motor by 1.25. The maximum permitted size of branch-circuit short-circuit and ground-fault protective device rating is determined from *430.52* and *Table 430.52*. The maximum permitted rating of circuit breaker for the circuit is stated in *430.52(C)(1), Exception 2(c)* as 400% of the motor full-load current. This is still smaller than the smallest circuit breaker rating available and would tend to indicate that the circuit breaker rating is still too high for the motor circuit. But this situation is covered in *Exception 1* to *430.52(C)(1)*. Standard ratings for circuit breakers smaller than 15 amperes are not available; therefore, this is the smallest standard circuit breaker.

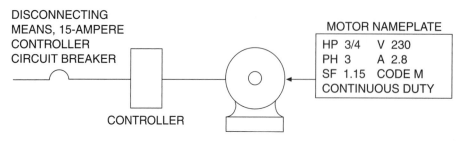

Figure 7.11   A circuit breaker provides the branch-circuit short-circuit and ground-fault protection for a ³/4 horsepower, 3-phase, 230-volt design B, squirrel cage-motor.

The answer to the question is yes. The 15-ampere circuit breaker is permitted to be used for this motor circuit.

$$2.5 \times 2.8 \text{ A} = 7.0 \text{ A}$$

$$4.0 \times 2.8 \text{ A} = 11.2 \text{ A}$$

Example 7.6   A specific fixed motor load consisting of 10-, 7.5-, and 5-horsepower, 3-phase, 230-volt design B motors is supplied by a specific purpose feeder. Fuses are used as short-circuit and ground-fault protection for each motor branch-circuit, and the ratings of these time-delay fuses for the branch-circuits are as follows: 10-horsepower motor, 45-ampere; 7.5-horsepower motor, 30-ampere; and 5-horsepower motor, 25-ampere. The circuit is shown in Figure 7.12. Determine the following:

1.   The minimum copper, Type THWN feeder wire size permitted assuming 75°C terminations
2.   The maximum feeder time-delay fuse size
3.   The minimum tap wire sizes for each motor branch-circuit assuming 75°C terminations

Answer: (1) Look up the motor full-load current from *Table 430.250*. Then determine the minimum size wire permitted using the rule in *430.24*. Look up the minimum wire size permitted from *Table 310.16*. The minimum feeder wire size is 4 AWG.

$$15.2 \text{ A} + 22 \text{ A} + 28 \text{ A} + 0.25 \times 28 \text{ A} = 72.2 \text{ A}$$

(2) The maximum permitted feeder time-delay fuse size is determined from *430.62*. Start with the maximum branch-circuit fuse rating permitted (50A) and add to it the full-load currents of the other motors supplied by the feeder and get 87.2 amperes. It is not permitted to exceed this value calculated; therefore, an 80-ampere time-delay fuse would seem to be the maximum permitted for the feeder. In this case, *430.62(A)* would require overcurrent protection on the feeder that was less than the ampacity of the feeder conductor. *NEC® 430.62(B)* permits the feeder conductor to be sized larger than required for the specific motor load, and the overcurrent protection to be based upon the ampacity of the conductors. This means the rules of *240.4* will apply with respect to selecting the overcurrent device rating and conductor size. In this case the size 4 AWG copper feeder conductor is permitted to be protected with a 90-ampere overcurrent device. Another example of this procedure is given in *Example D8, Annex D* of the Code.

$$50 \text{ A} + 22 \text{ A} + 15.2 \text{ A} = 87.2 \text{ A}$$

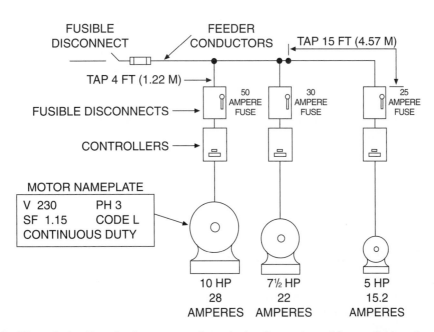

Figure 7.12   Three design B squirrel-cage motor branch-circuits are tapped from a feeder where these motors are the only loads on the feeder.

(3) The minimum branch-circuit tap sizes permitted are determined using the tap rule of *430.28*. With the 10-ft (3 m) rule, the minimum tap conductor ampacity is one-tenth of the rating of the feeder overcurrent device. In this case, the feeder overcurrent device is a set of 80-ampere time-delay fuses. One-tenth of that rating is 8 amperes. In this case, the tap conductors will be simply sized adequate to serve the individual motor branch-circuits.

$$10\text{-horsepower motor: } 28\text{ A} \times 1.25 = 35\text{ A, size 10 AWG}$$

$$7.5\text{-horsepower motor: } 22\text{ A} \times 1.25 = 27.5\text{ A, size 10 AWG}$$

For the 5-horsepower motor, the branch-circuit tap is more than 10 ft (3 m); therefore, the 25-ft (7.5-m) tap rule shall be used. The tap conductor shall have an ampacity of not less than one-third that of the feeder conductors, which is 85 amperes for a size 4 AWG copper Type THWN conductor with 75°C terminations. Also, be sure to check the minimum size wire required to supply the motor. A size 14 AWG wire is the minimum permitted to supply the motor, but it is required to use a size 10 AWG wire to satisfy the minimum tap permitted.

$$5\text{-horsepower motor: } 15.2\text{ A} \times 1.25 = 19\text{ A, size 14 AWG}$$

$$\frac{85\text{ A}}{3} = 28.3\text{ A, size 10 AWG}$$

Grounding Motors and Controllers

Grounding of equipment in a motor circuit is discussed in *Part XIII* of *Article 430* and in *250.122(D)*. *Part XIII* specifies what equipment is required to be grounded and how the grounding is to be accomplished. Frequently, metal raceway acts as the equipment grounding conductor for components of the circuit such as the disconnecting means enclosure, controller, and control devices. A typical motor installation has a flexible section of raceway between the controller and the motor. This flexible section is necessary for motor alignment to the driven machine and to minimize vibration from being transmitted to other equipment. A bonding jumper can be installed across this flexible section of raceway or an equipment grounding wire can be run inside the raceway with the circuit conductors. When supplied by flexible cord, an equipment grounding conductor is also required to be installed. Determining the minimum size of equipment grounding conductor for a motor circuit is based upon the size of overcurrent device protecting the branch-circuit. The minimum size bonding jumper or equipment grounding conductor is determined according to the rule in *250.122(A)*. *Table 250.122* is used to determine the minimum wire size using the branch-circuit overcurrent device rating. Since the branch-circuit overcurrent device is only intended to protect for ground faults and short circuits, it often has a rating much higher than the ampacity of the motor-circuit conductors. The following example will illustrate how the minimum size of an equipment grounding conductor is determined for a motor circuit:

Example 7.7: Determine the minimum size copper equipment grounding conductor required for a motor branch-circuit where the design B motor is rated 20 horsepower, 460 volts, 3 phase with a full-load current of 27 amperes. Assume the circuit is protected with time-delay fuses rated at 50 amperes, and the circuit conductors are size 10 AWG copper. Determine the minimum size copper equipment grounding conductor permitted for this circuit.

Answer: Look up the minimum size grounding electrode conductor for this motor circuit in *Table 250.122* using the 50 ampere circuit overcurrent device rating. The minimum size grounding conductor for this motor circuit is 10 AWG copper which is the same size as the ungrounded circuit conductors.

MOTOR-CONTROL CIRCUIT WIRING

Motor-control circuit wiring can be confusing, but after the general concept is understood, it can be easily performed. Common types of control circuit wiring for a magnetic motor starter are the 3-wire control and the 2-wire control. A typical example of the 3-wire control circuit is a start-stop station that operates the solenoid of a magnetic motor starter. A 2-wire control is any device that opens and closes a switch to operate the motor starter. A pressure switch, limit switch, or thermostat are typical examples of a 2-wire con-

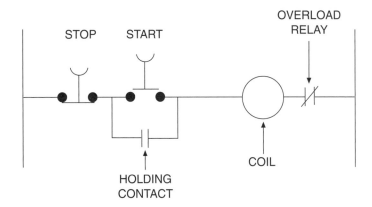

Figure 7.13   Ladder diagram for a start-stop station controlling a magnetic motor starter.

trol device. Ladder diagrams are frequently used to provide a means of visualizing the control circuit and how it works. Figure 7.13 shows a ladder diagram for a start-stop station operating a motor starter.

A schematic diagram of a magnetic motor starter operated with a start-stop station is shown in Figure 7.14. Compare the diagram of Figure 7.14 with Figure 7.13 to see how a ladder diagram represents the actual control circuit. With the ladder diagram, it is easy to see how the circuit works, but with Figure 7.14 it is easy to see how the wiring is installed.

A 2-wire control device is frequently used to open and close the motor-control circuit. Figure 7.15 shows a motor starter operated with a thermostat using power from two of the supply lines of the 3-phase source. Note that the thermostat in Figure 7.15 simply completes the circuit from line L1 to the coil. In the case of a 2-wire control circuit, the holding contact in the magnetic motor starter is not needed. Some physical action opens and closes the contacts of the 2-wire control device. In the case of Figure 7.15, that physical action is change in temperature. Other types of physical action are pressure, fluid level, flow rate, mechanical pressure, proximity, and many other physical quantities that can be detected by some type of sensor. It is important to remember that when power is restored after a power interruption, the motor will start immediately if the 2-wire control device is still in the closed position. This is the importance of *455.22* in

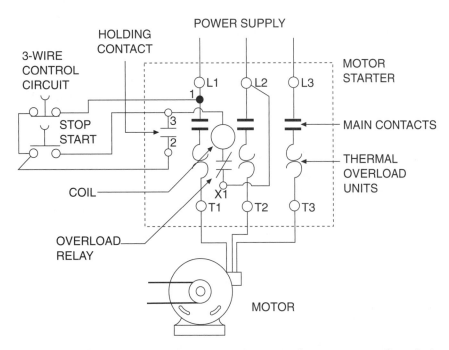

Figure 7.14   The wiring of a start-stop station to control a magnetic motor starter for a 3-phase design B squirrel-cage motor.

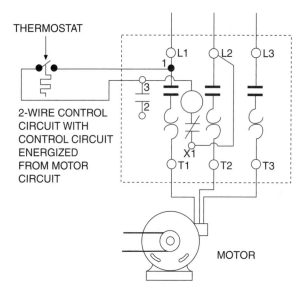

THERMOSTAT

2-WIRE CONTROL
CIRCUIT WITH
CONTROL CIRCUIT
ENERGIZED
FROM MOTOR
CIRCUIT

MOTOR

Figure 7.15 **The wiring of a simple switch device such as a thermostat to control a magnetic motor starter.**

the case of a power interruption of a rotary phase converter circuit. It is necessary to make sure that the phase converter is restarted before the loads are started. Safety may be a factor in the case of a power interruption, and automatic restarting of loads may not be desirable. Note that in the case of the 3-wire control circuit of Figure 7.14, the motor will not restart after a power interruption because the holding contact is now open.

A 2-wire control circuit is used when an external device, such as a programmable controller, is used to operate the motor starter as illustrated in Figure 7.16. In this case, the power source to operate the motor starter solenoid may be from a source other than the motor branch-circuit. It will be necessary to make sure there is no connection between the two power sources.

In the case of a control circuit power source separate from the motor branch-circuit, *430.74(A)* requires that all power sources be capable of being disconnected from the motor and the controller. It is permitted to have a separate disconnecting means for the motor branch-circuit and the control circuit. The dotted line in Figure 7.16 shows the wire that must be removed to make sure the control circuit power source is separated from the motor branch-circuit power source. The solenoid in the motor starter must match the control circuit voltage, and the control circuit must be separated from the motor power supply by removing a factory-installed wire inside the motor starter.

When a motor is operated at 480 volts, a control circuit is permitted to reduce the 480-volt supply to a lower voltage such as 120 volts by installing a control transformer inside or adjacent to the motor starter. Figure 7.17 shows a typical installation of a control transformer for a motor-control circuit. Note in Figure 7.17 that a factory-installed wire inside the motor starter must be removed when a control transformer is supplying the control circuit power.

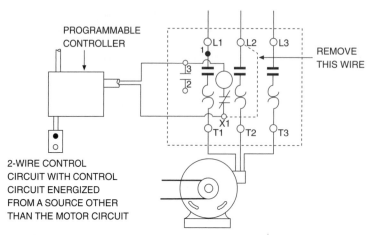

PROGRAMMABLE
CONTROLLER

REMOVE
THIS WIRE

2-WIRE CONTROL
CIRCUIT WITH CONTROL
CIRCUIT ENERGIZED
FROM A SOURCE OTHER
THAN THE MOTOR CIRCUIT

Figure 7.16 **An external power source such as a programmable controller is used to control the motor at 120 volts.**

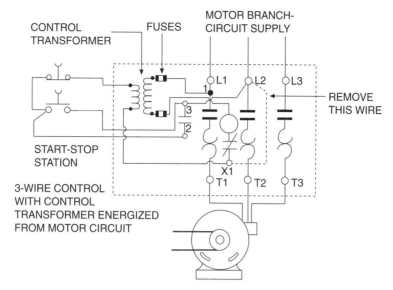

CONTROL TRANSFORMER

FUSES

MOTOR BRANCH-
CIRCUIT SUPPLY

REMOVE
THIS WIRE

START-STOP
STATION

3-WIRE CONTROL
WITH CONTROL
TRANSFORMER ENERGIZED
FROM MOTOR CIRCUIT

Figure 7.17   A control transformer steps 480 volts down to 120 volts for the control circuit.

Adjustable-Speed-Drives

The majority of the electric motors in use are induction motors that operate at a nearly constant speed. By applying 3-phase power to the windings in the stator of the motor, a magnetic field rotates in space about the axis of the motor. This rotating magnetic field induces a current into the aluminum squirrel cage of the rotor. That induced current in the rotor creates a magnetic field that tries to follow the stator magnetic field. The rotor of an induction motor will always turn at a speed slightly slower than the stator magnetic field. In a 4-pole induction motor energized with 3-phase, 60-Hz power, the stator magnetic field will rotate at a constant 1800 rpm. The rotor will turn at somewhere between 1725 rpm to 1750 rpm, depending upon the load being powered. If the frequency of the supply to the motor can be increased, the rotor will turn at a faster speed. If the frequency of the supply to the motor is decreased, the rotor will turn at a slower speed. This is how an adjustable speed drive works. The power conversion unit of the drive rectifies the 60-Hz input power into direct current, then creates an output that is not a sine wave, but it does have a repetitive pattern. If the output frequency of the conversion unit is less than 60 Hz, the rotor of the motor will turn at a slower speed. An adjustable-speed drive makes it possible to vary the rotor speed of what would normally be a fixed-speed motor. A common method of creating a variable frequency output to power a 3-phase induction motor is by a method called pulse width modulation (PWM). The induction motor is supplied with a repetitive pattern of positive and negative pulses with varying width to the pulses. Figure 7.18 is a simplified representation of a pulse width modulated output of one of the phases supplied to an induction motor from an adjustable-speed drive.

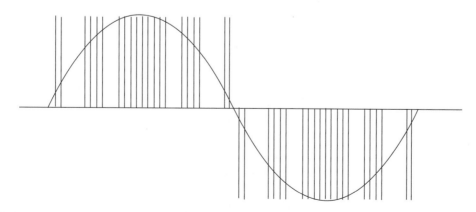

Figure 7.18   The output power from an adjustable-speed drive conversion unit is typically a series of positive and negative pulses with varying spacing that to a motor acts similar to a sine wave.

The power developed by an electric motor is proportional to the torque developed by the shaft times the revolutions per minute (rpm). Horsepower developed at the shaft of a motor can be determined using Equation 7.5 where torque is in lb-ft and rotor speed is in rpm. Motors are frequently used on loads that do not require a constant horsepower. A centrifugal pump may be required to maintain a constant pressure, but the flow is variable. If the motor operates at a constant speed when the flow is variable, some means must be devised to prevent the motor from over pressurizing the system. This type of regulation is a waste of energy. Using an adjustable-speed drive, the motor rpm can be reduced when the flow decreases, and still maintain the desired pressure. Note in Equation 7.5 that a reduction in rpm results in a reduction in horsepower and yet the torque is maintained at a constant level. This is how adjustable-speed drives are used to save energy.

$$\text{Horsepower} = \frac{2 \times \pi \times \text{torque} \times \text{rpm}}{33,000} \qquad \text{Eq. 7.5}$$

MAJOR CHANGES TO THE 2005 CODE

These are the changes to the 2005 *NEC®* that correspond to the Code sections studied in this unit. The following analysis explains the significance of the changes from the 2002 to the 2005 Code only and this analysis is not intended to be used in place of the Code. Refer to the actual section of the 2005 Code for the exact wording and meaning of each section discussed. Changes are indicated in the Code with a vertical line in the margin. If material was deleted or moved to another location in the Code, the location of the deletion is indicated with a dark dot in the margin.

Article 409 **Industrial Control Panels**

This is a new article. An industrial control panel is defined in *409.2*, and it is a single enclosure or group of attached sectional enclosures that direct the control of equipment as well as contain controllers. The key difference between a motor-control center and an industrial control panel is that the industrial control panel contains actual controllers and carries the main current to the loads. An industrial control panel and a motor-control center are illustrated in Figure 7.19.

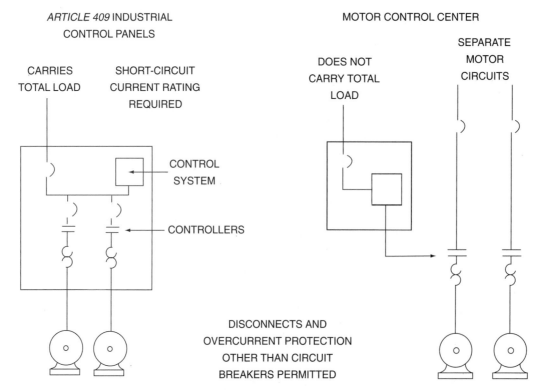

Figure 7.19   The principle difference between an industrial control panel and a motor control center is that the industrial control panel contains load controllers and carries the total load current.

409.20: The conductors supplying an industrial control panel are sized in the same manner as a feeder supplying a specific motor load as described in *430.24*.

409.21(C): The overcurrent protection for an industrial control panel is determined the same as the overcurrent protection for a feeder supplying a specific motor load as described in *430.62*.

409.110(3): Another key difference between an industrial control panel and a motor control center is that the
industrial control panel is required to have a short-circuit current rating labeled on the assembly.

Article 430   Motors, Motor Circuits, and Controllers

Design E motor references deleted: The design E high-efficiency motor was never built for production and
instead the premium efficiency design B motor has achieved the characteristics intended for the design
E motor. All references to the design E motor were deleted.

430.2: A new definition was added for an **adjustable-speed drive,** which consists of auxiliary devices and
sensors, an induction motor, and a power converter that changes normal alternating current into a
power form and frequency that causes a normally fixed-speed induction motor to operate as a variable-
speed motor.

430.2: A new definition was added for an **adjustable-speed drive system,** which is an adjustable-speed drive
plus auxiliary electrical components that are required to make the equipment function for a specific
application.

430.2: A new definition of **system isolation equipment** was added, which is a magnetically operated switch
that can have control power disconnected remotely from multiple locations so the switch can be locked
in the open position from multiple locations.

430.7(A)(15): Resistance heaters are sometimes installed inside motors to keep the inside of the motor warm
enough to prevent condensation when not in operation. Accumulation of moisture can lead to internal
corrosion and deterioration of winding insulation. This is a new requirement that, if a motor is
equipped with an internal condensation prevention heater, the heater voltage, phase, and rated power
is required to be marked on the motor. Condensation-prevention electric resistance heaters are usually
installed at one or both ends of the stator windings as shown in Figure 7.20.

430.8: A motor controller is now required to be marked with a short-circuit current rating. There are several
exceptions where a short-circuit current rating is not required to be marked on the controller.

430.8 Exception 1: A short-circuit current rating is not required to be marked on a controller for stationary
motors rated 2 horsepower or less and not over 300 volts controlled by a general-use snap switch, and
portable motors rated 1/3 horsepower or less.

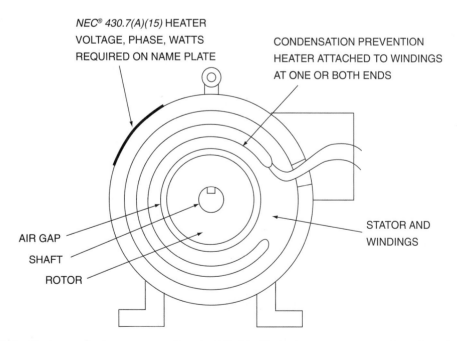

**Figure 7.20   Resistance heaters are sometimes installed inside motors to prevent condensation when the motor
is not in operation.**

430.8 Exceptions 2 and 3: These exceptions are similar. A short-circuit current rating is not required to be actually marked on the controller if a short-circuit current rating is marked elsewhere on the equipment or enclosure that also includes the controller.

430.8 Exception 4: A short-circuit current rating is not required to be marked on a controller listed for motors 2 horsepower or less and not over 300 volts and intended only for use on general-purpose branch-circuit.

430.52(C)(6) FPN: *Table 430.52* lists four types of overcurrent devices that can be used to provide short-circuit and ground-fault protection for a motor branch-circuit. This subsection deals with a listed combination controller that has an adjustable instantaneous-trip device. This new fine print note points out that an individual pole in the controller has a lower current-interrupting rating during a ground fault than it does for a short circuit. The difference is that only one pole must interrupt the current during a ground fault, while two poles in series are available to interrupt current during a short circuit. This same fine print not appears in *240.85* for circuit breaker interrupting ratings.

430.53(C)(3): A circuit breaker installed to provide branch-circuit short-circuit and ground-fault protection for a group motor installation is now only required to be listed. It is no longer required that the circuit breaker be listed for group motor installation. A standard listed inverse-time circuit breaker is now permitted to be installed for a group motor application.

430.52(C)(6): This new paragraph is only a clarification, but since the meaning in the previous editions of the Code were not always clear, this may seem like a new rule. An overcurrent device installed for protection of a group installation is not permitted to serve as the overcurrent protection for nonmotor loads of the group. Each nonmotor load, such as resistance heaters, is required to be provided with individual overcurrent protection in accordance with the rules in *Article 240* and any other appropriate article of the Code.

430.83(E): This section explains the meaning of voltage markings on controllers. The rule is the same as *240.85* for circuit breakers. The change involves controllers with slash voltage ratings such as 480Y/277 volts. This rating means the controller is capable of interrupting a locked-rotor condition or short-circuit condition between two circuit conductors with up to 480 volts line-to-line. The 277 volts means that the controller is only capable of interrupting a short-circuit, ground-fault, or locked-rotor condition not operating in excess of 277 volts. This means the controller marked 480Y/277 volts is only required to be installed to control a load on a solidly grounded wye electrical system. It is not permitted to be installed to control a load on an ungrounded system, a corner-grounded system, or high-impedance grounded system.

430.102(B) Exception: A disconnecting means is required to be located in-sight-from the motor and driven machine and the disconnecting means for the controller is permitted to serve this function. The only exception is for industrial installations and where location of a disconnect is impractical or increases hazards to personnel. When this exception applies, the disconnect is required to be capable of being locked in the open position. That locking means was in the previous edition of the Code required to be permanently installed. Now the only stipulation is that the locking means is required to be of a type that remains in place whether the lock is installed or removed.

430.109(A)(6): A manual motor starter listed as suitable as a motor disconnect is permitted to be installed between the last branch-circuit overcurrent device and the motor. A new last sentence was added to deal with what amounts to an exception to this rule where there needs to be a supplementary overcurrent device between the manual starter listed as suitable to serve as a disconnect and the motor. This supplementary overcurrent device is typically of the semiconductor fuse-type, installed as permitted in *430.52(C)(5)*, and is intended for the protection of electronic equipment in the circuit.

430.109(A)(7): This is a new paragraph the provides rules for the installation of system isolation equipment. This is a magnetically operable switch that can be controlled from multiple remote locations and locked in the open position at those remote locations as illustrated in Figure 7.21. System isolation equipment is required to be listed for disconnection purposes. It is also required to be installed on the load side of the circuit or feeder overcurrent protection and disconnecting means. The disconnecting means permitted on the line side of system isolation equipment is a motor-circuit switch, circuit breaker, or molded-case switch.

Part X: This is a new part in *Article 430* that provides rules for the installation of adjustable-speed drives for the operation of an electric motor. There are considerations related to the installation of circuit components for an adjustable-speed drive that are different from a typical motor circuit that were not cov-

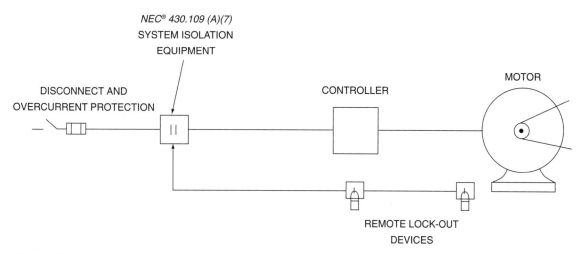

NEC® 430.109 (A)(7)
SYSTEM ISOLATION
EQUIPMENT

DISCONNECT AND
OVERCURRENT PROTECTION

CONTROLLER

MOTOR

REMOTE LOCK-OUT
DEVICES

Figure 7.21 **System isolation equipment is permitted to be installed on the load side of the circuit or feeder disconnect and overcurrent protection and can be controlled from multiple locations with a lockout to prevent the system from being energized**

ered in the Code. A schematic diagram of an adjustable-speed drive circuit is shown in Figure 7.22. The drive unit is a power converter and controller. The drive unit provides overload protection for the motor. The drive unit may also be capable of providing thermal protection for the motor by connecting to a temperature sensor in the motor. Depending upon the specific conditions involved, the adjustable-speed drive may be mounted in an enclosure that requires forced ventilation. Installation requirements and safeguards for the ventilation circuit are not covered in the Code. The adjustable-speed drive output power to the motor is not usually a sine wave, and a true rms reading ammeter is required to accurately measure current to the motor. A representation of the output power from the adjustable-speed drive is shown in Figure 7.18.

430.122: The previous edition of the Code provided a rule for sizing the conductors supplying a power conversion unit in *430.22 Exception 2*. That same rule is now in *430.122(A)* and sets the minimum conductor ampere rating at 1.25 times the rated input current of the power conversion unit. There is no rule stated in *Part X* for sizing the conductors between the power conversion unit and the motor. Those

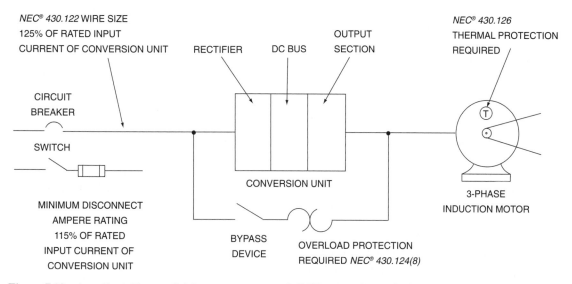

NEC® 430.124(A) OVERLOAD PROTECTION
PERMITTED IN CONVERSION UNIT

NEC® 430.122 WIRE SIZE
125% OF RATED INPUT
CURRENT OF CONVERSION UNIT

RECTIFIER   DC BUS

OUTPUT
SECTION

NEC® 430.126
THERMAL PROTECTION
REQUIRED

CIRCUIT
BREAKER

SWITCH

MINIMUM DISCONNECT
AMPERE RATING
115% OF RATED
INPUT CURRENT OF
CONVERSION UNIT

CONVERSION UNIT

BYPASS
DEVICE

OVERLOAD PROTECTION
REQUIRED *NEC® 430.124(8)*

3-PHASE
INDUCTION MOTOR

Figure 7.22 **An adjustable-speed drive converts normal 60-Hz power into a form of power at varying frequencies for the purpose of adjusting the rotor speed of an induction motor.**

wires should then be sized using *430.22,* which is 1.25 times the full-load current of the motor in most cases using the current from *Table 430.250* for the rated voltage and horsepower of the motor.

430.122(B): Some adjustable-speed drive systems are equipped with a bypass device to shunt power around the adjustable-speed drive unit. In this case, normal 3-phase power is supplied directly to the motor, which will then only run at a fixed speed. If a bypass device is provided, then the conductors supplying the power conversion unit and bypass device are required to be sized at not less than 1.25 times the motor full-load current or 1.25 times the power conversion unit rated current, whichever is greater.

430.126: Depending upon the way in which an adjustable-speed drive system operates, the motor may experience overheating that cannot be detected by current sensors in the power conversion unit. Motor over-temperature protection is required for an adjustable-speed drive system. There are a number of methods of providing over-temperature protection for a motor.

430.128: A disconnecting means is required to be provided to disconnect all ungrounded conductors supplying a variable-frequency drive system and is permitted to be installed in the line supplying the conversion unit. The disconnecting means is required to have an ampere rating not less than 115% of the rated input current of the conversion unit. There is no mention in this section about a bypass device. It is presumed that in the case where a bypass device is installed, the disconnecting means is required to have a current rating not less than 115% of the motor full-load current or the conversion unit rated current, whichever is greater. If the adjustable-speed drive unit is out of the circuit due to a bypass device, then the circuit becomes a normal motor branch-circuit. In this case, a motor switch used as a disconnect is required to have a horsepower rating not less than the rating of the motor. There is no mention in this section of a switch required to be rated in horsepower.

430.245(B): The junction box for making connections between the branch-circuit wires and the motor leads is permitted to be located up to 6 ft (1.8 m) from the motor. The wiring methods permitted for the motor leads between the motor and the junction box are specified. Added to the list of wiring methods is Type MC cable with interlocking metal tape.

Tables: The tables listing the full-load current for different types and sizes of motors were renumbered. For example, *Table 430.150* gave the full-load current of 3-phase motors. Now the table is numbered *Table 430.250.* The other tables at the end of *Article 430* are numbered in a similar manner.

Article 440   **Air-Conditioning and Refrigerating Equipment**

440.4(B): Refrigeration equipment and air-conditioning equipment is typically multimotor equipment. In addition to a compressor motor, there will often be several smaller horsepower motors to operate air-circulating fans. There may also be electric heaters. Now the nameplate of this multimotor and combination load equipment is required to provide the short-circuit current rating of the motor controllers or industrial control panel operating the equipment. The short-circuit current rating is not required to be on the nameplate of some equipment as stated in *Exception 3.*

440.4(B) Exception 3: Multimotor and combination load air-conditioning and refrigeration equipment that is cord- and plug-connected, installed in one- and two-family dwellings, or supplied by a branch-circuit rating not exceeding 60 amperes is now required to have a short-circuit current rating marked on the nameplate.

440.14 Exception 1: This exception permits the disconnecting means to be located out of sight of air-conditioning and refrigeration equipment that is necessary for the operation of an industrial process. The disconnecting means is required to be capable of being locked in the open position. The change in this exception is that now the means for locking the switch or circuit breaker must be permanent and not removable when the lock is not in place.

440.32: A 3-phase delta connected motor that has all six leads available can be reduced-voltage started by connecting the windings in a wye configuration during starting then switching the windings to a delta configuration when the motor reaches nearly full speed. This is called wye-delta starting. Since there are six leads between the controller and the motor, each lead does not carry the full-load current of the motor. Once the motor is running at full speed and the windings are in delta configuration, each lead carries 58% of the full-load current of the motor. The previous edition of the Code permitted the leads between the controller and the motor to be sized at 58% of the motor full-load current. In *430.22* the motor-circuit conductors are required to be sized at 125% of the current, and this was not factored into the 58%. Multiplying 58% by 1.25 gives 72%. Now the leads between the controller and the motor of a wye-delta motor are required to be sized not smaller than 72% of the motor full-load current.

Wye-delta motor starting is permitted for applications other than air conditioning and refrigeration. The same rule applies for sizing the conductors between the controller and the motor for other applications in *430.22(C)*. The same change that was made in *440.32* was not made to *430.22(C)*. The conductors between the controller and the motor for wye-delta operation in *430.22(C)* should also be sized with an ampacity not less than 72% of the motor full-load speed. Part-winding motor starting involves two parallel sets of phase conductors between the controller and motor as indicated in *430.22(D)*. The wires between the controller and the motor each carry half of the motor full-load current. It is necessary to multiply the 50% full-load current by 1.25 for part-winding motors so that each winding is required to have an ampacity not less than 63% of motor full-load current. A wye-delta start, 3-phase motor controller is shown in Figure 7.23 connected for starting and for running.

Article 675 Electrically Driven or Controlled Irrigation Machines

675.8(A) Exception: A molded-case switch is permitted to serve as a controller for an irrigation machine, and it is not required to have a horsepower rating. The molded-case switch is only required to have a current rating as determined by *675.7*.

675.8(B) Exception 2: A listed fusible molded-case switch is permitted to serve as the disconnecting means for an irrigation machine.

Article 685 Integrated Electrical Systems

685.1(2): An integrated electrical system is one where automatic shutdown may create a hazardous condition and an orderly shutdown is required. This is only permitted if qualified persons service the system. What constitutes a qualified person is now specifically described. The actual names of the qualified persons are required to be kept on file at the office of the establishment including documentation of training of the person. A qualified person is required to have the knowledge and skills to operate and service the system. The qualified person must have documentation of training related to the hazards involved with the system.

NEC® *440.32* AND *430.22(C)* PERMIT CONDUCTORS BETWEEN CONTROLLER AND MOTOR TO BE SIZED AT 72% OF MOTOR FULL-LOAD CURRENT.

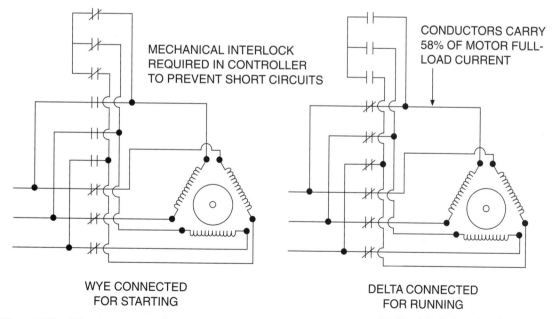

MECHANICAL INTERLOCK REQUIRED IN CONTROLLER TO PREVENT SHORT CIRCUITS

CONDUCTORS CARRY 58% OF MOTOR FULL-LOAD CURRENT

WYE CONNECTED FOR STARTING

DELTA CONNECTED FOR RUNNING

Figure 7.23 **When a motor compressor is operated with windings in a wye configuration for starting and a delta configuration for running, there will be six wires between the controller and motor and each wire will not carry more than 58% of the motor full-load current. The conductors are required to have an ampacity not less than 72% of the motor full-load current.**

WORKSHEET NO. 7—BEGINNING MOTOR-CIRCUIT WIRING

Mark the single answer that most accurately completes the statement based upon the 2005 Code. Also provide, where indicated, the Code reference that gives the answer or indicates where the answer is found, as well as the Code reference where the answer is found.

1. A 3-phase, 460-volt, 30-horsepower, design B, squirrel-cage induction motor is supplied with size 8 AWG copper conductors with 75°C insulation and terminations. The branch-circuit is protected from short circuits and ground faults with 70-ampere rated time-delay fuses. A short section of Liquidtight Flexible Nonmetallic Conduit is installed in the raceway between the controller and the motor to prevent transmission of vibration as shown in Figure 7.24 and a copper equipment grounding conductor is run from a grounding lug on the controller enclosure to a grounding lug in the motor terminal housing. The minimum size equipment grounding conductor permitted is:

   A. 14 AWG.   C. 10 AWG.   E.   6 AWG.
   B. 12 AWG.   D.   8 AWG.

                          Code reference _____

2. According to *430.102(B)* in the Code, a disconnecting means is required to be located "in sight from" a motor and driven machinery location. In addition to being directly in the line of sight, the term "in sight from" means the disconnect must not be located from the motor and driven machine a distance greater than:

   A. 10 ft (3 m).   D. 25 ft (7.5 m).
   B. 15 ft (4.5 m).   E. 50 ft (15 m).
   C. 20 ft (6 m).

                          Code reference _____

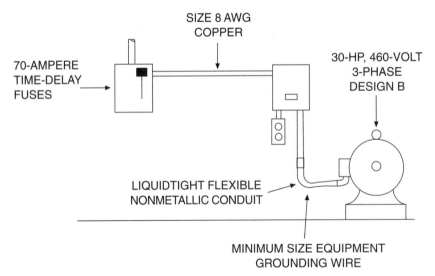

Figure 7.24 Determine the minimum size copper equipment grounding conductor required from the controller to the motor with the circuit protected with 70-ampere time-delay fuses.

3. A flexible cord attachment plug and receptacle are permitted to serve as the controller for a 230-volt, single-phase, portable induction motor that has a rating not greater than:

    A. $^{1}/_{8}$ horsepower.
    B. $^{1}/_{4}$ horsepower.
    C. $^{1}/_{3}$ horsepower.
    D. $^{1}/_{2}$ horsepower.
    E. $^{3}/_{4}$ horsepower.

    Code reference _____

4. A 3-phase, 5-horsepower, 230-volt, design B, squirrel-cage induction motor has a nameplate full-load current of 14 amperes and a service factor of 1.15. The branch-circuit short-circuit and ground-fault protection is provided with a 40-ampere inverse-time circuit breaker. The circuit is shown in Figure 7.25. The minimum branch-circuit copper conductor size with 75°C insulation and terminations permitted for this motor is:

    A. 16 AWG.
    B. 14 AWG.
    C. 12 AWG.
    D. 10 AWG.
    E. 8 AWG.

    Code reference _____

5. Each controller shall be capable of starting and stopping the motor and shall be capable of:

    A. interrupting the locked-rotor current of the motor.
    B. interrupting the full-load current of the motor.
    C. interrupting 125% of the full-load current of the motor.
    D. sensing overload currents.
    E. sensing short circuits and ground faults.

    Code reference _____

6. A 10-horsepower, 230-volt, 3-phase, design B, squirrel-cage induction motor has a nameplate full-load current of 26 amperes and a service factor of 1.15. The branch-circuit conductors are size 10 AWG copper with 75°C insulation and terminations. The motor has no difficulty starting the load. The circuit is similar to Figure 7.25. If

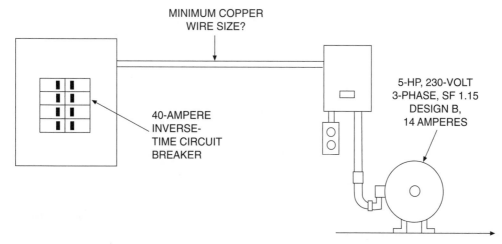

Figure 7.25 Determine the minimum size copper conductor for a motor branch-circuit that is short-circuit and ground-fault protected by a 40-ampere inverse-time circuit breaker and supplies a 5-horsepower, 230-volt, 3-phase motor.

an inverse-time circuit breaker is used as the branch-circuit short-circuit and ground-fault protection, the maximum standard rating permitted for this circuit is:

A. 30 amperes.   C. 50 amperes.   E. 70 amperes.

B. 40 amperes.   D. 60 amperes.

Code reference _____

7. A 230-volt, 3-phase, 10-horsepower, design B, squirrel-cage induction electric motor has a nameplate full-load current of 26 amperes and a service factor of 1.15. The branch-circuit conductors are size 10 AWG with 75°C insulation and terminations. If a set of time-delay fuses is installed to provide both running overload protection and branch-circuit short-circuit and ground-fault protection, the maximum standard rating permitted is:

A. 30 amperes.   C. 40 amperes.   E. 50 amperes.

B. 35 amperes.   D. 45 amperes.

Code reference _____

8. A motor and driven machine are located within sight of the controller and a fusible switch capable of being locked in the open position is located within sight of the controller, but not within sight of the motor or driven machine as shown in Figure 7.26. This commercial installation is permitted:

A. in any type of occupancy.

B. only in commercial occupancies where there are qualified personnel on duty to service the installation.

C. if there is an additional disconnect located between the controller and the motor within sight of the motor and driven machine.

D. even if the disconnect cannot be locked in the open position as long as a warning label can be attached to the disconnect during servicing.

E. because the fuses can be removed from the disconnect during servicing.

Code reference _____

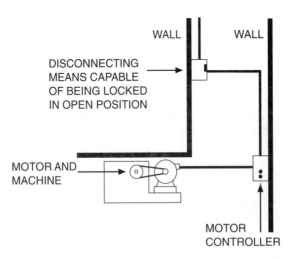

Figure 7.26 The single disconnect is capable of being locked in the open position and is in sight from the controller, but not in sight from the motor.

9. A 3-phase, 230-volt, 7$\frac{1}{2}$-horsepower squirrel-cage induction motor is required to be free to move during operation and is supplied power with a Type SO flexible cord. The cord is copper with three insulated ungrounded conductors and a green insulated equipment grounding conductor. The minimum size cord permitted for this motor is:

A. 14 AWG.   C. 10 AWG.   E. 6 AWG.
B. 12 AWG.   D. 8 AWG.

Code reference _____

10. A 3-phase, 230-volt, design B squirrel-cage induction motor has branch-circuit short-circuit and ground-fault protection provided by a 60-ampere circuit breaker. The magnetic motor starter is operated by a start-stop station located 10 ft (3 m) from the controller as shown in Figure 7.27. Separate overcurrent protection for the control circuit is not provided and the 60-ampere circuit breaker is the only protection for the control circuit. This is permitted provided the copper control-circuit wires with 75°C insulation and terminations are not smaller than size:

A. 18 AWG.   C. 14 AWG.   E. 10 AWG.
B. 16 AWG.   D. 12 AWG.

Code reference _____

11. An air-conditioning unit is located on the roof of a commercial building. The disconnecting means for the equipment is:

A. permitted to be a switch capable of being locked in the open position and located not in sight of the equipment.
B. required to be located within sight of the equipment and readily accessible from the equipment.
C. is required to be capable of being locked in the open position even when located within sight of the equipment.
D. not permitted to be attached to the outside of the air-conditioning equipment.
E. not permitted to be installed within the air-conditioner enclosure.

Code reference _____

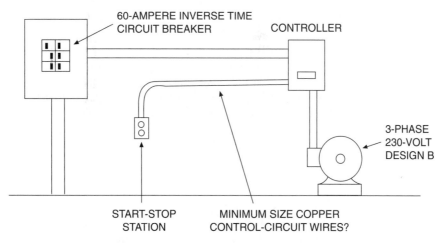

Figure 7.27 Determine the minimum size class 1 copper control-circuit conductors to the start-stop station where the motor circuit is protected with a 60-ampere inverse-time circuit breaker.

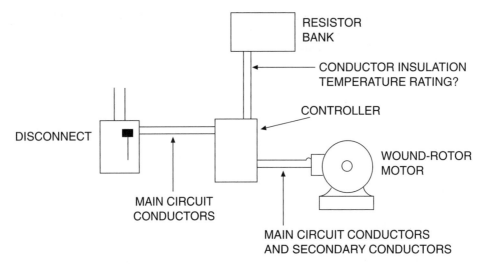

Figure 7.28   **The wound-rotor motor circuit has a resistor bank separate from the controller. Determine the minimum temperature rating of the conductors from the controller to the resistor bank.**

12. A wound-rotor motor circuit is installed with a resistor bank separate from the controller as shown in Figure 7.28. For this application the resistors may be in the circuit during normal operation. The insulation temperature rating for the conductors between the controller and the resistor bank is required to be not less than:
    A. 75°C.   C. 105°C.   E. 200°C.
    B. 90°C.   D. 150°C.

    Code reference _____

13. A 125-volt room air conditioner is permitted to be supplied by a 15- or 20-ampere, 125-volt general lighting branch-circuit in a dwelling if the air-conditioner load is not:
    A. more than 50% of the rating of the branch-circuit.
    B. more than 80% of the rating of the branch-circuit.
    C. large enough to overload the branch-circuit.
    D. more than 12 amperes.
    E. sharing the circuit with general-use receptacles.

    Code reference _____

14. A 3-phase bank of capacitors is installed on an electrical system for power factor correction. The nameplate full-load current of the capacitor bank is 50 amperes. The minimum size copper conductors with 75°C insulation and terminations permitted to connect to the capacitor bank is:
    A. 8 AWG.   C. 4 AWG.   E. 2 AWG.
    B. 6 AWG.   D. 3 AWG.

    Code reference _____

15. The main disconnecting means that is separate from the control panel for a center pivot irrigation machine, as illustrated in Figure 7.29, is required to be located:
    A. not more than 10 ft (3 m) from the center pivot point of the machine.
    B. at any distance from the center pivot point provided the disconnect is capable of being locked in the open position.
    C. where it will not be exposed to falling water.

D. at any distance from the center pivot point of the machine provided it is in direct line of sight of the center pivot point.

E. within 50 ft (15 m) of the machine and capable of being locked in the open position.

Code reference_____

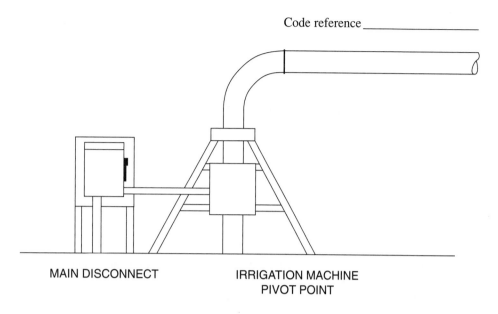

MAIN DISCONNECT   IRRIGATION MACHINE
PIVOT POINT

Figure 7.29 **What is the requirement for the location of the main disconnect for a center pivot irrigation machine operating at 460 volts.**

WORKSHEET NO. 7—ADVANCED MOTOR-CIRCUIT WIRING

Mark the single answer that most accurately completes the statement based upon the 2005 Code. Also provide, where indicated, the Code reference that gives the answer or indicates where the answer is found, as well as the Code reference where the answer is found.

1. A 75-horsepower, 3-phase, 460-volt design B squirrel-cage induction motor has a service factor of 1.15 and a nameplate full-load current of 92 amperes. The *motor* current to be used to determine the minimum size branch-circuit conductor is:

    A. 92 amperes.   C. 104 amperes.   E. 130 amperes.
    B. 96 amperes.   D. 110 amperes.

    Code reference _____

2. A 15-horsepower, 3-phase, 230-volt design B squirrel-cage induction motor has a service factor of 1.15 and a nameplate full-load current of 39 amperes. If the branch-circuit conductors are copper with 75°C insulation and terminations, the minimum size conductor permitted is:

    A. 10 AWG.   C. 6 AWG.   E. 3 AWG.
    B. 8 AWG.   D. 4 AWG.

    Code reference _____

3. A 25-horsepower, 3-phase, 460-volt design B squirrel-cage induction motor has a service factor of 1.15 and a nameplate full-load current of 30 amperes. The branch-circuit conductors are size 8 AWG copper with 75°C insulation and terminations. The overload sensing elements in the controller are set at not more than 38 amperes. The motor circuit illustrated in Figure 7.30 is powering an easy starting load. If the motor branch-circuit is protected from short circuits and ground faults by time-delay fuses, the maximum standard rating permitted is:

    A. 35 amperes.   C. 45 amperes.   E. 60 amperes.
    B. 40 amperes.   D. 50 amperes.

    Code reference _____

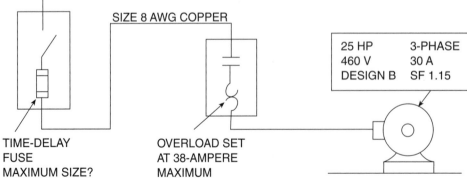

Figure 7.30 Determine the maximum rating of the time-delay fuse permitted for branch-circuit short-circuit and ground-fault protection of the 25-horsepower, 3-phase, 460-volt motor.

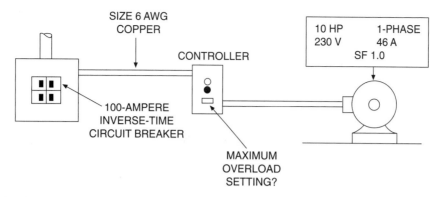

Figure 7.31   Determine the maximum setting in amperes for the running overload protection for a 10-horse-power, single-phase, 230-volt motor with a service factor of 1.0.

4. A flexible cord attachment plug not horsepower-rated and receptacle are permitted to be used as the disconnect for a portable electric motor that is:
   A. single-phase, 115 volts, rated not over $1/2$ horsepower.
   B. 3-phase, 460 volts, rated not over 1 horsepower.
   C. 3-phase, 208 volts, rated not over $1/2$ horsepower.
   D. 3-phase, 230 volts, rated not over $3/4$ horsepower.
   E. 3-phase, 230 volts, rated not over $1/3$ horsepower.

   Code reference _____

5. A 10-horsepower, single-phase, 230-volt, squirrel-cage induction motor has a service factor of 1.0 and a nameplate full-load current of 46 amperes. The branch-circuit conductors are size 6 AWG copper with 75°C insulation and terminations. The short-circuit and ground-fault protection for the branch-circuit is provided by a 100-ampere rated inverse-time circuit breaker as illustrated in Figure 7.31. The maximum current setting permitted for the running overload protection of this motor is:
   A. 40 amperes.   C. 53 amperes.   E. 63 amperes.
   B. 46 amperes.   D. 58 amperes.

   Code reference _____

6. A thermal protector is integral with a $7^1/2$-horsepower, 230-volt, single-phase, squirrel-cage motor to act as the running overload protection. The nameplate full-load current is 40 amperes, and the service factor is 1.15. The circuit is shown in Figure 7.32. If an

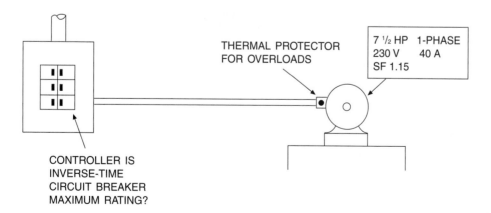

Figure 7.32   Determine the maximum rating permitted for a circuit breaker to act as a controller and disconnect for a thermally protected $7^1/2$-horsepower, single-phase, 230-volt motor.

inverse-time circuit breaker is used as the controller for the motor, the maximum standard rating permitted is:

A. 50 amperes.   C. 80 amperes.   E. 100 amperes.
B. 70 amperes.   D. 90 amperes.

Code reference_____

7. A 3-phase, 230-volt, 10-horsepower, design B, squirrel-cage induction motor is supplied with size 8 AWG copper conductors with 75°C insulation and terminations. The nameplate full-load current is 28 amperes and the service factor is 1.15. The branch-circuit is tapped from a feeder to a combination disconnect/motor controller containing a 200-ampere rated instantaneous-trip circuit breaker as the short-circuit and ground-fault protection. A short section of Liquidtight Flexible Nonmetallic Conduit is installed in the raceway between the controller and the motor to prevent transmission of vibration as shown in Figure 7.33 and a copper equipment grounding conductor is run from a grounding lug on the controller enclosure to a grounding lug in the motor terminal housing. The minimum size equipment grounding conductor permitted is:

A. 14 AWG.   C. 10 AWG.   E. 6 AWG.
B. 12 AWG.   D. 8 AWG.

Code reference_____

8. A 3-phase, 460-volt, 40-horsepower continuous-duty wound-rotor induction motor has a nameplate primary full-load current of 45.5 amperes and a nameplate secondary full-load current of 82 amperes. There is a resistor bank remote from the controller, as shown in Figure 7.34, and is classified as heavy-starting duty. The motor has a temperature rise of 40°C marked on the nameplate. All circuit conductors are copper with 75°C insulation and terminations. The minimum size secondary conductors permitted between the motor and the controller is:

A. 6 AWG.   C. 3 AWG.   E. 1 AWG.
B. 4 AWG.   D. 2 AWG.

Code reference_____

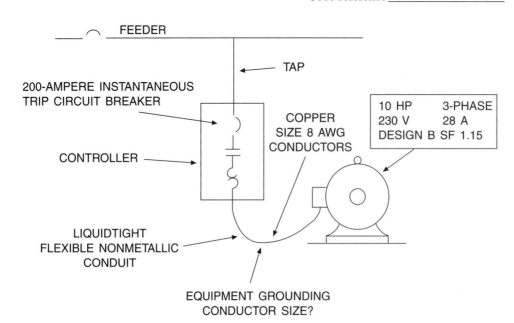

Figure 7.33 Determine the minimum size permitted for a copper equipment grounding conductor installed between a controller and a motor when the instantaneous-trip circuit breaker has a 200-ampere rating.

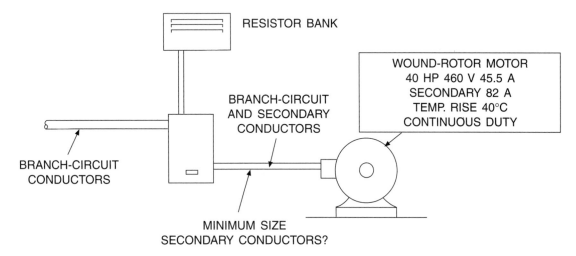

RESISTOR BANK

WOUND-ROTOR MOTOR
40 HP 460 V 45.5 A
SECONDARY 82 A
TEMP. RISE 40°C
CONTINUOUS DUTY

BRANCH-CIRCUIT
AND SECONDARY
CONDUCTORS

BRANCH-CIRCUIT
CONDUCTORS

MINIMUM SIZE
SECONDARY CONDUCTORS?

Figure 7.34   Determine the minimum size permitted for the secondary conductors between a wound-rotor motor and the controller where the secondary current is listed on the motor nameplate as 82 amperes.

9. A controller for an electric motor is within sight of a circuit breaker panel where a circuit breaker acts as the disconnect for the controller. The electric motor and machine are not located in sight from the controller or the disconnect for the controller. Due to the nature of the machine, power is supplied to the motor with a Type SO flexible cord. An acceptable disconnect for the motor and machine is:
    A. a disconnect switch located on the machine next to the motor.
    B. a lock on the door of the circuit breaker panel where the controller disconnect is located.
    C. a removable locking mechanism for the circuit breaker acting as the controller disconnect.
    D. a permanently installed locking mechanism for the circuit breaker acting as the controller disconnect where another disconnect within sight would be practical.
    E. an attachment plug and connector installed in the cord within 3 ft (900 mm) of the motor.

    Code reference_____

10. A 3-phase bank of capacitors is installed at a main service to a building to provide power factor correction. The rating of the capacitor bank is 55 kVAR, at 480 volts, 60 hertz. A fusible switch is provided as the disconnect and overcurrent protection for the conductors and the capacitors. If all conductors are copper with 75°C rated insulation and terminations, the minimum size permitted for this capacitor bank is:
    A. 6 AWG.   C. 3 AWG.   E. 1 AWG.
    B. 4 AWG.   D. 2 AWG.

    Code reference_____

11. A ladder diagram is shown for a start-stop station operating a magnetic motor starter in Figure 7.35 on the next page. The control circuit is supplied from a transformer in the motor controller enclosure with a 120-volt secondary that has one conductor grounded. The Code violation for the control circuit is that:
    A. a pilot light is required to be installed in parallel with the solenoid coil.
    B. an accidental ground fault can energize the coil and operate the motor.
    C. control circuits are required to be operated at motor line voltage.
    D. a manual disconnect switch is required to be installed in series with the start-stop station.
    E. one conductor is not permitted to be grounded.

    Code reference_____

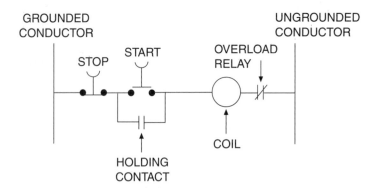

Figure 7.35   Find the Code violation in the way this control circuit of an electric motor is connected.

12. A refrigeration motor compressor is rated 208-volt, 3-phase, and has a branch-circuit selection current of 32 amperes. If the branch-circuit conductors are copper with 75°C insulation and terminations, the minimum size permitted is:

   A. 10 AWG.   C. 6 AWG.   E. 3 AWG.

   B. 8 AWG.   D. 4 AWG.

Code reference _____

13. A center pivot irrigation machine is operated with twelve design B, $3/4$-horsepower, 460-volt, 3-phase, squirrel-cage motors located on towers along the machine. The motors operate drive wheels that propel the machine in a circle around the center pivot point. These drive motors are tapped to a branch-circuit conductor that runs the entire length of the machine. The equivalent continuous-current rating for the selection of branch-circuit conductors for this machine is:

   A. 12.6 amperes.   D. 19.6 amperes.

   B. 15.8 amperes.   E. 24.0 amperes.

   C. 19.2 amperes.

Code reference _____

14. In an industrial installation with a trained and documented qualified maintenance staff on duty at all times, if an automatic shutdown of a motor in a system due to an overload would create a safety hazard greater than continued operation of a motor, it is:

   A. permitted to connect the overload sensing devices to a supervised alarm.

   B. permitted to eliminate motor overload relays form the control system.

   C. required to supply all motors of the system with an ungrounded electrical supply.

   D. permitted to eliminate motor overload relays if all motors are double insulated.

   E. required to ground all metal equipment in reach of personnel with an insulated copper equipment grounding conductor.

Code reference _____

15. A single-phase, 240-volt feeder consisting of size 2 AWG copper conductors with 75°C insulation and terminations supplies four induction motors rated one at $7^1/2$ horsepower, one at 5 horsepower, one at 3 horsepower, and one at 2 horsepower as shown in Figure 7.36. Each motor has a service factor of 1.15. The branch-circuit short-circuit and ground-fault protection for each motor is provided with time-delay fuses and the fuse ratings for each motor circuit is 70 amperes for the $7^1/2$-horse-

power motor, 50 amperes for the 5-horsepower motor, 30 for the 3-horsepower motor, and 25 amperes for the 2-horsepower motor. If a circuit breaker is used as the short-circuit and ground-fault protection for the feeder, the maximum standard rating permitted is:

A. 90 amperes.

B. 100 amperes.

C. 110 amperes.

D. 125 amperes.

E. 150 amperes.

Code reference _____

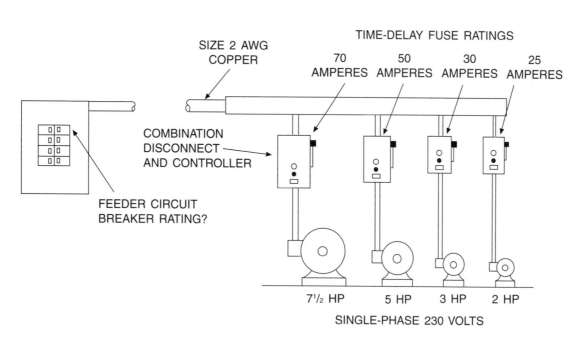

Figure 7.36 Determine the maximum rating permitted for the feeder overcurrent protection supplying a single-phase, 230-volt, 7^1/2-, 5-, 3-, and 2-horsepower motors.

UNIT 8

Transformers

OBJECTIVES

After completion of this unit, the student should be able to:

- define insulating transformer and autotransformer.
- determine the voltage of a transformer winding if turns ratio and the other winding voltage are given.
- determine the full-load current of a transformer winding if the voltage and kilovolt-amperes of the transformer are given for single-phase or 3-phase transformers.
- determine the minimum permitted kilovolt-amperes required for a specific application.
- draw the proper connections of the windings of a dual-voltage single-phase transformer for the desired voltage.
- explain the purpose of the taps on the primary winding of some transformers.
- state a specific example of the use of a boost and buck transformer.
- determine the maximum permitted overcurrent protection for a specific transformer.
- determine the minimum permitted primary and secondary conductors for a specific transformer application.
- determine the maximum distance permitted from a feeder tap to the transformer overcurrent protection.
- determine the maximum permitted input overcurrent protection for a boost or buck transformer application.
- explain how to ground the transformer and the secondary electrical system of the transformer.
- answer wiring installation questions relating to *Article 450*.
- state a significant change that occurred in Code *Article 450* from the 2002 to the 2005 edition.

CODE DISCUSSION

Article 450 deals with transformers and transformer vaults. *NEC® 450.1* gives the exceptions for transformers that are covered in other sections of the Code. *NEC® 450.3* deals with overcurrent protection of transformers. Some specific rules for autotransformer overcurrent protection are covered in *450.4*. The remainder of *Part I* of this article covers general installation requirements. *Part II* covers requirements for specific types of transformers. *Part III* deals with transformer vaults that are enclosures of specific construction for the purpose of housing transformers and other equipment such as switchboards and panelboards.

Article 450 of the Code applies only to the transformer itself, and not to the conductors leading to or away from the transformer. The branch-circuit, feeder, and tap conductors must be protected according to the rules of *Article 240*. Grounding must be accomplished according to the rules of *Article 250*. Here are Code sections helpful in working transformer circuit problems: *240.4(F), 240.21(B)(3), 240.21(C), 250.30, 250.104(D), 408.36(D), 430.72(C), 600.21, 600.23, 600.31, 600.32, 680.23(A), 725.66, 725.120(A)(1),* and *725.124*. These are in addition to information found in *Article 450*.

TRANSFORMER FUNDAMENTALS

The purpose of a transformer is to change electrical voltage to a different value. For example, a large 480-volt, 3-phase motor is powering a well pump. The motor is in a building, and one 120-volt circuit for a few lights and a receptacle outlet is needed. A transformer is used to lower the voltage from 480 to 120 for the lighting circuit. The controls for furnaces and air-conditioning units are often operated at 24 volts. A small transformer inside the equipment lowers the line voltage to 24 volts for the control circuit. Transformers are frequently used inside electronic equipment.

Types of Transformers

Transformers are of the dry type or oil-filled. Two to five percent of the electrical energy is lost in a transformer, mostly due to the resistance of the windings. Large transformers circulate oil through the windings to remove the heat. Dry transformers use air for cooling. Heat is moved from the windings to the case by conduction in smaller sizes of the dry type. Large dry-type transformers actually allow air to circulate through the windings. Oil-cooled transformers are used by the electric utility and for industrial or large commercial applications.

Common two-winding transformers are often called insulating transformers. The primary winding and the secondary winding are separate from each other, and they are not electrically connected. An autotransformer has the windings interconnected so that the primary and the secondary share the same winding. These transformers, therefore, have an electrically connected primary and secondary. A major advantage of the autotransformer over the insulating type is its lighter weight and compact size. An insulating transformer and an autotransformer are compared in Figure 8.1. A common application of an autotransformer is for electric discharge lighting ballasts.

A special type of autotransformer called a grounding autotransformer, or zig-zag transformer, is occasionally used to create a neutral wire or a ground for an ungrounded 480-volt 3-phase electrical system. These transformers are found occasionally in industrial wiring. The name zig-zag is derived from the shape of the schematic diagram. Standard insulating transformers can be used to make a zig-zag transformer. The wiring of these transformers is covered in *450.5*.

Voltage and Turns Ratio

The input winding to a transformer is called the primary winding. The output winding is called the secondary winding. If there are more turns of wire on the primary winding than on the secondary winding, the output voltage will be lower than the input voltage.

It is important to know the ratio of the number of turns of wire on the primary winding as compared with the secondary winding. This is called the turns ratio of the transformer. The actual number of turns is not important, just the turns ratio. The turns ratio of a transformer can be determined with Equation 8.1 if the actual number of turns on the transformer windings is known.

$$\text{Turns Ratio} = \frac{\textbf{Number of Turns on the Primary}}{\textbf{Number of Turns on the Secondary}} \qquad \text{Eq. 8.1}$$

The step-down transformer of Figure 8.2 has 14 turns on the primary winding and 7 turns on the secondary winding; therefore, the turns ratio is 2 to 1, or just 2. The step-up transformer has 7 turns on the pri-

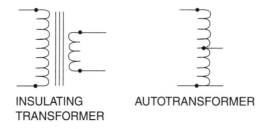

INSULATING TRANSFORMER   AUTOTRANSFORMER

Figure 8.1   Two basic types of transformers are the insulating transformer and the autotransformer.

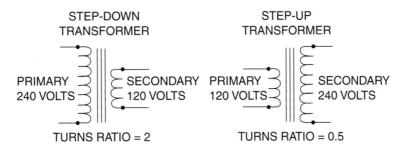

Figure 8.2 Schematic diagrams of step-down and step-up transformers.

mary and 14 on the secondary; therefore, the turns ratio is 1 to 2, or 0.5. If the voltage of one winding and the turns ratio are known, the voltage of the other winding can be determined using either Equation 8.2 or Equation 8.3.

$$\text{Primary voltage} = \text{Secondary Voltage} \times \text{Turns Ratio} \qquad \textbf{Eq. 8.2}$$

$$\text{Secondary Voltage} = \frac{\text{Primary Voltage}}{\text{Turns Ratio}} \qquad \textbf{Eq. 8.3}$$

Transformer Ratings

Transformers are rated in volt-amperes (VA) or kilovolt-amperes (kVA). This means that the primary winding and the secondary winding are designed to withstand the VA or kVA ratings stamped on the transformer nameplate. The primary and secondary full-load current usually are not given. The installer must be able to calculate the primary and secondary full-load current from the nameplate information. If the volt-ampere rating is given along with the primary voltage, then the primary full-load current can be determined using Equation 8.4 or Equation 8.5 for the case of a single-phase transformer. Equation 8.6 is used for determining the full-load current of either the primary or the secondary winding of a 3-phase transformer.

Single-phase:

$$\text{Full-Load Current} = \frac{\text{VA}}{\text{Voltage}} \qquad \textbf{Eq. 8.4}$$

$$\text{Full-Load Current} = \frac{\text{kVA} \times 1000}{\text{Voltage}} \qquad \textbf{Eq. 8.5}$$

Three-phase:

$$\text{Full-Load Current} = \frac{\text{kVA} \times 1000}{1.73 \times \text{Volts}} \qquad \textbf{Eq. 8.6}$$

An example will help to show how the previous equations are used to determine the full-load current of a transformer winding. It may be a good idea to write Equations 8.5 and 8.6 into a copy of the Code for easy reference when making transformer installations.

Example 8.1   A single-phase transformer is connected for a 480-volt primary and a 120-volt secondary. The transformer has a rating of 2 kVA. Determine the primary winding and the secondary winding full-load current of the transformer.

Answer: This is a single-phase transformer rated in kilovolt-amperes; therefore, Equation 8.5 is used to determine the full-load current for both windings. The full-load current of the primary winding of the transformer is 4.17 amperes, and for the secondary winding, the full-load current is 16.67 amperes.

$$\text{Primary Full-Load Current} = \frac{2\,\text{kVA} \times 1000}{480\,\text{V}} = 4.17\,\text{A}$$

$$\text{Secondary Full-Load Current} = \frac{2\,\text{kVA} \times 1000}{120\,\text{V}} = 16.67\,\text{A}$$

Connecting Transformer Windings

Transformer wiring diagrams are printed on the transformer nameplate, which may be affixed to the outside of the transformer or printed inside the cover to the wiring compartments. The lead wires or terminals are marked with Xs and Hs. The Hs are the primary leads, and the Xs are the secondary leads.

Some transformers have two primary and two secondary windings so they can be used for several applications. These are called dual-voltage transformers. Connections must be made correctly with dual-voltage transformers. If connected improperly, it is possible to create a short-circuit that will usually damage or destroy the transformer when it is energized.

Consider a dual-voltage transformer rated 240/480 volts on the primary and 120/240 volts on the secondary. Each of the two primary windings is, therefore, rated 240 volts. Each secondary winding is rated 120 volts. The transformer must be connected so each primary winding receives the proper voltage. Figure 8.3 shows the transformer with the primary windings connected in series with H1 and H4 connected to a 480-volt supply. The voltage across H1 and H2 is 240, and the voltage across H3 and H4 is 240. Each winding is receiving the proper voltage. With each primary winding receiving the proper 240 volts, each secondary winding will have an output of 120 volts. Connecting the secondary windings in series produces 240 volts across X1 and X4.

Next, consider a case where the primary voltage available is 480, but the desired output is 120 volts, single-phase. In this case, the primary windings are connected in series, while the secondary windings are connected in parallel, as shown in Figure 8.4.

Three-Phase Transformers

Changing the voltage of a 3-phase system can be done with a 3-phase transformer, or it can be done with single-phase transformers. Three-phase transformers are generally designed and constructed for specific voltages. For example, a transformer may have a 480-volt delta primary and a 208/120-volt wye secondary.

The 3-phase transformer has one core with three sets of windings. A primary and secondary winding are placed one on top of the other on each of the three legs of the core. Single-phase transformers can be used to form a 3-phase transformer bank. It is important that single-phase transformers are identical when connecting them to form a 3-phase system. They should be identical in voltage, kilovolt-amperes, impedance, manufacturer, and model number. Transformer impedance is the combined effect of resistance and inductance and is given in percent.

Connecting single-phase transformers to form a 3-phase bank must be done with extreme caution. The windings can only be connected in a certain way. Reversing a winding can damage the transformer. Figure 8.5 shows three individual single-phase transformers connected to step down from 480 volts delta to 240

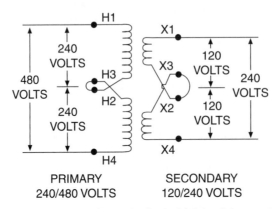

PRIMARY
240/480 VOLTS

SECONDARY
120/240 VOLTS

Figure 8.3   The windings are connected in series to obtain the higher of the rated transformer voltages.

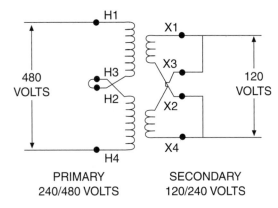

Figure 8.4 The secondary windings are connected in parallel for an output of 120 volts.

volts delta. It may be advisable to obtain a 3-phase transformer rather than connecting single-phase transformers. To illustrate the complexity, standard dual-voltage single-phase transformers are used to change 480-volt 3-phase delta to 208/120-volt wye, as shown in Figure 8.6.

Winding Taps

Transformers, except for small sizes, are often supplied with winding taps to compensate for abnormally low or high primary voltage. Assume, for example, that a transformer is rated 480 volts primary and 240 volts secondary. This means that 240 volts will be the output if the input is 480 volts. But what if the input is only 444 volts? The turns ratio for this transformer is 2-to-1; therefore, the output will be 222 volts. Equation 8.3 is used to determine the output, which would be 222 volts.

$$\text{Secondary Voltage} = \frac{444 \text{ Volts}}{2} = 222 \text{ Volts}$$

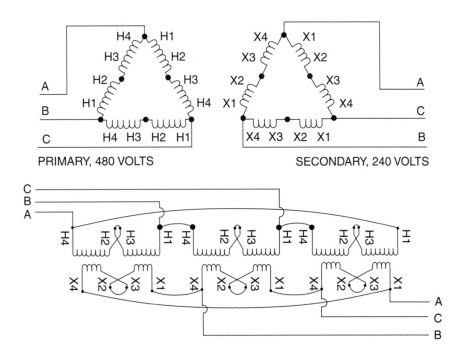

Figure 8.5   Dual-voltage single-phase transformers with 240/480-volt primary windings and 120/240-volt secondary windings are shown connected to form a 480-volt delta to a 240-volt delta 3-phase step-down transformer bank.

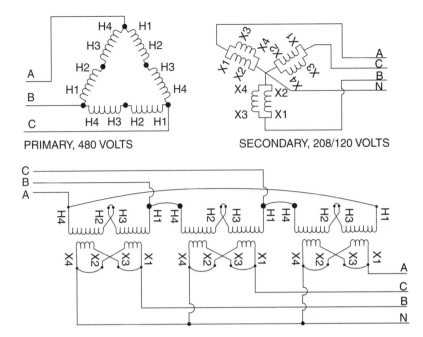

Figure 8.6 Dual-voltage single-phase transformers with 240/480-volt primary windings and 120/240-volt secondary windings are shown connected to form a 480-volt delta to a 208/120-volt wye 3-phase step-down transformer bank.

To get an output of 240 volts with an input of only 444 volts, the turns ratio will have to be changed to 1.85-to-1. The 1.85 was determined by dividing the 444 volts by 240 volts. The purpose of the tap connections, usually on the primary, is to easily change the transformer turns ratio. A typical single-phase transformer nameplate with primary taps is shown in Figure 8.7.

Consider an example in which the desired output voltage from a single-phase step-down transformer is 120/240 volts, but the available input is only 450 volts rather than 480 volts. A standard step-down transformer with a 2-to-1 turns ratio will give an output of 112.5/225 volts with a 450-volt input. By changing the primary taps, as shown in Figure 8.8, the turns ratio of the transformer is changed, and the output is now close to the desired 120/240 volts.

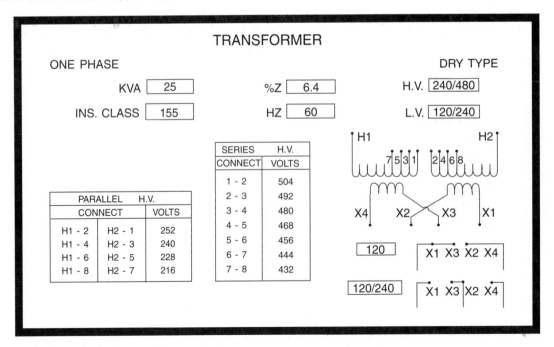

Figure 8.7 Transformer nameplate showing primary taps.

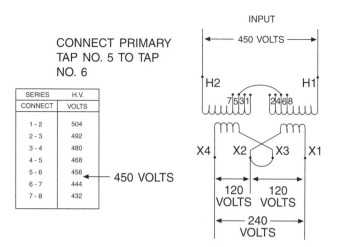

Figure 8.8 Single-phase transformer showing tap connection for an input of 450 volts and a 3-wire, 120/240-volt output.

Winding taps each make a 2.5% change in the voltage. A transformer will often have two taps above normal voltage and four taps below normal voltage. A transformer usually comes preconnected for normal voltage. If an abnormal voltage is present, it is up to the installer to change the tap connections.

Input and Output Current

The primary kilovolt-amperes (kVA) of a transformer will be equal to the secondary kVA less any small losses. If the primary voltage is reduced from 240 to 120 volts, this is a voltage ratio reduction of 2-to-1. If the primary and secondary kVA are to remain equal, the current must be higher on the secondary than on the primary by a factor of two. For example, assume the 240-volt primary current of the previous transformer is 5 amperes. The primary volt-amperes will be 1200 VA. For the 120-volt secondary, 10 amperes must flow to keep the volt-amperes at 1200 VA. If the primary and secondary voltages are known and if either the primary or secondary current is known, the other current may be determined using Equation 8.7 or Equation 8.8. These are useful equations especially when working with autotransformers.

Find primary input current when secondary output current is known:

$$\textbf{Input Current} = \textbf{Output Current} \times \frac{\textbf{Output Voltage}}{\textbf{Input Voltage}} \qquad \textbf{Eq. 8.7}$$

Find secondary output current when primary input current is known:

$$\textbf{Output Current} = \textbf{Input Current} \times \frac{\textbf{Input Voltage}}{\textbf{Output Voltage}} \qquad \textbf{Eq. 8.8}$$

Boost and Buck Transformers

A boost and buck transformer is an insulating transformer that can be connected as an autotransformer. The boost and buck transformer is used to make small adjustments in voltage, either up or down. For example, a machine has an electric motor that requires 208 volts, but the electrical supply is 240 volts. If ordering the machine with a 240-volt motor is expensive, a less costly solution to the problem may be to buck the voltage from 240 down to 208 with a boost and buck transformer.

Low voltage resulting from voltage drop can be corrected with a boost and buck transformer. This practice may not be energy-efficient, but it may be the best solution in unusual circumstances. Voltage drop on wires is wasted energy and should be avoided.

Boost and buck transformers for single-phase applications have a dual-voltage primary rated 120/240 volts. There is a choice of two sets of secondary voltages depending on the amount of boosting or bucking

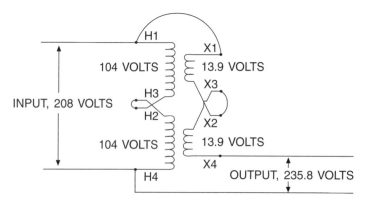

Figure 8.9   A boost and buck transformer is connected to boost a 208-volt supply to approximately 236 volts.

required: 12/24 volts and 16/32 volts. Three-phase applications from 380 to 500 volts require the use of a boost and buck transformer with a 240/480-volt primary and a 24/48-volt secondary. Figure 8.9 shows a single-phase boost and buck transformer connected to boost 208 volts to approximately 240 volts. Refer to manufacturer literature for other combinations of boosting and bucking.

Boost and buck transformers may be connected for 3-phase applications, but all 3-phase combinations are not possible. The commonly used 3-phase boost and buck transformer connections are:

- Wye (4-wire)   to   wye (3- or 4-wire)
- Wye (3- or 4-wire)   to   open delta (3-wire)
- Delta (3-wire)   to   open delta (3-wire)

A confusing aspect of boost and buck transformers is the kVA rating of the transformer required to supply a load. For some applications, the kVA rating of the load may be several times larger than the kVA rating of the boost and buck transformer. A boost and buck transformer, when used as an autotransformer for boosting or bucking, can supply a load several times the kVA rating of the transformer. The maximum kVA rating of the load supplied depends on the full-load current rating of the transformer secondary winding and the operating voltage of the load. Each manufacturer supplies load current and kVA data for boost and buck transformers for all combinations of input and output voltages.

K-Rated Transformers

The K-rating marked on some transformers is an indication of the ability of the transformers to supply loads, which produce harmonic currents. A transformer designated K-1 is one that has not been modified to supply loads that produce nonlinear or harmonic currents. Standard transformer ratings are K-4, K-9, K-13, K-20, K-30, K-40, and K-50. The higher the K-rating number, the greater is the ability of the transformer to supply loads that have a higher percentage of harmonic current-producing equipment without overheating. It is most important to select a transformer with a K-rating for the specific harmonic frequencies present and their magnitude in relation to the total 60-hertz current. A number that is probably of little use when selecting transformer K-rating is the percent of total harmonic distortion (THD). Another value that can give an indication that harmonics are present but is of little use in determining proper transformer K-rating is the value of root-mean-square (rms) current. When viewed on an oscilloscope, the current sine wave will distort when large numbers and significant levels of harmonic currents are present. This will usually result in an increase in the rms line current. Transformer overheating due to harmonics is primarily the effect of specific frequencies of harmonics and their magnitudes.

Electronic equipment that draws current for only a portion of the cycle is the type that produces harmonic currents in the electrical system. Electronic dimmer switches and some other electronic controllers switch on for only a portion of the cycle. As a result, current flows to the loads in pulses. Other electronic equipment that produces nonlinear currents includes personal computers, video display terminals, copiers, fax machines, uninterruptible power supplies (UPS), variable speed drives, electronic high-efficiency ballasts, some medical electronic monitors, welders, mainframe computers, data processing equipment, and induction heating systems. Most electronic office equipment has switching-mode power supplies (SMPS) that draw current in pulses for only part of the cycle. These pulsing input currents produce harmonic currents in the electrical system. These harmonic currents can result in overheating of a transformer and other distribution equipment and wiring.

One way of dealing with loads that consist of a high percentage of harmonic-producing electronic equipment is to oversize the transformer for the load. But a transformer designed to supply 60-hertz loads does not perform the same when currents are at a higher frequency than 60 hertz. Core saturation can occur when a standard transformer is subjected to loads with a high percentage of harmonic currents. This can occur even when the transformer is supplying less than rated full-load current. Transformer heating is different for a given level of rms current at a different frequency.

There are industry recommendations for transformer K-ratings for particular loads. Dealing with loads that produce harmonic currents is sometimes complex, and even these K-rating recommendations may not be correct for all applications. A harmonic analysis may be necessary in some cases with the transformer specifically matched to the load. Here are some general recommendations for matching transformers to loads:

- Use nonrated K-l transformers when the loads producing harmonic currents are less than 15% of the total load.
- Use K-4 rated transformers when the loads producing harmonic currents are 15 to 35% of the total load.
- Use K-13 rated transformers when the loads producing harmonic currents are 35 to 75% of the total load.
- Use K-20 rated transformers when the loads producing harmonic currents are 75 to 100% of the total load.
- Use K-30 and higher rated transformers for specific equipment where the load and transformer are matched for harmonic characteristics.

In the case where a transformer is supplying specific loads known to be producers of harmonic currents, some general guidelines can be used to help avoid transformer overheating. The transformer can be sized to the load kVA to be supplied when the proper K-rating is selected. Not only do different brands of the same equipment produce different harmonic currents, the identical equipment can produce different harmonic currents when supplied by different electrical systems. The harmonic current problem, when it exists, may be complex, sometimes requiring experienced personnel for analysis and design. The following is a general industry recommendation of approximate current K-ratings for different types of loads. Actual specifications for a specific installation made by an experienced engineer or technician may be different.

- K-4   welders and induction heaters
- K-4   electric discharge lighting
- K-4   solid-state controls
- K-13   telecommunications equipment
- K-13   branch-circuits in classrooms and in health care facilities
- K-20   mainframe computers and data processing equipment
- K-20   variable-speed drives

On a balanced 3-phase wye electrical system, odd-numbered harmonics in multiples of the third harmonic (3rd, 9th, 15th, 21st, and so on) if present will not cancel and will increase the neutral current. Examples are the 208/120-volt and the 480/277-volt 4-wire 3-phase electrical systems serving line-to-neutral loads. When a 3-phase wye transformer is supplying balanced line-to-neutral loads where harmonics are present, it is possible for the neutral current to be as high as twice the level of line current. This can result in neutral conductor and conductor termination overheating. A K-rated transformer will not eliminate this type of problem. If the harmonic currents cannot be reduced, then it will be necessary to increase the size of the neutral bus and neutral conductor.

OVERCURRENT PROTECTION FOR TRANSFORMERS

Wiring a transformer circuit is one of the most difficult wiring tasks unless the installer understands transformer fundamentals. Rules for sizing overcurrent protection for a transformer operating at not more than 600 volts are covered in *450.3(B)* and *Table 450.3(B)*. It must be noted that these rules apply only to the transformer itself, and not necessarily to the input and output circuit wires. If one or both of the windings operate at more than 600 volts, *450.3(A)* and *Table 450.3(A)* will apply.

According to *Table 450.3(B)*, both the primary and secondary windings are permitted to be protected from overcurrent by one overcurrent device located on the primary side of the transformer and sized at not more than 125% of the transformer full-load current. If that primary overcurrent device has a rating greater than 125% of the transformer full-load primary current rating, then overcurrent protection not greater than 125% of the transformer secondary full-load current rating is required to protect the transformer secondary winding. Both situations are explained in the following two sections.

Overcurrent Protection Only on the Primary (600 volts or less)

A transformer is permitted to be protected by one overcurrent device on the primary side rated not more than 1.25 (125%) times the primary full-load current, as shown in Figure 8.10. If the calculation does not correspond with a standard rating of overcurrent device as listed in *240.6*, it is permitted to round up to the next standard fuse or circuit breaker rating. This overcurrent device may be a set of fuses in a panelboard or fusible switch, or a circuit breaker. The secondary winding is not required have separate protection from overcurrent in this situation. When the secondary of the transformer is not single-voltage 2-wire, the secondary conductors are required to be protected according to *240.4(B)*, but the secondary winding is not required to be protected by other than the primary overcurrent device. *NEC® 240.21(C)(1)* applies to the secondary conductors, and not to the winding of the transformer. Example 8.2 will show how overcurrent protection is selected to protect the transformer using this rule, and how overcurrent protection for the conductors must be considered separately.

Example 8.2   A 37.5-kVA 3-phase transformer is to be installed to supply a 100-ampere 208Y/120-volt panelboard from a 480-volt, 3-phase power panel. The conductors are copper with 75°C insulation and terminations. The panelboard has a 100-ampere main circuit breaker, and it is located not more than 25 ft (7.5 m) from the transformer. The feeder and transformer are protected with time-delay fuses, and the feeder supplies only this transformer and panelboard. A diagram of the circuit is shown in Figure 8.10. Determine the maximum rating of the fuses permitted to protect the transformer when using the minimum size conductors supplying the primary of the transformer.

Answer: First calculate the primary full-load current of the transformer and obtain 45 amperes.

$$\text{Primary full-load current} = \frac{37.5 \text{ kVA} \times 1000}{1.73 \times 480 \text{ V}} = 45.1 \text{ amperes}$$

The rule is found in the first row of *Table 450.3(B)* for determining the maximum permitted rating of the primary overcurrent device. The transformer primary winding is permitted to be protected at 1.25

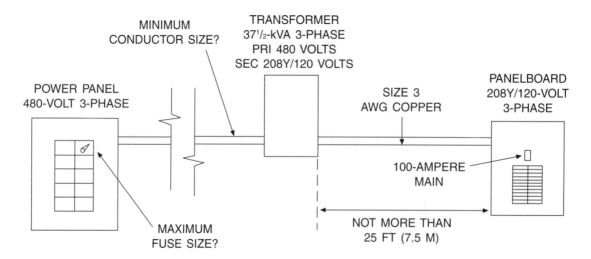

Figure 8.10   A 37.5-kVA, 3-phase, transformer supplies a 208Y/120-volt, 4-wire, 100-ampere panelboard from a 480-volt supply. Determine the minimum copper primary conductor size and the maximum permitted rating of fuse protecting the transformer and circuit.

times the full-load current of the transformer, which is 56 amperes. This value is permitted to be rounded up to the next standard rating, which is 60 amperes (see *Note 1*). If the conductors supplying the transformer are protected at 60 ampere, then *240.4* requires the conductor to have an allowable ampacity of 60 amperes, which is size 6 AWG. But if the primary full-load current of the transformer is only 45 amperes, it would seem more reasonable to use 50-ampere fuses and size the conductors at 8 AWG. The maximum fuse rating permitted assuming the minimum primary conductor size is 50 amperes.

The previous example points out the choices that can be made when sizing the overcurrent protection and the conductors for a transformer installation. For the previous example, the secondary full-load current is 104 amperes, and there is a 100-ampere overcurrent device installed on the secondary side of the transformer, which meets the requirement of protecting the transformer secondary within 125% of the transformer secondary full-load current. What this means is that it is actually permitted to size the primary overcurrent protection in Example 8.2 at 250% of the primary full-load current. This calculates to be 113 amperes, which must be rounded down to a 110-ampere overcurrent device rating. So actually, the maximum permitted rating of primary overcurrent protection for the transformer in Example 8.2 is 110 amperes. But this would not be practical since the primary conductors supplying the transformer would be required to be size 2 AWG when size 8 AWG is adequate to supply the load.

When there are several different secondary windings supplying the secondary conductors, an overload could occur on one or more secondary conductors and go undetected by the primary overcurrent protection. This can be prevented by properly sized overcurrent protection installed on the secondary side of the transformer. The primary overcurrent device can protect the secondary conductors, if properly sized, when the transformer secondary is single voltage and there are only two conductors leaving the transformer. This rule is found in *240.4(F)*, and an example is shown in Figure 8.11. The overcurrent device on the primary side of the transformer is required to be sized to prevent an overload on the secondary conductors. Example 8.3 describes this process.

Example 8.3 Consider the case of a 3-kVA transformer used to step down 480 to 120 volts to supply one 20-ampere single-phase 2-wire circuit. There will be no secondary overcurrent protection as shown in Figure 8.11. Determine the maximum permitted overcurrent device rating on the primary circuit to protect the transformer and the secondary circuit wires.

Answer: First, determine the primary and secondary full-load current rating of the 3-kVA transformer using Equation 8.5.

$$\text{Primary Current} = \frac{3 \text{ kVA} \times 1000}{480 \text{ V}} = 6.25 \text{ A}$$

$$\text{Secondary Current} = \frac{3 \text{ kVA} \times 1000}{120 \text{ V}} = 25 \text{ A}$$

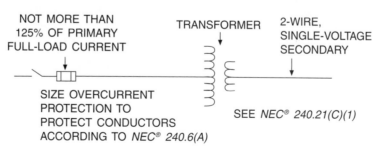

TABLE 450.3(B) TRANSFORMER PROTECTED ONLY ON THE PRIMARY

NOT MORE THAN 125% OF PRIMARY FULL-LOAD CURRENT

TRANSFORMER

2-WIRE, SINGLE-VOLTAGE SECONDARY

SIZE OVERCURRENT PROTECTION TO PROTECT CONDUCTORS ACCORDING TO *NEC®* 240.6(A)

SEE *NEC®* 240.21(C)(1)

Figure 8.11 A transformer may be protected with one overcurrent device on the primary sized at not more than 125% of the primary full-load current.

Next, determine the maximum permitted overcurrent device rating for the primary using the rules of *Table 450.3(B)*. The primary overcurrent device rating is not to exceed 167% of the transformer primary full-load current.

$$\text{Maximum Overcurrent Device Rating} = 6.25 \text{ A} \times 1.67 = 10.4 \text{ A}$$

Note 1, which permits rounding up, does not apply in this case. Therefore, it will be necessary to round down to a 10-ampere standard overcurrent device rating as listed in *240.6(A)*. Circuit breakers are not generally available with small ratings. Therefore, time-delay fuses will be used to protect the primary of the transformer. The primary circuit conductors will be size 14 AWG copper.

The transformer is considered protected with a 10-ampere fuse on the primary, but are the conductors on the secondary side of the transformer protected? The purpose of the installation was to provide a 20-ampere circuit at 120 volts. This requires size 12 AWG copper conductors. Use Equation 8.8 to determine the current flow on the secondary circuit wires required to blow the 10-ampere fuses on the primary.

$$\text{Secondary current} = 10 \text{ A} \times \frac{480 \text{ V}}{120 \text{ V}} = 40 \text{ A}$$

If 10-ampere fuses are installed on the primary, the secondary conductors will be overloaded to 40 amperes. Use Equation 8.7 to determine the current on the primary side of the transformer when there is a 20-ampere current flow on the secondary. The maximum rating fuse that is permitted to be installed on the primary side of the transformer to prevent overloading of the secondary conductors is 5 amperes.

$$\text{Primary current} = 20 \text{ A} \times \frac{120 \text{ V}}{480 \text{ V}} = 5 \text{ A}$$

Primary and Secondary Overcurrent Protection (600 volts or less)

The overcurrent device protecting the primary of a transformer is permitted to be rated as large as 2.50 (250%) times the primary full-load current, provided the transformer secondary winding is protected. The transformer secondary overcurrent device rating is not permitted to be greater than 1.25 (125%) of the secondary full-load current. Protection for both the primary and the secondary of the transformer is illustrated in Figure 8.12.

This rule is very useful when a feeder is available in the area where the transformer is to be installed and has sufficient capacity to supply the load served by the transformer. The transformer circuit is permitted

TABLE 450.3(B) OVERCURRENT PROTECTION RATING
WHEN TRANSFORMER IS PROTECTED ON THE PRIMARY
AND ON THE SECONDARY

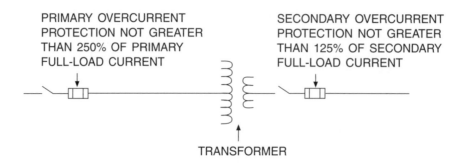

Figure 8.12   If the transformer is protected on the secondary with an overcurrent device rated not more than 125% of the secondary current, the primary overcurrent device is permitted to be sized as large as 250% of the primary current.

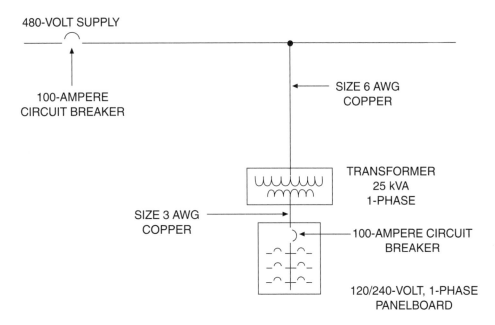

480-VOLT SUPPLY

100-AMPERE
CIRCUIT BREAKER

SIZE 6 AWG
COPPER

TRANSFORMER
25 kVA
1-PHASE

SIZE 3 AWG
COPPER

100-AMPERE CIRCUIT
BREAKER

120/240-VOLT, 1-PHASE
PANELBOARD

Figure 8.13   The primary overcurrent protection for a transformer is permitted to have a rating not exceeding 2.5 times the transformer primary full-load current provided overcurrent is provided on the secondary rated not higher than the transformer secondary full-load current.

to be tapped directly to the feeder and the feeder overcurrent device can serve as the transformer overcurrent protection provided the rating is not greater than 2.5 times the full-load current of the transformer primary. But the secondary winding of the transformer is now required to be provided with overcurrent protection not greater than 1.25 times the transformer secondary full-load current. This is illustrated in Figure 8.13 where a 25-kVA single-phase transformer is tapped to a 480-volt feeder protected by a 100-ampere circuit breaker. Example 8.4 shows how to determine the maximum rating of overcurrent protection permitted for the primary and secondary windings of the transformer. Rules in *Article 240* specify the minimum size of conductors and the locations of overcurrent devices. Those rules will be discussed later.

Example 8.4   A 25-kVA single-phase transformer is tapped to a 480-volt feeder protected with a 100-ampere circuit breaker. The transformer is supplying a 120/240-volt, 3-wire panelboard with a 100-ampere main circuit breaker. The circuit is illustrated in Figure 8.13. The total distance from the point of tap at the feeder, through the transformer to the 100-ampere circuit breaker in the panelboard is 25 ft (7.5 m). Determine the maximum rating of overcurrent device permitted to protect the primary and secondary of the transformer.

Answer: First determine the full-load current of the primary winding, which is 52 amperes. According to *Table 450.3(B)*, if overcurrent protection is provided on the secondary of the transformer at not more than 1.25 times the secondary full-load current, the overcurrent protection on the primary is permitted to be increased to 2.5 times the primary full-load current, which is 130 amperes (52 A × 2.5 = 130 A). There is no provision to round this value up to the next standard rating. Therefore, it is necessary to round the value down to the next standard rating as listed in *240.6*, which is 125 amperes. The feeder is protected with a 100-ampere circuit breaker, therefore, the feeder overcurrent device is less than the maximum permitted, and can serve as the transformer primary overcurrent protection.

$$\text{Primary full-load current} = \frac{25 \text{ kVA} \times 1000}{480 \text{ V}} = 52 \text{ A}$$

Next, determine the maximum permitted rating of the secondary overcurrent protection. The secondary full-load current is 104 amperes. The secondary overcurrent protection is permitted to have a rating of 1.25 times the full-load secondary current of the transformer, which is 130 amperes. *Note 1* permits this value to be rounded up to the next standard rating overcurrent device as listed in *240.6*, which is 150 amperes. Therefore, the secondary of this transformer is permitted to have an overcurrent device rated at 150 amperes. The panelboard has a 100-ampere main circuit breaker, which is well within the maximum permitted.

$$\text{Secondary full-load current} = \frac{25 \text{ kVA} \times 1000}{240 \text{ V}} = 104 \text{ A}$$

If the overcurrent protection for the 480-volt feeder of Figure 8.13 had a rating greater than 125 amperes, then the situation is handled as a feeder tap, *240.21(B)(3)*, ending in an overcurrent device similar to Figure 8.14. Then the overcurrent device is sized as small as practical to handle the load as was the case with Example 8.2.

In the case where the primary overcurrent protection for the transformer is greater than 1.25 times the full-load current of the primary winding, the overcurrent protection for the secondary winding is permitted to consist of up to six individual overcurrent devices grouped together at the end of one set of feeder conductors. This is *Note 2* and applies in the case of *Table 450.3(A)* for transformers operating at more than 600 volts and *Table 450.3(B)* for transformers operating at 600 volts or less. The sum of the ratings of the overcurrent devices is not permitted to total more than the rating permitted for a single overcurrent device. This is illustrated in Figure 8.15 with the overcurrent device ratings worked out in Example 8.5.

Example 8.5   A 37.5-kVA 3-phase transformer is tapped to a 480-volt feeder and supplies three 208-volt 3-phase motors. The individual motor branch-circuit time-delay fuse ratings are 30, 50, and 60 amperes. Is it permitted to omit the single secondary overcurrent device and terminate the secondary conductors from the transformer at the three motor circuit disconnects grouped together at one location?

Answer: *Note 2 of Table 450.3(B)* permits this practice if the three motor circuit fuse ratings do not total more than the highest rated single overcurrent device permitted for the transformer. First, determine the transformer secondary full-load current of 104 amperes. Then multiply that value by 1.25 to get 130 amperes (104 A × 1.25 = 130 A). *Note 1* permits this value to be rounded up to the next standard rating which according to *240.6* is 150 amperes. The three individual motor branch-circuits total to 140 amperes (30 A + 50 A + 60 A = 140 A). This is less than 150 amperes. Therefore, the installation is permitted as shown in Figure 8.15. If the motor circuit overcurrent devices had totaled more

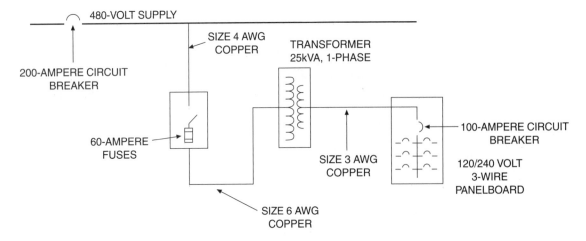

Figure 8.14   **If a transformer circuit is tapped to a feeder with feeder overcurrent protection rated more than 2.5 times the transformer primary full-load current, then overcurrent protection will be required on the primary specifically sized for the transformer circuit.**

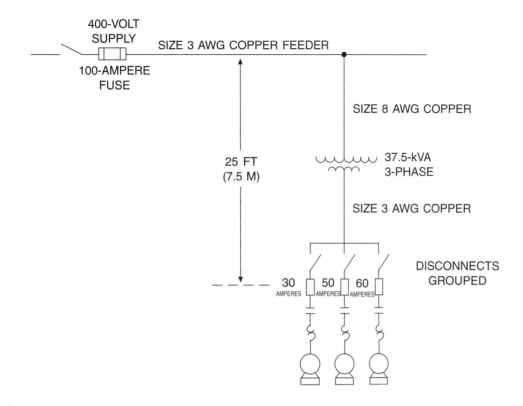

Figure 8.15 When overcurrent protection is required on the secondary of the transformer, such as this case of a feeder tap, the secondary is permitted to end in up to six overcurrent devices with ratings that do not total more than the rating required for a single overcurrent device.

than 150 amperes, then the feeder would have been required to terminate at a single overcurrent device rated not greater than 150 amperes.

$$\text{Secondary full-load current} = \frac{37.5 \text{ kVA} \times 1000}{1.73 \times 208 \text{ V}} = 104 \text{ A}$$

Boost and Buck Transformer Overcurrent Protection

The specific transformer, winding connections, kVA rating, and maximum load current for an application of voltage boosting or voltage bucking are determined with information from specific boost and buck transformer manufacturers. These are insulating transformers connected as autotransformers. *NEC® 450.4* gives the rules for installing autotransformers. Overcurrent protection for insulating transformers (*450.3*) is based upon the full-load current of the primary and secondary windings. For an autotransformer, the overcurrent protection is based upon the full-load **input** current. Figure 8.16 shows a boost and buck transformer connected to boost 208 volts up to 236 volts to power a 3-horsepower, single-phase, 230-volt motor. The boost and buck transformer suitable for this application is rated 0.75 kVA, with a primary rated 120/240 volts and the secondary rated 16/32 volts. When connected for maximum boosting, this transformer has a rated full-load output current of 23.4 amperes.

The first step in selecting the overcurrent protection for the transformer and the input and output conductor size is based upon knowing the input and output current. From *Table 430.248,* a 3-horsepower, 208-volt motor draws 18.7 amperes. Connected for maximum boosting, the transformer is capable of a maximum continuous current of 23.4 amperes. Overcurrent protection is based upon the full-load input current. Using Equation 8.7, determine the input current when the motor is drawing 18.7 amperes and the maximum input current when the transformer is delivering an output current of 23.4 amperes. The input current will be 21.2 amperes when the motor is operating at full-load. This transformer connection is capable of a maximum input current of 26.6 amperes.

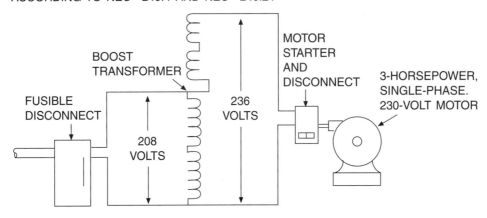

NEC® 450.4 OVERCURRENT PROTECTION FOR
AUTOTRANSFORMER; ALSO PROTECTS CONDUCTORS
ACCORDING TO NEC® 240.4 AND NEC® 240.21

Figure 8.16 **A boost and buck transformer is used to boost the 208-volt supply to 236 volts for a 3-horsepower, single-phase motor.**

$$\text{Input current} = 18.7\ A \times \frac{236\ V}{208\ V} = 21.2\ A$$

$$\text{Input current} = 23.4\ A \times \frac{236\ V}{208\ V} = 26.6\ A$$

It is recommended the overcurrent protection and conductors be based upon the full-load current of the load, which is 18.7 amperes. The full-load input current will then be 21.2 amperes. NEC® 450.4 permits the boost and buck transformer overcurrent protection to be rated not more than 1.25 time the full-load input current, which gives an overcurrent device rating of 26.5 amperes. In this case, it is permitted to round this value up to the next standard rating of overcurrent device listed in 240.6, which is a 30-ampere fuse or circuit breaker in each ungrounded input conductor. Because this transformer is slightly oversized for the load, the maximum input overcurrent device rating could be up to 35 amperes.

The circuit conductor size in most cases is based upon the rating of the overcurrent device. The input and output current are not the same. Therefore, an adjustment for voltage is necessary to properly size the conductors. If the transformer input overcurrent protection is 30 amperes, then using Table 310.16 and 240.4(B), size 10 AWG copper conductors are required for the input circuit to the transformer. Use Equation 8.8 to determine the output current that corresponds to an input current of 30 amperes. This value will be 26.4 amperes. Using Table 310.16, this also is a size 10 AWG copper conductor if insulation and terminations are assumed to be rated 75 °C.

$$\text{Output current} = 30\ A \times \frac{208\ V}{236\ V} = 26.4\ A$$

The conductors for the circuit in Figure 8.16 are permitted to be sized using the rules of 430.22. The conductor is required to have an allowable ampacity not less than 1.25 times the full-load current. The full-load current to the motor is 18.7 amperes and the full-load input current to the transformer is 21.2 amperes. Multiply each of these values by 1.25 and select the minimum conductor size using Table 310.16. The minimum size copper input current to the transformer in Figure 8.16 is size 10 AWG, and the minimum size copper output conductor to the motor is 12 AWG.

TRANSFORMER TAPS

The overcurrent protection rules for a transformer in 450.3 do not provide for the protection of the conductors supplying a transformer or the conductors leading away from the transformer. The protection of con-

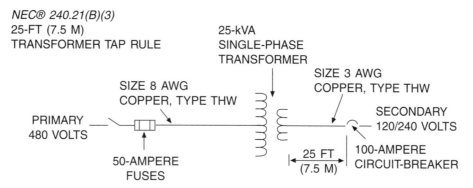

Figure 8.17   The secondary overcurrent protection is permitted to be located 25 ft (7.5 m) from the secondary winding when the primary circuit wire has overcurrent protection not exceeding the ampacity of the primary wire.

ductors is covered in *240.4* and *240.21*. The most basic situation is where a circuit is installed to supply a specific load, such as shown in Figure 8.17, and a transformer is required to change the supply voltage to match the requirement of the load. In Figure 8.17, the purpose is to supply a 120/240-volt, 3-wire panelboard with a 100-ampere main circuit breaker from a 480-volt electrical supply. Assume for the purpose of this example that all conductors are copper with 75°C insulation and terminations. The conductors from the secondary of the transformer to the 100-ampere circuit breaker are size 3 AWG. The voltage is changed from 480 volts to 240 volts. Therefore, there will be 50 amperes flowing in the primary conductors when 100 amperes is flowing in the secondary conductors (Equation 8.7). The 480-volt conductors supplying the transformer are size 8 AWG protected with 50-ampere time-delay fuses. The overcurrent protection for this circuit is well within the limits set by *Table 450.3(B)*. The conductors supplying the transformer primary are protected, but the secondary conductors leading away from the transformer are only protected at their load end similar to service entrance conductors. A short circuit or ground fault that may occur between the transformer and the secondary overcurrent device can create a hazardous condition without opening the primary overcurrent device. As a result, special rules were developed to minimize the risk of a hazardous condition developing. These are the transformer tap rules and they are located in *240.21*. For an inside installation, the maximum distance permitted from the transformer to the overcurrent device is 25 ft (7.5 m). This rule is found in *240.21(C)(6)*.

   NEC® *240.21(C)(6)* does not restrict the number of conductors leading away form the transformer. Therefore, several sets of conductors are permitted to supply different loads. In the case of Figure 8.18, a

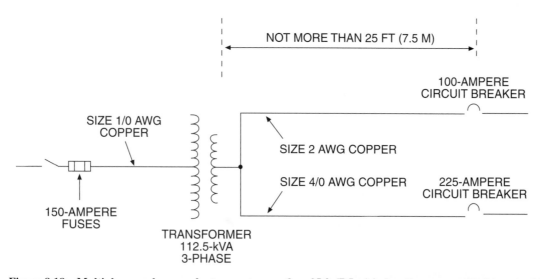

Figure 8.18   Multiple secondary conductors, not more than 25 ft (7.5 m) in length, are permitted to extend from a transformer and end in a single overcurrent device rated for the conductor, but the conductor is required to have an ampacity not less than one-third the rating of the primary overcurrent device adjusted for the voltage change.

112.5-kVA, 3-phase transformer supplies two 208Y/120-volt panelboards, one with a 225-ampere main circuit breaker and the other with a 100-ampere main circuit breaker. The only restriction is that each set of conductors ends at a single circuit breaker or set of fuses, and there is a minimum size conductor specified. The secondary conductors leading away from the transformer are required to have an ampacity not less than one-third the ampacity of the primary overcurrent device when converted to equivalent current at the secondary voltage (Equation 8.8). The transformer shown in Figure 8.18 is connected to a 480-volt, 3-phase supply and is protected on the primary with 150-ampere fuses. This fuse rating is small enough, according to *Table 450.3(B)*, that additional overcurrent protection for the secondary windings is not required. Divide the 150-ampere fuse rating by three and use Equation 8.8 to find the equivalent current at the secondary voltage, which is 115 amperes. This is the minimum ampacity permitted for any set of conductors leading away form this transformer. Note that instead of a size 3 AWG copper conductor to supply the 100-ampere panelboard, a size 2 AWG conductor is required.

$$\text{Secondary current} = 50 \text{ A} \times \frac{480 \text{ V}}{208 \text{ V}} = 115 \text{ A}$$

NEC® 240.21(C)(2) permits multiple sets of conductors leading away from a transformer secondary with only a small restriction upon the ampacity of the conductors relative to the primary overcurrent device, provided the length of conductor from the transformer to the overcurrent device does not exceed 10 ft (3 m). This is illustrated in Figure 8.19. The conductors are required to be run in raceway from the transformer to the panelboard or other device containing the overcurrent device. The 10 ft (30 m) tap conductors are required to have an ampacity not less than 10% of the rating of the primary overcurrent device corrected for the change in voltage using Equation 8.8, which in this case is only 10 amperes.

It is frequently desirable to tap a transformer from a feeder as shown in Figures 8.13 and 8.20 without overcurrent protection between the tap point and the transformer. This is permitted by *240.21(B)(3)*. The restriction in this case is that the total distance from the point where the primary conductor taps the feeder to the overcurrent device on the secondary is not permitted to exceed 25 ft (7.5 m). The feeder overcurrent device protects the primary and secondary tap conductors and the primary winding of the transformer. If the feeder overcurrent device has a rating higher than 1.25 times the transformer full-load current, then a single overcurrent device is required to protect the transformer secondary subject to the conditions of *Note 2* of *Table 450.3(A)* and *Table 450.3(B)*. The transformer installation of Figure 8.20 has a primary full-load current of 104 amperes, and a secondary full-load current of 208 amperes.

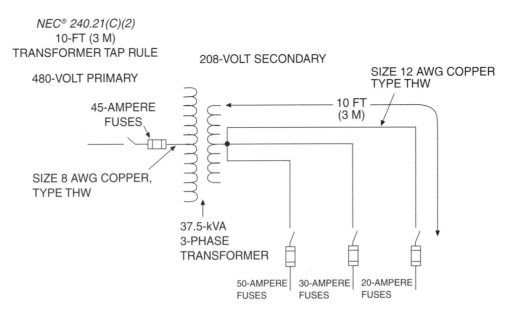

Figure 8.19   The size of the secondary tap conductors is permitted to be sized to supply the load with no relation to the rating of the primary overcurrent device when the tap length does not exceed 10 ft (3 m).

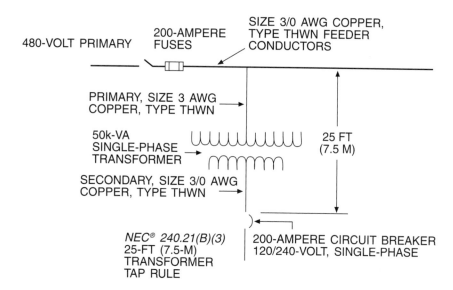

Figure 8.20 When the primary circuit wire is tapped from a feeder, the secondary winding overcurrent protection shall be located not more than 25 ft (7.5 m) from the point of the primary tap.

$$\text{Primary full-load current} = \frac{50 \text{ kVA} \times 1000}{480 \text{ V}} = 104 \text{ A}$$

$$\text{Primary full-load current} = \frac{50 \text{ kVA} \times 1000}{240 \text{ V}} = 208 \text{ A}$$

Table 450.3(B) permits the primary overcurrent device to have a rating of up to 250 amperes and a secondary overcurrent device of up to 300 amperes. The secondary overcurrent device rating is only 200 amperes. The conductor between the transformer and the 200-ampere circuit breaker is sized according to *240.4* which requires a minimum size 3/0 AWG copper conductor. The primary tap conductor must have an allowable ampacity not less than required to supply the load on the secondary of the transformer. Use Equation 8.7 to determine how much current will flow on the primary conductor when 200 amperes is flowing on the secondary conductors. The result will be 100 amperes, which requires a size 3 AWG copper conductor to supply the transformer.

Primary overcurrent device = 104 A × 2.5 = 260 A (must round down to 250 A)

Secondary overcurrent device = 208 A × 1.25 = 260 A (permitted to round up to 300 A)

NEC® 240.21(C)(3) is a special case of the 25-ft (7.5-m) transformer secondary tap rule that applies only for industrial installations. The difference between this tap rule and the others is that the secondary conductor is permitted to end is multiple overcurrent devices provided they are grouped together in one location. There are several conditions that must be satisfied. The secondary conductor is not permitted to have an allowable ampacity less than the secondary full-load current of the transformer. For example, assume a 75-kVA, 3-phase transformer supplies 208Y/120-volt power from a 480-volt supply. The secondary full-load current is 208 amperes. The secondary conductor is not permitted to be smaller than size 4/0 AWG copper. According to *Table 310.16*, this conductor is rated for 230 amperes. The secondary conductor is permitted to terminate at multiple overcurrent devices, but in this case, they are not permitted to total to more than 260 amperes (208 A × 1.25 = 260 A).

NEC® 240.21(C)(4) is a transformer secondary conductor tap rule that only applies if the conductors are run outside a building. Multiple tap conductors are permitted to be run outdoors from a transformer secondary to individual loads such as disconnects in individual buildings. The tap conductors are permitted to

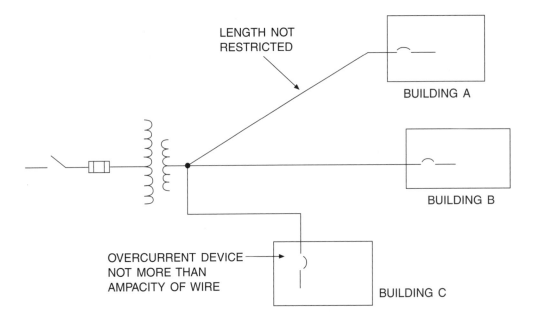

LENGTH NOT
RESTRICTED

BUILDING A

BUILDING B

OVERCURRENT DEVICE
NOT MORE THAN
AMPACITY OF WIRE

BUILDING C

Figure 8.21   Multiple outside secondary taps are permitted to a transformer with no restriction in length or conductor size except that the taps are to end at a single overcurrent device with a rating not exceeding the ampacity of the conductors.

be sized for the load to be served and they are permitted to be of any length. The conductors are required to end at a single circuit breaker or set of fuses with a rating not higher than the ampacity of the conductors. This type of installation is illustrated in Figure 8.21.

TRANSFORMER GROUNDING

An electrical system derived from a transformer is required to have noncurrent-carrying metal parts and equipment grounded the same as any other part of the electrical system. In addition, the electrical system produced by the transformer also may be required to be grounded according to *250.20(B). NEC® 250.26* specifies the wire that shall be grounded. The method of grounding the grounded circuit conductor is specified in *250.30* for separately derived systems. A separately derived system is defined in *Article 100*. A transformer installed at some point in an electrical system is considered to be a separately derived system if a derived grounded-circuit conductor is not electrically connected to a grounded-circuit conductor of the building electrical system. Further information about separately derived systems is found in *250.20(D)*. Most transformer installations are considered to be separately derived systems, and the grounding is covered in *250.30*. The rules of grounding are illustrated in Figure 8.22.

It is important to understand the purpose for grounding a separately derived system. If a ground fault should occur, the current attempts to return to the source, which in this case is the separately derived system. An adequate low impedance path must be provided back to the grounded conductor of the separately derived system in order for fault current to flow. In the case of an ungrounded system, a low impedance path must be provided back to the grounding point at the main disconnect for the separately derived system. Therefore, it is important the grounding electrode be located as close as practical to the separately derived system. Most insulating transformer installations are considered to be separately derived systems.

The requirement for a grounding electrode for a separately derived system is found in *250.30(A)(7)*. The grounding electrode is required to be the closest of either an effectively grounded structural metal member in the building or an effectively grounded metal water pipe. If the transformer is located near or adjacent to the service equipment, then it is probably most convenient and effective to ground the separately derived system to the same grounding electrode as the service equipment. Assume a transformer is installed in a building a considerable distance from the service equipment. Also, assume there is a metal water pipe in the building but the transformer is located a distance from the point where the water pipe enters the building. If the steel frame of the building is effectively grounded and closer to the transformer, then the transformer

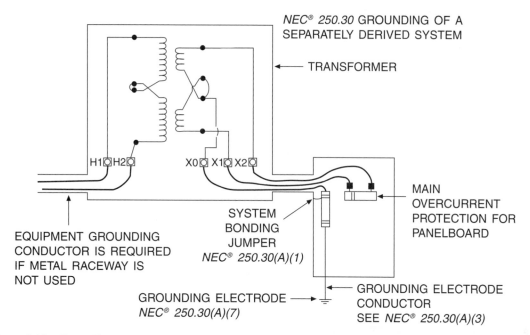

NEC® 250.30 GROUNDING OF A
SEPARATELY DERIVED SYSTEM

TRANSFORMER

H1 H2   X0 X1 X2

MAIN
OVERCURRENT
PROTECTION FOR
PANELBOARD

SYSTEM
BONDING
JUMPER
NEC® 250.30(A)(1)

EQUIPMENT GROUNDING
CONDUCTOR IS REQUIRED
IF METAL RACEWAY IS
NOT USED

GROUNDING ELECTRODE
NEC® 250.30(A)(7)

GROUNDING ELECTRODE
CONDUCTOR
SEE *NEC® 250.30(A)(3)*

Figure 8.22   Grounding and bonding for a transformer are permitted to be at the first disconnect with overcurrent protection, or it can be at the transformer.

should be grounded to the building steel as close to the transformer as practical. In this case it is not required to run a grounding electrode conductor back to a point within 5 ft (1.5 m) of the point where the metal water pipe enters the building. But if the metal water pipe serves the same area as the separately derived system, it is required according to *250.104(D)(1)* to bond to the metal water pipe in the local area. This is a bonding requirement not a grounding requirement, although the conductors are the same size. If there is no metal building steel available that is effectively grounded, then the transformer is required to be grounded to the metal water pipe within 5 ft (1.5 m) of the point where it enters the building. For industrial or commercial buildings, the connection to the water pipe may be permitted to be made a distance from the point where it enters the building if the conditions of the *Exception* to *250.52(A)(1)* are satisfied. If there is no effectively grounded metal water pipe or effectively grounded building structural steel, then any one of the other grounding electrodes described in *250.50* or *250.52* is required to serve as the grounding electrode for the separately derived system.

For a building where a number of transformers will be installed and finding a satisfactory grounding electrode may be difficult, it is permitted to run a common grounding electrode conductor through the building to which several transformers can be grounded. This method of grounding separately derived systems is permitted in *250.30(A)(4)* and is illustrated in Figure 5.18 in *Unit 5*. This common grounding conductor can be connected to the grounding electrode for the building. The size of the common grounding conductor is required to be not smaller than 3/0 AWG copper according to *250.30(A)(4)(a)*. If there is an exposed metal building frame and metal water pipe in the area served by the circuits from the separately derived system, they are to be bonded to the common grounding electrode conductor as required by *250.104(D)*.

The size of the grounding electrode conductor for any one transformer is based upon the largest size ungrounded secondary conductor from the transformer using *Table 250.66*. Any bonding conductor required by *250.104(D)* is also sized using *Table 250.66*. The grounding electrode conductor connection to the grounded conductor is permitted to be made at the transformer or at the first overcurrent device supplied by the transformer. In Figure 8.22, the grounding electrode conductor connection is made at the first disconnect. It is also required to bond to the disconnect enclosure in a manner similar to that of a service. If this bond is made at the transformer, usually a wire will be used as a bonding jumper from the grounded conductor terminal in the transformer to the enclosure. A grounding electrode conductor must be run from the transformer enclosure to the grounding electrode. Grounding at the transformer and at the first disconnecting means is not permitted if a parallel path is created for grounded conductor current. It is required that the transformer enclosure and the enclosure of the first disconnecting means be effectively bonded together.

MAJOR CHANGES TO THE 2005 CODE

These are the changes to the 2005 *NEC®* that correspond to the Code sections studied in this unit. The following analysis explains the significance of the changes from the 2002 to the 2005 Code only and this analysis is not intended to be used in place of the Code. Refer to the actual section of the 2005 Code for the exact wording and meaning of each section discussed. Changes are indicated in the Code with a vertical line in the margin. If material was deleted or moved to another location in the Code, the location of the deletion is indicated with a dark dot in the margin.

Article 450   **Transformers and Transformer Vaults**

450.5: A new sentence was added to deal with the installation of a zig-zag grounding autotransformer installed for the purpose of creating a neutral conductor on an electrical system that is grounded, but does not have a neutral conductor. The new sentence requires that the zig-zag grounding autotransformer be installed and connected on the supply side of any system grounding connection.

450.6: Editorial changes were made to this section to improve clarity and to help point out the differences between this section and *450.7*. Secondary ties are sometimes used for large industrial plants where transformers are installed at strategic locations throughout the plant to serve major loads. The secondaries of these transformers are then permitted to be connected together with secondary ties to form a network distribution system for the industrial plant. All transformers feed into the secondary network, and all loads are fed from the secondary network. To increase reliability, the individual transformer locations are often supplied by two primary systems, each capable of supplying the load. If one primary goes out of service, the transformers can be switched to the other primary system. Such systems are designed by engineers and installed under engineering supervision. This section places basic requirements on the secondary ties that connect the different power supply points. Secondary ties receive power from two ends and must be protected at both ends. The secondary tie can be protected from overcurrent according to the ampere rating to the tie conductors. The secondary tie is also permitted to be protected only from faults and short circuits with devices called limiters, in which case requirements are placed on the minimum ampacity of the secondary ties depending upon the arrangement of supply and load points.

A new last sentence was added to the first paragraph pointing out that secondary ties are different than parallel transformer operation as described in *450.7*. A network system connected with secondary ties is supplied with transformers that operate independent of each other. With parallel operation, two or more transformers operate as a single unit, each sharing a portion of the load and switched as a unit. Transformers supplying a network system connected with secondary ties can go out of service independent of each other, in which case a reverse-current relay is required to sense the outage and isolate the transformer from the system.

A new subsection *(C)* was added requiring that, when the secondary is a grounded system, the individual transformers are to be treated as separately derived systems and grounded according to the rules of *250.30*. The rules for grounding and bonding a separately derived system are different from the rules for grounding and bonding service equipment. The transformers connected with secondary ties are services supplied from a primary system. Separately derived systems are supplied from the secondary network in the plant. The rules for grounding and bonding a separately derived system are different from the rules that apply to a service.

WORKSHEET NO. 8—BEGINNING TRANSFORMERS

Mark the single answer that most accurately completes the statement based upon the 2005 Code. Also provide, where indicated, the Code reference that gives the answer or indicates where the answer is found, as well as the Code reference where the answer is found.

1. A single-phase transformer has a rating of 25 kVA with a primary winding rated 480 volts and the secondary winding rated 120/240 volts. The full-load current of the primary winding of the transformer is:
   A. 30 amperes.
   B. 52 amperes.
   C. 104 amperes.
   D. 130 amperes.
   E. 208 amperes.

   Code reference _No code reference_

2. A single-phase, 5-kVA transformer is connected to a 480-volt supply to provide power at 120 volts. The current draw in the 120-volt, 2-wire secondary circuit supplied by the transformer is 40 amperes as shown in Figure 8.23. The current flowing in the 480-volt circuit supplying the transformer, assuming minimal losses, is approximately:
   A. 10 amperes.
   B. 20 amperes.
   C. 40 amperes.
   D. 80 amperes.
   E. 160 amperes.

   Code reference _No code reference_

3. Information that is not required to be provided on the nameplate of a dry-type transformer from the following list is:
   A. full-load current of the primary and secondary.
   B. frequency.
   C. rated kilovolt-amperes.
   D. primary and secondary voltage.
   E. impedance if rated 25 kVA or larger.

   Code reference _____

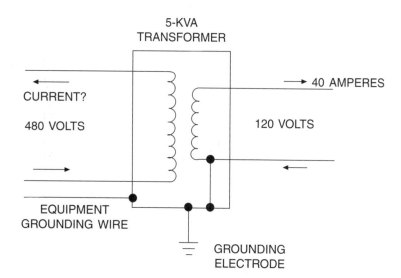

Figure 8.23 Determine the current flowing in the primary conductors of this insulating transformer.

4. Transformers are required to be installed where they are readily accessible. In the case of a 10-kVA dry-type transformer operating at not over 600 volts and installed in the open on the wall of an industrial space, the transformer is permitted to be installed:

A. not more than 6 ft 7 in. (2 m) above the floor.

B. not more than 8 ft (2.5 m) above the floor.

C. not more than 10 ft (3 m) above the floor.

D. not more than 20 ft (6.1 m) above the floor.

E. at any height as long as a ladder or other portable elevating device is capable of providing direct access to the transformer without removal of obstacles.

Code reference _____

5. A transformer installed within a building to change the voltage to create a new electrical system such as a transformer connected to a 480-volt supply to create a 120/240-volt 3-wire system, shown in Figure 8.24, is considered to be:

A. a secondary service.

B. an alternate power source.

C. an isolated electrical system.

D. a separately derived system.

E. a premises electrical system.

Code reference _____

6. A 5-kVA dry-type transformer, with both windings operating below 600 volts, has an outer metal covering in direct contact with the core and no ventilation openings to the inside of the transformer. The transformer is installed in a room with walls constructed of a combustible material. If no fire resistant, thermal insulating material is installed between the transformer and the wall, the minimum separation distance permitted between the transformer and the wall, as shown in Figure 8.25, is:

A. of no concern and the transformer can be mounted directly to the wall.

B. 3 in. (75 mm).

C. 6 in. (150 mm).

D. 12 in. (300 mm).

E. 18 in. (450 mm).

Code reference _____

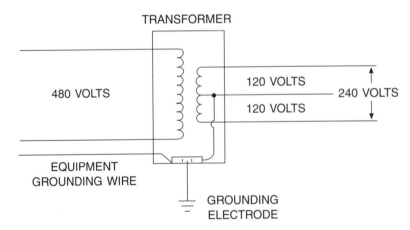

Figure 8.24   What is the name used in the Code for a transformer that changes the voltage and creates a new electrical system within a building?

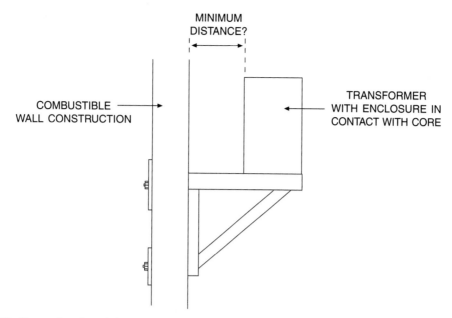

MINIMUM
DISTANCE?

COMBUSTIBLE
WALL CONSTRUCTION

TRANSFORMER
WITH ENCLOSURE IN
CONTACT WITH CORE

Figure 8.25   Determine the minimum distance between a wall of combustible construction and a transformer with the core in contact with the enclosure and no ventilation openings.

7. A 480-volt, single-phase circuit protected with time-delay fuses supplies only one 50-kVA dry-type transformer. The purpose of the transformer is to supply a single-phase, 120/240-volt panelboard which is provided with a 200-ampere main circuit breaker as shown in Figure 8.26. The distance from the fuses protecting the transformer primary to the circuit breaker in the panelboard is 40 ft (12.19 m), and the distance from the transformer to the panelboard is 15 ft (4.57 m). All conductors are copper with 75°C insulation and terminations. The conductor supplying the primary is size 2 AWG and the conductor on the secondary to the panelboard is size 3/0 AWG. The maximum rating time-delay fuse permitted to protect this transformer circuit is:

A. 100 amperes.   C. 150 amperes.   E. 250 amperes.
B. 125 amperes.   D. 200 amperes.

Code reference _____

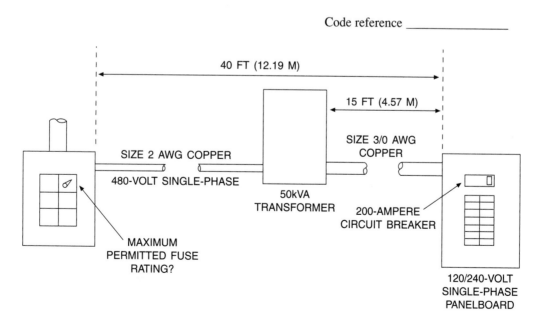

40 FT (12.19 M)

15 FT (4.57 M)

SIZE 2 AWG COPPER

480-VOLT SINGLE-PHASE

SIZE 3/0 AWG
COPPER

50kVA
TRANSFORMER

200-AMPERE
CIRCUIT BREAKER

MAXIMUM
PERMITTED FUSE
RATING?

120/240-VOLT
SINGLE-PHASE
PANELBOARD

Figure 8.26   Determine the maximum permitted primary fuse rating for this transformer and the maximum distance permitted from the transformer to the 120/240-volt panelboard.

8. Refer to the transformer installation of Figure 8.26 which is in a commercial building. The distance from the transformer to the panelboard is permitted to be increased from 15 ft to a maximum distance of:
   A. 25 ft (7.5 m).
   B. 50 ft (15 m).
   C. 75 ft (22.5 m).
   D. 100 ft (30 m).
   E. not limited to a specific distance.

   Code reference _____

9. A 15-kVA, single-phase dry-type transformer is installed to supply a 240-volt load from a 480-volt supply. All conductors are copper with 75°C insulation and terminations. The primary circuit conductor is size 10 AWG 2-wire at 480 volts, and the secondary conductor is size 6 AWG 2-wire at 240 volts. The primary is protected with 30-ampere time-delay fuses, and a disconnect switch with 60-ampere time-delay fuses is installed on the secondary circuit as shown in Figure 8.27. The maximum distance from the transformer to the secondary disconnect switch is:
   A. 25 ft (7.5 m).
   B. 50 ft (15 m).
   C. 75 ft (22.5 m).
   D. 100 ft (30 m).
   E. not limited to a specific distance.

   Code reference _____

10. An autotransformer:
    A. has two separate windings not electrically connected to each other.
    B. has a winding common to both the primary and secondary circuits.
    C. is only permitted to be used in electric cars.
    D. is required to have overcurrent protection installed on both the primary and secondary.
    E. installation is not permitted to any input conductor connected directly to an output conductor.

    Code reference _____

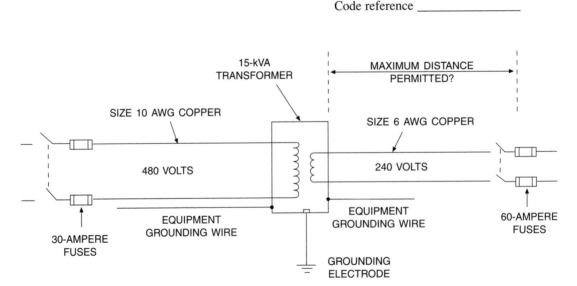

Figure 8.27 For this transformer installation with a single-voltage 2-wire secondary, determine the maximum distance permitted from the transformer to the secondary overcurrent device.

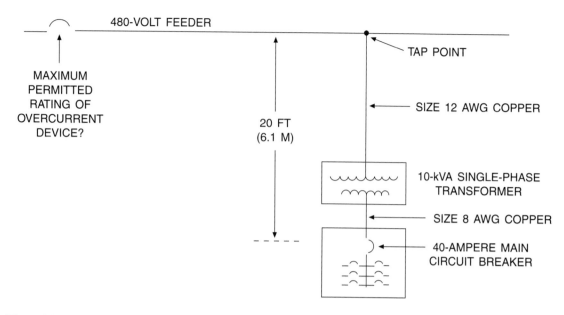

Figure 8.28 Determine the maximum rating feeder overcurrent device permitted where the overcurrent device also serves as the transformer overcurrent protection.

11. A 10-kVA single-phase transformer supplies a 120/240-volt, 3-wire panelboard with a 40-ampere main circuit breaker. The conductors between the transformer and the panelboard are size 8 AWG copper and the tap conductors supplying the primary of the transformer are size 12 AWG copper. The primary conductor is tapped to a 480-volt feeder with the distance from the tap point to the feeder being 20 ft (6.1 m) as illustrated in Figure 8.28. If the feeder overcurrent device serves as the primary protection for the transformer, the maximum rating feeder overcurrent device permitted is:
    A. 20 amperes.
    B. 40 amperes.
    C. 50 amperes.
    D. 125 amperes.
    E. 160 amperes.

    Code reference _____

12. A 75-kVA, 3-phase transformer supplies a 200-ampere, 208Y/120-volt, 4-wire panelboard from a 480-volt supply. The panelboard is supplied with size 3/0 copper conductors terminating at a 200-ampere circuit breaker within 10 ft of the transformer. The primary is supplied with size 3 AWG copper conductors protected by a 100-ampere circuit breaker. All conductors have 75°C insulation and terminations, and are run in Rigid Metal Conduit. The installation is shown in Figure 8.29. A metal water pipe enters the building and is used as a grounding electrode for the service, but the water pipe is not visible at all points between the point of entry to the building and the area of the transformer installation. The building has a structural steel frame, which is considered effectively grounded. The grounding electrode required for this installation is:
    A. the structural steel frame of the building and the metal water pipe.
    B. the metal water pipe with a connection made within 5 ft of the point where it enters the building.
    C. the metal water pipe but permitted to have the connection made at the point nearest the transformer.
    D. the structural steel because it is closer to the transformer than the point where the water pipe enters the building.
    E. a ground rod driven to earth at the point closest to the transformer.

    Code reference _____

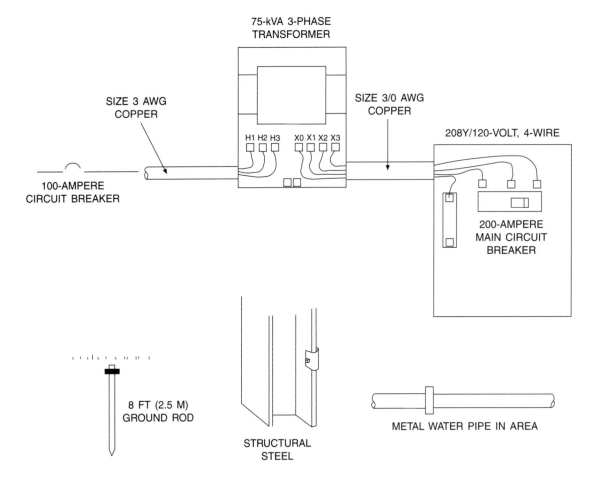

Figure 8.29 Determine the grounding electrode required for this transformer installation as well as the minimum size copper grounding electrode conductor. Also determine the permitted location for the bonding jumper, and the minimum size bonding jumper and the minimum size bonding jumper to the metal water pipe serving the area served by the transformer circuits.

13. Refer to the transformer installation of Figure 8.29. If the grounding electrode conductor is copper, the minimum size permitted is:

   A. 6 AWG.   C. 3 AWG.   E. 2/0 AWG.

   B. 4 AWG.   D. 2 AWG.

Code reference _____

14. The system bonding jumper for the transformer installation shown in Figure 8.29, which connects the grounded circuit conductor to the equipment grounding conductor, is:

   A. required to be at the transformer.

   B. required to be at the first disconnect supplied by the transformer.

   C. required to be made at both the transformer and the first disconnect supplied by the transformer.

   D. required to be made back at the main service equipment.

   E. permitted to be at the first disconnect or overcurrent device supplied by the transformer.

Code reference _____

15. The metal water pipe in an area supplied power from a transformer is not effectively bonded back to the point where the metal water pipe enters the building. Assuming the same transformer installation as Figure 8.29, the minimum size copper conductor required to bond the metal water pipe in the area to the grounded conductor of the separately derived system is:

A. 8 AWG.

B. 6 AWG.

C. 4 AWG.

D. 3 AWG.

E. 1 AWG.

Code reference _____

WORKSHEET NO. 8—ADVANCED TRANSFORMERS

Mark the single answer that most accurately completes the statement based upon the 2005 Code. Also provide, where indicated, the Code reference that gives the answer or indicates where the answer is found, as well as the Code reference where the answer is found.

1. A 3-phase transformer has a rating of 112.5 kVA with a primary winding rated 480 volts and the secondary winding rated 208Y/120 volts. The full-load current of the secondary winding of the transformer is:
    A. 243 amperes.   C. 313 amperes.   E. 541 amperes.
    B. 293 amperes.   D. 390 amperes.

    Code reference __No code reference__

2. A 3-phase transformer is connected for 480 volts on the primary and 208 volts on the secondary. If 200 amperes are flowing on the secondary conductors, the current in the primary conductors, assuming minimal losses, is approximately:
    A. 87 amperes.   C. 200 amperes.   E. 462 amperes.
    B. 108 amperes.   D. 400 amperes.

    Code reference __No code reference__

3. A transformer, which is intended to be connected to a 3-phase 3-wire ungrounded electrical system for the purpose of creating a 3-phase 4-wire grounded electrical system, is called a:
    A. boost and buck transformer.   D. shielded transformer.
    B. insulating transformer.   E. zig-zag transformer.
    C. isolation transformer.

    Code reference _____

4. The door sill at the entrance of a vault, as shown in Figure 8.30, for an oil-insulated transformer is required to have sufficient height to contain all of the liquid in the largest transformer and in no case shall the height be less than:
    A. 4 in. (100 mm).   D. 10 in. (250 mm).
    B. 6 in. (150 mm).   E. 12 in. (300 mm).
    C. 8 in. (200 mm).

    Code reference _____

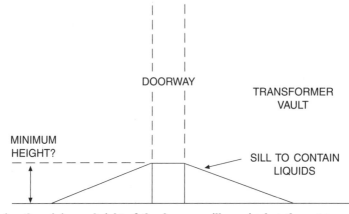

Figure 8.30   Determine the minimum height of the doorway sill required at the entrance to a liquid-insulated transformer vault.

5. A 500-kVA oil-insulated transformer is installed in a vault inside a building. The vault is ventilated to the outside of the building using natural circulation of air. After deducting the area for screens, gratings, and louvers, the minimum cross-sectional area of the ventilation opening shall be not less than:
    A.   375 in² (237,500 mm² or 2375 cm²).
    B.   500 in² (316,500 mm² or 3165 cm²).
    C.   750 in² (475,000 mm² or 4750 cm²).
    D.   1000 in² (633,300 mm² or 6333 cm²).
    E.   1500 in² (950,000 mm² or 9500 cm²).

    Code reference _____

6. A building contains one large 480-volt, 3-phase pump motor. One 20-ampere, 120-volt general-purpose circuit is installed to provide lighting outlets and receptacles for maintenance. The circuit is provided with a 3-kVA dry-type transformer and connected to the 480-volt supply. The conductors are copper with 75°C insulation and terminations. The conductor supplying the transformer is size 14 AWG and the conductors on the secondary side of the transformer are size 12 AWG. The installation is shown in Figure 8.31. If time-delay fuses in a disconnect switch protect all circuit conductors and the transformer, the maximum rating fuses permitted is (see Table 7.2 for typical fuse ratings):
    A. 3 amperes.   C. 10 amperes.   E. 20 amperes.
    B. 5 amperes.   D. 15 amperes.

    Code reference _____

7. A 45-kVA 3-phase transformer supplies a 208Y/120-volt, 4-wire panelboard with a 125-ampere main circuit breaker. The transformer is tapped from a 480-volt, 3-phase feeder with the feeder fuses also protecting the tap conductors and the transformer. The distance from the tap point through the transformer to the panelboard is 25 ft (7.5 m) with size 4 AWG primary and size 1 AWG secondary copper conductors run in Rigid Metal Conduit as shown in Figure 8.32. This installation is only permitted if the fuses protecting the feeder have a standard rating not exceeding:
    A. 125 amperes.   D. 225 amperes.
    B. 150 amperes.   E. 400 amperes.
    C. 200 amperes.

    Code reference _____

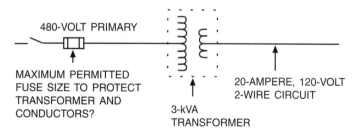

Figure 8.31   Determine the minimum rating primary overcurrent device required to protect the transformer and the secondary circuit conductors for a 2-wire, single-voltage secondary.

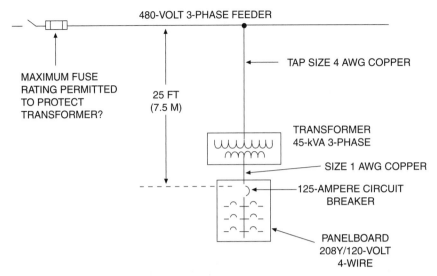

Figure 8.32 Determine the maximum rating feeder overcurrent device permitted where the overcurrent device also serves as the transformer overcurrent protection and the distance from the tap point to the secondary overcurrent device does not exceed 25 ft (7.5 m).

8. A 50-kVA dry-type single-phase transformer is installed outdoors on a concrete pad and supplies 120/240-volt single-phase power to panelboards with 100-ampere main circuit breakers in each of two separate buildings. The secondary conductors are size 1 aluminum Type USE run underground to each building. The conductors terminate at the panelboards immediately upon entering the buildings. The transformer is connected to a 480-volt supply with size 1 copper conductors protected by 150-ampere time-delay fuses. The maximum length of secondary conductors permitted to each buildings is:

A. 10 ft (3 m).

B. 25 ft (7.5 m).

C. 50 ft (15 m).

D. 100 ft (30 m).

E. is not restricted.

Code reference _____

9. A 45-kVA 3-phase transformer supplies a 208Y/120-volt, 4-wire panelboard with a 125-ampere main circuit breaker. The distance from the transformer to the panelboard is 25 ft (7.5 m) with size 1 AWG copper conductors run in Rigid Nonmetallic Conduit. The transformer circuit is tapped from a 400-ampere, 480-volt, 3-phase feeder with the tap ending at a fusible disconnect switch as shown in Figure 8.33. The primary conductors supplying the transformer are 6 AWG copper. The maximum standard rating time-delay fuses permitted for this transformer circuit is:

A. 60 amperes.

B. 70 amperes.

C. 100 amperes.

D. 125 amperes.

E. 175 amperes.

Code reference _____

10. The 45-kVA 3-phase transformer in Figure 8.33 with size 1 AWG copper secondary conductors feeding a panelboard with a 125-ampere main circuit breaker is grounded to a steel building support. The minimum size copper grounding electrode conductor permitted to be run from the panelboard to the steel support is:

A. 8 AWG.

B. 6 AWG.

C. 4 AWG.

D. 2 AWG.

E. 1 AWG.

Code reference _____

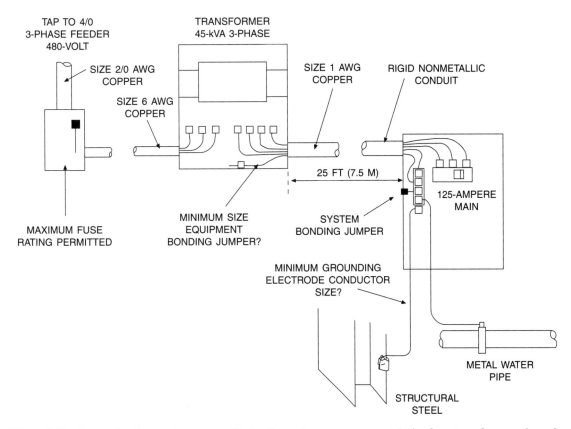

Figure 8.33 Determine the maximum permitted rating primary overcurrent device for a transformer where the primary conductors are size 6 AWG copper. Also determine the minimum size copper grounding electrode conductor for the transformer installation as well as the equipment bonding jumper in the section of Rigid Nonmetallic Conduit between the transformer and the panelboard.

11. The 45-kVA 3-phase transformer in Figure 8.33 with size 1 AWG copper secondary conductors feeding a panelboard with a 125-ampere main circuit breaker has the grounding electrode conductor connection to the grounded-circuit conductor at the panelboard. Because the conductors between the transformer and the panelboard are run in Rigid Nonmetallic Conduit, a bonding jumper is run from the panelboard grounding bus back through the conduit to the transformer enclosure. The minimum size equipment bonding jumper permitted for this purpose is:

   A. 12 AWG.   C. 8 AWG.   E. 4 AWG.

   B. 10 AWG.   D. 6 AWG.

Code reference _____

12. A 75-kVA 3-phase transformer is supplied with size 3 AWG copper conductors and protected on the primary by 100-ampere time-delay fuses. The transformer is connected to a 480-volt supply and provides power for two 208Y/120-volt, 4-wire panelboards as shown in Figure 8.34. One panelboard has a 150-ampere main circuit breaker and the other has a 50-ampere main circuit breaker. All conductors are copper and have 75°C insulation and terminations, and the maximum length of conductors from the transformer to either panelboard is not more than 10 ft (3 m). The minimum size conductors permitted to supply the panelboard with the 50-ampere main circuit breaker is:

   A. 8 AWG.   C. 4 AWG.   E. 3/0 AWG.

   B. 6 AWG.   D. 1/0 AWG.

Code reference _____

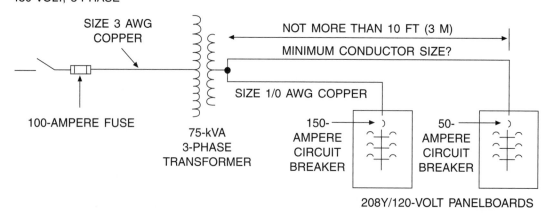

480-VOLT, 3-PHASE

SIZE 3 AWG COPPER

NOT MORE THAN 10 FT (3 M)

MINIMUM CONDUCTOR SIZE?

SIZE 1/0 AWG COPPER

100-AMPERE FUSE

75-kVA 3-PHASE TRANSFORMER

150-AMPERE CIRCUIT BREAKER

50-AMPERE CIRCUIT BREAKER

208Y/120-VOLT PANELBOARDS

Figure 8.34   Determine the minimum size copper conductors supplying the panelboard with the 50-ampere main circuit breaker when the length of the conductor is 10 ft (3 m), and when the length of the conductor is 25 ft (7.5 m).

13. A 75-kVA 3-phase transformer in a commercial building is supplied with size 3 AWG copper conductors and protected on the primary by 100-ampere time-delay fuses. The transformer is connected to a 480-volt supply and provides power for two 208Y/120-volt, 4-wire panelboards as shown in Figure 8.34. One panelboard has a 150-ampere main circuit breaker and the other has a 50-ampere main circuit breaker. All conductors are copper and have 75°C insulation and terminations. The installation is the same as shown in Figure 8.34 except the length of conductors from the transformer to both panelboards is more than 10 ft (3 m) but not more than 25 ft (7.5 m). The minimum size conductors permitted to supply the panelboard with the 50-ampere main circuit breaker is:

A. 8 AWG.   C. 4 AWG.   E. 3/0 AWG.
B. 6 AWG.   D. 1/0 AWG.

Code reference _____

14. A 5-horsepower single-phase motor is to be operated at 240 volts, but the building only has 208 volts available. A 1-kVA boost and buck transformer with a primary winding rated 120/240 volts and the secondary winding rated 16/32 volts is selected to boost from 208 volts to 240 volts. This transformer, connected in this manner, has a maximum continuous load current rating of 31.2 amperes (1000 VA/32 V = 31.2 A). The circuit is shown in Figure 8.35. It will be assumed the motor is powering an easy starting load, and the circuit conductors are properly sized. The maximum standard rating time-delay fuses permitted to provide overcurrent protection for the transformer and circuit is:

A. 25 amperes.   C. 35 amperes.   E. 45 amperes.
B. 30 amperes.   D. 40 amperes.

Code reference _____

15. A 5-horsepower single-phase motor is to be operated at 240 volts, but the building only has 208 volts available. A 1-kVA boost and buck transformer with a primary winding rated 120/240 volts and the secondary winding rated 16/32 volts is selected to boost from 208 volts to 240 volts as shown in Figure 8.35. The transformer, connected in this manner, has a maximum continuous load current rating of 31.2 amperes. All conduc-

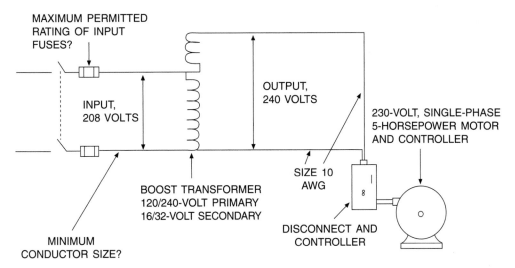

MAXIMUM PERMITTED
RATING OF INPUT
FUSES?

INPUT,
208 VOLTS

OUTPUT,
240 VOLTS

230-VOLT, SINGLE-PHASE
5-HORSEPOWER MOTOR
AND CONTROLLER

SIZE 10
AWG

BOOST TRANSFORMER
120/240-VOLT PRIMARY
16/32-VOLT SECONDARY

DISCONNECT AND
CONTROLLER

MINIMUM
CONDUCTOR SIZE?

Figure 8.35   Determine the maximum permitted standard rating of the input conductor fuses, and the minimum size input conductors for this boost and buck transformer installation supplying a single-phase 230-volt motor from a 208-volt supply.

tors are copper with 75°C insulation and terminations, and the conductors from the transformer to the motor are size 10 AWG. The minimum size conductors permitted to supply the transformer from the 208-volt source is:

A. 12 AWG.   C. 8 AWG.   E. 4 AWG.
B. 10 AWG.   D. 6 AWG.

Code reference _____

UNIT 9

Hazardous Location Wiring

OBJECTIVES

After completion of this unit, the student should be able to:

- explain the difference between a Division 1 and a Division 2 (or a Zone 0, Zone 1, and Zone 2), Class I hazardous location.

- explain the difference between a Division 1 and a Division 2, Class II hazardous location (or Zone 20, Zone 21, and Zone 22).

- describe the conditions that constitute a Class III hazardous location.

- explain the function of an explosionproof enclosure.

- select the proper atmospheric group when given the name of a common flammable vapor or dust.

- explain a wiring method permitted for use in a Class I hazardous location.

- explain a wiring method permitted for use in a Class II, Division 1 and Division 2 hazardous location.

- explain the special bonding requirements when double locknuts are used at an enclosure feeding a circuit in a hazardous location.

- state the minimum number of threads required on a piece of Rigid Metal Conduit or IMC to be used in a Class I hazardous location for field-made fittings.

- answer wiring installation questions relating to electrical installations in hazardous locations, as described in *Articles 500, 501, 502, 503, 504, 505,* and *506,* including the special occupancies described in *Articles 510, 511, 513, 514, 515,* and *516.*

- state at least four significant changes that occurred from the 2002 to the 2005 Code for *Articles 500, 501, 502, 503, 504, 505, 506, 510, 511, 513, 514, 515,* or *516.*

CODE DISCUSSION

An area is considered hazardous because of the highly flammable nature of a vapor, gas, dust, or solid material that may be easily ignited to cause fire or an explosion. Classification of the type of hazardous material, and the boundaries of the hazardous area, is covered in *Article 500,* with general requirements for wiring in these areas. *Articles 501, 502, 503, 504, 505,* and *506* cover specific wiring requirements for the type of hazardous area involved. *Article 504* covers the installation of intrinsically safe wiring systems in any classified location. Intrinsically safe wiring systems are designed such that even if a fault or a short circuit occurs, sufficient energy is not available to ignite the hazardous vapor, dust, or material in the surrounding area. *Article 505* describes an alternate method to *Article 501* of classifying Class I hazardous locations into zones, and describes the wiring methods permitted to be used for Zone 0, Zone 1, and Zone 2. In the case of flammable dusts, fibers, and flyings, wiring and equipment is permitted to be installed according to the rules of *Article 506* where materials and extent of hazard are organized into Zone 20, Zone 21, and Zone 22. This is an alternative using *Article 502* or *Article 503* for the installation of equipment and wiring. *Articles 510* through *516* deal with specific types of occupancies where hazardous conditions exist. These articles bring about uniformity of installations. Many occupancies are unique, and therefore, it is necessary to depend on *Articles 500* through *506* to make decisions about the wiring. There are other National Fire Protection Association (NFPA) documents that may be necessary to determine the wiring requirements or to decide the boundaries of the hazardous area. It may be necessary to involve an expert, such as a registered professional engineer, with knowledge of the process and materials to decide the boundaries of the hazardous area, as well as to determine when a material constitutes a hazard.

Article 500 covers some explanation of the nature of the conditions in a hazardous location as far as electrical equipment is concerned. The types of hazardous vapors, gases, and dusts are categorized into Groups A through G. Flammable vapor and gas can also be categorized into Groups IIA, IIB, and IIC when installed under the zone system of *Article 505*. The markings required on electrical equipment for use in a hazardous location are covered. Hazardous locations are arranged into Class I, Class II, and Class III. Class I hazardous locations are described in *500.5(B)*. Class II locations are described in *500.5(C)*, with Class III locations described in *500.5(D)*.

Article 501 covers the specific requirements and wiring methods permitted in a Class I hazardous location, which is one where flammable vapors or gas are or may be present in sufficient concentration to be ignited. The wiring methods permitted are covered in *501.10*, and some common methods are illustrated in Figure 9.1. The wiring methods permitted in a Class I, Division 1 location are threaded Rigid Metal Conduit, threaded Intermediate Metal Conduit, and Type MI Cable. Some additional cables and raceways are permitted in a Class I, Division 2 location. These are covered in *501.10(B)*. The requirements for providing seals in the wiring system are covered in *501.15*. The types of motors permitted to be installed in a Class I area are covered in *501.125*, and luminaire (lighting fixture) markings and installation are covered in *501.130*. NEC® *501.30* covers the requirements for grounding and bonding, which in some ways are different from the general requirements of *Article 250*. Refer to *500.8(D)* for a caution about preventing sparking during fault conditions by making sure threaded fittings are wrench-tight. This, as well as adequate grounding and bonding of the equipment, can reduce the chances of sparking.

Article 502 gives the wiring requirements for a Class II hazardous location where combustible dust is present in the air or may become suspended in the air. The wiring methods permitted are given in *502.10*. It should be noted there are differences in requirements for a Division 1 and a Division 2 hazardous area. Sealing the wiring system is covered in *502.15*. The type of motors permitted for use in a Class II location and the ventilation of the motor are covered in *502.125* and *502.128*. Luminaire (lighting fixture) installation is covered in *502.130*. The bonding and grounding requirements are similar to those of the Class I hazardous location and are covered in *502.30*.

Article 503 deals with wiring in areas where there will be flammable fibers and flyings. The key here is flammable. A woodworking area where wood flyings that are easily ignitable and collect on surfaces and equipment is a typical example. A sawmill, on the other hand, usually involves wood particles that are heavy in weight and have a high moisture content. This material is generally not considered to be easily ignitable, and thus would not be considered to be a Class III hazardous location. Wiring methods permitted for use in a Class III area are stated in *503.10*. Motors are covered in *503.125*, and *503.128* and luminaires (lighting fixtures) in *503.130*. Grounding and bonding are covered in *503.30*.

Article 504 covers the requirements for the installation of intrinsically safe wiring systems in hazardous classified locations. NEC® *504.2* provides definitions important for the application of this article, such as the definition of intrinsically safe circuits. NEC® *504.20* states that any wiring method suitable for similar conditions in unclassified locations is permitted to be used as a wiring method for intrinsically safe wiring systems in classified locations. NEC® *504.30* specifies separation requirements for intrinsically safe wiring from power and light wiring systems. The minimum separation is 2 in. (50 mm) if the wiring of both systems is fixed in place. There are exceptions, stated in *504.30*. It is necessary to maintain the 2 in. (50 mm) separation everywhere in the building, not just in the classified area. This is illustrated in Figure 9.2. NEC® *504.80* specifies that terminals shall be identified to prevent accidental interconnection of normal power cir-

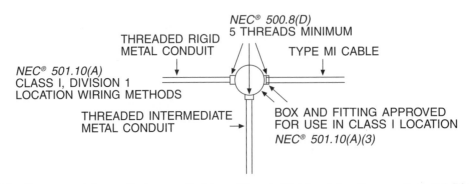

Figure 9.1 Common wiring methods permitted in a Class I, Division 1 location are threaded Rigid Metal Conduit, threaded Intermediate Metal Conduit, and Type MI Cable with approved fittings.

NEC® 504.30 MINIMUM SEPARATION OF OPEN CONDUCTORS

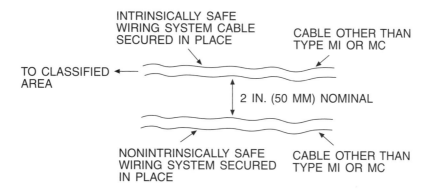

Figure 9.2   The intrinsically safe wiring shall be separated a minimum distance of 2 in. (50 mm) from lighting, power, and Class I circuits everywhere in the building—not just in the classified location.

cuits and intrinsically safe wiring systems. The intrinsically safe wiring system shall be identified with a permanent label affixed to the wiring at intervals with a minimum spacing of 25 ft (7.5 m). If a label is not visible in every separated section of the building, a label must be attached. For example, if intrinsically safe wiring passes through a room and there is no label visible in that room, a label must be attached. The label shall read "Intrinsic Safety Wiring," *504.80(B)*.

Article 505 provides an alternative method of classifying Class I hazardous locations as well as specifying the wiring methods. This article separates the Class I area into three zones, Zone 0, Zone 1, and Zone 2. As described in *505.5(B)(1)*, in a Zone 0 of a Class I location, an ignitable concentration of gas or vapor is present continuously or is normally present for extended periods of time. The rules for wiring in Zone 0 are provided in *505.15(A)*, but essentially the only wiring permitted is intrinsically safe wiring, nonincendive circuits, nonconductive optical fiber cable, or similar systems with an energy-limited electrical supply that has been approved for the purpose. *NEC® 505.5(B)(2)* describes Zone 1 of a Class I location as one where an ignitable concentration of gas or vapor is likely to be present for limited periods of time during normal operation, system breakdown, or repair. In general, a Class I, Division 1 location, as described in *500.5(B)(1)*, is separated into Zone 0 and Zone 1 when the rules of *Article 505* are applied. The wiring methods permitted in Zone 0 are permitted in Zone 1 are in *505.15(B)*.

In Zone 2 of a Class I location, an ignitable concentration of gas or vapor is not likely to exist during normal operation, and if an ignitable mixture becomes present, it will only exist for a short period of time. Usually the short-duration ignitable gas or vapor concentration in Zone 2 is the result of an accident or an unexpected system component rupture or failure. For Zone 2, the wiring methods used in more hazardous Class I, Division 1 or Zone 0 or 1 are permitted, as well as those wiring methods described in *505.15(C)* location.

Typical equipment intended for use in classified hazardous locations is required to be marked to show the class, group, and operating temperature. But equipment to be installed in a Class I location in accordance with *Article 505* is required to be marked with the class, zone, and symbol indicating the equipment is built to American standards (AEx), the type of protection designation, gas classification group, and temperature classification. An example is shown in Figure 9.3. The types of protection system designations are listed in Code *Table 505.9(C)(2)* and are described later. Temperature classifications are described in *Table 505.9(D)*. The gas groups as applied to *Article 505* are labeled differently than as applied to *Article 501*. As applied to *Article 505*, an ignitable concentration of gas or vapor is labeled Group II. But Group II is subdivided into Group IIA (same as Group D), Group IIB (same as Group C), and Group IIC (same as Groups A and B).

Different methods of preventing ignition of a flammable vapor in the area are used when wiring is installed according to *Article 505*. Intrinsically safe equipment cannot release enough energy to ignite a flammable vapor, and it would have a protection system designation of **ia** or **ib**. The term **explosion proof** is not generally used with respect to wiring installed according to *Article 505*. The corresponding term is **flameproof,** which carries a protection system designation **d.**

Flameproof protection is identified by the letter **d.** Type **d** equipment is permitted to be installed in Zone 1 areas and is designed to withstand an internal explosion without causing the ignition of the vapors surrounding the equipment.

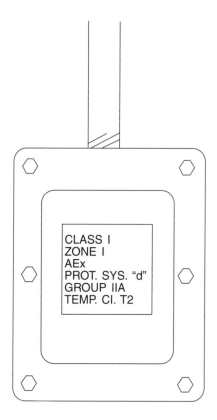

Figure 9.3   Markings on equipment suitable to be installed in a Class I hazardous location under the zone system shall be marked with the class, zone, United States specification symbol, protection system designation employed, vapor group, and temperature classification.

Equipment type known as **p** is purged and pressurized. This type of equipment reduces the chance of the possible ignition of the surrounding flammable gases such as with the ventilation of air from an unclassified location. Purged and pressurized equipment is permitted to be installed in Zone 1 and 2 hazardous locations.

Intrinsic safety equipments are identified by two different designations. Type **ia** is an intrinsically safe equipment, which is intended for use in a Zone 0 location, while type **ib** is the intrinsically safe equipment but is designed for Zone 1 locations. Equipment identified with the designation **ia** or **ib** is an associated apparatus such as a barrier which is connected to their respective intrinsically safe equipment but is generally not installed in the classified area.

Equipment that is provided with type **n** protection is permitted to be installed in Zone 2 locations. This type of equipment is designed in a manner such that it is not capable of igniting a surrounding hazardous vapor. Furthermore, type **n** equipment is broken down into three subcategories—**nA**, **nC**, and **nR**. The subcategories represent different design characteristics of the equipment or enclosures.

Protection provided by oil immersion is identified by the letter **o**. Through the use of this type of protection, the electrical equipment or part of the equipment is immersed in a nonflammable liquid. Any spark or arc from this type of equipment cannot ignite the surrounding vapors or gases. This type of equipment is intended to be installed in a Zone 1 location.

Type **e** equipment is known as increased safety equipment. This type of equipment can be installed in Zone 1 hazardous locations. This type of equipment is characterized by those not producing any arcs or sparks under normal operating conditions. Also, measures are taken to reduce the likelihood of surface temperatures of the equipment being of a level that would ignite the surrounding hazardous vapors or gases.

Encapsulation, type **m** equipment, is approved for Zone 1 locations. With this type of protection, arc producing contacts are encased in a compound so that any resultant arc could in no way ignite gases or vapors surrounding the enclosure.

Powder-filled protection, type **q**, is very similar to encapsulation except that a filling powder surrounds the spark producing contacts rather than a rigid compound. Type **q** equipment is for installation in Zone 1 locations.

NEC® 501.1 is the key for wiring in a Class I hazardous location. An area described as Class I, Division 1 or 2 according to *500.5(B)* is to be wired in accordance with the rules of the Code and as modified by *Article 501*. As pointed out in the fine print note to *501.1*, if the Class I area is subdivided into zones rather than divisions, then the rules of *Article 505* shall apply. Therefore, wiring in a Class I location is permitted to be selected and installed in accordance with either *Article 501* or *Article 505*. *NEC® 505.7(B)* is a precaution for instances when a structure or location is classified by both the zone and division system. Class I, Zone 0 or Zone 1 classified areas are not permitted to border Class I, Division 1 or Division 2 locations. It is permitted for a Class I, Zone 2 location to border, but not overlap, a Class I, Division 2 location. This is illustrated in Figure 9.4. There is an additional requirement if the class location is to be installed according to *Article 505*. A qualified registered professional engineer is required to supervise the definition of zone boundaries, equipment selection, and wiring system selection according to *505.7(A)*.

Article 506 is an alternate method of installing wiring and electrical equipment in a Class II or a Class III hazardous location. This is an installation method that can be used in place of *Article 502* for Class II wiring or *Article 503* for Class III wiring. When using this alternate method, flammable dusts, fibers, and flyings are treated the same with respect to rules for wiring. Combustible metallic dusts (Group E) are not covered by this alternate method, and for those areas it will be necessary to apply the rules of *Article 502*. Types of materials are not subdivided into groups as they are with dusts of different characteristics for *Article 502*. With *Article 506*, dusts, fibers, and flyings are categorized according to degree of hazard. If a condition will occur where a combustible concentration of the material exists all of the time or for extended periods of time, that condition is considered a Zone 20. If a combustible accumulation or suspension is likely to occur, then those areas are considered Zone 21. In areas where a condition of combustible accumulation is not likely to occur, but is possible under unusual or accidental circumstances, then that area is considered to be a Zone 22. Classification of Zones with examples is discussed in *506.5*.

Wiring and electrical equipment installation in areas where there is a combustible dust or a combustible fiber and flying accumulation is treated differently in *Article 502* from *Article 503*. Equipment such as motor controllers is required to have a marking similar to that shown in Figure 9.3 with the difference that the Class I is not present and a safe temperature operating range is required to be listed on the equipment, such as $-20°C$ to $+40°C$. The ambient temperature of the area where the equipment is to be installed is required to be within

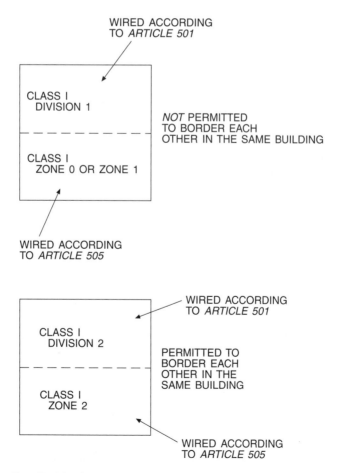

Figure 9.4 When some Class I wiring in a facility is wired according to *Article 501* and other areas are wired to *Article 505*, Division 1 and Zone 0 and 1 locations are not permitted to border each other, but Division 2 and Zone 2 areas are permitted to border each other.

the listed range. It is required in *506.6(A)* that design of the equipment layout and wiring system, selection of equipment and materials, installation, and inspection is to be performed by qualified personnel. In the case of a Zone system applied to a Class I area, the design, selection of equipment and materials, and installation are to be performed under the supervision of a Registered Professional Engineer. Wiring and equipment installed in a facility using *Article 502* or *Article 503*, according to *506.6(B)*, are not permitted to overlap with or even adjoin wiring installed according to *Article 506* except that a Class II, Division 2 or a Class III, Division 2 area is permitted to be located next to a Zone 22 area. This is a similar rule as shown in Figure 9.4.

The rules for the selection and installation of equipment such as motors, luminaires (lighting fixtures), controllers, heaters, transformers, and the like are much more detailed in *Article 502* and *Article 503* than in *Article 506,* and the rules are not necessarily equivalent. Wiring methods and materials are also different. For example, Electrical Metallic Tubing is permitted to be used as a wiring method in a Class III, Division 1 location using *Article 503*, but is not permitted in that same location if classified into Zone 20 or Zone 21 using *Article 506*. If sealing of the wiring system is judged to be necessary, the seal shall be identified. No such requirement applies when installing wiring using *Article 503*. The rules for grounding and bonding are the same for wiring installed using *Article 502*, *Article 503*, or *Article 506*.

Article 510 is a short article that states that *Articles 511* through *516* apply to special occupancies where hazardous areas exist. It also states that the general provisions of the Code apply, unless specific requirements are given in the article that apply to the type of occupancy.

Article 511 applies to commercial garages where repair is done to automobiles and other combustion motor equipment. The hazardous locations are defined, and the wiring methods permitted in those hazardous areas are described in *Article 501*. Special requirements are placed on the installation of luminaires (lighting fixtures) above a hazardous area. When wiring is installed in these facilities, as much wiring as possible is located outside of the hazardous area to reduce the cost of the installation.

Article 513 covers the requirements for installing wiring in an aircraft hangar. There are similarities in the requirements for a commercial garage. The hazardous area includes a larger portion of room because of the height of aircraft and the location of fuel tanks in the wings. The wiring methods permitted for the hazardous areas are described in *Article 501*. Raceways within or beneath a hangar floor are considered to be a part of the Class I area above the floor by *513.8* and *514.9*. This means conduits emerging up through the floor into the Class I area above are not considered to be crossing a classification barrier and are, therefore, not required to be sealed. If, however, those conduits leave the classified area and emerge into an area of a different division or an unclassified area, those conduits are required to be sealed.

Article 514 covers the classification and specific wiring requirements for gasoline dispensing and service stations. It is important to be familiar with the type of equipment used in these areas. *NEC® 514.3(B)(1)* describes the boundaries of the hazardous areas for Class I, liquids. Flammable gases such as compressed nat-

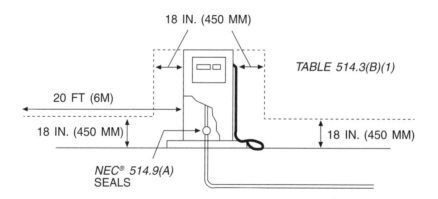

Figure 9.5   The limits of the hazardous locations for a gasoline dispensing unit are found In Code *Table 514.3(B)(1).*

ural gas or liquified petroleum have the limits of the boundaries described in *514.7(B)(2)*. For example, in the case of a gasoline dispenser, the space within the dispenser is considered to be a Class I, Division 1 hazardous location. The space in all directions horizontally from the dispenser and down to the grade level is considered to be a Class I, Division 2 location. This Division 2 location extends outward from the dispenser 20 ft (6 m) in all directions and up to a height of 18 in. (450 mm) above grade level. This is illustrated in Figure 9.5. For a dispensing unit, *514.11(A)* requires that each circuit conductor be wired so that the disconnecting means opens all of the wires, including a neutral conductor.

Article 515 covers wiring installed in hazardous areas around bulk storage plants. This is an area where gasoline or other volatile flammable liquids are stored in tanks or distributed by container, rail car, tank vehicle, or ship. The boundaries of the hazardous area are described in *Table 515.3*.

Article 516 applies to the wiring installed in the area of spray application, dipping, and coating processes. A paint spray area is a typical application. The boundaries of the hazardous location are described in *516.3*. All dimensions are described in this section and shown in *Figure 516.3(B)(1), Figure 516.3(B)(2)*, and *Figure 516.3(B)(4)* of this *Article 516*. In the case of a dipping process, it is important to make a determination of the point where the object that was dipped is no longer considered to be a vapor source. The definition of a vapor source is found in *516.3(B)(4)*. The limits of the hazardous location for an open tank dipping process are shown in an illustration of an open tank dipping process in *Figure 516.3(B)(5)*. Frequently, it is possible to locate wiring, controls, motors, and lighting fixtures outside of the hazardous area.

HAZARDOUS LOCATION WIRING FUNDAMENTALS

The first step for wiring in a hazardous location is to determine the type of material that makes the area hazardous. Next, determine if the type of facility is considered to be a special occupancy, which is described in one of the articles of the Code, for example, a paint spray booth or a motor vehicle fuel-dispersing station. Then it will be possible to determine the areas classified as hazardous. Achieving a safe and low-cost wiring system is the result of locating as much electrical equipment as possible out of the hazardous area. In the case of an industrial process, chemicals may be involved that require the assistance of officials or engineers other than the electrical inspector to determine the type of hazard. The National Fire Protection Association has additional publications that deal with most materials that would result in an area being classified as a hazardous location.

Types of Hazardous Locations

The types of hazardous materials are described in *Article 500, Article 505,* or *Article 506* with specific wiring requirements for each class covered in *Articles 501, 502, 503, 504, 505,* and *506*.

- Class I   Groups A, B, C, and D, ignitable gas or vapor

- Class I   Groups IIC, IIB, and IIA, ignitable gas or vapor

- Class II   Groups E, F, and G, combustible dust

- Class III   Easily ignitable fibers or flyings

Explosion characteristics of different air mixtures of flammable vapors are different; therefore, flammable vapors are separated into Groups A, B, C, and D when the wiring is installed according to *Article 501*, and the same vapors are separated into Groups IIA, IIB, and IIC when the wiring is installed according to *Article 505*. The Code does not provide a list of typical vapors that fall into each group. Instead, the Code gives one typical example of a vapor in each group. For Groups B, C, and D, vapors are placed according to ignition performance tests. The Code uses maximum experimental safe gap (MESG) and minimum ignition current (MIC) ratio to place vapors into the groups. Pressure of the vapor air mixture is an important factor with respect to flammability. One criteria is the maximum experimental safe gap under test conditions between adjacent metal parts. Another criteria is the minimum current to cause ignition. The ignition characteristics of some common materials such as gasoline are well-known, and specific requirements can be established for the installation of wiring. For other materials, special training may be needed.

Table 9.1   Group designations for flammable vapor air mixtures that apply when wiring is installed according to *Article 501*.

Group	MESG	MIC ratio	2005 Code vapor list	1996 Code vapor list
A or IIC	NA	NA	Acetylene	Acetylene
B or IIC	Less than or equal to 0.45 mm	Less than or equal to 0.4	Hydrogen	Gases containing more than 30% hydrogen by volume: butadiene, ethylene oxide, propylene oxide, and acrolein
C or IIB	Greater than 0.45 mm, but less than or equal to 0.75 mm	Greater than 0.4, but less than or equal to 0.8	Ethylene	Ethyl, ether, ethylene
D or IIA	Greater than 0.75 mm	Greater than 0.8	Propane	Acetone, ammonia, benzene, butane cyclopropane, ethanol, gasoline, hexane, methanol, methane, natural gas, naphtha, and propane.

A summary of the Groups A, B, C, and D is shown in Table 9.1, along with the MESG and MIC ratio for each group.

Once the hazardous material has been identified, the degree of hazard must be determined. A Division 1 area is one in which the hazardous material is likely to be present. A Division 2 area is one in which, under normal operating conditions, a hazardous material is present in dangerous quantities only under accidental spills or other unusual circumstances. The final condition is the nonhazardous area. The division line between hazardous and nonhazardous may be a physical barrier, or it may simply be a distance limit. Class I zones were discussed earlier in this unit.

Ratings of Equipment

Electrical equipment other than conduit, wire, and some fittings will be marked if suitable for hazardous locations. The equipment shall be marked with the class and the group of hazardous material. The National Electrical Manufacturers Association (NEMA) has established a numbering system for different types of enclosures. Typical motor and control enclosure designations for hazardous locations are listed as follows:

- Class I   Hazardous gas or vapor
  NEMA 7 explosion-proof enclosure
- Class II   Flammable dust
  NEMA 9 dust-ignition-proof enclosure
  Motors are permitted to be totally enclosed pipe-ventilated
- Class III Flammable flyings and fibers
  NEMA 4, and NEMA 12 dusttight enclosure
  Motors are permitted to be rated totally enclosed, totally enclosed pipe-ventilated, or totally enclosed fan-cooled

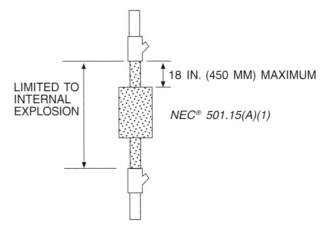

Figure 9.6   Seals confine combustion in the wiring system to a limited portion of the electrical system.

Objectives of Explosionproof Enclosures

An explosionproof wiring system is installed in Class I hazardous locations. The wiring system is installed with the assumption that it is impossible to prevent the entry of the hazardous gas or vapor into the wiring system. A flammable mixture of gas or vapor and oxygen may accumulate in a heat-producing or an arc-producing portion of the wiring system, such as a switch. The distance from the enclosure to the seal is not permitted to be greater than 18 in. (450 mm). If the raceway is trade size 2 (53) or larger in diameter, and the enclosure contains splices, taps, or terminations, then it is also required to provide a seal within 18 in. (450 mm) of the box or enclosure. Seals installed in conduit entries to enclosures limit the internal combustion to a small portion of the wiring system, as shown in Figure 9.6.

It is, therefore, assumed that an internal explosion cannot be prevented. The explosion produces extreme internal pressure. The enclosure, seals, conduit, and fittings shall have sufficient strength to withstand the pressure of the internal explosion. This is why Rigid Metal Conduit or Intermediate Metal Conduit (IMC) is required, and why joints must have a minimum of five threads fully engaged. This is also why explosionproof enclosures are so massive. Therefore, it is absolutely necessary to install all bolts and screws on enclosure covers.

Heat is produced during an internal explosion, and this heat will raise the outside surface temperature of the enclosure. This surface temperature shall be kept below the ignition temperature of the gas or vapor on the outside, or an explosion is likely to occur. This is another reason why explosionproof enclosures are so massive. There must be enough metal mass to absorb the heat of the internal explosion.

The high pressure developed during an internal explosion will cause the ignition gases to eventually escape through joints and threads. The covers and threads shall be tight enough to retard the leakage of products of combustion. If leakage is too fast, the escaping gas will be hot enough to ignite the vapor on the outside. Therefore, slowly escaping gas will cool before it gets to the outside, as shown in Figure 9.7. All threaded joints must be tight, with at least five threads engaged. Machined metal surfaces must be clean with no

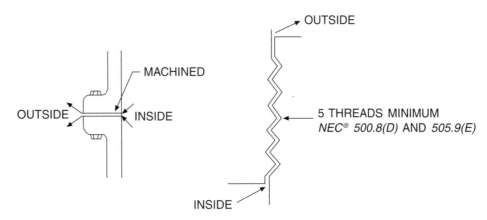

Figure 9.7   Explosionproof equipment is designed to absorb the heat from escaping combustion vapors through the threads or a metal surface machined to a specified clearance.

scratches. A grain of sand on the machined surface of a cover, or a scratch, will allow hot gas to escape, and an external explosion may occur.

Sealing Fittings

The thickness of the compound in a Class I sealing fitting is specified in *501.15(C)(3)*. The minimum thickness of the sealing compound is not permitted to be less than the trade size of the sealing fitting. The minimum thickness under any circumstances is $^5/_8$ in. (16 mm). If a sealing fitting is made for trade size $^3/_4$ (21) conduit, then the minimum sealing compound thickness in the sealing fitting is trade size $^3/_4$ (21). This is illustrated in Figure 9.8. If a trade size 1 (27) sealing fitting is used with trade size $^3/_4$ (21) conduit, using a reducer at the fitting, the minimum thickness of sealing compound in the fitting is now trade size 1 (27).

It is also important to make sure the conductors are separated so that the sealing compound will flow around each conductor, leaving no voids for vapor to pass. To help in the separation of conductors in the sealing fitting, standard fittings are listed for only 25% fill. This means that the total cross-sectional area of the conductors is not permitted to exceed 25% of the cross-sectional area of the sealing fitting based on the cross-sectional area of Rigid Metal Conduit as listed in *Table 4* in *Chapter 9*. This would mean that the IMC or Rigid Metal Conduit is permitted to be filled to only 25% of the conduit cross-sectional area. There are some ways around this problem. There are sealing fittings rated for 40% fill. Another solution is to use an oversized sealing fitting with reducers. The following example will show how to determine the minimum size of sealing fitting required for a particular installation.

Example 9.1   Four size 6 AWG, Type THWN conductors are run to an explosion-proof enclosure where a sealing fitting is required to be installed within 18 in. (450 mm) of the enclosure. Determine the minimum trade diameter sealing fitting required for this installation.

Answer: Look up the cross-sectional area of size 6 AWG, Type THWN conductors in *Table 5* in *Chapter 9* and find 0.0507 sq. in. Next, multiply the individual conductor cross-sectional area by 4 to get the total cross-sectional area of the conductors (4×0.0507 in.^{2} = 0.2028 in.^{2}). There is no 25% cross-sectional area column in *Table 4* in *Chapter 9,* so divide the cross-sectional area of the wires by 0.25 to get the minimum total cross-sectional area of the fitting required, which in this case is 0.8112 sq. in. (523.2 mm^2) (0.2028 ÷ 0.25 = 0.8112 in.^{2}). Now go to the 100% cross-sectional area column of the Rigid Metal Conduit section of *Table 4* in *Chapter 9* and find a size that is not smaller than 0.8112 sq. in. (523.2 mm^2), which is trade size 1 (27).

$$4 \times 0.0507 \text{ in.}^2 \quad = \quad 0.2028 \text{ in.}^2$$
$$(4 \times 32.71 \text{ mm}^2 \quad = \quad 130.8 \text{ mm}^2)$$

$$0.2028 \text{ in.}^2 / 0.25 \quad = \quad 0.8112 \text{ in.}^2$$
$$(130.8 \text{ mm}^2 / 0.25 \quad = \quad 523.2 \text{ mm}^2)$$

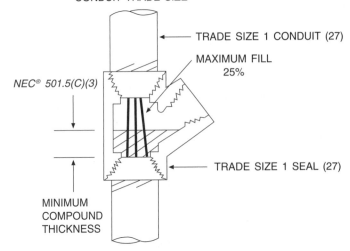

NEC® 501.15(C)(6) MAXIMUM CONDUCTOR FILL
FOR A SEALING FITTING IS 25%
OF EQUIVALENT RIGID METAL
CONDUIT TRADE SIZE

TRADE SIZE 1 CONDUIT (27)

MAXIMUM FILL
25%

NEC® 501.5(C)(3)

TRADE SIZE 1 SEAL (27)

MINIMUM
COMPOUND
THICKNESS

Figure 9.8   Standard Class I sealing fittings are rated for 25% fill and the thickness of the sealing compound is not permitted to be less than $^5/_8$ in. (16 mm) or the trade size of the sealing fitting.

Table 9.2   Allowable wire fill area for standard sealing fittings (based on RMC trade size area)

| Sealing Fitting Trade Size | Cross-Sectional Area | | | |
| | 25% | | 100% | |
	sq. in.	sq. mm	sq. in.	sq. mm
³/₄ (21)	0.137	88	0.549	353
1 (27)	0.222	143	0.887	573
1¹/₄ (35)	0.382	246	1.526	984
1¹/₂ (41)	0.518	333	2.071	1333
2 (53)	0.852	550	3.408	2198
2¹/₂ (63)	1.217	784	4.866	3137
3 (78)	1.875	1210	7.499	4840
3¹/₂ (91)	2.503	1615	10.010	6461
4 (103)	3.221	2079	12.882	8316
5 (129)	5.053	3263	20.212	13050
6 (155)	7.290	4705	29.158	18821

The conduit is permitted to be sized according to a 40% fill. The total cross-sectional area of the conductors in this example is 0.2028 sq. in. Look up the minimum trade size conduit from the 40% fill column of *Table 4* and find trade size ³/₄ (21). It is permitted to use trade size ³/₄ (21). Rigid Metal Conduit in this case with a trade size 1 (27). sealing fitting and an explosion-proof reducer from trade size 1 to ³/₄ (27 to 21).

Since the maximum permitted cross-sectional area of the wires passing through a sealing fitting is based upon the cross-sectional area of Rigid Metal Conduit of the same trade size, *Table 9.2* was created showing 25% of the cross-sectional area of the sealing fitting. Referring to Example 9.1, the four wires have a total cross-sectional area of 0.2028 sq. in. Using *Table 9.2,* it is easy to see that a trade size 1 in. sealing fitting is required.

Equipment Grounding and Bonding

Grounding is extremely important in hazardous locations. Extra care must be taken to make sure the grounding path is of low impedance. An adequate grounding system will conduct current without allowing the enclosure of the equipment to develop a voltage above ground. This is important to prevent the case of the electrical equipment from arcing to ungrounded metal equipment and structural supports. A normally harmless static or small fault-induced arc can cause an explosion in a hazardous location.

It is also important that all threaded connections be made wrenchtight to prevent arcing at the joint in the event the conduit and metal enclosures are needed to conduct fault current. An arc at a loosely joined conduit connection can cause an explosion. Special bonding requirements are also required at certain areas of hazardous locations, as shown in Figure 9.9. Double locknuts are not permitted to serve as the equipment grounding for a circuit in a hazardous location. It is necessary to bond directly from conduit entries into an

BONDING REQUIREMENTS FOR WIRING IN CLASSIFIED LOCATIONS ARE FOUND IN *NEC®* 501.30(A), 502.30(A), AND 503.30(A), 505.25(A), AND 506.25(A)

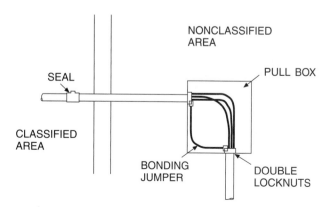

Figure 9.9   Double locknuts are not considered adequate bonding for conduit systems supplying wiring in a hazardous location and, therefore, special bonding is required even in the nonhazardous area.

enclosure even if the enclosure is not in the hazardous area. Code *Articles 501.30(A), 502.30(A), 503.30(A), 505.25(A)* and *506.25(A)* cover the special bonding requirements for wiring to hazardous locations.

Dust-Ignitionproof Equipment

The principle of installing wiring in a Class II area is to prevent the entry of flammable dust. Dust is a solid material that is heavier than air. Gaskets will prevent the entry of dust particles. In calm air, dust will settle. If dust is prevented from entering heat- and arc-producing parts of the wiring system, a fire or explosion will be prevented.

Enclosures for Class II areas are not required to be of equally heavy construction as Class I enclosures. Seals are not required to isolate arc- and heat-producing components. Sealing is required simply to prevent the entry of dust. The same extra care taken to ensure a good grounding system is required because arcs, due to poor bonding, can set off a general explosion if a fault occurs.

Flammable Fibers and Flyings

The primary problem in these types of facilities is to prevent the entry of heavier-than-air fibers and flyings such as cotton, textile fibers, and dry wood fibers. Enclosures generally have hinged doors with a latch and a gasket around the edges of openings. Enclosures and equipment must be selected to prevent the surface temperature from rising above the ignition temperature of the fibers or flyings that may collect on the enclosure. This temperature limitation must be achieved even when fibers and flyings have accumulated in layers on wiring and equipment.

Special Occupancies

The general wiring requirements for hazardous locations are provided in Code *Articles 501, 502, 503, 504,* and *505.* Later articles provide special wiring requirements and rules for determining the hazardous classified area. An example of a special occupancy with rules to define the extent of the hazardous locations is an automobile repair facility that also serves motor vehicles that run on compressed natural gas or other lighter-than-air fuels. In such an area, according to *511.3(B)(4),* the area 18 in. (450 mm) down from the ceiling, as well as the area 18 in. (450 mm) above the floor, is considered to be a Class I, Division 2 hazardous location as illustrated in Figure 9.10.

The local electrical inspector, building code, or other organizations or agencies will provide additional requirements for types of hazardous facilities not specifically covered in the Code. Some manufacturers of electrical equipment and wiring materials for installation in hazardous locations publish excellent literature that is helpful when installing wiring in hazardous areas.

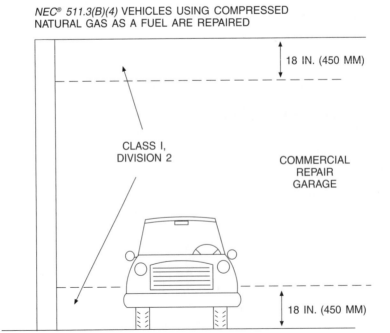

NEC® 511.3(B)(4) VEHICLES USING COMPRESSED NATURAL GAS AS A FUEL ARE REPAIRED

18 IN. (450 MM)

CLASS I, DIVISION 2

COMMERCIAL REPAIR GARAGE

18 IN. (450 MM)

Figure 9.10 Commercial garages that repair vehicles using compressed natural gas or other lighter-than-air fuels have the area 18 in. (450 mm) down from the ceiling in the service area classified as a Class I, Division 2 location.

MAJOR CHANGES TO THE 2005 CODE

These are the changes to the 2005 *NEC*® that correspond to the Code sections studied in this unit. The following analysis explains the significance of the changes from the 2002 to the 2005 Code only, and this analysis is not intended to be used in place of the Code. Refer to the actual section of the 2005 Code for the exact wording and meaning of each section discussed. Changes are indicated in the Code with a vertical line in the margin. If material was deleted or moved to another location in the Code, the location of the deletion is indicated with a dark dot in the margin.

Article 500   Hazardous Locations, Classes I, II, and III, Divisions 1 and 2

500.1 FPN No. 1: A new fine print note was added to the scope to point out that *Article 500* does not address the installation of wiring in facilities where pyrotechnics, explosives, and other blasting agents are manufactured, stored, or handled.

500.1 FPN No. 3: A new *Article 506* was added that is an alternative to the wiring methods described in *Article 502* and *Article 503*. This new article contains rules for the installation of wiring in Zones 20, 21, and 22 where fire or explosive hazards exist with flammable dusts and ignitible flyings.

500.2: The definition of purged and pressurized was changed. The past edition of the Code described purged and pressurized as requiring flow as well as pressurization to prevent the entry of flammable vapors in sufficient concentration to create an ignition. The new definition describes purging and pressurization separately. Purging is a flow under sufficient positive pressure. Pressurization may or may not involve a flow of fluid. An enclosure, equipment, or wiring system can simply be pressurized with no flow.

500.7(K): In some situations, an acceptable protection technique permitted for electrical installations in Class I hazardous locations is the presence of a combustible gas detection system. A new paragraph was added that requires the detection equipment to be listed for the specific application and documentation is required with regard to location of the equipment, calibration, type of alarm utilized, and the system shut-down procedure.

500.8(A)(2): Equipment that is identified for use in a Division 1 location is permitted to be installed in a Division 2 location of the same class and group of flammable material. An additional requirement was added that when equipment approved for a Division 1 location is installed in a Division 2 location, the equipment temperature class must also be acceptable. There is another new requirement relating to intrinsically safe wiring. If the wiring being installed in the Division 2 location is intrinsically safe wiring, it is to be installed in the same manner as required in the Division 1 location.

500.8(D): A new sentence was added to this section that establishes the requirements for threaded connections for wiring in classified locations. Connections into explosionproof equipment are to be made up with 5 threads fully engaged. This 5-thread requirement in the previous edition of the Code was in *501.4(A)* and *505.9(E)*. The 5-thread requirement only applied in Class I, Division 1 locations. Now the requirement specifies connections to explosionproof equipment. If explosionproof equipment is installed in a Class I, Division 2 location, the 5-thread connection requirement will apply. An exception permits manufacturers of listed equipment to only be required to make hubs and other threaded entries with 41⁄2 threads. The change was made because of a manufacturing problem involving adapters to metric threads.

Article 501   Class I Locations

The sections were renumbered to create a similarity of numbering and sections between this and *Article 502* and *Article 503*. For example, wiring methods are discussed in *501.10*, sealing is discussed in *501.15*, and grounding and bonding are discussed in *501.30*.

501.10(A)(1) Exception: This exception permits Rigid Nonmetallic Conduit to be installed encased in 2 in. (50 mm) of concrete and buried not less than 24 in. (600 mm) below grade level in a Class I, Division 1 location. In the case of wiring supplying a gasoline-dispensing unit and bulk storage areas, the 2 in. (50 mm) encasement of concrete is not required. What appears to be a change in this exception is the deletion of a reference to wiring installed beneath the floor of a service area of a commercial garage. This seems to imply that a 2 in. (50 mm) encasement of concrete is required when run under the floor in Rigid Nonmetallic Conduit, but that is not the case. (See discussion of *511.4(A)(1)*).

501.10(B)(6): Single-conductor Medium Voltage Cable, Type MV, when run in cable tray, is required to be shielded or be provided with a metallic armor.

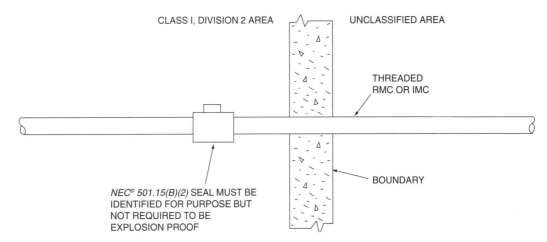

CLASS I, DIVISION 2 AREA   UNCLASSIFIED AREA

THREADED
RMC OR IMC

BOUNDARY

NEC® 501.15(B)(2) SEAL MUST BE
IDENTIFIED FOR PURPOSE BUT
NOT REQUIRED TO BE
EXPLOSION PROOF

Figure 9.11   The purpose of the seal between a Class I, Division 2 location and an unclassified location is to prevent the movement of combustible vapors through the raceway into the unclassified area, not to contain an explosion.

501.15(B)(2): An explosionproof seal is not required at the boundary between a Class I, Division 2 area and an unclassified area as illustrated in Figure 9.11. The seal needs only to be identified for the purpose. The wording of this paragraph was changed to make this clear. An explosionproof seal was usually required at this location because the meaning was not clear. The purpose of this seal is to prevent vapor from passing from the classified area into the unclassified area, not to contain an explosion. Seals are available for this purpose that permit the conductors to be removed at a later date for remodeling and resealed. The sealing compound is not required to be permanent, but it must be identified for the purpose.

501.15(C) Exception: This section describes how an explosionproof seal is to be installed. A new exception was added that does not require a seal to be made in this manner for the situations described in *501.15(B)(2)* and *504.70.*

501.15(F)(3): This section deals with process instrumentation and similar equipment connected to a system primarily for measurement purposes where a failure of a seal or barrier can result in flammable vapors or liquids entering the wiring system. The wiring installer is required to provide some seal or approved barrier as a backup prevention in case of leakage beyond the connection to the process. Equipment is now available that does not require an additional field-installed seal or barrier. This section permits equipment and instrumentation that is marked or listed as "Dual Seal" to be connected to a process, and an additional field installed seal is not required.

501.25: This was *501.15* in the previous edition of the Code, and it was a rule that stated there shall be no exposed live parts in a Class I, Division 1 or Division 2 location. That rule prevented the termination of intrinsically safe circuits and nonincendive circuits. Now these types of circuits are permitted to have exposed live parts provided 30 volts is not exceeded in dry locations and 15 volts is not exceeded in wet locations.

501.35: This section was numbered *501.17* in the previous edition of the Code. The section places requirements on the installation of surge suppressors. Now transient voltage surge suppressors (TVSS) devices are also required to be installed in the same manner as surge suppressors.

501.130(B)(2): This section contains rules for the installation of luminaires (lighting fixtures) in a Class II, Division 2 hazardous location. This is a new section dealing with protection form physical damage. There was no such requirement in the past. The luminaire (lighting fixture) is now required to be protected from physical damage by a suitable guard or by where it is mounted. If there is a danger of hot metal from damage to the lamps falling into an area where there is a potential hazard from a material such as combustible vapor, steps are required to be taken to prevent such a hazard.

Article 502   **Class II Locations**

The sections were renumbered to create a similarity of numbering and sections between this and *Article 501* and *Article 503*. For example, wiring methods are discussed in *502.10*, sealing is discussed in *502.15*, and grounding and bonding are discussed in *502.30.*

502.10(A)(2): This subsection describes permitted means of making flexible wiring connections in a Class II, Division 1 hazardous location. The change is that now Type MC Cable with interlocked armor and an overall jacket of polymeric material is permitted if listed terminations are used. Type MC Cable with interlocked armor has adequate flexibility to be used to make flexible connections.

502.15: There is an addition to this section to clarify the intent of the case where a run of raceway is permitted to act in place of a seal between an enclosure that is required to be dust-ignitionproof and one that is not required to be dust-ignitionproof. Taking the language in the previous edition of the Code literally results in different interpretations of the sealing requirements. A seal is required if a conduit extends downward from an enclosure required to be dust-ignitionproof for 1 ft (300 mm) then makes a bend and runs horizontally 10 ft (3.05 m) to an enclosure that is not required to be dust-ignitionproof. If the raceway runs 10 ft (3.05 m) horizontally from the dust-ignitionproof enclosure then makes a bend downward for 1 ft (305 m) and terminates at the enclosure not required to be dust-ignitionproof, a seal is not required. A new item *(4)* was added to make it clear the intent was to establish, somewhere in the run, a 10 ft (3.05 m) horizontal section or a 5 ft (1.52 m) vertical section. It would also seem that other combinations are equivalent; a run that is 7 ft (2.13 m) horizontal and 3 ft (1.52 m) vertically down is definitely equivalent to a 10 ft (3.05 m) horizontal run.

502.25: This was *502.15* in the previous edition of the Code, and it was a rule that stated there shall be no exposed live parts in a Class II, Division 1 or Division 2 location. That rule prevented the termination of intrinsically safe circuits and nonincendive circuits. Now these types of circuits are permitted to have exposed live parts provided 30 volts is not exceeded in dry locations and 15 volts is not exceeded in wet locations.

502.35: This section was numbered *502.17* in the previous edition of the Code. The section places requirements on the installation of surge suppressors. Now transient voltage surge suppressors (TVSS) devices are also required to be installed in the same manner as surge suppressors.

Article 503   **Class III Locations**

The sections were renumbered to create a similarity of numbering and sections between this and *Article 501* and *Article 502*. For example, wiring methods are discussed in *503.10* and grounding and bonding are discussed in *503.30*.

503.10(A)(3): Nonincendive field wiring was added as a wiring method permitted in a Class III, Division 1 location. The installation rules are the same as in *Article 502*.

503.25: This was *503.15* in the previous edition of the Code, and it was a rule that stated there shall be no exposed live parts in a Class III, Division 1 or Division 2 location. That rule prevented the termination of intrinsically safe circuits and nonincendive circuits. Now these types of circuits are permitted to have exposed live parts provided 30 volts is not exceeded in dry locations and 15 volts is not exceeded in wet locations.

Article 504   **Intrinsically Safe Systems**

504.10(B): When installing an intrinsically safe control system, components are required to be listed. Some components are believed to be so low-power producing or power using that they do not develop high enough surface temperature to create a hazard. These components are called simple apparatus, and they are not required to be listed. Whether simple apparatus can create a hazard has been called into question. Listing of simple apparatus is still not required, but there is now a rule that only permits simple apparatus to be installed if the surface temperature developed does not exceed the ignition temperature of the flammable vapors, flammable liquids, combustible dust, or ignitible fibers or flyings present. A formula for calculating surface temperature and a table for evaluating acceptability are provided, but without experience they are of not much value. Simple apparatus are passive devices such as LEDs, resistance temperature devices, and switches; and active devices that generate energy such as photocells and thermocouples.

504.30(B)(3): A different intrinsically safe circuit is an intrinsically safe circuit that is not installed in such a manner that the intrinsically safe quality can be guaranteed to be maintained. When terminating these circuits there is a new requirement that the terminals be not closer to each other than 1/4 in. (6 mm) unless a lesser distance is permitted by the control diagram.

Article 505   **Class I, Zone 0, 1, and 2 Locations**

505.8(I): A type of protection system permitted is a combustible gas detection system. With such a system it may be permitted to install equipment suitable for a Class I, Zone 2 in a Class I, Zone 1 area. Or it may be permitted to treat a Class I, Zone 2 area as unclassified. There are some new requirements for the combustible gas detection system. Detection equipment installation location, type of alarm utilized, calibration frequency, and system shutdown criteria are required to be documented.

505.9(E) Exception: If explosionproof or flameproof equipment is installed in a Class I, Zone 0, Zone 1, or Zone 2 location the 5-thread connection requirement will apply. The new exception permits manufacturers of listed equipment to only be required to make hubs and other threaded entries with 4 1/2 threads. The change was made because of a manufacturing problem involving adapters to metric threads.

505.15(C)(1)(b): Single-conductor Medium Voltage Cable, Type MV, when run in cable tray, is required to be shielded or be provided with a metallic armor.

505.16(B)(1): When an enclosure of protection type "d" or "e" is installed in Class I, Zone 1, a seal is now required to be installed in each conduit entry within 2 in. (50 mm) of the enclosure. Two new exceptions were added where the seal may be at a greater distance or omitted.

505.16(E)(3): This section deals with process instrumentation and similar equipment connected to a system primarily for measurement purposes where a failure of a seal or barrier can result in flammable vapors or liquids entering the wiring system. The wiring installer is required to provide some seal or approved barrier as a backup prevention in case of leakage beyond the connection to the process. Equipment is now available that does not require an additional field-installed seal or barrier. This section permits equipment and instrumentation that is marked or listed as "Dual Seal" to be connected to a process, and an additional field installed seal is not required.

505.19: In the previous edition of the Code, this was a rule that stated there shall be no exposed live parts. That rule prevented the termination of intrinsically safe circuits and nonincendive circuits. Now these types of circuits are permitted to have exposed live parts provided 30 volts is not exceeded in dry locations and 15 volts is not exceeded in wet locations.

Article 506   Zone 20, 21, and 22 Locations for Combustible Dusts, Fibers, and Flyings

The following is a discussion of the key points covered in this new article to assist in finding information relative to an installation. This article does not apply to metal dusts.

506.1: This scope statement points out this is an alternative method to the methods of *Article 502* and *Article 503* for installing wiring and electrical equipment in areas where a hazard exists due to the presence in sufficient quantity of combustible dusts, ignitible fibers, or ignitible flyings. Depending upon the degree of hazard, the locations are subdivided into Zone 20, Zone, 21, and Zone 22. A Class II or Class III, Division 1 area is subdivided into Zone 20 or Zone 21 depending upon whether the hazard exists or is likely to exist for extended periods of time. A Class II or Class III, Division 2 area is equivalent to Zone 22.

506.5: This section contains a detailed description of each zone with examples. Zone 20 is described in *506.5(B)(1)*, Zone 21 is described in *506.5(B)(2)*, and Zone 22 is described in *506.5(B)(3)*.

506.6(A): Qualified personnel are required to design the wiring system and select equipment. Qualified personnel are also required to supervise the installation of wiring and electrical equipment as well as inspect the installation.

506.6(B): Areas installed using the methods of *Article 506* are not permitted to directly adjoin an area with wiring installed using *Article 502* or *Article 503* except a Zone 22 area is permitted to adjoin a Class II, Division 2 or a Class III, Division 2 area.

506.6(C): An area with wiring previously installed using the rules of Class II or Class III is permitted to be reclassified using the Zone system.

506.8: Various types of protection systems, such as dust-ignitionproof enclosures, are listed in this section with the zones for which each is permitted to be installed.

506.9: This section provides the required markings for equipment that is to be installed in Zone 20, Zone 21, or Zone 22. This is similar to Figure 9.3 except that the Class is not required and an ambient temperature range for which the equipment is suitable must be provided on the equipment.

506.15: The wiring methods that are permitted in the three zones are described in this section. Rules that are acceptable for a Class II area are required to be applied to Class III areas. There are some differences between the wiring methods permitted using the Zone system as compared to using the rules of *Article 502* and *Article 503*.

506.16: Sealing requirements are in this section and seals and materials used are required to be identified. Explosionproof and flameproof seals are not required, and seals are only required where necessary to prevent entry of dust, fibers, or flyings. There is considerable room for judgment and interpretation with respect to installation of seals.

506.20: Installation of equipment such as lighting, motors, transformers, and other equipment is covered in this section. Surface temperature requirements in *506.20(E)* are essentially the same as for equipment installed according to *Article 502* and *Article 503*. The equipment requirements are much more extensively described in *Article 502* and *Article 503* than they are in *506.20*.

506.25: Grounding and bonding requirements are specified in this section, which is essentially the same as the rules in *502.30* and *503.30*.

Article 511 Commercial Garages, Repair and Storage

There was a complete reorganization of the section dealing with classification of areas in a commercial garage or a storage garage. There is no longer a reference to *Table 514.3(B)(1)* for specifications on ventilation for an underfloor service area where only minor repairs are conducted. That material in *Article 514* was moved to this section. *Section 511.3(A)* describes areas that are unclassified, and *511.3(B)* describes the areas that are classified.

511.3(A)(3): An area adjacent to a classified area is not necessarily required to have ventilation at a rate of four air changes per hour in order to be considered an unclassified area. The area is now permitted to be designed to have a positive air pressure with respect to the surrounding classified areas.

511.4(A)(1): This section in the previous edition of the Code was deleted. This paragraph stated that raceway embedded in a masonry wall adjacent to a Class I area, or buried under a concrete floor above which is a Class I area, is considered to also be in the Class I area. Since this statement no longer exists in *Article 511*, those areas are no longer considered to be in the Class I area. As a result there was no need for an exception, and the *Exception* to *511.4(A)(1)* was also deleted. Rigid nonmetallic conduit is permitted to be run under the floor of the auto repair area of a commercial garage and there is no depth requirement, as illustrated in Figure 9.12. Rigid Nonmetallic Conduit is not permitted to be installed in a Class I, Division 2 hazardous area. If a run of Rigid Nonmetallic Conduit emerges up through the concrete floor above which is a Class I, Division 2 area, that portion of the conduit run must be Rigid Metal Conduit or Intermediate Metal Conduit.

511.7(A)(1): Now, Type AC cable is permitted to be installed in spaces above a Class I area of a commercial garage.

Article 513   **Aircraft Hangars**

513.12: All 125-volt, 15- and 20-ampere receptacles located in a hangar and intended for powering diagnostic equipment, electrical hand tools, or portable lighting are now required to be ground-fault circuit-interrupter protected.

Article 514   **Motor Fuel-Dispensing Facilities**

Table 514.3(B)(1) Footnote 1: A new footnote was added to the table to establish the equivalent of grade level in the case of fuel-dispensing equipment for water craft. Grade level is the surface of the pier down to the water level.

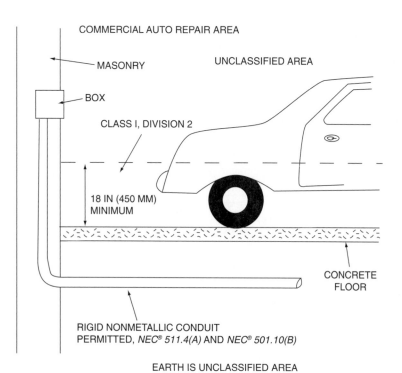

COMMERCIAL AUTO REPAIR AREA

MASONRY

UNCLASSIFIED AREA

BOX

CLASS I, DIVISION 2

18 IN (450 MM) MINIMUM

CONCRETE FLOOR

RIGID NONMETALLIC CONDUIT PERMITTED, *NEC*® *511.4(A)* AND *NEC*® *501.10(B)*

EARTH IS UNCLASSIFIED AREA

Figure 9.12   The earth below the floor of the Class I, Division 2 hazardous area of a commercial service garage is considered unclassified and any wiring method suitable for direct burial is permitted to be installed in the area.

514.8: The earth below a Class I hazardous location of a motor fuel-dispensing facility is no longer considered to be a Class I hazardous location. There is, however, a requirement that any conduit emerging from the ground be sealed within 10 ft (3.05 m) of the point of emergence, as illustrated in Figure 9.13. Also, there shall be no coupling, fitting, box, or union between the point of emergence and the seal, except a listed explosionproof reducer at the seal is permitted. The wiring is required to be threaded Rigid Metal Conduit or threaded steel Intermediate Metal Conduit. *Exception 2* does permit Rigid Nonmetallic Conduit for the portion of the run that is at least 2 ft (600 mm) below grade, the same as permitted by the previous edition of the Code.

514.13: It is required in *514.11(A)* that all circuit conductors to a dispensing unit be capable of being disconnected simultaneously from the source of supply. This includes the grounded conductor if one exists. The new rule in *514.13* requires that this means of disconnection of circuit conductors supplying the dispensing unit be capable of being locked in the open position.

Article 516 Spray Application, Dipping, and Coating Processes

The major change is the addition of Zone designations to classification of areas in *516.3*. Now Class and Division as well as Class and Zone are shown for all of the areas covered. There are also several references to *Article 505*, which contains the rules for installing wiring and electrical equipment utilizing the Zone designations. There are no significant changes in the requirements.

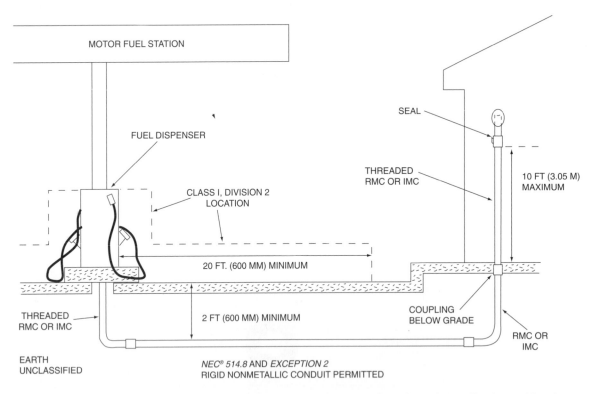

Figure 9.13   The earth beneath the Class I, Division 2 area of a motor fuel dispensing station is considered an unclassified area; however, there are requirements placed on the sealing of any conduit passing through the earth beneath the classified area.

WORKSHEET NO. 9—BEGINNING HAZARDOUS LOCATION WIRING

Mark the single answer that most accurately completes the statement based upon the 2005 Code. Also provide, where indicated, the Code reference that gives the answer or indicates where the answer is found, as well as the Code reference where the answer is found.

1. In a Class I, Division 1 area, Rigid Metal Conduit or Intermediate Metal Conduit is required to be installed into hubs or fittings with the number of threads fully engaged not less then:

    A. 3.   C. 7.   E. 10.
    B. 5.   D. 8.

    Code reference _____

2. A duplex receptacle is installed on a masonry wall in the service area of a commercial garage to provide power for portable tools. The commercial garage is used for the repair of gasoline and diesel powered engines. The receptacle is installed in a surface-mounted metal masonry box and supplied with EMT using set screw connectors. This installation is permitted provided the distance from the service area floor to the bottom of the box is not less then:

    A. 12 in. (300 mm).   D. 3 ft (900 mm).
    B. 18 in. (450 mm).   E. 5 ft (1.5 m).
    C. 24 in. (600 mm).

    Code reference _____

3. Current-interrupting contacts of a motor starter for an installation are not enclosed within a hermetically sealed chamber or immersed in oil. In a Class I, Division 1 hazardous location as shown in Figure 9.14, conduit seals are required to be installed in each conduit within a distance from the motor starter of not more then:

    A. 6 in. (150 mm).   D. 18 in. (450 mm).
    B. 10 in. (200 mm).   E. 24 in. (600 mm).
    C. 12 in. (300 mm).

    Code reference _____

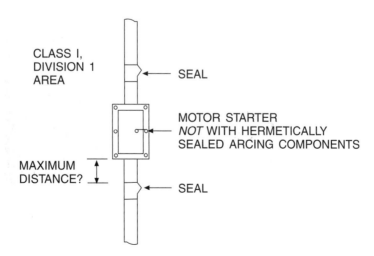

Figure 9.14   This motor starter is installed in a Class I hazardous location.

4. The wiring method permitted to be used in a Class I, Division 1 industrial hazardous location with restricted access to the public is threaded Rigid Metal Conduit, Type MI Cable, Type MC-HL Cable, Type ITC Cable, or:
   A. Electrical Metallic Tubing.
   B. enclosed gasketed busways.
   C. threaded Steel Intermediate Metal Conduit.
   D. Liquidtight Flexible Metal Conduit.
   E. Flexible Metal Conduit.

   Code reference_____

5. An area within a textile manufacturing building containing electrical operated equipment where combustible cotton fibers are frequently in the air is considered to be a:
   A. Class III, Division 1 location.
   B. Class II, Division 2 location.
   C. Class I, Division 2 location.
   D. Class II, Division 1 location.
   E. Class I, Division 1 location.

   Code reference_____

6. An area of a feed processing facility is considered a Class II, Division 1 location. A dust-ignitionproof motor starter is supplied with Rigid Metal Conduit where the last 10 ft (3.05 m) is a horizontal run with no fittings as shown in Figure 9.15. The fittings in the run of conduit do not contain splices and are listed as dusttight. A seal in the conduit run is:
   A. shall be placed within 18 in. of the dust-ignitionproof enclosure.
   B. required to be listed as explosionproof.
   C. required to be installed within 10 ft (3 m) of the dust-ignitionproof enclosure.
   D. not required because the horizontal conduit is considered an adequate seal.
   E. required to by any permanent compound installed in either end of the last portion of the conduit run.

   Code reference_____

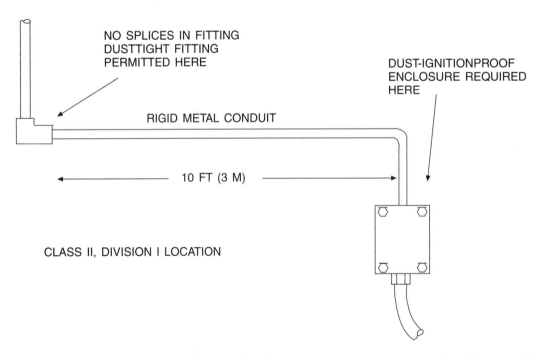

NO SPLICES IN FITTING
DUSTTIGHT FITTING
PERMITTED HERE

DUST-IGNITIONPROOF
ENCLOSURE REQUIRED
HERE

RIGID METAL CONDUIT

10 FT (3 M)

CLASS II, DIVISION I LOCATION

Figure 9.15   A 10-ft (3.05 m) horizontal length of Rigid Metal Conduit connects an unclassified fitting to a NEMA 9 enclosure in a Class II hazardous location.

7. Intrinsically safe wires in route to a Class I, Division 1 classified hazardous location are installed in Electrical Metallic Tubing. A power circuit not serving a hazardous location is run in Rigid Nonmetallic Conduit parallel with the intrinsically safe circuits. The minimum separation distance required to be maintained between the two raceway runs is:

    A. not specified in the Code.
    B. $^1/_2$ in. (13 mm).
    C. $^3/_4$ in. (19 mm).
    D. 1.97 in. (50 mm).
    E. 6 in. (152 mm).

    Code reference _____

8. A conduit run supplies wiring in a Class III, Division 1 classified hazardous location. In route to the Class III area the conduit enters and leaves a metal pull box. Bonding at the pull box is considered adequate:

    A. if a locknut is installed on the inside and outside and made up tight with no concentric knockouts in place.
    B. if a locknut is installed on the inside and outside and is made up tight.
    C. only if the pull box has threaded hubs.
    D. if bonding bushings are installed on the conduit entries and a bonding conductor connects the two bonding bushings together.
    E. if a locknut is installed on the inside and a metal bushing is installed on the outside.

    Code reference _____

9. To prevent sparking when fault current flows, all threaded conduit installed in classified hazardous locations shall have:

    A. all threaded connections treated with a conductive sealing compound.
    B. a copper bonding conductor run either inside or on the outside of the conduit.
    C. all threaded connections made up wrench tight.
    D. bonding jumpers installed when circuits operate at less than 250 volts.
    E. exposed threads coated with corrosion protection after installations in wet locations.

    Code reference _____

10. Oil immersion is a protection technique that is permitted to be used in areas classified as:

    A. Class II, Zone 1 or 2.
    B. Class I, Zone 0, 1, or 2.
    C. Class I, Zone 1 only.
    D. Class I, Zone 0 or 1.
    E. Class I, Zone 1 or 2.

    Code reference _____

11. Examine the seal shown in Figure 9.16. Unless specifically approved for a higher fill, the cross-sectional area of the conductors in a sealing fitting installed in a Class I hazardous location shall not exceed the cross-sectional area of a Rigid Metal Conduit of the same trade size by more then:

    A. 20%.
    B. 25%.
    C. 40%.
    D. 60%.
    E. 80%.

    Code reference _____

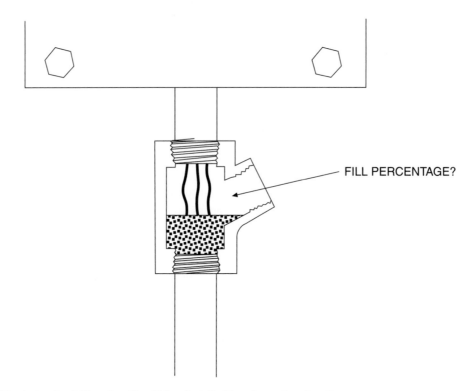

FILL PERCENTAGE?

Figure 9.16   A standard Class I sealing fitting installed in a hazardous location.

12. Conductors supplying a gasoline-dispensing unit are run underground in Rigid Metal Conduit from the service station building to the dispenser. A seal is required to be installed in this conduit run:
   A. only at the dispenser where the conduit emerges from the concrete.
   B. only if the service station area where the conduit emerges from the earth is not under positive ventilation.
   C. only at the service station where the conduit emerges from the concrete.
   D. within 10 ft (3 m) of the boundary where the conduit leaves the Class I, Division 2 area.
   E. both at the dispenser and at the service station where the conduit emerges from the concrete.

   Code reference _____

13. Equipment installed in classified hazardous locations shall be marked to show the class, operating temperature or temperature class, and the flammable material:
   A. name.   C. division.   E. group.
   B. classification. D. zone.

   Code reference _____

14. A roof is installed above an area containing gasoline dispensers. High-intensity discharge luminaires (lighting fixtures) are connected to boxes with a short length of flexible wiring. A wiring method not permitted to be installed for this purpose in this location is:
   A. Flexible Metal Conduit.
   B. Type MC Cable.
   C. Type AC Cable.
   D. Electrical Nonmetallic Tubing.
   E. Nonmetallic-Sheathed Cable, Type NM.

   Code reference _____

15. Paint spraying is conducted in a factory in an open-front booth that is equipped with a ventilation system that is interlocked with the spray application equipment. Luminaires (lighting fixtures) are installed outside of the spray booth and 2 ft (600 mm) in front of the opening to illuminate the interior of the booth. Luminaires (lighting fixtures) suitable for installation in an unclassified area are permitted for this application provided they are mounted a minimum height above the top edge of the booth opening a distance not less than:

A. 3 ft (914 mm).

B. 5 ft (1.525 m).

C. 8 ft (2.438 m).

D. 10 ft (3.048 m).

E. 12 ft (3.658 m).

Code reference _____

WORKSHEET NO. 9—ADVANCED HAZARDOUS LOCATION WIRING

Mark the single answer that most accurately completes the statement based upon the 2005 Code. Also provide, where indicated, the Code reference that gives the answer or indicates where the answer is found, as well as the Code reference where the answer is found.

1. Equipment that is supplied having metric threaded entries:
   A. shall not be permitted to be installed.
   B. shall be provided with an adapter for connection to NPT threads.
   C. shall be installed only with conduits with metric threads.
   D. is permitted to be installed with conduits having metric threads.
   E. is only required to be identified as having metric threads.

   Code reference _____

2. Luminaires (lighting fixtures) installed in a Class II, Division 1 location where grain dust that can dehydrate and carbonize is present shall have a surface temperature classification not exceeding the ignition temperature or:
   A. T1.   C. 165°C.   E. T3A.
   B. 500°C.   D. T2C.

   Code reference _____

3. A gasoline dispenser at a service station is supplied with 120-volt power from a circuit breaker panelboard where the circuit breaker will act as the disconnecting means for the dispenser. A locking mechanism that stays in place with or without a lock is installed at each circuit breaker supplying a fuel dispenser. The circuit breaker permitted to serve as the dispenser disconnect is:
   A. a single-pole circuit breaker.
   B. a single-pole, ground-fault circuit-interrupter breaker.
   C. a single-pole, arc-fault interrupter.
   D. an instantaneous-trip circuit breaker.
   E. a switched-neutral circuit breaker.

   Code reference _____

4. Unless specifically stated otherwise, intrinsically safe wiring installations are:
   A. permitted as open conductors without a cable jacket or raceway protection.
   B. required to be installed in Rigid Metal Conduit or steel Intermediate Metal Conduit.
   C. required to be installed in metal raceway or as metallic sheathed cable.
   D. permitted to be run in Electrical Metallic Tubing.
   E. required to be Type MI or Type MC when run as cable wiring.

   Code reference _____

5. Rigid Metal Conduit or steel Intermediate Metal Conduit is required to be installed into hubs or fittings with 5 threads fully engaged for fuel wiring in a hazardous location classified as:
   A. Class I, Division 1 or 2.
   B. Class I, Division 1.
   C. Class I, Division 1 or 2, or Class II, Division 1.
   D. Class I, Division 1 or 2, or Class II, Division 1 or 2.
   E. Class I, Class II, or Class III, both Divisions 1 or 2.

   Code reference _____

6. A conduit run supplies wiring for a Class III, Division 1 hazardous location, as shown in Figure 9.17. The conduit contains size 8 AWG, Type THW copper conductors, which are protected by a 50-ampere overcurrent device in a 200-ampere panelboard. The minimum size of copper bonding jumper required to be installed in this pull box is:

A. 12 AWG.   C. 8 AWG.   E. 4 AWG.

B. 10 AWG.   D. 6 AWG.

Code reference _____

7. A magnetic motor starter in a NEMA 1 enclosure is to be installed on a wall near the open end of a paint spray booth. The paint spray booth is equipped with ventilation that is interlocked with the sprayer such that the sprayer will not operate unless the ventilation is operating. The minimum horizontal distance permitted from the edge of the paint spray booth opening to the NEMA 1 enclosure is:

A. 3 ft (900 mm).   D. 20 ft (6 m).

B. 5 ft (1.5 m).   E. not specified.

C. 10 ft (3 m).

Code reference _____

8. A parking garage with no mechanical ventilation provided is used for the parking and storage of gasoline and other fuel powered vehicles. If no maintenance is performed in this garage, the space up to a height of 18 in. (457 mm) above the floor is considered to be:

A. an unclassified location.

B. a Class I, Division 1, location.

C. a Class I, Division 2, location.

D. a Class I, Zone 1, location.

E. a nonspecified area.

Code reference _____

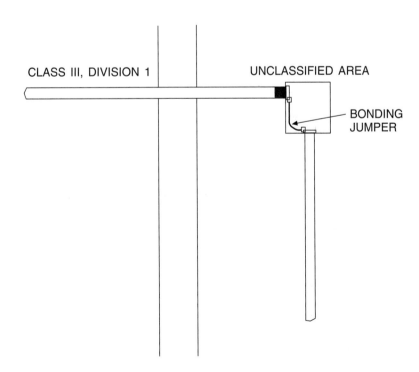

Figure 9.17   **A pull box placed outside of a Class III Division 1 hazardous location.**

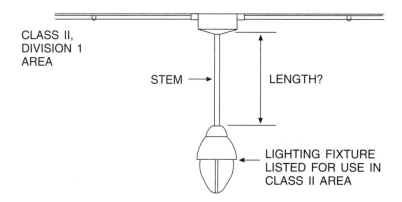

Figure 9.18 The Class II listed luminaire (lighting fixture) is installed in a Class II hazardous location with a Rigid Metal Conduit stem from the box to the luminaire (lighting fixture).

9. A luminaire (lighting fixture) is installed in a Class II, Division 1 area of a grain elevator, as shown in Figure 9.18. The box is properly supported and listed for use in a Class II location. A Rigid Metal Conduit stem is threaded into the cover of the box and is used as the sole support for the luminaire (lighting fixture). There is no flexible connection between the stem and the box, and there is no lateral support for the stem. The maximum length of stem permitted is:
A. 6 in. (150 mm).   D. 24 in. (600 mm).
B. 12 in. (300 mm).   E. 3 ft (900 mm).
C. 18 in. (450 mm).

                           Code reference _____

10. Refer to the illustration of the wiring to the outdoor gasoline dispenser in Figure 9.19. The Class I, Division 2, hazardous location is from the ground level to a height of 18 in. (450 mm) and extends outward from the edge of the gasoline dispenser in all directions a minimum distance of:
A. 3 ft (900 mm).   D. 25 ft (7.5 m).
B. 10 ft (3 m).   E. 50 ft (15 m).
C. 20 ft (6 m).

                           Code reference _____

11. Refer to Figure 9.19 for the wiring supplying the gasoline dispenser where condensed gasoline vapors may collect on the conductors. For the wires inside of the dispenser and run in the underground conduit to the dispenser, electrical conductors:
A. shall not be smaller than size 10 AWG.
B. are only required to have a moisture-resistant type of insulation.
C. shall be marked on the insulation suitable for use in a Class I location.
D. are required to have lead covering over the insulation.
E. shall be identified for these conditions such as being gasoline resistant.

                           Code reference _____

12. Fixed wiring installed above a Class I hazardous area of a bulk storage plant is not permitted to be run as:
A. Intermediate Metal Conduit.
B. Electrical Metallic Tubing.
C. Schedule 80 PVC Rigid Nonmetallic Conduit.
D. Liquidtight Flexible Nonmetallic Conduit.
E. Metal Clad, Type MC Cable.

                           Code reference _____

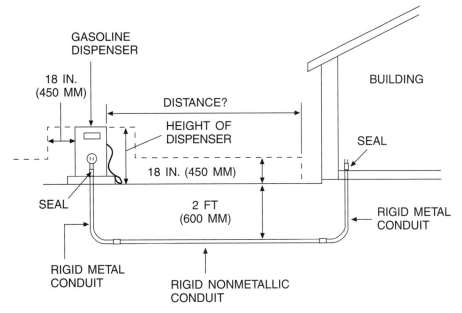

Figure 9.19   Wiring is run underground from the service station to a gasoline dispenser using Rigid Nonmetallic Conduit in direct contact with the earth.

13.   A trade size $^3/4$ (21), Rigid Metal Conduit passes from a Class I, Division 1, hazardous location to an unclassified location. A sealing fitting listed for only a 25% fill is installed on the nonhazardous side of the boundary. The maximum number of size 10 AWG, Type THWN copper conductors that are permitted to be installed in the fitting is:

A.   4.   C.   6.   E.   8.
B.   5.   D.   7.

Code reference _____

14.   For instances when a structure or location is classified by both the Zone and Division system, it is permitted to have Class I, Division 2 locations border, but not overlap a:

A.   Class I, Zone 1 or 2 location.
B.   Class I, Zone 0 location.
C.   Class I, Zone 1 location.
D.   Class I, Zone 2 location.
E.   any location installed using the Zone system.

Code reference _____

15.   During the construction process of an aircraft hangar it is deemed necessary to install a conduit under the floor of the hangar. The proposed conduit would be buried 24 in. (600 mm) below grade level and would be encased in concrete with a thickness of 2 in. (50 mm). The conduit would pass under the hangar from one unclassified location to another unclassified location, as shown in the Figure 9.20. The conduit:

A.   is permitted to be Rigid Nonmetallic Conduit for the below grade installation and Rigid Metal Conduit or steel Intermediate Metal Conduit where it emerges from the earth.
B.   is only permitted to be installed in this manner by use of Rigid Metal Conduit or Steel Intermediate Metal Conduit.
C.   is required to be relocated so it does not pass under the hangar.

D. system is not permitted to be installed in this location if the raceway is Rigid Nonmetallic Conduit.

E. is permitted to be installed as Rigid Nonmetallic Conduit for the entire run including the points where the conduit emerges from the ground in the unclassified locations.

Code reference_____

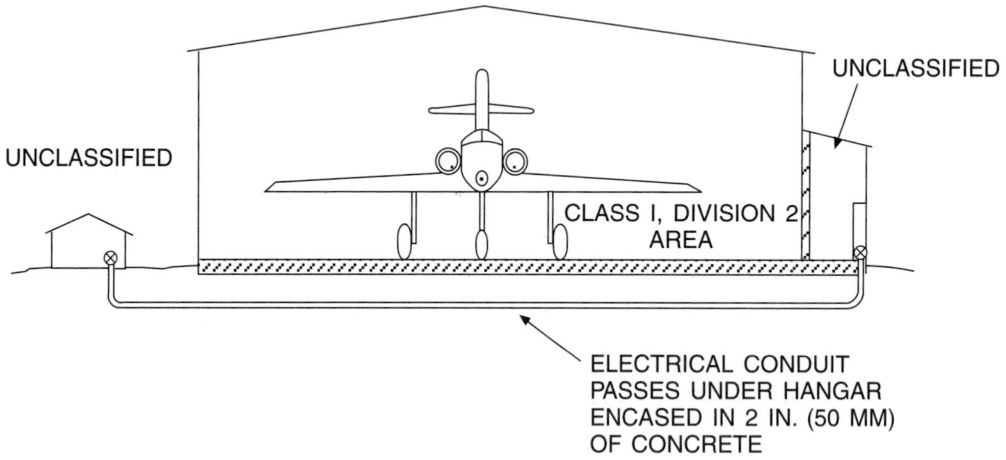

Figure 9.20   An electrical conduit passes completely under an aircraft hanger and emerges from the ground outside of the hazardous location.

UNIT 10

Health Care Facilities

OBJECTIVES

After completion of this unit, the student should be able to:

- name two types of patient care areas of a health care facility.

- describe when an anesthetizing location is also considered to be a hazardous location.

- give examples of the types of electrical equipment to be supplied by the life safety branch of the essential electrical system of a health care facility.

- give an example of the type of electrical equipment to be supplied by the critical branch of the essential electrical system of a health care facility.

- explain the purpose of a reference grounding point for a patient care area of a health care facility.

- name a specific location of a health care facility where a receptacle outlet is required to be listed for hospital use.

- describe how equipment grounding is required to be installed from a receptacle outlet at a bed location of a critical care area to the service panel that supplies the receptacle.

- state the minimum number of receptacle outlets required at the bed location of a general care area of a hospital.

- state the minimum number of circuits required at a patient bed location of a critical care area to be supplied by the critical branch of the essential electrical system.

- answer wiring questions about installations in health care facilities from Code *Articles 517* or *660*.

- state at least three significant changes that occurred from the 2002 to the 2005 Code for *Articles 517* or *660*.

CODE DISCUSSION

An essential step to understanding the wiring requirements of a health care facility is knowledge of the definitions at the beginning of *Article 517*. This article addresses the following types of health care facilities: (1) clinics, medical offices, dental offices, and outpatient facilities; (2) nursing homes and residential custodial care facilities; and (3) hospitals. Other articles also are essential to the wiring installation for health care facilities. These are *Article 700,* which covers emergency systems and illumination for building egress, *Article 701,* covering legally required standby electrical systems, and *Article 760* on fire alarm systems.

Article 517 covers specific electrical wiring requirements for health care facilities that are not covered elsewhere in the Code. *Part I* of this article provides definitions necessary for the understanding and uniform application of the Code to health care facilities. In a multifunction building, the appropriate part of this article shall apply to an area with a specific function.

Part II specifies the wiring methods that can be used in health care facilities. *NEC®517.13(A)* specifies that branch-circuits supplying patient care areas shall be run in metal raceway such as Rigid Metal Conduit, Intermediate Metal Conduit, Electrical Metallic Tubing, or cable assemblies such as Type AC, Type MI Cable, or Type MC where the outer metal jacket is an approved grounding means. Grounding and bonding requirements for receptacles and equipment in patient care areas are also covered in *Part II. NEC® 517.13(B)* requires that all receptacles supplying power at over 100 volts to patient locations shall be grounded with an insulated copper equipment grounding conductor run with the circuit conductors in metal raceway. The insulated copper equipment grounding conductor is also permitted to be run in Type AC, Type MC, or Type MI

343

Cable that has a metal sheath that is approved for grounding. With electrical equipment frequently attached to patients, redundant grounding is important. The insulated copper conductor acts as a path for fault current as well as the metal conduit or approved metal sheath of the cable. This redundancy in the grounding also reduces the resistance of a ground fault path which helps keep voltages during a fault at extremely low levels.

Generally a patient location is served by branch-circuits from more than one panelboard. It is likely that a patient could be connected to electrical equipment that is supplied from at least two separate panelboards. It is necessary that the equipment grounding bus in each panelboard be at the same potential. This is accomplished by bonding together of the equipment grounding bus of any panelboard that serves receptacles in the same patient location. *NEC® 517.14* specifies a minimum size 10 AWG copper bonding conductor be used between equipment grounding terminals in panelboards serving the same patient location.

The patient vicinity is defined in *517.2* as the area within 6 ft (1.8 m) of the patient bed. In a general care area, a minimum of four receptacles is required to serve the patient bed location. These may be single receptacles, or they may be multiple receptacles on the same yoke. For example, a duplex receptacle would count as two receptacles. The receptacles are required to be listed as hospital grade. A minimum of two branch-circuits is required to serve a patient bed location in a general care area. At least one of the branch-circuits is required to be supplied by the normal electrical system and one from the emergency system, as stated in *517.18*, unless the room is supplied from two separate emergency power sources. These branch-circuit and receptacle requirements are illustrated in Figure 10.1. *Exception 3* of *517.18(A)* permits both required circuits of a general care area to be supplied from the emergency system. The stipulation is that the two circuits be supplied from two completely separate emergency systems with separate transfer switches and panelboards. An installation meeting these requirements for a general care area is illustrated in Figure 10.2.

A critical care area is required to have six receptacles at the patient bed location, as shown in Figure 10.3. At least two circuits are required, one of which is required to be supplied from the emergency electrical system, as stated in *517.19*. The receptacles are required to be listed as hospital grade. At least one receptacle at the patient bed location is required to be supplied from the normal power system or other emergency system.

Part III provides the requirements for the types of electrical systems for a health care facility. There is a normal power system providing power for various circuits throughout the health care facility. An emergency electrical system is required in certain types of health care facilities, such as hospitals. This emergency electrical system is permitted to consist of a life safety branch and a critical branch. These systems are illus-

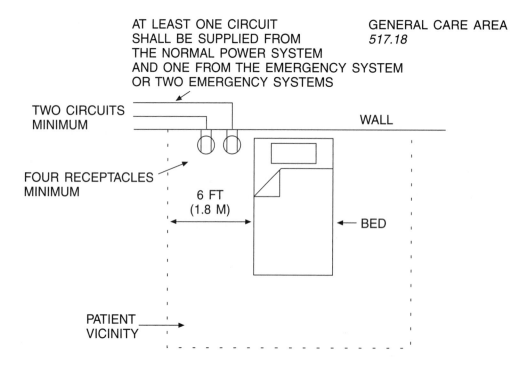

Figure 10.1 **Four receptacles and at least two circuits are required to serve a patient bed location of a general care area.**

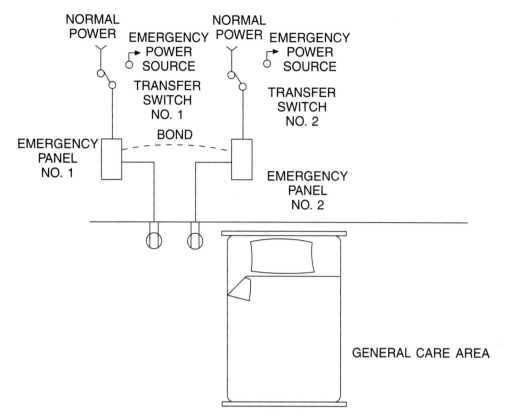

NEC® 517.18(A) EXCEPTION 3 BOTH REQUIRED CIRCUITS
PERMITTED FROM EMERGENCY SYSTEM IF SUPPLIED
THROUGH SEPARATE TRANSFER SWITCHES

Figure 10.2   The two required bed location circuits in a general care area are permitted to be from the emergency system if the circuits are supplied by separate transfer switches.

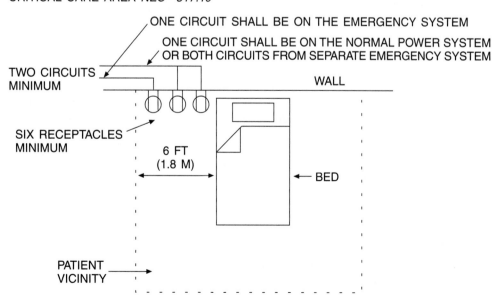

Figure 10.3   Six receptacles are required to serve the patient bed location of a critical care area. These are served by a minimum of two branch circuits, one of which is supplied from the emergency electrical system.

ESSENTIAL ELECTRICAL SYSTEM
OF A HOSPITAL *NEC® 517.30*

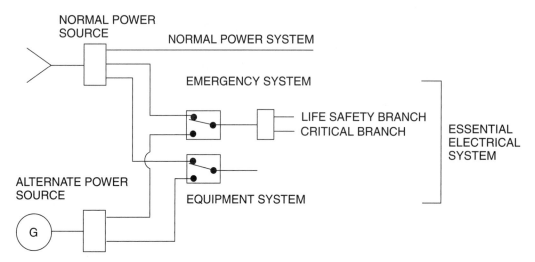

Figure 10.4   The essential electrical system consists of the emergency system and the equipment system. If the essential electrical system demand load is greater than 150 kVA, then each branch is required to be served by a separate transfer switch.

trated in Figure 10.4 for a hospital with an essential electrical system demand load not exceeding 150 kVA. It is permitted to serve both the emergency system and the equipment system from the same transfer switch. If the demand load for the essential electrical system exceeds 150 kVA, then a separate transfer switch is required for the equipment system, the life safety branch, and the critical branch. In addition, an equipment system shall be provided in most types of health care facilities. All three of these systems make up the essential electrical systems. For a hospital, these systems are covered in *517.30* through *517.35*. For nursing homes and limited care facilities, these systems are covered in *517.40* through *517.44*.

Part IV deals with wiring in inhalation anesthetizing locations. Where flammable anesthetics are administered, the area is considered a classified hazardous location. An isolated power system is required where flammable anesthetics are used. Wiring requirements for other than hazardous anesthetizing locations are found in *517.61(C)*.

Part V deals with X-ray installations. Rating of supply conductors, disconnect, and overcurrent protection is covered by this part. Wiring of the control circuit and grounding is also covered. X-ray units have a momentary current rating and a long-time current rating. *NEC® 517.72(A)* requires the disconnecting means to have a current rating not less than 50% of the momentary input rating of the X-ray equipment or 100% of the long-time rating, whichever is greater. *NEC® 517.73(A)(1)* requires the supply conductors to be sized in the same manner as the disconnecting means.

Part VI covers requirements for communications, signaling systems, data systems, fire alarm systems, and systems operating at less than 120 volts. The key issue here is grounding. It is important that a patient not be exposed to hazard through the grounding of electrical equipment.

Isolated power systems are covered in *Part VII*. Isolated power systems are required to supply circuits within areas where flammable anesthetizing agents are used. The requirements on which isolated systems are required are found in *517.61(A)(1)*. The installation of the isolated power system is covered in *517.160*. Circuits serving an anesthetizing location where flammable anesthetics are used are required to be isolated from the electrical distribution system in the building. This can be accomplished by operating the circuits ungrounded. A common means of establishing an ungrounded electrical system is by use of an isolation transformer that has no connection between the primary and secondary windings. *NEC® 517.160(A)(1)* requires all isolated circuits to be controlled by a switch that opens each ungrounded conductor. Each operating room is required to have the circuits supplied by a separate isolation transformer. An induction room that serves an operating room is permitted to have the circuits supplied from the same isolation transformer as the operating room. The isolated circuits are generally 125-volt, single-phase, 2-wire. The conductors are required to be identified by the colors orange and brown. When supplying 125-volt, 15- or 20-ampere receptacles, the orange conductor shall be connected to the silver-colored terminal where a grounded conductor would normally have been terminated. This rule found in *517.160(A)(5)* is illustrated in Figure 10.5.

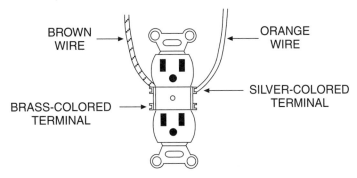

NEC® 517.160(A)(5) FOR 125-VOLT, 15- OR 20-AMPERE
RECEPTACLES ON ISOLATED POWER SYSTEMS,
CONNECT THE ORANGE WIRE TO THE SILVER TERMINAL

BROWN
WIRE

ORANGE
WIRE

SILVER-COLORED
TERMINAL

BRASS-COLORED
TERMINAL

Figure 10.5   The orange conductor of an isolated 125-volt circuit connected to a 15- or 20-ampere receptacle shall be connected to the silver-colored terminal intended for connection of the grounded conductor in a normal power system.

Article 660 covers the installation of X-ray equipment for nonmedical and nondental use. The article provides specifications for minimum circuit rating and wire size, and the disconnect for the equipment. Wiring of the control system and grounding of the equipment are also covered.

Article 680, Part IV covers the wiring to and in the area around therapeutic pools and tubs in health care facilities. Specifications are placed on receptacles and luminaires (lighting fixtures) in the area. Grounding and bonding of equipment is the main emphasis of this part. Bonding and grounding of metal parts within 5 ft (1.5 m) of the inside edge of the hydromassage unit is important for safety. The permitted methods of grounding and bonding are given in *680.62(C)* and *(D)*. Therapeutic tubs and hydromassage tubs are required by *680.62(A)* to be ground-fault circuit-interrupter protected. *NEC® 680.62(E)* requires that any receptacle within 5 ft (1.5 m) is also required to be ground-fault circuit-interrupter protected.

WIRING IN HOSPITALS

Several very important issues deal with wiring in health care facilities and hospitals in particular. Reliability of electrical supply is necessary for equipment and lighting needed for human life support. Lighting and exit marking are needed for egress from the building in case of an emergency. Special measures must be taken to ensure that equipment in a patient area is grounded in such a way that differences in voltage will not be present between equipment such that a hazard to the patient will be created.

Definition of Areas Within a Hospital

Patient vicinity: The area within 6 ft (1.8 m) horizontally from the perimeter of the bed, and within 7^1/$_2$ ft (2.3 m) above the floor. Figure 10.1 and Figure 10.3 illustrate the patient vicinity.

Patient bed location: The intended location of the patient bed, or the procedure table of a critical care area.

General care area: An area where the patient is not generally connected to electrical equipment. If so, the connection is basically external to the body, and there is no apparent hazard of electrical current affecting life-supporting organs of the body.

Critical care area: This is an area where a patient may be put in intimate contact with electrical equipment that may produce a real hazard of electrical shock to essential body organs, or where the reliability of the equipment is necessary for life support. See the Code for specific locations that come within this definition.

Wet location: An area that is made intentionally wet for some specific purpose. Wet locations are covered in *517.20*. Refer to *Article 680, Part VI* for installation requirements for therapeutic pools and tubs in health care facilities.

Anesthetizing location: An area intended for administration of flammable or nonflammable inhalation anesthetic agents in the course of examination or treatment.

Essential electrical system: This electrical system is required to have an alternate source of power. The emergency electrical system and the equipment electrical system are part of the essential electrical system. The building electrical system is made up of the normal power system and the essential electrical system. This system includes lighting circuits and equipment considered necessary for minimal operation and life safety. Power is provided for the life safety branch and the critical branch. See Figure 10.4. There is usually a delay before power is provided to the equipment systems branch.

Life safety branch: This system provides light and power for the emergency systems of *Article 700,* such as lighting for egress, exit signs, and alarm and communications systems. This is covered in *517.32* and is illustrated in Figure 10.4.

Critical branch: This system supplies power in selected areas for illumination and receptacles considered essential for protection of life. For hospitals, specific locations are specified in *517.33.* A patient bed location in a critical care area shall have at least one branch-circuit supplied from the critical branch, as shown in Figure 10.4.

Hospital Receptacle Requirements

Hospital grade receptacles are required to be installed in patient care areas, above the classified portion of an anesthetizing location, and in certain other locations. These devices are usually identified by a visible green dot, as shown in Figure 10.6. Some devices are made of a clear material to allow employees to determine visually if an electrical malfunction occurs. Required hospital grade receptacles for the general care patient area are found in *517.18(B),* for the critical care area in *517.19(B)(2),* and for hospital use above hazardous anesthetizing locations in *517.61(B)(5).*

Receptacles with equipment grounding terminals insulated from the yoke are permitted to be used in hospitals for such devices as electronic equipment in which electrical noise may be a problem. These receptacle outlets with the insulated grounding terminals are required to be distinctively identified from the front. They have an orange triangle visible on the receptacle, as shown in Figure 10.6. Rules on the installation of isolated ground receptacles are found in *250.146(D), 406.10(D),* and *517.16.* They shall also be identified as hospital grade. In the case of an insulated ground receptacle, grounding of the equipment plugged into the receptacle is dependant only on the insulated copper equipment grounding conductor. There is no redundancy for the equipment ground. These receptacles should only be used where there is a definite need for an isolated ground to prevent electrical noise of sensitive equipment.

In the case of a critical care area, there shall be at least one branch-circuit from the emergency system, and that receptacle or receptacles shall be identified. This is necessary to make sure essential equipment is attached to an emergency circuit if normal power is lost. It is also required that the location of the branch-circuit overcurrent device panelboard be indicated so the device can be reset quickly if necessary to maintain operation of life-saving equipment. These requirements are found in *517.19(A).*

Grounding in Hospitals

Proper installation of the equipment grounding system, particularly in hospitals, is extremely important. The goal, especially in critical care areas, is to prevent any two pieces of metal equipment that may be contacted simultaneously by a person from developing a voltage difference harmful to a patient. Receptacles

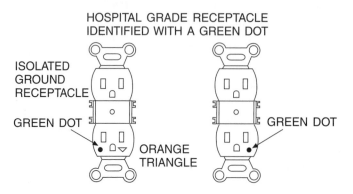

Figure 10.6   A receptacle identified as hospital grade may have the equipment grounding terminal insulated from the yoke of the receptacle and is identified by a green dot and an orange triangle.

PATIENT EQUIPMENT BONDING POINT *NEC® SECTION 517.19(C)*

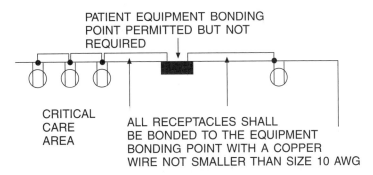

PATIENT EQUIPMENT BONDING
POINT PERMITTED BUT NOT
REQUIRED

CRITICAL
CARE
AREA

ALL RECEPTACLES SHALL
BE BONDED TO THE EQUIPMENT
BONDING POINT WITH A COPPER
WIRE NOT SMALLER THAN SIZE 10 AWG

Figure 10.7   If a patient equipment grounding point is provided in a critical care area, the receptacle outlets are required to be bonded to that point.

and fixed equipment in patient care areas shall be grounded with an insulated copper equipment grounding conductor. Several types of grounding points are discussed in the Code and are defined as follows:

- Patient equipment grounding point: This is a grounding bus with plug jacks available for the grounding of equipment to be operated in the patient bed location and listed for the purpose. The patient equipment grounding point is not required, as stated in *517.19(C)*. A patient equipment grounding point is illustrated in Figure 10.7.
- Reference grounding point: This is a terminal bus that is a convenient collecting point for equipment grounding conductors and bonding wires. A reference grounding point is required for circuits serving a critical care area. This requirement is found in *517.19(B)(2)*. A reference grounding point is illustrated in Figure 10.8. The reference grounding point is defined in *517.2* as the ground bus in a panelboard serving a patient care area. For an operating room it is the grounding bus of the isolated grounding system. Patient care areas are frequently served from two separate panelboards. This means the grounding bus in each panelboard is a reference grounding point. *NEC® 517.14* requires the grounding bus of any panelboard serving circuits in the same patient vicinity to be bonded together with an insulated copper conductor not smaller than size 10 AWG.

The patient equipment grounding point and the reference grounding point are frequently the same point for critical care areas. When the room is large, these grounding and bonding points may be separated. A patient equipment grounding point is permitted but not required. If installed, the patient equipment grounding point shall be connected directly to the reference grounding point by means of a continuous length of insulated copper conductor. The reference grounding point of the room is bonded to the panelboard grounding bus supplying power to the room. Circuits in the critical care patient area are supplied from a panelboard of the normal power system and a panelboard of the essential electrical system. The equipment grounding buses of each of these panelboards serving the patient care area are required to be bonded together by means of a continuous length of insulated copper conductor, as required by *517.14*. This is illustrated in Figure 10.9.

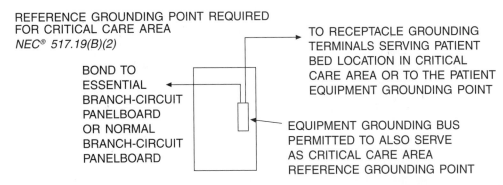

REFERENCE GROUNDING POINT REQUIRED
FOR CRITICAL CARE AREA
NEC® 517.19(B)(2)

TO RECEPTACLE GROUNDING
TERMINALS SERVING PATIENT
BED LOCATION IN CRITICAL
CARE AREA OR TO THE PATIENT
EQUIPMENT GROUNDING POINT

BOND TO
ESSENTIAL
BRANCH-CIRCUIT
PANELBOARD
OR NORMAL
BRANCH-CIRCUIT
PANELBOARD

EQUIPMENT GROUNDING BUS
PERMITTED TO ALSO SERVE
AS CRITICAL CARE AREA
REFERENCE GROUNDING POINT

Figure 10.8   A reference grounding point is required to be provided for circuits supplying the patient vicinity of a critical care area.

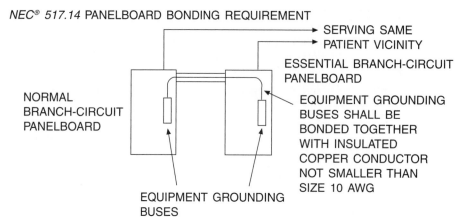

NEC® 517.14 PANELBOARD BONDING REQUIREMENT

SERVING SAME
PATIENT VICINITY

ESSENTIAL BRANCH-CIRCUIT
PANELBOARD

NORMAL
BRANCH-CIRCUIT
PANELBOARD

EQUIPMENT GROUNDING
BUSES SHALL BE
BONDED TOGETHER
WITH INSULATED
COPPER CONDUCTOR
NOT SMALLER THAN
SIZE 10 AWG

EQUIPMENT GROUNDING
BUSES

Figure 10.9   The equipment grounding bus of a panelboard serving circuits in a patient vicinity shall be bonded.

A patient bed location is permitted to have receptacles supplied from both the emergency and the normal power panelboards. The equipment grounding buses of these panelboards shall be bonded together to make sure a difference in voltage between equipment at the bed location cannot develop to a sufficient level to be harmful to the patient.

Hospital Required Electrical Systems

Essential electrical systems must be connected to the normal power system and the standby power source through a transfer switch. The essential electrical system may consist of a separate critical branch and a life safety branch. If there is a life safety branch, it is not permitted to occupy the same raceway or cable with other wiring. This requirement is found in *517.30(C)(1)*. The critical branch and the life safety branch are not permitted to share the same enclosures, raceways, or cables. The emergency wiring system in a hospital shall be in metal raceway. The types of loads permitted to be connected to the life safety branch of a hospital are limited to egress illumination, exit signs, alarm systems, emergency communication systems, and illumination at the generator location. It is important that receptacles supplied by the emergency electrical system be easily recognized. It is required that these receptacles have a distinctive color or be supplied with faceplates that have a distinctive color. Frequently the receptacles supplied by the emergency system are red.

Power circuits within an anesthetizing location classified as a hazardous location shall be isolated from the normal distribution system. The lower 5 ft (1.52 m) of an anesthetizing room where flammable agents are used shall be considered to be a Class I, Division 1 hazardous location, as stated in *517.60(A)(1)*.

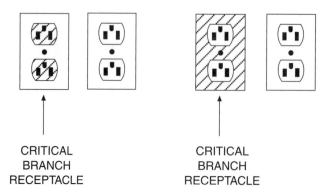

NEC® 517.30(E) CRITICAL BRANCH RECEPTACLE OR
FACEPLATE DISTINCTIVE COLOR MARKING REQUIRED

CRITICAL
BRANCH
RECEPTACLE

CRITICAL
BRANCH
RECEPTACLE

Figure 10.10   Receptacles supplied from the critical branch or the life safety branch of the emergency system are required to be either distinctively marked or have faceplates with a distinctive color.

MAJOR CHANGES TO THE 2005 CODE

These are the changes to the 2005 *NEC®* that correspond to the Code sections studied in this unit. The following analysis explains the significance of the changes from the 2002 to the 2005 Code only, and this analysis is not intended to be used in place of the Code. Refer to the actual section of the 2005 Code for the exact wording and meaning of each section discussed. Changes are indicated in the Code with a vertical line in the margin. If material was deleted or moved to another location in the Code, the location of the deletion is indicated with a dark dot in the margin.

Article 517   **Health Care Facilities**

517.14: When a patient location is supplied circuits from the critical branch panelboard and the normal power panelboard, the grounding bus of each panelboard is required to be bonded together with a continuous insulated copper wire not smaller than size 10 AWG. Some patient locations, such as a critical care area, may be supplied circuits from two separate emergency systems and no circuits from the normal power system. A new sentence was added to require the grounding bus of separate emergency panelboards to be bonded together when they both provide circuits to the same patient location.

517.17(A): A solidly grounded electrical system operating at more than 150 volts to ground and rated 1000 amperes or more is required to be provided with equipment ground-fault protection. For a hospital, it was also required to provide equipment ground-fault protection on the down-line feeders operating at more than 150 volts to ground. This is illustrated in Figure 10.11. The purpose of this procedure was to limit an outage due to a ground-fault to the feeder involved and not the entire hospital. This new subsection *(A)* points out that this requirement is intended to apply only to hospitals and similar facilities with critical care areas or electrical life-support equipment.

517.18(C): In a pediatric facility, the description of where receptacles are to be of the temper-resistant type or be equipped with tamper-resistant covers is more specific. The areas covered are patient rooms, bathrooms, playrooms, activity rooms, and patient care areas.

517.26: This is a new section that points out that circuits and wiring of the essential electrical system are required to meet the requirements of *Article 700* on emergency systems except where amended by rules in *Article 517*.

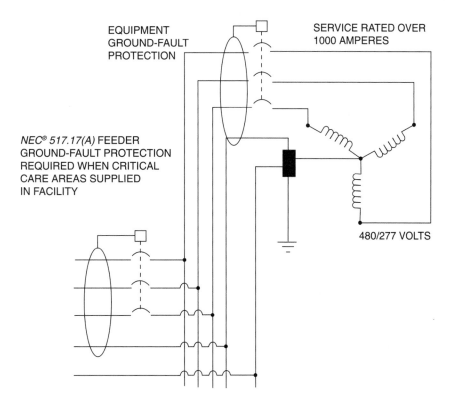

Figure 10.11   The requirement for down-line equipment ground-fault protection on feeders when the service is provided with equipment ground-fault protection only applies in the case where the health care facility supplies power to a critical care area or electrically powered life-support equipment.

517.30(C)(3): A new sentence was added to make it clear that the requirements of *517.13(A)* and *(B)*, with respect to grounding, also apply to emergency circuits supplying patient care areas. Raceway or the metallic sheath of cable is required to qualify as an equipment grounding conductor as well as the required addition of an insulated equipment grounding conductor.

517(C)(3)(2): Jacketed metallic cable assemblies listed for installation in concrete are now permitted to be installed encased in a minimum of 2 in. (50 mm) of concrete to supply emergency system circuits. Emergency system circuits in a hospital include circuits on the critical branch and the life safety branch. Jacketed Type MC cable that is marked for direct burial is also listed as suitable for encasement in concrete.

517.30 (C)(3)(3): Listed flexible metal raceway and listed metallic-sheathed cable is now permitted to be fished into existing walls or ceilings that are not otherwise accessible to supply outlets on emergency circuits. Grounding requirements of *517.13* must be met if the circuits supply outlets in patient care areas.

517.34(C): This is a new section and requirements in the Code, but it is not actually a new requirement since it was a requirement in another NFPA standard. AC electrically powered equipment necessary for the operation of the emergency electrical generating system is required to be connected to the emergency system without a time delay.

517.35(B)(4): An essential electrical system is required to be supplied by a minimum of two independent power sources. One source is usually the utility supply. Now, a battery system located on the premises is permitted to be one of the required sources of power for the essential electrical system.

WORKSHEET NO. 10—BEGINNING HEALTH CARE FACILITIES

Mark the single answer that most accurately completes the statement based upon the 2005 Code. Also provide, where indicated, the Code reference that gives the answer or indicates where the answer is found, as well as the Code reference where the answer is found.

1. For a general care area in a hospital, as illustrated in Figure 10.12, each patient bed location is required to be provided with two branch-circuits:
   A. one of which shall originate from the emergency electrical system.
   B. both of which shall originate from the emergency electrical system.
   C. both of which shall originate from the normal electrical system.
   D. with not less than two receptacles connected to each branch-circuit.
   E. that are permitted to originate from any electrical system desired.

   Code reference _____

2. The minimum number of receptacles serving the patient bed location of a critical care area of a hospital is:
   A. not specified.
   B. four.
   C. six.
   D. eight.
   E. six duplex receptacles for a total of twelve.

   Code reference _____

3. All emergency system receptacles at the patient bed location of a critical care area of a hospital:
   A. shall be supplied by a minimum of two emergency circuits.
   B. are not permitted to originate from the same panelboard.
   C. are required to be supplied by a 20-ampere branch-circuit.
   D. are required to be identified as to panelboard and circuit number from which they are supplied.
   E. are to be divided so at least one receptacle is on each side of the bed.

   Code reference _____

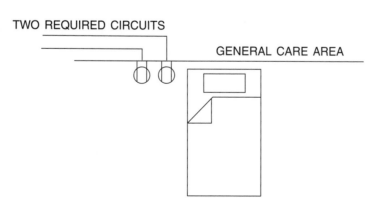

Figure 10.12   Which circuits are required to serve the receptacle outlets at the patient bed location of a general care area of a hospital?

4. Even though circuit wiring is run in metal raceway of a hospital, an insulated copper equipment grounding conductor is required to be run in the raceway and connected to the grounding terminal of:
    A. all equipment, receptacles and lighting fixtures in the hospital.
    B. receptacles installed in basements and other below grade areas.
    C. for every circuit protected with a ground-fault circuit-interrupter.
    D. ceiling luminaires (lighting fixtures) in patient care areas.
    E. all receptacles in patient care areas.

    Code reference _____

5. Circuits for patient bed locations of critical care areas of a hospital are supplied from a panelboard on the normal power system and a panelboard from the critical branch of the emergency system. The equipment grounding bus in each panelboard is required to be bonded together, as shown in Figure 10.13, with an insulated copper conductor not smaller than size:
    A. 12 AWG.   C. 8 AWG.   E. 4 AWG.
    B. 10 AWG.   D. 6 AWG.

    Code reference _____

6. All receptacle outlets rated 20 amperes or less at 125 volts installed in the patient vicinity of a hospital general care area are required to be:
    A. of the insulated ground type.
    B. listed as hospital grade.
    C. protected by a ground-fault circuit-interrupter.
    D. constructed of clear plastic to allow examination of the internal connections.
    E. listed as commercial grade.

    Code reference _____

7. A room of a hospital where patients are anesthetized with flammable anesthetics, as illustrated in Figure 10.14, is classified as a Class I, Division 1 location from the floor up to a height of:
    A. 2 ft (600 mm).   D. 5 ft (1.52 m).
    B. 3 ft (900 mm).   E. of the ceiling.
    C. 4 ft (1.22 m).

    Code reference _____

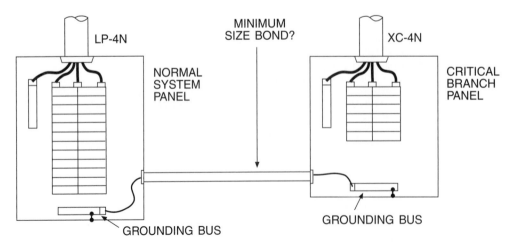

Figure 10.13 Determine the minimum size conductor required to bond the equipment grounding buses of the normal panelboard and the emergency panelboard serving the same patient vicinity.

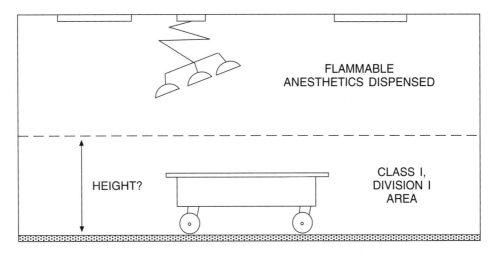

FLAMMABLE
ANESTHETICS DISPENSED

HEIGHT?

CLASS I,
DIVISION I
AREA

Figure 10.14   In the portion of an operating room from the floor up, what height is considered to be a Class I, Division 1 hazardous location when flammable anesthetics are dispensed in the room?

8. In a hospital, an exit sign circuit is a part of the:
   A. equipment system.
   B. critical branch of the emergency system.
   C. normal power system.
   D. alternate power branch of the normal power system.
   E. life safety branch of the emergency system.

   Code reference_____

9. A receptacle rated 15 or 20 amperes, 125 volts installed adjacent to the sink of a bathroom in a general care area patient room of a hospital is:
   A. not required to be ground-fault circuit-interrupter protected.
   B. required to be grounded to the room bonding point.
   C. required to be ground-fault circuit-interrupter protected.
   D. grounded with a size 10 AWG insulated copper equipment grounding wire.
   E. supplied from an isolated power system.

   Code reference_____

10. Circuit conductors of the life safety branch in a nursing home are:
    A. required to be kept entirely independent of other wiring and shall not share the same box, raceway, or cabinet with other wiring.
    B. permitted to share the same box, raceway, or cabinet with other wiring in the building.
    C. permitted to share the same box, raceway, or cabinet with circuits of the critical branch, but not other circuits in the building.
    D. permitted to share the same box, raceway, or cabinet with any other circuit of the essential electrical system but not other building circuit.
    E. limited to a maximum rating of 20 amperes.

    Code reference_____

11. Receptacles in a hospital supplied from the emergency system are:
    A. required to be kept 18 in. (450 mm) from other receptacles.
    B. required to be distinctively identified to indicate they are supplied by the emergency system.
    C. required to be equipped with pilot lights to indicate when they are energized.
    D. required to be red or their face plates are required to be red.
    E. required to be marked with a label indicating emergency power.

    Code reference_____

12. A wiring method that is permitted to be installed in a hospital to supply circuits serving a patient care area is:
    A. Rigid Nonmetallic Conduit, schedule 80.
    B. Electrical Nonmetallic Tubing, which is listed as fire rated.
    C. Type UF Cable.
    D. Type MC Cable where the metal covering is identified as a grounding return path.
    E. any listed cable that has a metal sheath for protection.

    Code reference _____

13. Receptacles installed within the general patient care area, bathroom, or playroom of a pediatric facility in a hospital are required to be equipped with a listed tamperesistant device or cover for circuits:
    A. rated 125 volts, 15 amperes.
    B. rated 125 volts and not over 20 amperes.
    C. rated 125 or 250 volts and not over 30 amperes.
    D. rated 125 volts up to 30 amperes.
    E. having any amperage or voltage rating.

    Code reference _____

14. If an outage of the normal power to a hospital occurs, the circuits supplied from the critical branch are required to have their power restored in not more than:
    A. 10 seconds.
    B. 20 seconds.
    C. 30 seconds.
    D. 60 seconds.
    E. whatever is considered a reasonable time.

    Code reference _____

15. Receptacles rated 20 amperes at 125 volts in an operating room of a hospital are supplied from an isolated power system (see Figure 10.15). The orange conductor of the isolated power system is to be:
    A. connected to the equipment grounding terminal of the receptacles.
    B. run without splices and looped at each receptacle terminal connection.
    C. connected only to the silver-colored screw terminal of the receptacles.
    D. connected only to the brass-colored screw terminal of the receptacles.
    E. connected to either the silver- or brass-colored terminal of the receptacles

    Code reference _____

ISOLATED POWER SYSTEM, 125-VOLT, SINGLE-PHASE

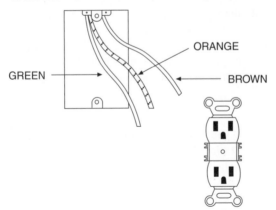

Figure 10.15 Which conductor of an isolated power system in a hospital is required to be connected to the silver-colored terminal of a receptacle?

WORKSHEET NO. 10—ADVANCED HEALTH CARE FACILITIES

Mark the single answer that most accurately completes the statement based upon the 2005 Code. Also provide, where indicated, the code reference that gives the answer, or indicates where the answer is found, as well as the Code reference where the answer is found.

1. The equipment grounding bus of a panelboard supplying receptacle circuits at the bed location of the critical care area of a hospital, shown in Figure 10.16, is considered to be:
   A. the patient equipment grounding point.
   B. a room bonding point.
   C. a remote bonding point.
   D. the reference grounding point.
   E. the emergency system grounding point.

   Code reference_____

2. For a critical care area in a hospital, each patient bed location is required to be provided with two branch-circuits:
   A. one of which shall originate from the emergency electrical system.
   B. both of which shall originate from the emergency electrical system.
   C. both of which shall originate from the normal electrical system.
   D. with not less than two receptacles connected to each branch-circuit.
   E. that are permitted to originate from any electrical system desired.

   Code reference_____

3. The patient bed location of a critical care area of a hospital is supplied by only one circuit from the critical branch of the emergency system. The circuit from the emergency system serving a receptacle at the patient bed location, as shown in Figure 10.17, is:
   A. also permitted to serve receptacles in the same room outside of the patient vicinity.
   B. not permitted to serve receptacles outside of the patient vicinity.

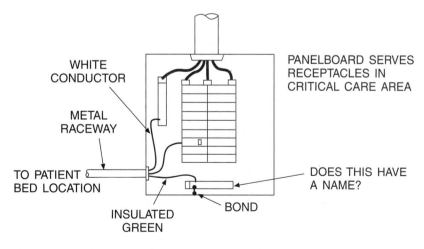

Figure 10.16   What is the name given to the equipment grounding bus of an emergency panel that serves receptacles in the patient bed location of a critical care area of a hospital.

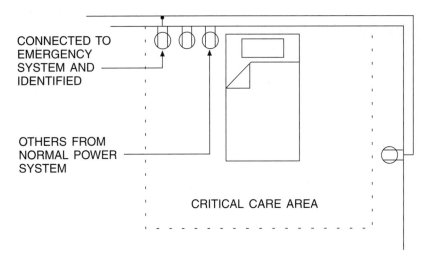

CONNECTED TO
EMERGENCY
SYSTEM AND
IDENTIFIED

OTHERS FROM
NORMAL POWER
SYSTEM

CRITICAL CARE AREA

Figure 10.17　There is only one emergency system circuit serving this patient bed location of a critical care area. Is it permitted to supply another receptacle outside of the patient bed location on that same circuit?

　　C.　permitted to supply fixed equipment in the room.
　　D.　permitted to supply a receptacle at an adjacent bed location.
　　E.　not permitted to supply a second receptacle at that same patient bed location.

Code reference_____

4.　The minimum number of receptacles serving the patient bed location of a general care area of a hospital is:
　　A.　not specified.
　　B.　four.
　　C.　six.
　　D.　eight.
　　E.　six duplex receptacles for a total of twelve.

Code reference_____

5.　An isolated power system in a hospital is required to be supplied power from the:
　　A.　critical branch of the emergency system.
　　B.　equipment system.
　　C.　life safety branch of the emergency system.
　　D.　normal power system but connected to the generator through an optional equipment transfer switch.
　　E.　normal power system.

Code reference_____

6.　All receptacles installed in wet locations and not served by an isolated power system shall be:
　　A.　supplied by an isolated power system.
　　B.　bonded to a room bonding point with a copper wire not smaller than size 10 AWG.
　　C.　located not less than 5 ft (1.5 m) above the floor.
　　D.　mounted in nonmetallic boxes with nonmetallic face plates.
　　E.　ground-fault circuit-interrupter protected for personnel.

Code reference_____

7. A hydromassage bathtub installed in a health care facility is required to have the metal piping, hand rails, metal parts of electrical equipment, and the pump motor bonded together. If bonded with a copper wire, it is required to be:
   A. not smaller than size 10 AWG.
   B. solid and not smaller than size 10 AWG.
   C. solid and not smaller than size 8 AWG either insulated, covered, or bare.
   D. solid or stranded and not smaller than size 8 AWG.
   E. not smaller than size 6 AWG.

   Code reference _____

8. Fixed room luminaires (lighting fixtures) are mounted at least 8 ft (2.5 m) above the floor of an anesthetizing location where flammable anesthetics are used and an isolated power system is installed. These luminaires (lighting fixtures) are permitted in this location:
   A. and powered by a normal grounded electrical system if the switch is located outside in another area.
   B. and powered by a normal grounded electrical system if the switch is located above the Class I, Division 1 hazardous location.
   C. if powered only by the isolated power system.
   D. if provided with explosionproof enclosures.
   E. if the fixtures have no exposed metallic parts.

   Code reference _____

9. The disconnecting means for permanently installed X-ray equipment shall be:
   A. operable from a location readily accessible from the X-ray control.
   B. located within sight of the X-ray machine.
   C. permitted to be a cord cap and attachment plug in the main circuit to the unit.
   D. capable of interrupting 125% of the momentary current of the unit.
   E. located anywhere in the room.

   Code reference _____

10. The wiring of circuits for exit signs and emergency lighting throughout patient care areas of a hospital are:
    A. permitted to be run in Type MC Cable where the metal sheath is approved as a grounding means.
    B. permitted to have short lengths of Flexible Metal Conduit not exceeding 6 ft (1.8 m).
    C. required to be run as nonflexible metal raceway or as Type MI Cable.
    D. required to be run as Rigid Metal Conduit or Intermediate Metal Conduit.
    E. permitted to be run using Type AC or Type MC Cable with an insulated equipment grounding conductor and metal sheath approved as a grounding means.

    Code reference _____

11. The power source for the essential electrical system for a hospital is an engine driven generator with a main disconnect and overcurrent device rated more than 1000 amperes, 3-phase, 480Y/277 volts solidly grounded as shown in Figure 10.18. Equipment ground-fault protection is:
    A. required between the generator and the essential electrical system transfer switches.
    B. required on the feeders on the load side of the essential electrical system transfer switches.
    C. permitted on the feeders between the generator and the essential transfer switches.
    D. permitted on the load side of the essential electrical system transfer switches.
    E. not permitted between the generator and the essential electrical system transfer switches or on the load side of the transfer switches.

    Code reference _____

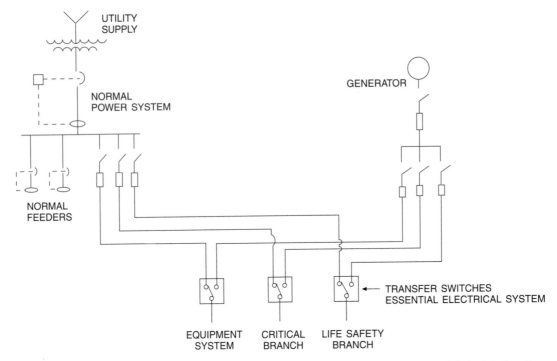

Figure 10.18   **Where is equipment ground-fault protection required relative to the essential electrical system of a hospital?**

12.   A critical care area of a hospital is equipped with a patient grounding point as shown in Figure 10.19. An equipment bonding jumper shall be used to connect the patient equipment grounding point and the grounding terminal of receptacles. The minimum size copper bonding jumper permitted is:

A.   14 AWG.

B.   12 AWG.

C.   10 AWG.

D.   8 AWG.

E.   not specified and depends upon the rating of the circuit and sized according to *Table 250.122.*

Code reference _____

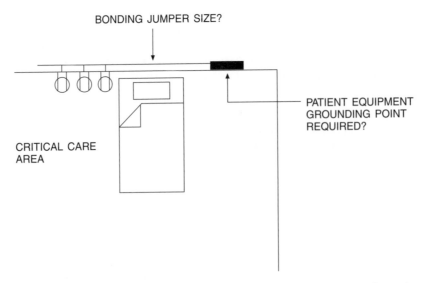

Figure 10.19   **Determine the minimum size bonding jumper required from the receptacles to the optional patient equipment grounding point in a critical care area of a hospital.**

13. A room in a hospital used for the storage of flammable anesthetics or disinfectants is considered to be a:
    A. Class I, Division 2 location for the entire room, floor to ceiling.
    B. Class I, Division 2 location from the floor up to a height of 18 in. (450 mm).
    C. Class I, Division 1 location from the floor up to a height of 18 in. (450 mm).
    D. Class I, Division 1 location for the entire room, floor to ceiling.
    E. an unclassified area, however, all wiring is to be in Rigid Metal Conduit using boxes with threaded hubs.

    Code reference _____

14. A hospital has a demand load on the essential electrical system of more than 150 kVA. A single standby generator supplies the equipment system, the critical branch, and the life safety branch. Upon a loss of normal utility power to the hospital, the equipment system is:
    A. required to be energized in not more than 10 seconds.
    B. permitted to be completely or partially transferred to the generator manually.
    C. required to be started simultaneously along with the emergency system.
    D. permitted to be started simultaneously along with the emergency system.
    E. required to have power restored automatically with an appropriate time lag.

    Code reference _____

15. A hospital has an equipment system, a life safety branch, and a critical branch all supplied from one standby generator. The demand load of the essential electrical system is more than 150 kVA. The minimum number of transfer switches required for this installation is:
    A. zero.
    B. one.
    C. two
    D. three.
    E. not specified in the Code.

    Code reference _____

UNIT 11

Emergency and Alternate Power Systems

OBJECTIVES

After completion of this unit, the student should be able to:

- determine the minimum size conductor permitted to connect a standby generator to a wiring system.

- explain the purpose of transfer equipment for a standby power system.

- explain ventilation and insulating needs for battery rooms.

- define terms used in the Code to describe solar photovoltaic systems.

- name the types of equipment and circuits that make up the emergency electrical system of a typical commercial building.

- explain how the wiring is to be run for emergency electrical circuits.

- explain the type of electrical equipment and circuits powered by a legally required standby electrical power system.

- describe a typical application where there is an optional standby electric power system.

- define an interconnected electric power production source.

- explain the difference between a supervised and an unsupervised fire alarm system.

- explain how the wires are to be attached to a fire alarm pull station.

- answer typical wiring installation questions from *Articles 445, 480, 690, 692, 695, 700, 701, 702, 705*, and *760*.

- state at least three changes that occurred from the 2002 to the 2005 Code for *Articles 445, 480, 690, 692, 695, 700, 701, 702, 705*, or *760*.

CODE DISCUSSION

Alternate electric power-producing systems are discussed in this unit, as well as the Code articles that cover their installation. These systems may be legally required for some buildings, such as a hospital, or they may be optional. Emergency electrical systems are discussed in this unit. The most simple emergency system consists of exit signs and lighting for building evacuation. Other buildings may have more extensive requirements. The electrical Code covers the installation of these systems, while other codes cover what systems and equipment are required in a particular building. The installation of fire alarm systems is covered in this unit.

Article 445 covers the requirements for generators to be connected to wiring systems. The markings for generators are covered in *445.11*. Even though the full-load rated current is required to be marked on the nameplate, there are many older generators in operation that list only the continuous rating in kilowatts or kilovolt-amperes. The Equations 8.5 and 8.6 in *Unit 8* can be used to determine the full-load current if the continuous rating is in kVA or kW. *NEC® 445.12(A)* requires constant voltage generators, such as the ones used for standby power, to be protected from overcurrent. This is generally done by the manufacturer with either a set of fuses or a circuit breaker as an integral part of the generator. The conductors between the generator and the point where the connection is made to the premises wiring system is required to be not less than 1.15 times the full-load continuous output current of the generator. This conductor is required by *240.4* to be protected according to its ampere rating, therefore, the conductor is generally sized to the rating of the output overcurrent device of the generator. The connection to the premises wiring system is usually made to a transfer switch.

A disconnecting means is required to be installed to disconnect the generator ungrounded conductors from the premises wiring system. The location of the disconnecting means is not specified. A switch or circuit breaker on the generator unit is a convenient means of disconnecting the ungrounded conductors from the generator. A transfer switch can also be used to disconnect the ungrounded conductors from the generator. *NEC® 445.18* does not require a disconnect for the conductors if the generator prime mover can readily be shut off and the generator is not operating in parallel with another electrical supply such as the utility.

Article 480 deals with stationary storage battery installations. The main emphasis of this article is the rigid mounting of the batteries, corrosion prevention, ventilation of the battery room, and guarding of live parts when the batteries are connected for the circuit to operate at 50 volts or more. The rules on guarding are found in *110.27*. It is also necessary to provide working clearances around battery installations according to *110.26*.

Specific requirements for the connection of storage batteries as a part of an electrical system are covered in various articles throughout the Code. Storage batteries can function as a stand-alone standby power system. An inverter can be used to convert the direct current to alternating current in the event of a power outage. Energy conversion efficiencies of more than 85% are common for inverters. The batteries can be kept charged as needed with a small automatically operated generator. Operating in this manner, generators are supplying a nearly constant load, and will operate more efficiently. Batteries are usually installed as a part of a photovoltaic power system. Batteries can also be used with wind generating systems and small hydroelectric turbines. There are no specific rules in the Code for general installation of battery power systems. The best source is *Part VIII* of *Article 690* on solar photovoltaic systems.

Article 690 applies to solar photovoltaic electric power systems including the dc-to-ac inverter, charge controller, sun tracking system, and wiring of components in the system. Several important definitions are covered in *690.2*. It is important to understand the basic photovoltaic terminology in order to apply the rules of *Article 690*. A typical silicon photovoltaic cell is about 4 in. (100 mm) in diameter and has an operating output of about 3 amperes at 0.5 volts dc. These photovoltaic cells are connected in series to obtain the desired output voltage and manufactured into a unit called a solar photovoltaic module. As many modules as needed are assembled into a solar photovoltaic array to achieve the desired output voltage and power. A typical solar photovoltaic power system is shown in Figure 11.1. Modules are usually connected in series to achieve the desired voltage. Each module or series of modules is called a photovoltaic source circuit. These source circuits are connected in parallel until the desired power is achieved. The wires connecting to a group of photovoltaic source circuits is called the photovoltaic output circuit. If an output circuit connects to ten source circuits each producing 3 amperes at 36 volts, the output circuit will be rated 30 amperes at 36 volts.

The rated output voltage of the solar photovoltaic array will be higher than the nominal operating voltage of the photovoltaic system because the array will not be able to maintain maximum output throughout

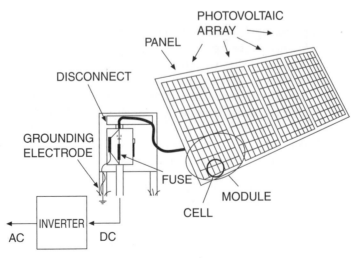

Figure 11.1   Photovoltaic cells are assembled into modules, which are then assembled into panels. A group of panels make up an array.

the day. The output of the array can be sent directly to the load or to a set of storage batteries. Output from the photovoltaic array goes to a charge controller which supplies the desired voltage and current to the batteries and the load. Solar photovoltaic systems are sometimes interconnected with building wiring and the utility simultaneously to form what is known as an interconnected electric power production system, the wiring of which is covered in *Article 705*.

Part II of *Article 690* covers circuit conductor sizing and overcurrent protection. *Part III* specifies the disconnecting means requirements. It is important to remember that if light strikes the photovoltaic array, a voltage is being produced. Also, when a system is interconnected with a utility system or run in parallel with another power production system, backfeeding may be a problem. When a switch is opened, both sides of the switch may still be energized. This is why placement of disconnects is important. Wiring methods are covered in *Part IV* and grounding is covered in *Part V*. A stand-alone solar photovoltaic power system is a separately derived system and grounding should meet the requirements of *250.30*. Connected as a standby power system by means of a transfer switch at the main service of a building or operated as an interconnected power system in parallel with a utility supply, the solar photovoltaic system connects to the service of a building and is not considered a separately derived system. In this case, the same rules for grounding are followed as would be the case for a utility supplied electrical system. This is explained a little more clearly in *Part V* of *Article 692* on fuel cell systems.

Article 692 provides the rules for the installation of fuel cell systems to supply power for premises wiring systems. Definitions important to understanding this article are in *692.2*. Fuel cells produce direct current. The direct current can be utilized for some loads, but for many loads the dc must be converted to ac with an inverter. The inverter may be a part of the fuel cell unit, or it may be connected as an auxiliary unit. *NEC® 692.8(B)* requires the feeder conductor from the fuel cell system to have an ampacity not less than the nameplate rated current or the rating of the overcurrent device on the fuel cell system. The output conductors from the inverter are sized according to the rated output current of the inverter. *NEC® 692.13* requires that a means be provided to disconnect all ungrounded conductors of the fuel cell system from the premises wiring system.

Fuel cells operated as a stand-alone system are considered a separately derived system and are required to be grounded following the rules of *250.30*. If the system is operated in parallel with another power system, such as the utility system, the fuel cell system is required to be grounded to the grounded circuit conductor terminal of the premises wiring system. This is also the case where the fuel cell system is operated as a standby power system and connected to the wiring by means of a transfer switch. The fuel cell system may also contain batteries. *NEC® 692.56* requires a warning to be placed at the main disconnect location calling attention to the energy storage system, but no rules are included for the installation of the batteries. The rules of *Article 480* must be followed for the battery installation but there are no rules for the installation of the complete battery system, including charge controller and inverter. The best place to find this information is in *Part VIII* of *Article 690* on solar photovoltaic power systems.

Article 695 deals with the electrical power, wiring, and controls for fire pumps and associated equipment. This article assembles the electrical installation requirements related to fire-pumping installations together in one location in the Code. Much of the article is taken directly from *NFPA20-1999, Standard for the Installation of Stationary Pumps for Fire Prevention*. The rules applying to fire pumping systems may seem contrary to normal safety rules in the Code, but it is important to note that the purpose of the system is to reduce a fire emergency to allow the building to be evacuated and to assist fire-fighting personnel in dealing with the emergency. Hopefully, the operation of the pumping system will put out the fire or limit the extent of damage. In any case, the electrical system is to be installed for maximum reliability under heavy load conditions. The system is to be installed to minimize its exposure to damage. Overcurrent protection is set high enough to prevent premature shutdown of the pumping system. With a building on fire, and human life in danger, damage or burnout of pumping system motors and associated equipment is of little concern. *NEC® 695.3* directs that the power supply be reliable. In the case where the pumping system is supplied power by a tap to the normal power system, the tap ahead of the service disconnect is not permitted to be made in the service disconnecting means enclosure, *695.3(A)(1)*.

The supply conductors for the pumping system are not permitted to be protected from overload, but only protected from short circuits as stated in *695.6(A)* and *240.4(A)*. When the system is supplied from a dedicated transformer, overcurrent protection is not permitted to be installed on the secondary of the transformer. The minimum setting of the transformer primary overcurrent protection shall be sufficient to continuously supply the sum of the locked-rotor currents of all motors of the system and the full-load current of the other loads of the system.

A separate service is required for a fire pump installation. The power source is required to be reliable and capable of supplying the locked-rotor current of all pump motors and pressure maintenance pump motors, and the full-load current of any associated equipment. Power source requirements are discussed in *695.3*. A separate service is permitted as shown in Figure 11.2. It is also permitted to tap ahead of the service disconnect. An on-site generator is also permitted to supply the fire pump and associated equipment. Rules for determination of the minimum size of generator are found in *695.3(B)(1)*. Obviously it is important to make sure the fire-pumping system is always ready for operation. Therefore, the system is required to be supervised to ensure that power is supplied to all system components, and a signal will indicate if any part of the system is not ready for operation. Disconnects must be capable of being locked in the "closed" position to make sure they are not inadvertently opened.

Article 700 deals with the installation of emergency electrical systems. The fine print notes to *700.1* provide information that helps understand the type of electrical circuits that may be included as part of an emergency electrical system. Emergency systems are required in buildings where there is public assembly, schools, hotels, theaters, sports arenas, and health care facilities. Typical equipment included as a part of the emergency system is illumination for safe exiting of the building, exit signs, fire detection and alarm systems, elevators, fire pumps, ventilation and other building utilities necessary for life safety. *NEC® 700.12* requires load transfer to a backup power source to occur in not more than 10 seconds.

The wiring of these systems is covered in *Part II*. Emergency circuits are required to be kept independent of other circuits used for light and power. If the building is wired with raceway, then the emergency circuits are required to be run in dedicated raceways that contain only emergency circuits. The sources of power for these systems are covered in *Part III*. Emergency illumination is the most common type of emergency system installed. The requirements for the circuits, equipment, and control of equipment are covered in *Parts IV* and *V*.

For some buildings requiring only safe evacuation of the building, unit equipment is satisfactory for safe exiting from the facilities using illuminated exit signs and adequate area lighting. The illumination units are not permitted to depend upon only one lamp for illumination in any particular area. Each area will require

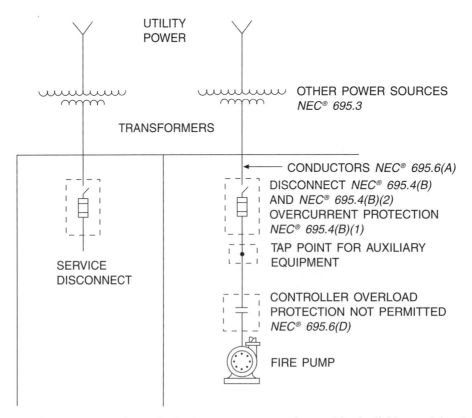

Figure 11.2   A fire pump system is required to have a separate service capable of reliably supplying the locked-rotor current of the fire pump and pressure maintenance pump and the full-load current of the other associated equipment. The conductors and equipment are to be protected from short circuits and ground faults only.

two single-lamp units or one two-lamp unit. Power for the illumination units is required to be from the same circuit that supplies the normal lighting in the area. The emergency light units are required to be connected to the circuit ahead of any local switches. These rules are found in *700.12(F)*. High intensity discharge lighting will generally not relight when power is restored and this needs to be considered when installing emergency lighting. The emergency lighting units will turn off when power is restored, but the HID lamps will not necessarily relight immediately. One way to solve this problem is to include fluorescent or incandescent lights as a part of the lighting in the area.

NEC® 700.12 specifies the power sources that are acceptable for emergency circuits. One source is battery-powered illumination units. More extensive systems may require more power. One alternative is a battery system that must be able to supply the load for not less than 1.5 hours. Deep cycle batteries are required and vented lead-acid batteries requiring periodic filling with water are required to have a transparent case to indicate electrolyte level. Another power source is an on-site generator which must have a fuel supply for the engine to last not less than two hours. The generator must be automatic starting with automatic load transfer. It is not permitted for the engine to operate on a utility natural gas supply. If the generator is not capable of starting and powering the load within 10 seconds, another temporary power system is required to fill the gap in time until the generator is capable of supplying the emergency load. An uninterruptible power supply is also permitted to supply emergency loads. A separate service in some situations may be permitted as an emergency power supply but only if approved by the local jurisdiction.

Article 701 on legally required standby systems do not include emergency systems. These are systems that are required to be installed because a system in a building may create a condition that becomes life threatening if normal power fails. Or it may be a system that would hamper rescue or fire fighting. It can include lighting, ventilation and smoke removal, and communications systems. *NEC® 701.11* requires the standby power system to be up and running within 60 seconds of a normal system power failure. The power source is permitted to be a generator, storage batteries, uninterruptible power supply, separate service if approved by the authority having jurisdiction, or a connection ahead of the service disconnect if acceptable to the authority having jurisdiction. *NEC® 701.10* permits the wiring of circuits for a legally required standby power system to occupy the same raceways, cables, boxes, and cabinets with normal wiring in the building. For illumination, unit equipment the same as specified for emergency systems is permitted to be used. *NEC® 701.7* requires transfer equipment to be automatic and it shall be arranged in such a way that there will not be an interconnection between the normal power supply and the legally required standby power source.

Article 702 applies to optional standby power systems intended to supply loads for industrial, commercial, farm, and residential buildings and facilities. The type of power source is not specified. It could be a generator, storage batteries, or any type of alternate power system such as solar photovoltaics or fuel cells. Portable generators and tractor driven generators are permitted for this type of installation. An example of a portable tractor driven generator is shown in Figure 11.3. *NEC® 702.9* permits the wiring of circuits for an

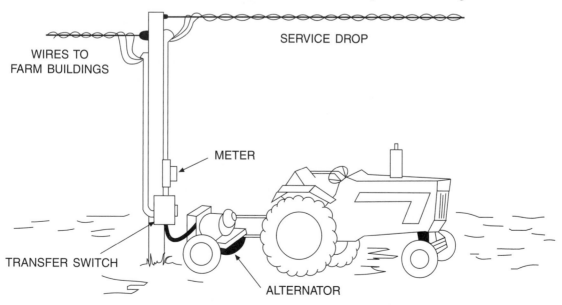

Figure 11.3 **A portable generator that has a permanent means of attachment to a premises wiring system, such as this tractor powered portable generator at a farm distribution pole, is considered to be an optional standby power system.**

optional standby power system to occupy the same raceways, cables, boxes, and cabinets with normal wiring in the building. *NEC® 702.6* allows manual transfer equipment but it shall be arranged in such a way that there will not be an interconnection between the normal power supply and the optional standby power source.

Article 705 specifies the method of interconnecting an alternate power system to operate in parallel with another power system such as a utility power supply. These systems are only permitted to be connected in parallel with a utility power system through a contractual agreement with the specific electric utility. Some states have a net power agreement law where the utility purchases power from the customer at retail price and in other states the utility purchases the power at a regulated wholesale price. *NEC® 705.12* generally requires the connection of the alternate power source to be at the service disconnecting means. The connection is permitted to be on the supply side of the service disconnect or it is permitted to be on the load side. *NEC® 705.14* requires the customer power production system to be compatible with the utility power supply. In particular, the customer power source frequency and voltage must be the same as the utility power supply and the waveshapes must be compatible. The customer power source making connection with the utility supply is usually a generator, the output from an inverter, or the output from a transformer. The connection must be controlled so that the customer power supply is synchronized with the utility supply when the connection is completed so the two wave forms are in phase with each other.

NEC® 705.21 requires a means be accessible to disconnect all ungrounded conductors of the customer power source. *NEC® 705.40* requires that the power production source be automatically disconnected from the utility supply in case of a failure of the utility supply. This can be in the form of a load transfer switch that will connect the customer source to the customer load while disconnecting the source from the utility supply.

EMERGENCY ELECTRICAL SYSTEMS

Emergency electrical systems must be installed to ensure the highest practical level of reliability. The wiring for these systems is required to be run completely independent from normal power and lighting circuits. These systems frequently are supplied power from a separate service or from a tap to the normal service plus a backup power source such as a generator. When a backup power source is installed, the transfer of load from normal power to the alternate power source must be automatic and occur in not more than 10 seconds. This transfer of load from one power source to another is accomplished using a transfer switch such as the one illustrated in Figure 11.4. *NEC® 700.6* requires this transfer of emergency load from normal to standby power to be done in such a way that there is never an interconnection between the standby power source and the normal power source. A legally required standby power source is intended to supply loads other than emergency systems that must be operated to protect personnel from injury or to aid fire fighters or rescue personnel. The load is permitted to be transferred to the standby source in not more than 60 seconds. In the case of an optional standby power system, the load transfer is permitted to be manual and there is no specified time in which the load transfer must occur.

The primary function of the emergency electrical system, covered in *Article 700,* is to provide automatic illumination for safe exiting of a building and panic control during an emergency. This would require illumination in places where people assemble in the event that power to the normal illumination is inter-

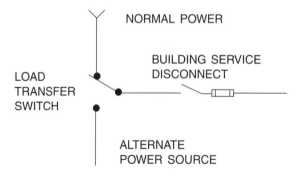

Figure 11.4   A load transfer switch is required when connecting a standby power source to a premises wiring system, and it prevents interconnection of the standby power source to the normal power source.

rupted. Illumination is required to safely exit the building, including power for the marking of exits from the building.

The emergency electrical system also may include fire alarm systems, power for fire pumps, ventilation where necessary to provide life safety, or other functions necessary to prevent other life-threatening hazards.

A self-contained, battery-operated exit sign may consist of a standard universal exit sign that is internally equipped with a battery, charger, and a means of automatically transferring to battery power on loss of normal operating power. Exit signs of this type, which are listed by an independent testing laboratory, are readily available with operational time ratings in excess of 4 hours. The following criteria are used in determining whether self-contained, battery-operated exit signs are acceptable in lieu of conventionally wired exit signs. Self-contained, battery-operated signs are permitted to be installed provided the following conditions are met:

- The power supply for the exit signs and illumination units shall be from the same branch-circuit as that serving the normal lighting in the area, and shall be connected ahead of any local switches, *700.12(F)*.
- The operational time rating of the units shall not be less than that required by the rules applicable to the facility in which they are installed.

Fire Alarm Systems

A fire alarm system, covered in *Article 760*, is an assemblage of components, acceptable to the authority having jurisdiction, that will indicate a fire emergency requiring immediate action. The system shall alert all occupants of a building in which it is installed when a fire emergency is present.

The authority having jurisdiction is the governmental, legally employed agency that can require the installation of a fire alarm system with specified features, characteristics, functions, and capabilities. The authority having jurisdiction may be a person, firm, or corporation with financial or other interest in the protected property and whose authority lies in contractual arrangements between the affected parties. The electrical inspection authority has jurisdiction over installation methods, materials, and some operational characteristics of all systems.

Fire alarm systems required to be installed by governmental agencies are designed and installed to save lives by alerting occupants to a fire emergency. Life safety systems may provide some property protection. These systems are often termed local protective signaling systems. Fire alarm systems installed to protect or limit property loss are not required by governmental agencies. These optional systems are installed to protect high-value properties or to reduce insurance premiums.

Fire alarm systems required for life safety shall be the supervised type. This requirement is not in the *NEC®*, but is required by other codes. A supervised circuit will indicate a ground fault or an open circuit with a distinctive audible signal at the control panel location and remote locations as required. Class A supervised circuits will continue to operate with a single open-circuit conductor. Class B supervised circuits will not operate with an open circuit. A Class A fire-alarm initiating circuit is shown in Figure 11.5. Supervision is provided by the design of the control panel and correct installation of the field wiring. Circuits required to be supervised are:

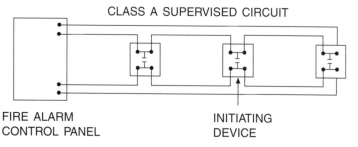

Figure 11.5   A Class A supervised fire-alarm initiating circuit completes a loop on each side of the initiating device and will allow the system to remain functional even with one open circuit.

- The main operating power
- The initiating circuits
- The sounding device circuits

The initiation circuit senses the fire emergency condition and permits initiation current to flow in the circuit. Initiating devices are of several common types: pull stations, products of combustion detectors, heat detectors, and water flow switches. Rules not a part of the electrical Code are provided for the proper location of these initiating devices.

The signaling circuit alerts building occupants to the fire emergency. Sounding devices installed on the signaling circuit are of several common types. Typical sounding devices are horns, buzzers, bells, chimes, electronic signals, and sirens. The type of sounding device used in a building must have a distinctive sound not used for any other purpose in the building. Sounding devices must be audible in all areas of the protected facility. Visual signals are now being provided for hearing-impaired persons by some building codes for barrier-free design. Visual signals, when installed, are required to be on a supervised circuit.

One type of initiating device is a heat detector. Several types of heat detectors are used for fire alarm systems. One type operates at a predetermined fixed temperature. Another operates at a predetermined rate of rise in temperature. Both types may be combined in one unit.

There are several types of products of combustion detectors. The photoelectric type is designed to detect visible particles in smoke by obscuring the light or by light scattering. Another is the ionization type, which is designed to detect invisible products of combustion by using the electrical conducting nature of air in the presence of low-level radioactivity. The conductivity of air is changed by the presence of combustion products or other gas or vapor.

Duct detectors are installed to prevent toxic vapors from being moved from one part of the building to another. There are basically two types of duct detectors. The heat type is used to detect abnormal temperatures in air-handling systems. They are permitted in systems under 15,000 cubic feet per minute (425 m³/min) air capacity by some codes. Some are similar to the fixed temperature detectors described earlier. This type is no longer acceptable for new construction by some state and local codes. Smoke-type duct detectors are adaptations of photoelectric or ionization-type detectors. They are installed to sample the air in a duct system to detect abnormal smoke in air-handling systems. All air-handling systems are not required to have automatic detectors. Refer to state and local codes for specific requirements.

Flame detectors are a type activated by visible light, light invisible to human eye, or flame flicker. These units are for special occupancies or hazards and are not required by codes, but they may be provided for an additional level of protection.

Wiring Fire Alarm Systems

It is important to understand how supervised electrical circuits operate to understand how the wiring is to be installed. The wiring for the initiation circuit or the sounding circuit shall be installed such that the system will function when an initiation action has been taken. If the initiation circuit is not properly installed, it is possible for an initiation station to develop an open circuit at a later time without activating the trouble indicator. Specific requirements for wiring the fire alarm system are found in *Article 760*. This discussion is intended to show the proper wiring of the fire-alarm circuits to make sure they will function properly.

The Class A initiating circuit requires four wires. A resistor is not required at the last initiating station. The current-limiting resistor is built into the alarm control unit. One wire on each side of the circuit can open, which sounds the trouble signal, and the initiating circuit will still function. Figure 11.5 shows a Class A initiating circuit properly installed.

The Class B initiating circuit has only two wires. An end-of-line resistor is installed to complete the supervisory circuit. A trickle current flows through the wires at all times. A short circuit, a ground fault, or an open wire will prevent proper operation, and a trouble signal will sound. A control unit with four terminals may be connected for a 2-wire Class B circuit. Figure 11.6 shows a properly wired Class B initiating circuit. An end-of-line resistor is used for a Class B supervised circuit.

The sounding circuit is also required to be supervised. One method is the series circuit, in which a trickle current passes through each sounding device connected in series. The fire-alarm control panel is adjusted for the number and type of sounding devices used. Figure 11.7 shows a series-type sounding circuit.

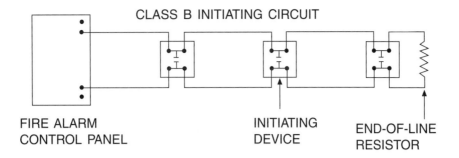

Figure 11.6 A Class B supervised fire-alarm initiating circuit has only one conductor for each side of the initiating devices and must have an end-of-the-line resistor to complete the supervisory circuit.

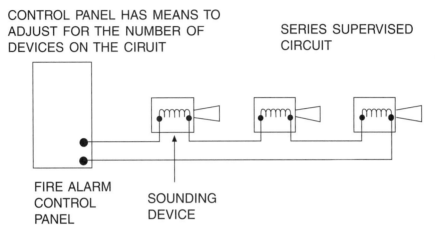

Figure 11.7 A series-type supervised fire-alarm sounding circuit passes a supervisory current through each sounding device.

A polarized sounding circuit is shown in Figure 11.8, and an end-of-line resistor is used. This system is supervised with a current flowing along one wire, then through the end-of-line resistor, and back to the alarm control panel on the other wire. Proper polarity must be observed for connecting to the sounding devices.

Wire splices and terminations are a source of problems in any wiring system. It is not uncommon to have a splice or termination fail and cause an open circuit. For this reason, it is not permitted to make a con-

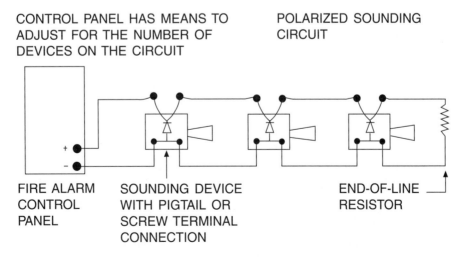

Figure 11.8 A parallel or polarized fire-alarm sounding circuit has two terminal connections to each side of the sounding device, and an end-of-the-line resistor is needed to complete the supervisory circuit.

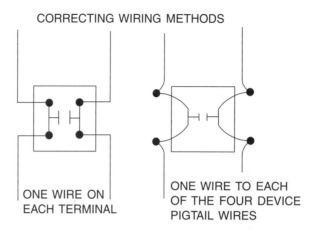

CORRECTING WIRING METHODS

ONE WIRE ON
EACH TERMINAL

ONE WIRE TO EACH
OF THE FOUR DEVICE
PIGTAIL WIRES

Figure 11.9   Only one conductor is attached to a terminal of a fire alarm device so that an open conductor will always stop the supervisory current flow and provide a trouble signal.

nection to an initiation station that depends on one splice or termination. If an open circuit should occur, it is possible for the supervisory circuit to function, but an initiation station may be inoperative. If a connection to an initiation station is made with two separate terminations or with a loop termination, an open circuit to an initiation station will most likely open the supervisory circuit. Proper connections to an initiating station are shown in Figure 11.9.

In the process of wiring in an initiating circuit, it may be necessary to make a tee tap. This procedure is a frequent source of error. If this is not done properly, the tee tap circuit will not be supervised. It is possible for an open to occur, rendering one or more initiating devices not operable without sounding a trouble signal. The proper method of making a tee tap for a Class A initiating circuit and a Class B initiating circuit is shown in Figure 11.10. Note that the initiating stations are all connected in a series circuit.

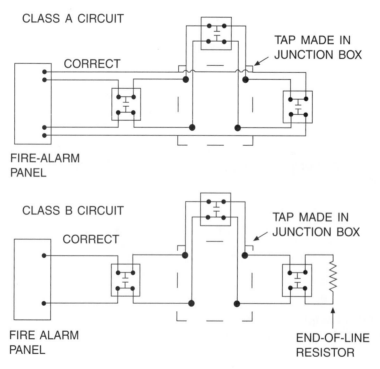

CLASS A CIRCUIT

CORRECT

TAP MADE IN
JUNCTION BOX

FIRE-ALARM
PANEL

CLASS B CIRCUIT

CORRECT

TAP MADE IN
JUNCTION BOX

FIRE ALARM
PANEL

END-OF-LINE
RESISTOR

Figure 11.10   Tapping a fire alarm circuit at a later time to install additional devices must be done in such a way that the supervisory current is forced to travel through every terminal in the circuit. A proper tap of a Class A and Class B initiating circuit is shown.

Computerized fire alarm systems are frequently used on large buildings. A remote terminal unit is placed in an area or on a floor. Each remote terminal unit has an initiating circuit and a sounding circuit. The central computer interrogates each remote terminal unit for a status check. If an alarm is initiated, the sounding signal is given in a predetermined strategy. For large buildings, evacuation of the areas of greater danger is initiated first. The rules for supervision and wiring discussed earlier must be followed for the initiating and sounding circuits of each remote terminal unit.

Fire Pumps

Power for a fire pump installation is required to be independent of the normal power system for a building. *NEC® 695.4(A)* permits the power source to connect directly to a listed fire pump controller, or a listed combination fire pump controller and transfer switch. *NEC® 695.4(B)* permits a disconnecting means with overcurrent protection to be installed ahead of the fire pump controller. Figure 11.11 shows a diagram of a fire pump installation with and without a transformer. A fire pump must be ready at all times. Therefore, there is a requirement that the disconnecting means must either be capable of being locked in the closed position or must be supervised to indicate if the switch is open. The disconnect is not sized the same as would be the case for a typical motor circuit. The overcurrent device is not permitted to have a rating less than the locked rotor current of the fire pump and the pressure maintenance pump. The next higher standard rating overcurrent device is selected from *240.6(A)*. The rating of the overcurrent device will dictate the minimum rating of switch.

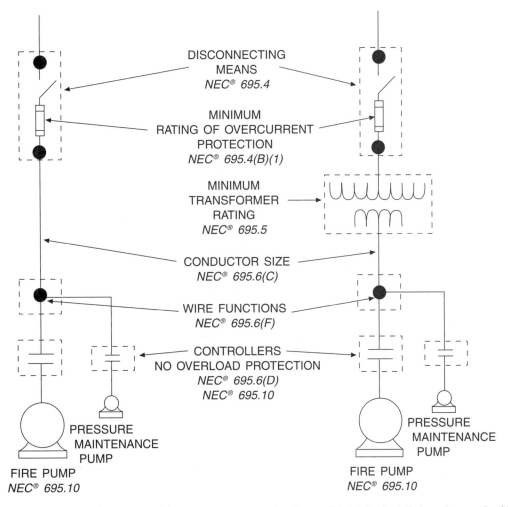

Figure 11.11   A disconnecting means with overcurrent protection is permitted to be installed on the supply side of a listed fire pump controller. The left diagram is supplied from a utility transformer and the right diagram is supplied with a customer owned transformer.

NEC® 695.6(C)(2) specifies the minimum rating of conductor required for the service to the fire pump installation and the individual circuit conductors to each component of the system. In Figure 11.11, there is a main service conductor supplying both motors, and a tap to the fire pump and another tap to the pressure maintenance pump (sometimes called a jockey pump). Overcurrent protection is not permitted on these individual taps. *NEC® 695.6(D)* permits the tap conductors to be sized according to the rules in the article with no restriction as to length of tap or size of conductor as would normally be the case if the rules in *240.21* applied. The conductor minimum ampacity is to be not less than 1.25 times the full-load current of the motor. If the conductor runs are long from the power supply to the fire pump controller, there is a maximum voltage drop permitted by *695.7*. When starting, the voltage at the controller terminals is not permitted to drop more than 15% below the nominal circuit voltage which, for a 480-volt motor, would be 408 volts (480 V × 0.85 = 408 V). The voltage drop is not permitted to be more than 5% of motor rated voltage when the motor is operating at 1.15 times the motor rated full-load current. For a 3-phase motor rated at 460 volts, the voltage is not permitted to drop below 437 volts. These voltage-drop criteria can be calculated based upon locked-rotor current and motor full-load current using equations 6.6 and 6.7 in *Unit 6* to determine the minimum permitted conductor size.

If a transformer is installed in the fire pump circuit, it is required to have a minimum rating of not less than 1.25 times the full-load current of the motors plus 100% of the rating of auxiliary equipment. The minimum kVA rating of the transformer needed to supply a fire pump load can be determined using Equation 11.1 for single-phase loads and Equation 11.2 for 3-phase loads. The next larger kVA rating of transformer would be chosen after making the calculation. Overcurrent protection for the transformer is installed only on the primary side of the transformer. The overcurrent device rating is not permitted to be more than the values given in either *Table 450.3(A)* or *Table 450.3(B)*. But the overcurrent device protecting the transformer is not permitted to be less than the locked-rotor current of the motors supplied after an adjustment in current is made for the primary to secondary voltage difference. This adjustment can be made using Equation 11.3. If this criteria cannot be met, then it will be necessary to increase the transformer rating in order to meet both the overcurrent protection rules of *450.3* and *695.4(B)(1)*.

Single-Phase:

$$\textbf{Fire Pump Transformer kVA} = \frac{\textbf{Voltage} \times \textbf{Current} \times \textbf{1.25}}{\textbf{1000}} \qquad \textbf{Eq. 11.1}$$

Three-Phase:

$$\textbf{Fire Pump Transformer kVA} = \frac{\textbf{1.73} \times \textbf{Voltage} \times \textbf{Current} \times \textbf{1.25}}{\textbf{1000}} \qquad \textbf{Eq. 11.2}$$

$$\textbf{Minimum Fuse Rating} = \textbf{Motor Locked-Rotor Current} \times \frac{\textbf{Secondary Voltage}}{\textbf{Primary Voltage}} \qquad \textbf{Eq. 11.3}$$

The following examples will help to understand the various rules for sizing conductors, disconnecting means, and overcurrent protection for a fire pump installation.

Example 11.1   A fire pump installation in a building consists of a 75-horsepower, 3-phase, 460-volt, design B motor and a pressure maintenance pump rated 1.5 horsepower, 3-phase, 460 volts, design B. There is a fusible disconnect installed ahead of the controller as shown in Figure 11.12. Determine the minimum size copper conductors with 75°C insulation and terminations permitted to supply these two motors assuming the voltage-drop limitations will be satisfied.

Answer: *NEC® 695.6(C)(2)* requires the conductors to be not less than 1.25 times the combined full-load current of the two motors. Look up the motor full-load currents in *Table 430.250*. The size the conductors using *Table 310.16*. The 75-horsepower motor has a full-load current of 96 amperes, and the 1.5-horsepower motor has a full-load current of 3.0. The minimum conductor ampacity is 124 amperes (99 A × 1.25 = 124 A). The minimum conductor size permitted is 1 AWG copper.

Example 11.2   With the same fire pump installation as Example 11.1, determine the minimum size fuses permitted to be installed in the disconnect.

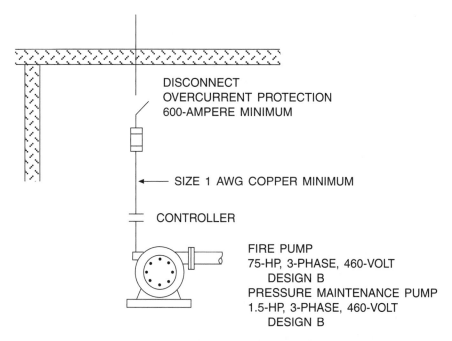

DISCONNECT
OVERCURRENT PROTECTION
600-AMPERE MINIMUM

SIZE 1 AWG COPPER MINIMUM

CONTROLLER

FIRE PUMP
75-HP, 3-PHASE, 460-VOLT
DESIGN B
PRESSURE MAINTENANCE PUMP
1.5-HP, 3-PHASE, 460-VOLT
DESIGN B

Figure 11.12   The minimum conductor size for a fire pump installation is required to be not less than 1.25 times the full-load current of the fire pump and pressure maintenance pump motors plus the full-load current of other associated loads.

Answer: *NEC® 695.4(B)* requires the rating of the overcurrent device in the disconnect to be not less than the sum of the locked currents of the two motors. Look up the locked rotor currents in *Table 430.251(B)* and find 543 amperes for the 75-horsepower motor and 20 amperes for the 1.5-horsepower motor. The combined locked rotor current is 563 amperes. The next higher standard overcurrent device rating listed in *240.6(A)* is 600 amperes.

Example 11.3   Assume the fire pump installation of Example 11.1 is supplied through a customer owned transformer as shown in the right-hand diagram of Figure 11.11. Determine the minimum rating of transformer required for the fire pump installation.

Answer: *NEC® 695.5(A)* requires the transformer to have a kVA rating not less than 1.25 times the full-load volt-amperes of the motor. This is determined using Equation 11.2. Look up the full-load current in *Table 430.250* and multiply by the voltage and by 1.25 and 1.73 to get 102.8 kVA (99 A × 480 V × 1.73 × 1.25 = 102,762 VA).

ALTERNATE ELECTRICAL POWER SYSTEMS

Alternate electric power systems can be used for a variety of applications. These systems can be used as a stand-alone system to provide the total power needs when the building is located in a remote area where extension of utility lines to the building is impractical. Alternate power systems can be used as standby power. In this case, the power system is connected to the normal power system with a transfer switch and operated only when utility power is not available. Another approach is to operate the alternate power system interactively in parallel with the utility power system. This must be done in such a way that the two power systems will be synchronized. Excess power leaves the property and is sold to the utility. The connection is made only by utility approval. The economics of this arrangement vary from state to state. Following is a discussion of the most popular alternate power sources and the major components of alternate power systems.

Solar Photovoltaic Systems

The photovoltaic process was first discovered in 1839. It was in 1954 when Bell Laboratories developed the first practical solar cell to produce dc power. Solid state electronic devices are made of semiconducting materials such as silicon crystals containing an impurity that gives them an apparent charge, either positive or negative. These crystals are called n-type semiconductors and p-type semiconductors. A solar cell consists of an n-type semiconductor in contact with a p-type semiconductor. When light penetrates these semiconducting materials, they absorb much of the energy of the photons of light, which dislodges electrons from their positions within the atomic crystal structure. Electrons from the p-type semiconductor cross the boundary (called a junction) and enter the n-type semiconductor. An electrical field between these two materials prevents the electrons from returning across the junction. If an electrical conductor is attached to each of the semiconductors, the electrons can return from the n-type semiconductor to the p-type semiconductor through an external electrical circuit. A cross-sectional view of a solar cell is shown in Figure 11.13.

A typical solar cell has an area of about 100 square centimeters (15.5 in.^{2}). The power output of a solar cell depends upon the intensity of light received. At noon in areas where the sun is very high and the sky is clear, the output of a solar cell is about 3 amperes or more at 0.5 volts which gives 1.5 watts. Later in the day, the power output will decrease as the sunlight must travel through the thick atmosphere. Located in space above the earth's atmosphere, solar cells produce about 35% more energy than at noon on earth. Solar cells are expensive to produce. Research continues to produce them at lower cost with as high an efficiency as possible. Single-crystal solar cells (mono-crystalline) have the highest practical efficiency which is about 20%. Polycrystalline solar cells (multicrystalline silicon) are less expensive to produce but have a lower efficiency. Amorphous silicon cells consist of a thin film deposited on a base material and are relatively inexpensive to produce, but their efficiency is lower than for single-crystal solar cells.

Solar cells are arranged into modules consisting typically of 36 solar cells connected in series to give a practical operating output of approximately 18 volts dc and a maximum operating current level of about 3 amperes. These are frequently called the maximum power point voltage and current (V_{mpp} and I_{mpp}). The open-circuit voltage of the module is the output voltage with no current being drawn. The open-circuit voltage (V_{OC}) will be higher than the operating output voltage. Wire insulation and all electrical components must be rated higher than the open-circuit voltage of the system. Another important characteristic of a photovoltaic module is the short-circuit current (I_{SC}). This is the maximum current that can be supplied by the photovoltaic module when the output terminals are shorted together. Ambient temperature affects the output of a solar cell. The lower the ambient temperature, the higher the rated open-circuit voltage of the solar cell. Required markings on a photovoltaic module are listed in *690.51* and include these values plus maximum power and maximum permitted system voltage.

Photovoltaic modules are connected together in series to obtain the desired operating voltage. Typical systems operate at 12, 24, and 48 volts. Modules can be connected together in series to obtain a higher sys-

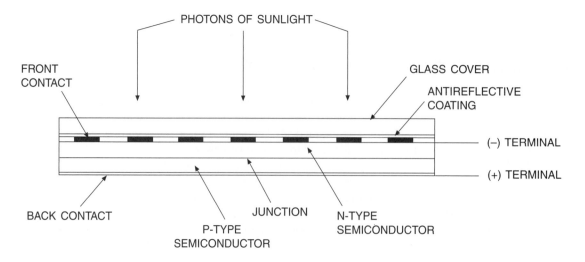

Figure 11.13   A solar photovoltaic cell consists of a p-type semiconductor on the bottom and an n-type semiconductor on top. When sunlight falls on the cell, electrons move across the junction from the bottom layer to the top layer producing about 3 amperes or more of direct current at 0.5 volts.

tem voltage. The maximum permitted system voltage marked on the module must not be exceeded. One module will supply 12 volts, two modules in series will supply 24 volts, and four modules in series will supply 48 volts. Panels are connected in parallel to increase the current and power output form an array. When connected in series, the voltage and power of the individual modules will add. The maximum operating current will remain a constant value. When connected in parallel, the power and current of modules or sets of modules will add. The operating output voltage will remain a constant value. A stand-alone photovoltaic system to provide only dc power will have a photovoltaic array, a charge controller, and a set of batteries. With the addition of an inverter both ac and dc loads can be supplied. A system of this type is shown in Figure 11.14.

The operating voltage of the photovoltaic array must be higher than the nominal operating voltage of the system. Maximum output will not be maintained at all times and the photovoltaic panels must produce a voltage high enough to charge the batteries. Output voltage to the batteries and the loads will be regulated by the charge controller. The charge controller monitors the status of the batteries and the load and controls the rate of charging. By supplying the batteries and loads with a series of dc pulses, the charge controller can maintain the average output voltage at the desired level. When the battery charge is high, the pulses are far apart and when the battery charge is low, the pulses are close together.

Article 690 deals with the installation of the total photovoltaic system. The photovoltaic system may be installed to provide only dc power as a stand-alone power source or fed into an inverter to produce ac 60-hertz power. To maintain a constant power supply throughout the day, this system must include batteries. Rules for the installation of batteries are found in *Article 480*. A typical example will help to understand how the total system ratings are determined and how different sections of the Code apply.

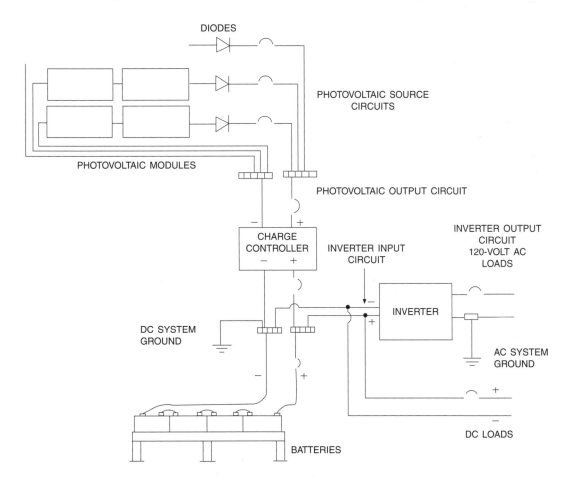

Figure 11.14   A solar photovoltaic power system that produces alternating current for use in a premises wiring system usually has a charge controller, batteries, and inverter along with the solar array. Disconnects are required to isolate each component of the system and overcurrent devices are required to protect all conductors and equipment.

Example 11.4   The individual modules of a mono-crystalline photovoltaic system to be installed for a dwelling is rated with an open-circuit voltage (V_{OC}) of 20.9 volts, a maximum short-circuit current (I_{SC}) of 4.2 amperes, operating voltage of 17.0 volts, maximum operating current of 3.82 amperes, and maximum power (P_{max}) of 65 watts. Each module is 18.5 by 40.5 in. (47 by 103 cm). The system will operate at a nominal 24 volts dc with two modules connected in series for each module set. The array will consist of 12-module sets connected in parallel. The installation is similar to Figure 11.14. The lowest expected winter temperature is 20°F (–7°C). For each module set, determine the following: open-circuit voltage, short-circuit current, operating voltage, maximum operating current, and maximum power output.

Answer: The short-circuit current (I_{SC}) will remain at 4.2 amperes. The maximum output current will remain at 3.82 amperes. There are two modules in series. Therefore, the open-circuit voltage (V_{OC}) will be 41.8 volts (20.9 V + 20.9 V = 41.8 V). The operating voltage will be 34.0 volts (17.0 V + 17.0 V = 34.0 V). The maximum power (P_{max}) will be 130 watts (65 W + 65 W = 130 W).

All wiring and equipment of the photovoltaic system are required to have a rating not less than the maximum open-circuit voltage of the system. *NEC® 690.7* describes the method of determining the open-circuit voltage of the system. The open-circuit voltage of the module sets must be corrected for the lowest expected ambient temperature. For Example 11.4, the open-circuit voltage of each module set is 41.8 volts and the correction factor from *Table 690.7* is 1.13 to give a maximum photovoltaic system voltage of 47.2 volts (41.8 V × 1.13 = 47.2 V). These values are used to size the conductors, overcurrent protection, and ratings of disconnects.

NEC® 690.15 requires that disconnecting means be provided so that all components can be isolated from all sources of power. In the case of a photovoltaic power source, a voltage will be present any time the sun is shining on the solar cells. Switches and circuit breakers installed in the dc circuits must be dc rated. *NEC® 690.8(B)* specifies that conductors shall have an ampacity not less than 1.25 times the computed maximum circuit current determined in *690.8(A)*. Photovoltaic systems are assembled in different ways, and care must be taken to identify the current that can flow on the various sections of the circuit.

Example 11.5   The photovoltaic array described in Example 11.4 supplies 60 hertz ac power at 120 volts, with a 2500-watt continuously rated true sine wave inverter. The inverter has a surge rating of 7500 watts. The continuous output current is 20.8 amperes. The nominal input is 24 volts and 123 amperes dc. The input voltage range is 14.9 to 30.7 volts dc with an input current at minimum voltage of 197 amperes dc. The efficiency of the inverter is 85%. There are 24 deep cycle, 6-volt lead-acid batteries with a rating of 350 ampere-hours. The batteries can deliver up to 75 amperes for three hours continuously on a charge or 17.5 amperes for 20 hours. Four batteries are connected in series to get 24 volts and six sets are connected in parallel. The ambient temperature of the wiring on the solar array could reach 130°F (54°C). Determine the minimum size copper conductors with 75°C rated terminations permitted for the following: (1) the photovoltaic source circuits to the solar modules and (2) the photovoltaic output circuit to the charge controller.

Answer: First determine the photovoltaic source circuit current, *690.8(A)(1)*. There are two modules connected in series for each source circuit with a maximum short-circuit current of 4.2 amperes. The source circuit current is the short-circuit current multiplied by 1.25 to get 5.25 amperes. *NEC® 690.8(B)(1)* requires the source circuit conductors to have an ampacity not less than 1.25 times the value calculated by *690.8(A)(1)*, which is 6.6 amperes (5.25 A × 1.25 = 6.6 A). Because the ambient temperature of the conductors is above 30°C, a temperature correction is required according to *690.31(C)*. The correction factor from *Table 690.31(C)* is 0.67. Divide the 6.6 amperes by 0.67 to get 9.9 amperes. From *Table 310.16*, the minimum conductor size permitted is 14 AWG. However, a smaller size may be permitted.

The photovoltaic output circuit current is the sum of the source circuit maximum currents, *690.8(A)(2)*. The source circuit current was determined to be 5.25 amperes. This solar array consists of six source circuits connected in parallel for an output current of 31.5 amperes. *NEC® 690.8(B)(1)* requires the conductor to have an ampacity not less than 1.25 times the output circuit current, which is 39.3 amperes. If the conductors are exposed to an ambient temperature greater than 30°C (86°F), then an adjustment must be made using the appropriate factor from *Table 690.31(C)*. Assuming no adjustment is needed, the minimum photovoltaic output circuit conductor size is 8 AWG copper. Single conductor cables listed as sunlight and moisture resistant are permitted to be installed for individual module interconnections, *690.31(D)*, in sizes 18 AWG copper and larger.

The maximum current output from the charge controller will not be greater than the maximum photovoltaic output circuit supplying the charge controller from the solar array. Therefore, the output dc conductors from the charge controller will be the same as the dc input conductors. Batteries will release current slowly for a long period of time or at a high ampere rate for a limited period of time. An overcurrent device near the batteries is important to minimize the length of unprotected circuit. The battery output conductors will depend upon the rating of the overcurrent device protecting the battery circuit. The rating of that overcurrent device will depend upon the maximum expected dc load to be drawn from the batteries. If the batteries supply an inverter, the maximum current will be the maximum draw of the inverter at the inverter minimum input voltage and producing the maximum continuous ac output current, *690.8(A)(4)*. If dc loads are supplied directly from the batteries, the sum of the ratings of the dc output circuit overcurrent devices is added to get the total dc load.

Example 11.6   Consider the photovoltaic system described in Examples 11.4 and 11.5 where all of the dc output of the system supplies the 2500-watt inverter that is 85% efficient and has a 2-wire 120-volt ac circuit that draws 20.8 amperes. The dc input to the inverter is nominal 24 volts with a minimum of 14.9 volts and a maximum of 30.7 volts. The maximum input current at minimum input voltage is 197 amperes. Determine the minimum size copper conductors between the batteries and the inverter and determine the minimum size 120-volt ac conductors from the inverter to the premises wiring system transfer switch.

Answer: The input dc current at minimum voltage is 197 amperes. *NEC® 690.8(B)(1)* requires the conductor ampacity to be not less than the value determined in *690.8(A)(4)*. The overcurrent device will be 250 amperes (200 A × 1.25 = 250 A). The minimum conductor size will be 4/0 AWG copper between the battery and the inverter applying the rule of *240.4(B)*. If the dc input current at minimum voltage is not given, it can be calculated by dividing the continuous output rating of the inverter by the minimum input voltage and the efficiency.

$$\text{dc inverter input current} = \frac{2500 \text{ W}}{14.9 \text{ V} \times 0.85} = 197 \text{ A}$$

The inverter ac output conductor minimum size is determined by multiplying the nameplate output rating by 1.25 as stated in *690.8(B)(1)* to get 26 amperes. The minimum conductor size is 10 AWG copper.

Fuel Cell Systems

A fuel cell is an electrochemical device that converts hydrogen and oxygen into electricity, water, and heat. It is like a battery except, rather than requiring periodic recharging, it runs continuously as long as it gets a constant supply of hydrogen gas and oxygen. The process was discovered in 1839 and remained a curiosity until NASA developed it as an energy source for space travel. Fuel cells were not practical until recently because they were expensive to build. In order for the process to work, hydrogen molecules had to be broken down into its basic components of electrons and protons. That can be accomplished by exposing the hydrogen molecules to a material called a catalyst. The catalyst that can break down hydrogen contains the metal platinum, which is very expensive. Advances in chemistry and engineering in recent years have reduced the amount of platinum needed in the catalyst. Advantages of fuel cells is they have no moving parts, they run at high efficiency, they produce virtually no noise, and if operated on pure hydrogen, they run pollution free. All of the major automobile manufacturers are testing fuel cell powered electric vehicles. Fuel cells are also being tested as electric power sources for dwellings. As a result, the Code makes provisions in *Article 692* for fuel cell connection to premises wiring systems.

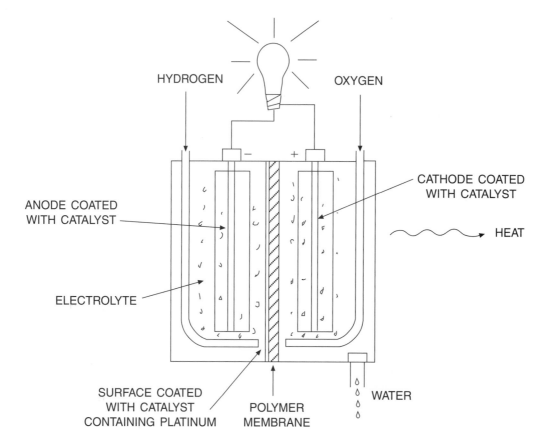

Figure 11.15　A fuel cell has two chambers containing an electrolyte and separated by a membrane. Hydrogen gas is injected into one chamber and oxygen into the other. A catalyst separates the hydrogen molecules into protons and electrons. The protons move through the membrane to the other side and combine with oxygen to form water and heat. The electrons must travel through an external circuit as direct current to get to the other chamber.

A simplified view of a fuel cell is shown in Figure 11.15. Fuel cells manufactured today are actually very thin, and they are assembled together in long stacks similar to slices of bread. The fuel cell has two chambers separated by a membrane that is porous to protons. There is an electrode in each chamber partially coated with a catalyst. Hydrogen molecules are fed into one changer, and the catalyst breaks down the molecules into protons and electrons. The protons pass through the membrane to the other side. The electrons are trapped in the chamber and build up a negative charge. The protons arriving in the second chamber create a positive charge. A little less than one volt will develop between the electrodes in each chamber. If a circuit is completed between the two electrodes, electrons can flow from the first to the second chamber. Oxygen is pumped into the second chamber which combines with the electrons and protons to form water. Heat is also produced during this chemical formation of water. Fuel cells produce direct current and many cells are assembled into stacks and connected in series to obtain the desired output voltage.

There are a number of different types of fuel cells available or being tested. Two types are being used as stationary units to produce power for premises wiring or interconnection to electric utility grids. One is the phosphoric acid fuel cell and the other is the proton-exchange membrane fuel cell. The latter type is being tested as a power source for electric vehicles.

Air contains plenty of oxygen to operate a fuel cell but hydrogen in the form of a gas is needed. Hydrogen gas will most likely be readily available in the future, but for now, it is not easy for the consumer to obtain. Petroleum fuels and alcohols are good sources of hydrogen. Fuel cells are available that run on natural gas, propane, ethanol, methanol, gasoline, and other fuels. Before the fuel cell can get the hydrogen it needs from these fuels, the fuel must first be reformed to free the hydrogen to become a gas. A device called a **reformer** must be installed between the fuel tank and the fuel cell. A reformer is not needed if pure hydrogen is used as the fuel. The reformer takes oxygen out of the air to combine with the carbon in the fuel to free the hydrogen. The waste product from the reformer is mostly carbon dioxide and a little carbon monoxide.

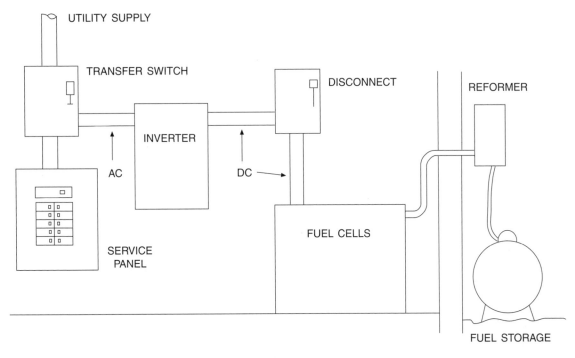

Figure 11.16   A fuel cell system operating as a standby power source uses an inverter to change the direct current to alternating current. The connection to the premises wiring system can be by means of a transfer switch. Fuels such as natural gas, ethanol, propane, and gasoline can be used as an energy source but a reformer is needed to free the hydrogen so it can be utilized by the fuel cell.

The direct current from the fuel cells can be used to power loads or it can be passed through an inverter to convert it to alternating current. A typical fuel cell power system produces a constant supply of electricity day and night and is illustrated in Figure 11.16. In order to handle temporary peak loads, the fuel cell power system can be combined with batteries to extend its capacity. The fuel cell power system can be interconnected to operate in parallel with the utility power grid. This may not be economical in many areas of the country. A fuel cell power system can act as a stand-alone system to power a building with no connection to the utility power grid. This may be especially desirable in remote areas where utility electrical lines are not available. It would be necessary to transport fuel to the location such as propane. Another alternative is to use the fuel cell power system as a standby power unit and connect it to a standard service through a transfer switch. Fuel cells produce heat as well as electricity. The heat can be reclaimed for water heating and space heating.

Storage Batteries

A stationary installation of storage batteries to provide electrical energy for premises lighting, power, and heating may be a separate system or it may be a part of a number of alternate power systems. Decisions relative to guarding of live parts in *110.27* or grounding of the direct current system in *250.162* depends upon the system voltage. *NEC® 480.2* under nominal battery voltage specifies that voltage is taken as a nominal value by cell. For a lead-acid battery it is 2 volts per cell and for an alkali battery it is 1.2 volts per cell. A 6-volt lead-acid battery has three cells and a 12-volt lead-acid battery has six cells.

Other important concerns are ventilation of the battery room, *480.8(A)*, working space around battery racks, *480.8(C)*, and containment of electrolyte leakage, *480.8(D)*. Ventilated lead-acid batteries release hydrogen when charging. If adequate ventilation is not provided, an explosive mixture can accumulate. The actual rules for determining working space about battery installations is found in *110.26. Table 110.26(A)* does not specify whether the voltages are ac or dc. Therefore, they apply to both ac and dc voltages. Minimum working space in front of equipment is given in *Table 110.26(A)* as 3 ft (900 mm). *NEC® 110.26(A)(1)(b)* permits a smaller working distance by special permission for systems operating at not more than 60 volts dc. Typical battery systems for alternate power production operate at 12, 24, and 48 volts.

Batteries used for home power should be of the deep cycle type. These are different from automotive batteries. Deep cycle batteries are built to be repeatedly heavily discharged. They give up their power at a slower rate than an automotive battery but they can take hundreds of deep discharging and recharging cycles. Their energy capacity is given in ampere-hours. Assume a deep cycle battery is rated 105 ampere-hours. Theoretically, the battery can give up its energy at a rate of 105 amperes for one hour or at a rate of one ampere for 105 hours. These batteries are generally rated for an even discharge over 20 hours. This would mean that a 105 ampere-hour battery is rated to deliver a steady current of 5.25 amperes for 20 hours. If more current is needed, multiple batteries can be connected in parallel. For example, if four 105 ampere-hour batteries are connected in parallel, they will be designed to deliver 21 amperes. The batteries will deliver current at a much faster rate, but batteries will deliver the maximum energy to a load if discharged slowly.

Golf cart batteries are generally used for home power applications. They deliver 6 volts and have a rating between 220 and 300 ampere-hours. They are unsealed and must be housed in a well ventilated area. Since most home applications are at 12 or 24 volts, they are required to be connected in series to obtain the required voltage and in parallel to obtain the desired current delivering capacity. Industrial stationary batteries deliver 2 volts and have ratings up to 3000 ampere-hours. They are generally too bulky and hard to handle for home applications.

Inverters

An inverter is an electronic device that converts direct current to alternating current. Alternate power sources such as batteries, photovoltaics, fuel cells, and some wind machines produce direct current. This direct current can be used for many applications but in the home there are appliances that require alternating current. Most inverters are made to accept 12, 24, or 48 volts dc. Inverters made for the U.S. market generally have a 2-wire output of 120-volt, 60-hertz ac. Some manufacturers list their inverters to be connected in parallel to the dc supply with their outputs connected in series to produce 120/240-volt, 3-wire ac. The Code recognizes that most inverters will have a 2-wire, 120-volt ac output. *NEC® 692.10(C)* permits a 120/240-volt, 3-wire panel to be supplied by a stand-alone 2-wire 120-volt power source provided there are no 240-volt loads operated or there are no multiwire branch-circuits.

The output of an inverter may be a true sine wave or it may be a modified sine wave. A true sine wave is more difficult to produce and the equipment is more expensive. The output from a true sine wave inverter and a modified sine wave inverter is shown in Figure 11.17. The modified sine wave inverter may result in interference with some electronic equipment.

Generators

Generators are available that produce alternating current and direct current. An ac generator is used for the purpose of supplying power to a premises wiring system. Rules applying to generators are found in

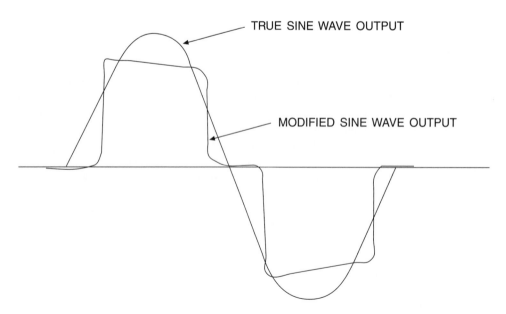

TRUE SINE WAVE OUTPUT

MODIFIED SINE WAVE OUTPUT

Figure 11.17   The alternating current waveform from an inverter is usually either a true sine wave or a modified sine wave.

Article 445. NEC® 445.12(A) requires that generators be provided with overload protection. No specific rules are provided as to how this is accomplished. It is, therefore, the responsibility of the manufacturer to provide generator overload protection. If overcurrent protection is not present and the generator is not marked as inherently protected, then overload protection is required to be installed at the generator end of the conductors supplying the premises wiring system. All the electrician can do is base the overcurrent protection on the full-load current rating of the generator.

The nameplate is required to provide the generator rated full-load output current. If that value is not given, the full-load current can be estimated from the continuous kW rating and the output voltage. Equations 11.4 and 11.5 can be used to estimate the full-load current of an ac generator if only the continuous kW rating and voltage are given.

$$\text{Approximate Full-Load Current}_{1\text{-Phase}} = \frac{kW \times 1000}{\text{Voltage}} \qquad \text{Eq. 11.4}$$

$$\text{Approximate Full-Load Current}_{3\text{-Phase}} = \frac{kW \times 1000}{1.73 \times \text{Voltage}} \qquad \text{Eq. 11.5}$$

NEC® 445.18 requires that a disconnecting means be installed at the generator on the load side of all overcurrent and control devices. Generally, this disconnecting means is an integral part of the generator. *NEC® 445.13* applies to the conductors from the generator terminals to the premises wiring system. The conductors between the generator and the point of connection to the premises wiring system are not permitted to have an ampacity less than 1.15 times the generator rated continuous current. The conductors must have an allowable ampacity not less than the rating of the overcurrent protection at the generator. For a permanent installation, the conductors are usually run in raceway and sized according to *Table 310.16,* as shown in Figure 11.18. For a portable generator installation where multiconductor flexible cord is used, the conductors are sized using *Table 400.5(A).* Large portable generator installations use single-conductor cables with pin and sleeve connectors for each conductor. These cables when used as individual conductors separated in air are sized using *Table 400.5(B).*

Connecting Alternate Power Sources

Connecting an alternate power source to a premises wiring system can be done in several ways depending upon the purpose of the installation. The most simple installation is where an alternate power source is the only electrical supply to a building. In this case, the building is wired as normal unless the power supply is 2-wire, 120-volt, 60-hertz ac. In this case, 240-volt circuits and multiwire branch-circuit with a common neutral are not permitted. A stand-alone power source such as a photovoltaic system or a fuel cell system is connected directly to the main disconnect or service panel for the building. A building panel

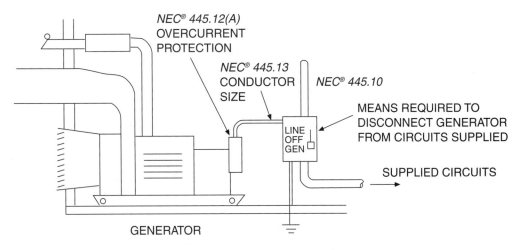

Figure 11.18   The output conductors from a standby generator are required to be sized based upon the rating of the overcurrent device on the generator, or if not provided at the generator, not less than 1.15 times the continuous full-load output of the generator.

TAP BOX

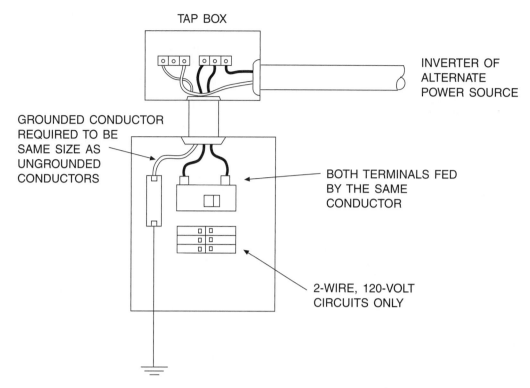

INVERTER OF
ALTERNATE
POWER SOURCE

GROUNDED CONDUCTOR
REQUIRED TO BE
SAME SIZE AS
UNGROUNDED
CONDUCTORS

BOTH TERMINALS FED
BY THE SAME
CONDUCTOR

2-WIRE, 120-VOLT
CIRCUITS ONLY

Figure 11.19   A 120-volt, 2-wire inverter output is permitted to supply power to a 3-wire panelboard provided there are no 240-volt loads or multiwire branch-circuits. For this stand-alone system, the connections are made in a tap box with both sides of the main circuit breaker supplied by the same undergrounded conductor from the inverter.

supplied from a 2-wire, 120-volt, 60-hertz inverter is shown in Figure 11.19. This type of connection is permitted by *690.10(C)* and *692.10(C)*.

Alternate power sources may be installed to act as backup systems to the normal utility power supply. A means must be installed to transfer the load or circuits from the utility supply to the alternate power source. This must be done with some type of transfer equipment in such a way that it prevents inadvertent interconnection of normal and alternate power sources. This is essential to ensure the safety of personnel at the premises and working on utility lines. Figure 11.16 shows a single-phase double-pole, double-throw transfer switch installed ahead of the main service panel so that all circuits can be supplied from the alternate power source. A transfer switch installed at this location will act as the service disconnecting means and must be rated as suitable for use as service equipment. The requirement for the installation of transfer equipment that is legally required is found in *701.7(A)*. The rule for the installation of optional standby power equipment is found in *702.6*. An optional standby power system is defined in *702.1*. An optional standby power system is one that is permanently installed including the generator and prime mover and also a portable generator connected to a premises wiring system through permanently installed transfer equipment such as shown in Figure 11.3.

For the connection of portable generators for temporary installations where the connection is to a premises wiring system normally supplied by utility power, the rules for connection of the alternate power source are found in *702.6(C)*. The connection is required to be made in such a manner that an inadvertent connection between utility power and alternate power cannot occur. The load transfer to the standby generator is permitted by *702.6* where a transfer switch is connected to each circuit that is to be switched to the alternate power source as illustrated in Figure 11.20. A fuel cell electrical power system may be used as a back-up system for utility power in which case the method of interconnection to the premises wiring system is covered in *692.59*.

Grounding dc Power Sources

Direct current electrical system grounding is covered in *Article 250, Part VIII*. Generally dc systems are operated as 2-wire systems, although there are applications for dc 3-wire electrical system. Standard convention for direct current systems is to ground the negative terminal or conductor. The ungrounded conductor is positive. Red terminals are ungrounded or positive, and black or white terminals are grounded or negative. *NEC® 250.162(A)* requires that 2-wire dc systems operating at more than 50 volts, but not more than 300 volts and serving a premises wiring system, are required to have one conductor grounded. Any 3-wire direct current system serving premises wiring is required to have one conductor grounded. *NEC® 250.160* makes reference to other sections of *Article 250* not specifically intended for ac power systems, and therefore, *250.52* will apply when selecting a grounding electrode for a dc power source.

The rules for sizing the grounding electrode conductor are found in *250.166*. The minimum size copper grounding electrode conductor in any case is 8 AWG. The minimum size grounding electrode conductor is not permitted to be smaller than the ungrounded conductor. If the grounding electrode is a rod, plate, or pipe, the grounding electrode conductor is not required to be larger than size 6 AWG. If a concrete-encased electrode is used, the maximum size required is 4 AWG. If a grounding ring is used, the minimum size conductor for the grounding ring is 2 AWG and the grounding electrode conductor is required to be of the same size. This grounding electrode conductor is required to be connected to the negative conductor either at the dc source or at the first overcurrent device or disconnecting means supplied by the source. Optional grounding of dc power systems intended to supply premises wiring is common.

NEC® 250.168 requires a bonding jumper to connect the grounded-circuit conductor, the grounding electrode conductor, and the equipment grounding conductor. This conductor is required to be of the same size as the grounding electrode conductor sized in accordance with *250.166*.

Even if a dc electrical system produced by a stand-alone power source is not required to have one conductor grounded to the earth, it is required to provide a grounding electrode and connect it to the metallic enclosures of the electrical systems with a equipment grounding electrode conductor. The rules for sizing the grounding electrode conductor are the same as described in *250.166*.

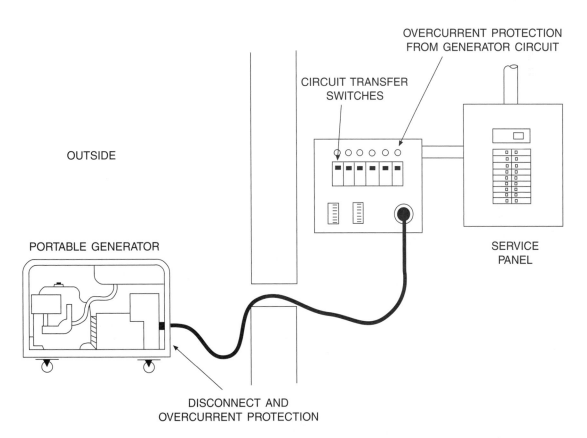

NEC® 702.1 OPTIONAL STANDBY POWER SYSTEMS NOW
INCLUDE PORTABLE GENERATOR CONNECTIONS TO PREMISES WIRING SYSTEMS

OVERCURRENT PROTECTION
FROM GENERATOR CIRCUIT

CIRCUIT TRANSFER
SWITCHES

OUTSIDE

SERVICE
PANEL

PORTABLE GENERATOR

DISCONNECT AND
OVERCURRENT PROTECTION

Figure 11.20   A portable generator connected to a premises wiring system is permitted to be connected through a load transfer device either on the supply side or the load side of the service disconnecting means.

Direct Current Rating

Direct current as discussed in *Unit 1* is a steady flow of current in one direction. The voltage maintains a constant level and the current never stops flowing. With an ac circuit, the voltage reverses polarity two times each cycle and reaches a zero value two times each cycle. The current builds to a maximum and decreases to zero and stops. Then the current reverses direction. The current actually stops flowing two times each cycle. It is easier to open a circuit when the current stops flowing two times each cycle as opposed to continuously flowing in one direction at a constant level. For this reason, devices intended to break current flow in a dc circuit must be rated for dc operation.

MAJOR CHANGES TO THE 2005 CODE

These are the changes to the 2005 *NEC®* that correspond to the Code sections studied in this unit. The following analysis explains the significance of the changes from the 2002 to the 2005 Code only and this analysis is not intended to be used in place of the Code. Refer to the actual section of the 2005 Code for the exact wording and meaning of each section discussed. Changes are indicated in the Code with a vertical line in the margin. If material was deleted or moved to another location in the Code, the location of the deletion is indicated with a dark dot in the margin.

Article 445   **Generators**

445.11: This section states the required markings for generators. It will now be necessary for the manufacturer to provide impedance values for the generator that can be used in calculating the short-circuit energy capabilities of the generator.

445.18: A generator is now permitted to have more than one disconnect so it can supply more than one feeder. Previous editions of the Code limited the generator disconnect to only one. This is illustrated in Figure 11.21.

Article 690   **Solar Photovoltaic Systems**

690.2: A new definition was added for a photovoltaic unit that is part of the protective outer surface of a building. The new term is "building integrated photovoltaics."

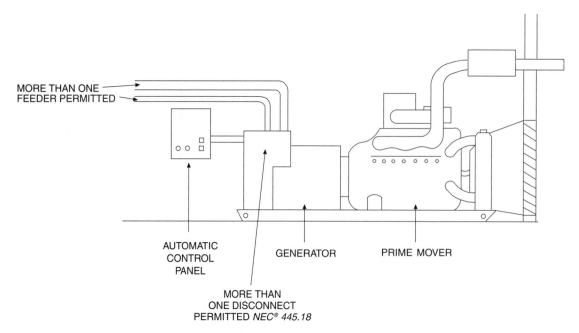

MORE THAN ONE FEEDER PERMITTED

AUTOMATIC CONTROL PANEL

MORE THAN ONE DISCONNECT PERMITTED *NEC® 445.18*

GENERATOR

PRIME MOVER

Figure 11.21   A standby generator output is permitted to have two or more disconnects, and it is now permitted to supply multiple feeders from the generator.

690.14(D): An inverter that is connected interactive with the utility is permitted to be mounted in a location that is not readily accessible such as on a roof or the exterior of a building. A disconnecting means is required to be located in sight of the inverter for the photovoltaic source and for the output conductors from the inverter.

690.17 Exception 1: This exception in the previous edition of the Code was deleted. A disconnect is required in the dc circuit between an inverter and the photovoltaic source to isolate the inverter from the source. This disconnect is no longer permitted to have a short-circuit rating less than the available short-circuit current that can be produced by the photovoltaic source.

690.31(E): This is a new subsection that requires direct current conductors from a photovoltaic source to be run in metal raceway or enclosures from the point of entry of a building or structure to the disconnecting means. These are the dc conductors supplying the inverter that is connected interactive with the utility and mounted inside a building or structure.

690.35: Photovoltaic power sources are now permitted to be operated ungrounded. The direct current output conductors from the photovoltaic array are required to be provided with a disconnect, overcurrent protection, and a ground-fault detection system that indicates a ground-fault condition has occurred, and automatically disconnects all conductors from the source. A diagram of an ungrounded photovoltaic system is shown in Figure 11.22. The dc output conductors are required to be run as a sheathed cable or run in raceway. Warnings that a shock hazard may exist are to be placed at all points where the conductors may be accessible for servicing. The inverter or charge controller is required to be listed to operate with an ungrounded direct current input.

690.41 Exception: This section required photovoltaic direct current output circuits operating at more than 50 volts to be grounded. This new exception recognizes an ungrounded photovoltaic circuit that is ungrounded as meeting the conditions of *690.35*.

690.47(C): This section specifies the rules for grounding the ac and dc circuits of a photovoltaic power system. Subsection *(C)* is new, and it deals with the usual case where there is a grounded ac system and a grounded dc system. This is a new requirement that permits the ac and dc circuits to be grounded to the same electrode. The rules for choosing a grounding electrode for a service are more restrictive. If there are a separate grounding electrodes for the ac and dc systems, then the two grounding electrodes are required to be bonded together using a conductor not smaller than the larg-

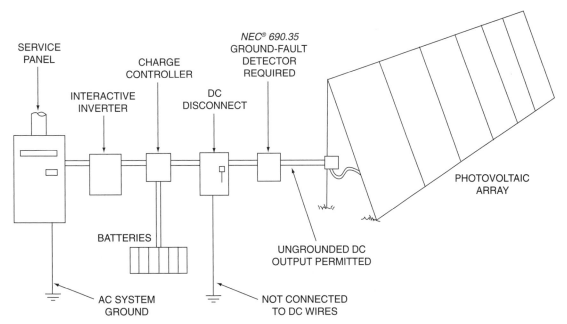

Figure 11.22   A photovoltaic source is permitted to be operated ungrounded, but it is required to install a ground-fault detector.

er of the grounding electrode conductors from either the ac system or the dc system as illustrated in Figure 11.23.

690.48: It is common practice to route an equipment bonding jumper for a photovoltaic system through raceways and equipment. With such a practice, removal of any component of the system will interrupt the continuity of the bonding jumper, potentially leaving components of the system ungrounded while illumination of the photovoltaic array produces voltage. This section requires a bonding jumper to be installed when removal of a component of the system will result in an interruption of the equipment grounding.

690.49: Grounding of the dc output circuit from the photovoltaic array is permitted by *690.42* to be at any single point on the output circuit. If the grounding connection is made at an inverter and the inverter is removed from the system for servicing, connection to the grounding electrode from the dc grounded conductor can be lost. This section requires a bonding jumper to be installed in the case where removal of any component of the system interrupts the bonding or grounding. By grounding the dc circuit at the first disconnecting means located adjacent to the photovoltaic array, the problem of loss of grounding of the dc circuit is avoided.

690.64(B)(5): A common method of connecting a photovoltaic system inverter output to operate interactively with the utility supply is to connect the inverter to backfeed a circuit breaker in the service panel. This circuit breaker is not required to be individually clamped to the panel as would be the case of a main circuit breaker feeding a panel. Clamping of the circuit breaker by the front cover is considered sufficient for this purpose. If the circuit breaker should separate from the panelboard bus, a listed interactive inverter will shut down automatically and the backfed circuit breaker will be de-energized.

690.71(B)(1): This paragraph limits dwelling battery systems associated with a photovoltaic array to 50 volts maximum. When lead-acid cells are used, nominal voltage is 2 volts per cell. There is now a new sentence that limits a dwelling lead-acid battery system to a maximum of 24 cells in series.

690.72(B)(3)(2): A photovoltaic power system operating interactive with the utility will typically have a battery system where the charge is maintained by excess power from the photovoltaic array or from the utility grid. Loss of utility power, failure of the inverter or switching off of the inverter can result in

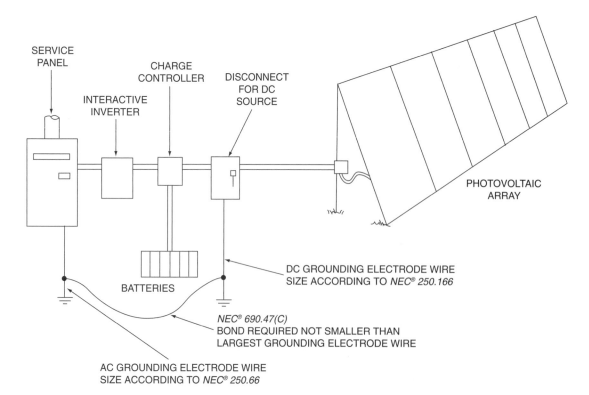

Figure 11.23   A photovoltaic system is permitted to have a common or a separate grounding electrode for the ac and dc systems and, if separate, the grounding electrodes shall be bonded with a conductor not smaller than the largest system grounding electrode conductor.

loss of charge control to the batteries. This new sentence requires an alternate means of charge control for the battery system if loss of the inverter can result in loss of charge control to the batteries.

Article 695 Fire Pumps

695.4(B)(1): This section requires that any overcurrent device in the circuit supplying a fire pump be set at not less than the locked-rotor current of the fire pump motor plus any other associated load current. A new last sentence was added, making it clear that this requirement applies only to the overcurrent device. Motor controllers, wires, and any other component in the circuit are not required to be sized to carry this level of current.

695.4(B)(2)(3): This new sentence makes it clear that the disconnecting means for a fire pump is not permitted to be located in the same enclosure that feeds other loads not associated with the fire pump.

695.5(B): This section prohibits an overcurrent device from being installed in the secondary of a transformer supplying a fire pump. The primary overcurrent device is required to be set higher than the locked-rotor current of the pump motor plus any other associated loads, and corrected for voltage change. A new last sentence was added that makes it clear that the transformer and conductors are not expected to be sized to carry a load equivalent to the locked-rotor current of the pump motor.

695.5(C)(2): When sizing the feeder for a fire pump, the conductors are not required to be sized to carry the locked-rotor current of the pump motor plus the other associated loads.

695.6(C)(2): This section deals with sizing of conductors to supply a fire pump and no other loads. A new last sentence was added that makes it clear that the voltage-drop requirements of *695.7* are a requirement that may result in the conductor size being increased.

695.6(E): Listed Type MC cable with an impervious covering is now permitted as wiring to supply a fire pump as illustrated in Figure 11.24.

695.6(H): This is a new section that makes it clear that equipment ground-fault protection is not permitted to be provided for a fire pump service even if it is supplied by a solidly grounded system operating at over 150 volts to ground and rated at more than 1000 amperes.

695.14(E): Fire pump control wiring is now permitted to be run as listed Type MC cable with an impervious covering.

Article 700 Emergency Systems

700.5(B): If an alternate power source has adequate capacity to supply the emergency and other necessary loads, now automatic load shedding equipment is not required.

700.9(D)(1)(1): If an emergency feeder is protected by a fire suppression system, the fire suppression system is now only required to be in the areas of the feeder, not necessarily the entire building.

700.12(E): A fuel cell system that is sized to power the full load of the emergency system for a period of not less than 2 hours is permitted to serve as the power source for the emergency system. Where a fuel cell system serves as a normal source of power for a building or group of buildings, it is not permitted to serve as the only power source for the emergency system.

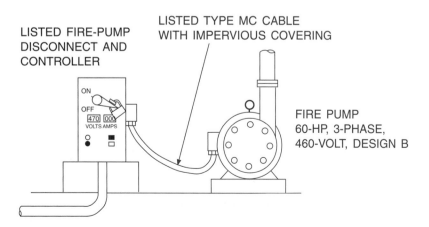

Figure 11.24 Type MG cable with an impervious covering is permitted to be used as wiring for a fire pump system.

700.27: The fine print note on overcurrent device coordination that was in *700.25* is now a requirement in this section. Coordination of overcurrent devices of the emergency system is now a requirement. This means that branch-circuit overcurrent devices are required to open before feeder overcurrent devices open on a ground fault or short circuit. It will now be necessary to examine the time-current characteristics of overcurrent devices in the emergency system, such as the layout in Figure 11.25, to make sure the down-line overcurrent devices open before the overcurrent devices near the source.

Article 701   Legally Required Standby Systems

701.11(F): A fuel cell system that is sized to power the total load for a period of not less than 2 hours is permitted to serve as a legally required standby power system. Where a fuel cell system serves as a normal source of power for a building or group of buildings, it is not permitted to serve as the only power source for the load required to be supplied.

701.18: Overcurrent devices in a legally required standby system are now required to be coordinated. This means that branch-circuit overcurrent devices are required to open before feeder overcurrent devices open on a ground fault or short circuit. It will now be necessary to examine the time-current characteristics of overcurrent devices in the emergency system to make sure the down-line overcurrent devices open before the overcurrent devices near the source.

Article 702   Optional Standby Systems

702.6 Exception: This section requires transfer equipment to be installed for the connection of an optional standby power source to a premises wiring system. This new exception permits, for an industrial facility, the temporary connection of an optional standby power source to the premises wiring system not through transfer equipment. The requirements are that this be done only by qualified personnel, that the normal power supply disconnect be capable of being locked in the open position, and that there be written safety procedures for connecting the standby power source.

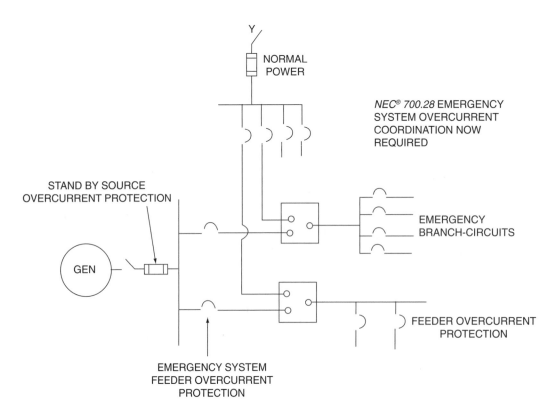

Figure 11.25   The overcurrent devices of an emergency system are now required to be coordinated such that a ground fault or short circuit on the branch-circuit or feeder will not open an up-line overcurrent device before opening the overcurrent device on the faulted circuit or feeder.

702.7 Exception: Signals are required to indicate when there is a malfunction with the optional standby power system, and when the system is carrying load. This new exception exempts portable standby power sources from this requirement.

702.11: If a disconnecting means is provided at an outside generator set that is located within sight of the building or structure served, another disconnecting means is not required at the building or structure served.

Article 760 **Fire Alarm Systems**

760.21: Non-power-limited fire alarm circuits are no longer permitted to be supplied by a circuit that is protected with an arc-fault circuit-interrupter (AFCI).

760.30(B)(2) Exception 3: Non-power-limited fire alarm cable rated for plenums with the circuit integrity marking (CI) is permitted to be installed in other spaces used for environmental air as a 2-hour fire-rated cable.

760.30(B)(3) Exception 3: Non-power-limited fire alarm cable rated for use as a riser with the circuit integrity marking (CI) is permitted to be installed as a riser between floors as a 2-hour fire-rated cable.

760.41: The power source for power-limited fire alarm circuits is no longer permitted to be supplied by a circuit that is protected with an arc-fault circuit-interrupter (AFCI).

760.56(D): This is a new section that prohibits power-limited fire alarm circuit conductors and Class 2 or Class 3 audio system circuit conductors to share the same cable or raceway. The issue is that a fault in the audio cable can have sufficient power to damage the fire alarm cable.

760.61(A): Type FPLP-CI cable is permitted to be installed in an environmental air plenum and is approved as a 2-hour circuit-integrity-rated cable.

760.61(B)(1): Type FPLR-CI cable is permitted to be installed as a riser between floors of a building and is approved as a 2-hour circuit-integrity-rated cable.

760.61(C): Type FPL-CI cable is permitted to be installed in all locations except in plenums or used as a riser and is approved as a 2-hour circuit-integrity-rated cable.

WORKSHEET NO. 11—BEGINNING EMERGENCY AND ALTERNATE POWER SYSTEMS

Mark the single answer that most accurately completes the statement based upon the 2005 Code. Also provide, where indicated, the Code reference that gives the answer or indicates where the answer is found, as well as the Code reference where the answer is found.

1. A 5.5-kW portable generator supplies 120/240-volt single-phase power to a generator panel in a building that is installed on the load side of the service disconnecting means. The panel contains single-pole double-throw switches installed on selected circuits from the main panel. The output of the generator has an overcurrent protector in each ungrounded conductor that does not serve as a disconnecting means at the generator. The generator is accessible and the engine can easily be shut off. A single disconnecting means is (see Figure 11.24):
   A. required to be installed either in the building or at the generator to disconnect all ungrounded conductors.
   B. required to be installed at the generator and also in the building.
   C. required to be installed only at the generator.
   D. required to be installed only at the termination of the conductors within the building.
   E. not required to be installed because the engine can readily be shut down.

   Code reference _____

2. A standby power system for sensitive equipment in a commercial building is supplied by a power system using single-cell stationary lead-acid batteries. The installation is shown in Figure 11.26. The nominal output voltage of 24 single-cell lead-acid batteries connected in series is:
   A. 24 volts.   C. 48 volts.   E. 144 volts.
   B. 28.8 volts.   D. 72 volts.

   Code reference _____

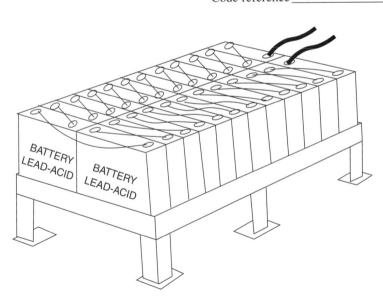

Figure 11.26   Twenty-four lead-acid, liquid electrolyte, single-cell batteries are connected in series as a part of an alternate power system.

3. Storage batteries are installed at a dwelling as a part of a solar photovoltaic power system. The cells of the storage batteries are not permitted to be connected in such a way that the system operates at:
   A. more than 24 volts.
   B. 50 volts or more.
   C. 67 volts or more.
   D. more than 120 volts.
   E. more than 150 volts.

   Code reference _____

4. A stand-alone photovoltaic system is shown in Figure 11.27. The photovoltaic source circuit in Figure 11.27 is labeled with the letter:
   A. A
   B. B
   C. C
   D. D
   E. E

   Code reference _____

5. The nameplate of a fuel cell electrical power unit, among other information, is required to include:
   A. maximum short-circuit current.
   B. minimum output feeder conductor size.
   C. maximum intermittent current rating.
   D. minimum safe operating voltage.
   E. continuous output current rating.

   Code reference _____

6. A fuel cell system is located outside and the feeder conductors run underground to the building and terminate at the distribution panel in the building. A disconnecting means for the feeder between the fuel cell unit and the building:
   A. shall be located at the fuel cell unit.
   B. shall be a fusible switch located on the outside of the building at the point of entry of the fuel cell feeder conductors.

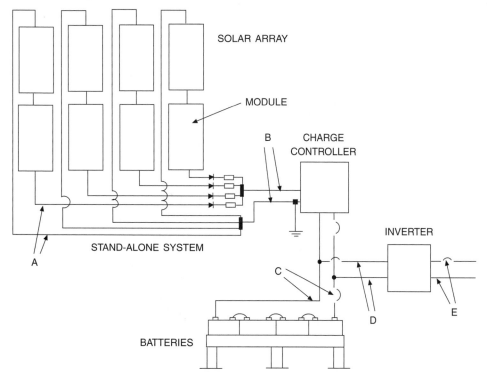

Figure 11.27   A photovoltaic power system consists of the solar array, a charge controller, a set of batteries, and an inverter, along with circuit breakers and fuses to provide safety and protection.

C. is permitted to be the main in the service panel.

D. is required to have integral overcurrent protection rated at not more than 1.15 times the continuous output current rating of the fuel cell unit.

E. shall open both conductors if the output is a 2-wire at 120 volts.

Code reference _____

7. The fire pump installation shown in Figure 11.28 receives the source of power directly from a utility transformer. There is no central station monitoring, local signaling, or sealing program with weekly inspections to supervise the disconnect to make sure it is maintained in the closed position. Supervision of the disconnect is:

A. not required when less than 1000 persons occupy the building.

B. not required if the fire pump installation is located in a dedicated 1-hour fire-rated room.

C. permitted to be a lock that prevents the switch to be turned off without use of a key.

D. required to be in the form of an audible intermittent sounding horn.

E. only required for certain facilities such as hospitals and assembly halls that are occupied by more than 1000 persons.

Code reference _____

8. A fire pump installation consists of only one motor that is rated 60-horsepower, 3-phase, 460-volt, design B as shown in Figure 11.28. Conductors are copper with 75°C rated terminations. The minimum size copper Type THWN conductor permitted to be run from the controller to the fire pump motor is:

A. 8 AWG.   C. 4 AWG.   E. 2 AWG.

B. 6 AWG.   D. 3 AWG.

Code reference _____

9. The circuit conductors from an emergency panelboard to exit signs in a building are:

A. required to be run only in Type MI, MC, or AC Cable, or metallic raceway.

B. required to be run in Rigid Metal Conduit.

C. permitted to be run in the same raceway with normal power conductors provided all conductors have 600-volt insulation.

D. required to be run in separate raceways even when two separate emergency circuits originate from the same panelboard.

E. not permitted to be run in the same raceway with other power or lighting circuit conductors.

Code reference _____

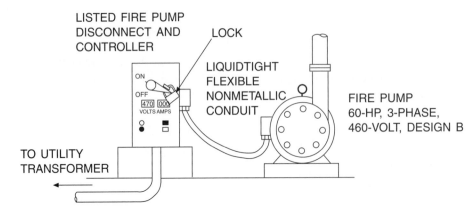

Figure 11.28   The fire pump system is powered with a single 60-horsepower, 3-phase, 460-volt design B motor with no other loads.

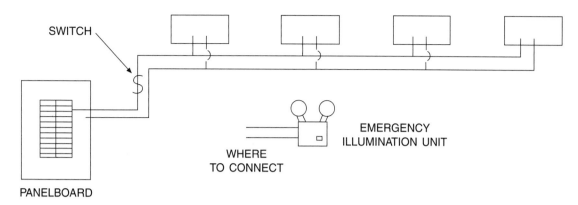

LIGHTING CIRCUIT IN AREA

SWITCH

WHERE
TO CONNECT

EMERGENCY
ILLUMINATION UNIT

PANELBOARD

Figure 11.29   How is the emergency illumination unit required to be connected in an area where emergency illumination is required.

10. An area requiring emergency illumination is provided with self-contained battery-powered automatically controlled lighting units with a build-in battery charger as shown in Figure 11.29. The illumination units are required to be connected to:
   A. a branch-circuit serving the lighting in the area covered by the emergency lighting unit and connected on the supply side of any switching.
   B. a dedicated circuit from the first normal power panelboard serving the building.
   C. any circuit serving the building.
   D. a dedicated circuit from the emergency panelboard serving the building.
   E. a branch-circuit serving the lighting in the area covered by the emergency lighting unit and connected on the load side of any switching.

   Code reference_____

11. A legally required standby power system is one that automatically supplies power to:
   A. emergency electrical systems.
   B. other than emergency electrical systems.
   C. batteries of unit emergency equipment such as area lighting and exit signs.
   D. exit signs and required building evacuation lighting units
   E. loads in not more than 10 seconds.

   Code reference_____

12. The wiring supplied by an optional standby power system is:
   A. required to be kept completely separate from circuits of the general wiring of the building.
   B. required to be energized in not more than 60 seconds of a power outage.
   C. required to be run in raceway.
   D. permitted to occupy the same cabinets, boxes, cables, and raceways with general building wiring.
   E. required to be run through the building concealed within spaces that have a minimum 15-minute fire rating.

   Code reference_____

13. An alternate power production system is to be connected to operate in parallel with the utility supply as an interconnected electric power production source. The alternate power production system connection to the building electrical system is:
   A. required to be on the supply side of the service disconnecting means.
   B. required to be on the load side of the service disconnecting means.

C. required to be at the service disconnecting means but permitted to be connected either on the supply side or on the load side of the disconnect.

D. permitted to be made at any location within the building.

E. required to be at the utility transformer location.

Code reference _____

14. Fire alarm cables that are abandoned and not tagged for future use:

A. are required to be removed in their entirety to prevent spread of fire.

B. shall have the accessible portion of the cables removed.

C. are required to be removed only over suspended grid-type ceilings.

D. are required to be removed only if the cables are not rated for installation in plenums.

E. are permitted to remain in place even if in accessible areas.

Code reference _____

15. A receptacle in the unfinished portion of the basement of a dwelling that is installed to provide power to a non-power-limited fire alarm system is:

A. required to be ground-fault circuit-interrupter protected.

B. required to be protected by a arc-fault circuit interrupter.

C. required to be controlled by a switch.

D. not permitted to be ground-fault circuit-interrupter protected.

E. not permitted and the installation is required to be supplied by a circuit near an exit door on the first level.

Code reference _____

WORKSHEET NO. 11—ADVANCED EMERGENCY AND ALTERNATE POWER SYSTEMS

Mark the single answer that most accurately completes the statement based upon the 2005 Code. Also provide, where indicated, the Code reference that gives the answer or indicates where the answer is found, as well as the Code reference where the answer is found.

1. An engine-driven standby generator set is located outside and is permanently installed to supply a building. The output of the generator is 3-phase, 208Y/120-volt, 4-wire and the generator is rated 60 kW with a 200-ampere circuit breaker protecting the output of the generator. The minimum size copper conductors with 75°C insulation from the generator to the distribution panelboard inside the building is:

   A. 1/0 AWG.   C. 3/0 AWG.   E. 250 kcmil.
   B. 2/0 AWG.   D. 4/0 AWG.

   Code reference _____

2. A standby power system for sensitive electrical equipment in a commercial building is supplied by single-cell stationary vented lead-acid batteries. Twenty-four of the batteries are installed in a rack that is 36 in. (0.91 m) wide and 72 in. (1.83 m) long. Each individual battery contains 2.2 gallons (8.3 L) of liquid electrolyte (see Figure 11.26). An electrolyte catchment below the batteries is:

   A. not required.
   B. required to be capable of retaining 2.2 gallons.
   C. required to be capable of retaining 13.2 gallons.
   D. required to be capable of retaining 52.8 gallons.
   E. required to be capable of retaining 66.0 gallons.

   Code reference _____

3. A solar photovoltaic array, as shown in Figure 11.30, consists of eight modules connected in parallel. The common connecting point with overcurrent protection and blocking diodes is located on the backside of the array. The photovoltaic output circuit runs from the back of the array to a disconnect switch with overcurrent protection then underground to an adjacent building which houses the charge controller, batteries, and inverter. The nameplate rating of the 36 cell modules is: open-circuit

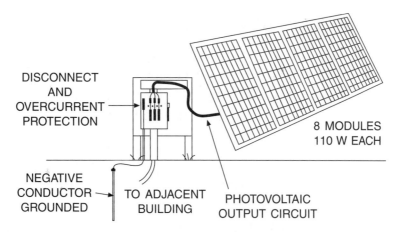

DISCONNECT AND OVERCURRENT PROTECTION

8 MODULES 110 W EACH

NEGATIVE CONDUCTOR GROUNDED

TO ADJACENT BUILDING

PHOTOVOLTAIC OUTPUT CIRCUIT

Figure 11.30 Eight solar modules are connected in parallel with one set of conductors between the solar panel and the disconnect and from the disconnect to an adjacent building.

voltage = 21.0 volts; short-circuit current = 7.22 amperes; operating voltage = 17.0 volts; operating current 6.74 amperes; and maximum power = 110 watts. The photovoltaic output circuit is run as copper conductors in underground raceway to the adjacent building and the conductors have 75°C insulation and terminations. The minimum size conductors is:

A. 10 AWG.   C. 6 AWG.   E. 3 AWG.
B. 8 AWG.   D. 4 AWG.

Code reference_____

4. A solar photovoltaic power system similar to the system shown in Figure 11.30 has parallel modules connected at a collecting point on the back of the array. The photovoltaic output circuit runs from the array to the disconnect then to an adjacent building housing the remaining components of the system. The photovoltaic output circuit conductors are size 4 AWG copper with 75°C insulation and terminations. The photovoltaic output dc circuit is grounded to a driven 8 ft (2.5 m) rod at the disconnect adjacent to the array. The minimum size grounding electrode conductor permitted for this solar photovoltaic power system is:

A. 8 AWG.   C. 4 AWG.   E. 2 AWG.
B. 6 AWG.   D. 3 AWG.

Code reference_____

5. A stand-alone fuel cell system is connected to a building wiring system through a transfer switch and serves as backup power for essential circuits. The installation is shown in Figure 11.31. The output from the inverter is 120 volts, 2-wire. This arrangement is permitted only if:

A. there are no 240-volt, or multiwire branch-circuits supplied by the fuel cell system.

B. the transfer switch is located ahead of the main disconnect for the service.

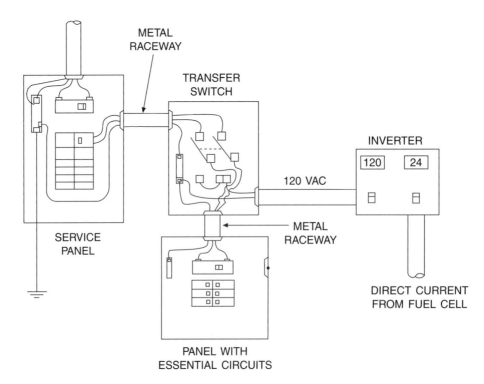

Figure 11.31   A transfer switch is installed on the load side of the service panel to a building with the essential circuits in a separate panelboard. The transfer switch is connected in such a way that the 120-volt, 2-wire output from the fuel cell system supplied both ungrounded busses in the essential circuit panel.

C. the inverter output is run through a transformer to produce a 120/240-volt, 3-wire electrical system.

D. the fuel cell system output is changed over to 120/240-volt, 3-wire.

E. the output of the inverter is not less than the sum of the overcurrent devices supplied.

Code reference_____

6. A fuel cell system is complete with inverter and transformer and has a nameplate output of 58 amperes continuous full-load alternating current at 120/240 volts, 3-wire. The output of the fuel cell unit is protected with a 2-pole, 60-ampere circuit breaker. The feeder conductors from the fuel cell to the premises wiring system are copper with 75°C insulation and terminations. The minimum size conductors permitted for the feeder is:

A. 10 AWG.   C. 6 AWG.   E. 3 AWG.
B. 8 AWG.   D. 4 AWG.

Code reference_____

7. A 50-horsepower, 460-volt, 3-phase, design B fire pump motor in a building is supplied with copper conductors size 4 AWG with 75°C insulation and terminations. The fire pump circuit is protected with a fusible disconnect as shown in Figure 11.32. The minimum size fuses permitted for the circuit is:

A. 90 amperes.   D. 350 amperes.
B. 100 amperes.   E. 400 amperes.
C. 300 amperes.

Code reference_____

8. The flexible conductor connection between the listed fire pump controller and the fire pump, as shown in Figure 11.32, is:

A. required to be Flexible Metallic Tubing.
B. required to be Liquidtight Flexible Nonmetallic Conduit
C. permitted to be Type MC Cable.
D. permitted to be Liquidtight Flexible Nonmetallic Conduit, Type B.
E. permitted to be Electrical Nonmetallic Tubing.

Code reference_____

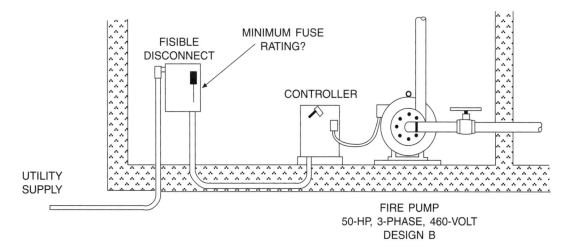

Figure 11.32 A fire pump is rated 50-horsepower, 3-phase, 460-volt, design B and is the only load supplied by the system.

9. The authority having jurisdiction may rule in some situations that the electrical supply for an emergency panel is permitted to be:
   A. a circuit in the first panelboard of the normal power system.
   B. a tap to the normal service conductors entering the building provided the tap is made ahead of the main service disconnect.
   C. a tap to the normal service made at the main lugs of the disconnect.
   D. a separate and independent service to the building supplying only the emergency panelboard.
   E. any circuit from the normal power system in the building.

   Code reference _____

10. An area requiring emergency illumination that has high-intensity discharge lighting such as high-pressure sodium, metal halide, or mercury vapor as the normal lighting, in the case of a power interruption, is required to:
    A. have emergency illumination that will remain operating until the normal illumination units are operating or provide other means of illumination until the luminaires (lighting fixtures) restrike.
    B. operate only until the power to the circuit has been restored.
    C. have several incandescent or fluorescent luminaires (lighting fixtures) that operate continuously controlled only by a locked-on circuit breaker.
    D. also have an incandescent or fluorescent luminaires (lighting fixtures) on the same circuits with the high intensity discharge lights.
    E. provide extra exit doors to the area.

    Code reference _____

11. Circuit wiring supplied by the legally required standby power system is:
    A. required to be kept completely independent of other circuits in the building.
    B. only permitted to be run in Type AC , MC, or MI Cable or run as single conductors in metal raceway.
    C. required to receive power after an outage with a time period not to exceed 10 seconds.
    D. is permitted to occupy the same raceways, cables, boxes, and cabinets with other general wiring.
    E. permitted to utilize manual transfer equipment if the building has qualified maintenance personnel on duty at all times.

    Code reference _____

12. A portable engine-generator set that is connected to a premises wiring system by means of a transfer switch is:
    A. permitted to serve on a legally required standby power system.
    B. only required to meet the rules of a temporary electrical system.
    C. a violation and not permitted to supply power to a premises wiring system.
    D. not permitted to be used to supply emergency circuits if rated less than 10 kW.
    E. considered to be an optional standby power system.

    Code reference _____

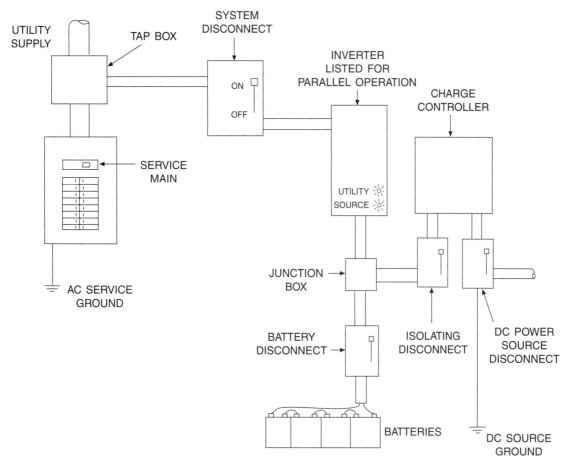

Figure 11.33   A dc power source is connected in parallel with a utility supply and tapped ahead of the service disconnect.

13. A dc power source is operated as an interconnected electric power production system in parallel with the utility supply through an inverter and connected as shown in Figure 11.33. The system disconnect is:
    A. required to be a manual switch.
    B. permitted to contain fuses that at times may be energized with the switch in the off position and marked with a warning.
    C. required to be kept locked in the open position.
    D. not permitted to be a circuit breaker.
    E. required to be power operable.

        Code reference_____

14. Fire alarm cables in a room with a non-fire-rated suspended ceiling grid of an existing building are not permitted to be supported by the ceiling grid:
    A. unless there are no more than three cables in any 10 ft by 10 ft (3 m by 3 m) area.
    B. except where there is a maintenance electrician.
    C. unless the cables have a diameter not more than $1/2$ in. (13 mm) and there are no more than three cables across any one ceiling tile.
    D. unless only one cable is across any one ceiling tile.
    E. in any type of building.

        Code reference_____

15. The initiation circuit conductors of a non-power-limited fire alarm system are stranded copper and run in Electrical Metallic Tubing throughout a building. The continuous supervisory current flow in the circuit conductors is 0.2 amperes. The minimum size conductor permitted for this installation is:

A.   24 AWG.

B.   20 AWG.

C.   18 AWG.

D.   16 AWG.

E.   14 AWG.

Code reference _____

UNIT 12

Industrial Electrical Applications

OBJECTIVES

After completion of this unit, the student should be able to:

- determine the minimum dimensions of cable tray permitted for given sizes and numbers of wires and cables.
- describe the conductor insulating method for integrated gas spacer cable.
- determine the maximum ampacity of wires and cables to be installed in cable tray.
- determine the minimum current rating of the supply conductors for an electric welder.
- determine the maximum ampacity for wires to be installed in cablebus.
- define the different methods of electrically heating pipelines and vessels.
- determine the maximum ampacity of branch-circuit conductors for a noncontinuous motor on a crane, hoist, or monorail hoist.
- explain the application of *Article 665* on induction heating and dielectric heating to the types of facilities.
- explain how dielectric heating works in comparison with induction heating.
- describe facilities to which *Article 668*, dealing with electrolytic cells, and *Article 669*, dealing with electroplating, shall apply.
- determine the minimum size supply conductor permitted for an industrial machine if the nameplate information is given.
- describe the different types of optical fiber cables and their applications.
- answer wiring installation questions from *Articles 326, 368, 370, 392, 427, 490, 610, 630, 665, 668, 669, 670, 727,* and *770*.
- state at least three changes that occurred from the 2002 to the 2005 Code for *Articles 326, 368, 370, 392, 427, 490, 610, 630, 665, 668, 669, 670, 727,* and *770*.

CODE DISCUSSION

This unit deals with wiring methods and materials commonly used in industrial installations but not necessarily limited to industrial use. Several wiring methods are discussed from *Chapter 3* of the Code that are used most frequently in industrial and commercial wiring. Installation of electric heating equipment for pipelines and vessels is also described. *Chapter 6* of the Code deals with the wiring of special equipment. Several types of special equipment for industrial applications are covered in this unit.

Article 326 deals with integrated gas spacer cable where pressurized sulfur hexafluoride gas and dry kraft paper tapes are used as the conductor insulation. The conductor or conductors are contained within a flexible nonmetallic conduit. The pressurized gas-filled cable assembly helps prevent contaminants from entering the assembly and causing cable failure. As stated in *326.104*, the minimum conductor size permitted is 250 kcmil solid aluminum. A single solid aluminum conductor not smaller than 250 kcmil is permitted in the cable, or the conductor is permitted to be formed using up to nineteen 250 kcmil solid aluminum rods laid parallel.

Article 368 covers busways, which is a form of an electrical distribution system that provides flexibility for future changes. Busway is especially well suited for manufacturing facilities where exact placement of equipment may not be known at the time of construction or where equipment is periodically rearranged. Busway is a factory-assembled metal enclosures containing electrical conductors, usually in the form of copper or aluminum bars insulated from the metal enclosure. Busway is frequently used in commercial and industrial work areas by attaching to the ceiling with a spacing not to exceed 5 ft (1.5 m), *368.30*. Circuit breaker or fusible tap boxes are attached to the busway, and various raceway or cable wiring methods listed in *368.56(A)* are permitted to extends down to a workstation, machine, motor control center, or electrical panelboard. *NEC® 364.56(B)* permits suitable cord and cable assemblies rated for extra-hard usage or hard usage to be used as drops from a busway tap box to equipment. There is also a listed bus drop cable for this purpose. A suitable tension take-up support device is required to be used with cord and cable assemblies. The maximum distance from the cord or cable termination to the strain relief take-up device or intermediate support is 6 ft (1.8 m) as shown in Figure 12.1. *NEC® 368.17(C)* requires that plug-in devices to tap for branch-circuits and feeders contain overcurrent protection. The plug-in tap device is required to contain a circuit breaker or a fusible switch, which shall by some means be operable for the floor level. Generally, these plug-in tap devices are located high in the room and out of reach of an operator. The busway will have a name-plate current rating and it is required to be protected in accordance with the busway ampacity. *NEC® 368.17(B) Exception* does permit the reduction in size of a busway without overcurrent protection at the point where the size is reduced. Under these circumstances, a length of reduced ampacity busway is required to be adequate to supply the load and not more than 50 ft (15 m) in length.

Article 370 covers the installation of cablebus. Cablebus is a factory-made cable and ventilated metal framework system that is usually field-assembled. *NEC® 370.3* permits cablebus to be used for services, feeders, and branch-circuits. The minimum permitted size of insulated conductors is size 1/0 AWG, which is covered in *370.4(C)*. The conductors are supported periodically on insulating blocks made for the cablebus. The maximum permitted support spacing for horizontal and vertical runs of cablebus is covered in *370.6*. Figure 12.2 shows a

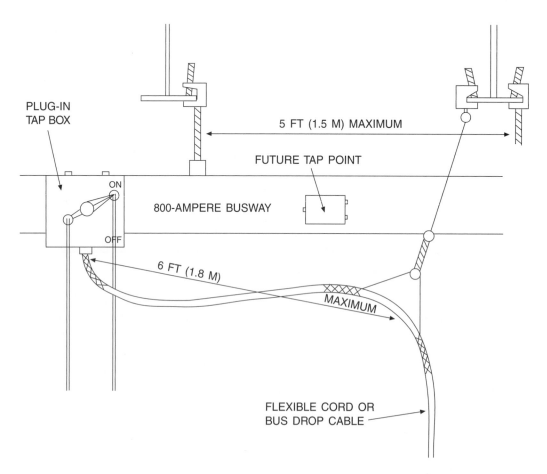

Figure 12.1   Flexible cord or cable drops from a busway are required to be supported at intervals of not more than 6 ft (1.8 m) and within a distance of 6 ft (1.8 m) from the strain relief take-up device.

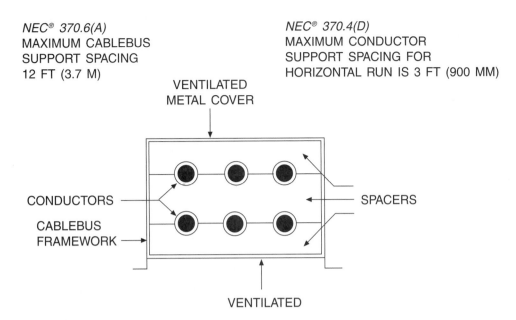

NEC® 370.6(A)
MAXIMUM CABLEBUS
SUPPORT SPACING
12 FT (3.7 M)

NEC® 370.4(D)
MAXIMUM CONDUCTOR
SUPPORT SPACING FOR
HORIZONTAL RUN IS 3 FT (900 MM)

VENTILATED
METAL COVER

CONDUCTORS

CABLEBUS
FRAMEWORK

SPACERS

VENTILATED

Figure 12.2 **Cross-sectional view of a cablebus which is permitted to be used for branch-circuits, feeders, and services.**

cross section of cablebus. The metal framework of cablebus is permitted to be used as an equipment grounding conductor, as stated in *370.3* and *250.118(B)*. The grounding requirements are covered in *365.9*.

The ampacity of the conductors is permitted to be determined using *Tables 310.17* and *310.19*, which are for single-insulated conductors in free air. For circuits operating at more than 600 volts, *Table 310.69* and *Table 310.70* shall be used. Also, it should be noted in *370.5* that the next standard rating overcurrent device is permitted to be used when conductor ampacity does not correspond to a standard rating overcurrent device provided the rating does not exceed 800 amperes. For example, a Type THWN copper wire size 700 kcmil with a rating of 755 amperes, found in *Table 310.17*, is permitted to be protected with an 800-ampere overcurrent device.

Article 392 deals with cable trays, which are defined in *392.2*. A cable tray has a rectangular cross section and is a form of support system for conductors, cables, conduit, and tubing. A typical ladder-type cable tray is illustrated in Figure 12.3. Single-conductor cable installed in cable tray is not permitted to be smaller than size 1/0 AWG, as stated in *392.3(B)(1)*. All single-conductor cable installed in cable tray is required to be marked for use in cable tray on the exterior of the cable. Generally single-conductor cables size 1/0 AWG and larger of types such as XHHW, THHN, THWN, THW, RHH, and RHW are listed for use in cable trays. Multiconductor cables Type TC and Type MC with three or four Type XHHW insulated conductors plus an equipment grounding conductor are in common use. Individual conductor sizes range from size 8 AWG copper to 1000 kcmil. Cables rated over 600 volts are usually Type MV or Type MC with conductors appropriate for the circuit voltage.

Standard widths of ladder-type and trough-type cable trays are 6, 9, 12, 18, 24, 30, and 36 in. (150, 225, 300, 450, 600, 750, and 900 mm). Typical lengths are 10 ft (3 m) and 12 ft (3.7 m). There is no specific support spacing required in the Code. In addition to the ladder type, cable tray is available as a solid bottom and a ventilated trough. Channel-type cable tray is available in widths of 2, 3, 4, and 6 in. (50, 75, 100, and 150 mm).

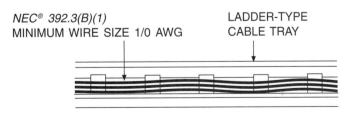

NEC® 392.3(B)(1)
MINIMUM WIRE SIZE 1/0 AWG

LADDER-TYPE
CABLE TRAY

Figure 12.3 **A ladder-type cable tray is permitted to support single-conductor cables, marked for use in cable trays, and multiconductor cables as listed in** *392.3(A)(1)*.

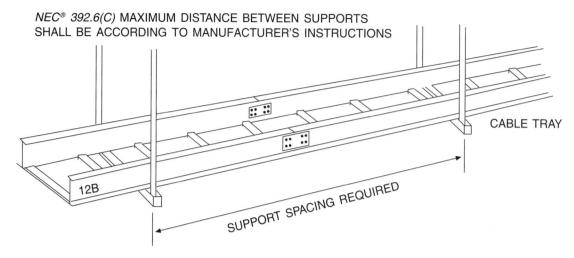

NEC® 392.6(C) MAXIMUM DISTANCE BETWEEN SUPPORTS
SHALL BE ACCORDING TO MANUFACTURER'S INSTRUCTIONS

CABLE TRAY

12B

SUPPORT SPACING REQUIRED

Figure 12.4 Cable tray is required to be supported at regular intervals according to manufacturer's instructions.

Channel type cable tray is often used for conductor and cable drops from the main cable tray to loads. It is available with a solid bottom and ventilated. *NEC® 392.6(C)* requires cable tray support to be according to the manufacturers instructions as shown in Figure 12.4. For a particular support spacing, cable tray is rated by the manufacturer for a maximum load on a pounds-per-linear-ft basis (1 pound per linear foot = 1.49 kilograms per linear meter) for cables and raceways supported by the cable tray and any other environmental loads. Standard support spacings for cable trays are 8, 12, 16, and 20 ft (2.5, 3.7, 4.9, and 6 m). A manufacturer will provide a load class designation for a particular cable tray consisting of a number and a letter. The number is the recommended maximum support spacing. The letter is the working load category, which is A for a 50-pound-per-linear-ft (75 kg/m) load, B for 75 pounds per linear ft (112 kg/m), and C for 100 pounds per linear ft (150 kg/m). Assume an aluminum ladder-type cable tray has a load classification 8B. The recommended maximum support spacing is every 8 ft (2.5 m), and with that spacing the load is not permitted to exceed 75 pounds per linear ft (112 kg/m) Figure 12.4.

When installed outside, it may be necessary to calculate the side pressure on a linear-ft basis for wind or a vertical loading expected from ice or snow. In northern climates, ice and snow load is added to the conductor loading. Support spacing may be determined on the availability of points within a building from which to anchor supports. When spanning open areas, the cable is usually supported in a trapeze style. When run adjacent to a wall, a cantilever support to the wall is frequently used. These supports are illustrated in Figure 12.5. If the desired support spacing is not practical to achieve, cable trays are available with

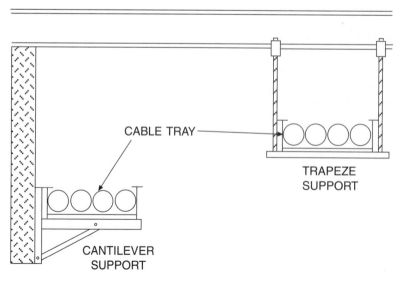

CABLE TRAY

TRAPEZE
SUPPORT

CANTILEVER
SUPPORT

Figure 12.5 Typical means of support for cable trays are the cantilever support when attaching to side walls, and the trapeze support when attaching to structural members above the cable tray.

stronger side rails to support the weight for longer spans. Cable trays are not required to be mechanically continuous, but bonding is required. Thermal expansion and contraction may also need to be considered in some locations where the cable tray will be installed in an area where the temperature will change.

NEC® 392.7 covers the situation in which the cable tray is permitted to serve as the equipment grounding conductor. *Table 392.7(B)* gives the maximum circuit rating for a given cable tray cross-sectional area for which the cable tray may act as the equipment grounding conductor. Much of the remainder of the article deals with the number of wires and cables permitted to be installed in cable trays. Examples of conductor installation and determination of the minimum permitted width of cable tray are explained later in this unit.

Methods for determining the width of cable tray for the types and sizes of cables to be installed are explained in *392.9* for multiconductor cables in the cable tray, and in *392.10* when single-conductor cables are installed. How the cables are installed in the cable tray influences the method to be used to determine the conductor ampacity. If multiconductor cables are installed so that the cables have a space between them of at least one cable diameter, then conductor ampacity can be determined using the method of *392.11(A)(3)*. If the minimum spacing between conductors is not maintained, the ampacity is determined according to *392.11(A)(1)*. A similar situation exists for single-conductor cables supported by cable tray. Two different methods are used to determine the ampacity of the same conductors. If the minimum spacing is maintained between conductors, the ampacity will be higher than if the spacing is not maintained. However, when the minimum spacing is maintained, a wider cable tray is frequently required.

A portion of a circuit run in cable tray is frequently also run in raceway. *NEC® 392.11(A)* and *(B)* is a reminder that the ampacity of the conductors run in cable tray is to be determined based upon *310.15(A)(2)*. When a set of conductors is subject to two different conditions for the purpose of determine conductor ampacity, the most restrictive condition shall be used. This is illustrated in Figure 12.6. For example, a set of single copper conductors with 75°C insulation and terminations size 1/0 AWG run in ladder-type cable tray with a maintained spacing between the conductors has an allowable ampacity based upon *Table 310.17* of 230 amperes. The portion of the circuit run that is in raceway, based upon *Table 310.16*, only has an allowable ampacity of 150 amperes.

Article 427 covers the installation of fixed electric heating equipment for pipelines and vessels. Several types of heating are defined in *427.2*. One type is a resistance heating element attached to or inserted into a pipeline or vessel. An impedance heating system is one where electrical current flows through the wall of the pipeline or vessel, and the heat is produced by the impedance of the wall and the current flow. Skin-effect heating uses a ferromagnetic envelope attached to a pipeline or vessel, and electrical current flow through the envelope produces the heat. The impedance of the envelope and the current flow produce the heat. Induction heating is accomplished by using an external induction coil. Electric current is caused to

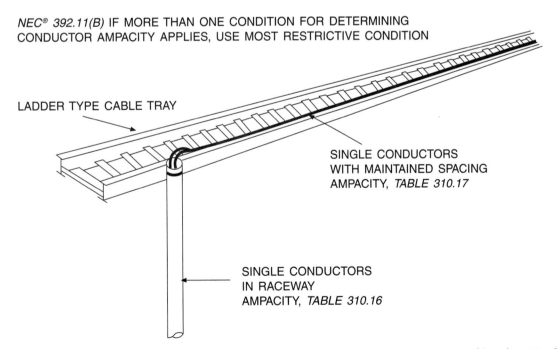

NEC® 392.11(B) IF MORE THAN ONE CONDITION FOR DETERMINING CONDUCTOR AMPACITY APPLIES, USE MOST RESTRICTIVE CONDITION

LADDER TYPE CABLE TRAY

SINGLE CONDUCTORS WITH MAINTAINED SPACING AMPACITY, *TABLE 310.17*

SINGLE CONDUCTORS IN RACEWAY AMPACITY, *TABLE 310.16*

Figure 12.6 **When determining the ampacity of a set of conductors run in cable tray, consideration must be given to the entire circuit and other conditions that may result in a different ampacity for the same conductors.**

flow in a pipeline or vessel wall by electromagnetic induction similar to a transformer. Induced wall current flow and the impedance of the wall produce the heat.

The minimum permitted ampacity of circuit conductors is required to be 125% of the total load of the heaters. This requirement is covered in *427.4,* which points out that it is permitted to round up the conductor ampacity to the next standard size of overcurrent device as permitted by *240.4(B).* The type of disconnecting means permitted for the heating is covered in *427.55.*

Special consideration is required for equipment grounding depending on the type of heating involved. In the case of resistance heating, the grounding requirements of *Article 250* shall apply. Grounding for impedance heating is covered in *427.29. NEC® 427.48* covers grounding for skin-effect heating. It is pointed out that the grounding requirement of *250.30* for a separately derived system does not apply in the case of skin-effect heating.

Article 490 provides basic rules for sizing, selection, and installation of equipment and conductors operating at more than 600 volts. The latter portion of the article provides rules for some specific high-voltage applications. Interrupting current at high voltage is much more difficult than interrupting current for loads operating at 600 volts or less. Current-interrupting devices are covered in *490.21* and consist of circuit breakers, power fuses, expulsion-type cutouts, and oil-filled cutouts. When working with equipment that operates at more than 600 volts, safety is of the highest importance and isolation of equipment for inspection, maintenance, and repair is essential. Equipment isolation is covered in *490.22.* Rules for the construction and installation of metal-enclosed power switchboards and industrial control assemblies are in *Part III.* Guarding of live parts and clearances between live parts and between live parts and ground is an important consideration for equipment construction, and when performing maintenance procedures near live parts. Some heavy industrial equipment is mobile and operates from a high-voltage power source. This creates the need for mobile or portable high-voltage distribution equipment, the rules for which are covered in *Part IV.* A unique high-voltage application is the electrode-type boiler where electrodes are placed in a liquid with heat generated by the passage of current through the liquid. The rules for applying high-voltage power for such an application are found in *Part V.*

Article 610 deals with the wiring methods and wiring requirements for cranes, hoists, and monorail hoists. Motors powering this equipment may have a duty cycle rating. That is, the motor may only be permitted to operate for 15 minutes, 30 minutes, or 60 minutes until it is required to cool before being operated again. If the motor has a duty cycle, then the conductor to the motor is permitted to carry a higher level of current than would be typical for a general circuit or a continuous load. *Table 610.14(A)* is used to determine the minimum wire size for time-rated motors. The article deals with the installation, support, and grounding of open and insulated conductors on hoists, cranes, and monorail hoists. All exposed noncurrent-carrying metal parts are required to be grounded.

Article 630 provides the requirements for wiring electrically powered welders. A nameplate or rating plate is required to be provided on the welder. Electric welders are of the arc-type and the resistance-type. Arc welders are of the ac transformer-type and the motor-generator type. With the arc-type, the metal parts to be joined are brought together and an arc is struck between the metal parts and a metal electrode. The metal electrode is melted away by the heat of the arc to provide extra metal for joining the parts. The actual load on a transformer-type arc welder is intermittent, which is taken into consideration when sizing the circuit components. Even during a continuous weld, current flows in pulses. Current flow is on for several cycles and off for several cycles. The duty cycle for a welder is determined based upon the number of cycles the welder is operating in a one-hour period of time. At the end of *630.31* is a FPN called explanation of terms. Duty cycle is explained in the third paragraph of that FPN.

The rules for determining the size of conductors for an individual arc welder or a feeder supplying a group of arc welders are found in *Part II.* Overcurrent protection is required for the welder and for the conductors supplying the welder. One overcurrent device may provide protection for both the conductors and the welder, in which case the minimum value is determined according to *630.12.*

Rules for sizing components for a resistance-type electric welder are found in *Part III.* With a resistance-type, the metal pieces to be joined are clamped between two electrodes and current flows from one electrode to the other through the metal. Because the area of contact is small and the current level is high, the area is heated to the melting point due to the resistance of the metal. The weld is actually accomplished with only a few cycles but at very high current. Voltage drop is an important consideration when sizing conductors for a resistance welder. The number of cycles the resistance welder delivers depends upon the particular weld. Duty cycle for a resistance welder is determined by multiplying the number of cycles required for a weld times the number of welds per hour and dividing by 216,000, which is the total number a 60-hertz cycles per hour.

Information on the welder rating plate or nameplate needed for determining conductor size, overcurrent rating, and disconnect rating is based on the effective rated primary current, or the rated primary current times a multiplying factor obtained from a table in either *630.11* or *630.31*. The welder duty cycle is also needed to find the multiplier from the table. Duty cycle is permitted to be calculated based upon the weld to be performed, and the FPN to *630.31(B)* explains how duty cycle can be determined. Here is how duty cycle and the rated primary current on the welder rating plate is used to determine the minimum supply conductor ampacity.

Example 12.1   One transformer arc welder has a primary current marked on the rating plate as 40 amperes. The welder is to be operated at a duty cycle of 70%. Determine the minimum current rating of the supply conductors for the welder.

Answer: This is a transformer-type arc welder. Therefore, use the nonmotor generator column of *Table 630.11(A)* to find the multiplying factor, which is 0.84 for a 70% duty cycle. Next multiply the factor and the 40-ampere primary current to get the minimum conductor ampacity of 33.6-amperes (40 A × 0.84 = 33.6 A).

Welders draw some current when running idle (not welding) and a higher current when welding. Effective rated primary current (I_{1eff}) combines the conductor heating due to these two levels of current. The effective rated primary current (I_{1eff}) may be given on the welder rating plate. The FPN to *630.12(B)* gives a formula for determining the effective rated primary current. It is based on the rated supply current for the welder, the no-load supply current, and the duty cycle. Duty cycle for a welder may be fixed, or it may be adjustable. If it is adjustable, then effective primary current can be calculated. Because heat in a conductor is proportional to the square of the current (Equation 1.16), the rated supply current with load is squared and multiplied by the duty cycle. That value is added to the square of the no-load current times the percentage of time remaining, which is one minus the duty cycle. The square root of the sum of those two values is taken to get the effective primary current. The heating of the supply conductor is considered to be proportional to this value. The rated supply current under load and the rated no-load supply current can be measured, but proper instrumentation is needed because during welding the current is flowing in pulses not continuously.

Determining the minimum size of conductor for a welder and the maximum rating of overcurrent protection permitted is based upon the maximum rated supply current and the effective rated supply current (rated primary current and duty cycle may also be used). The welder is required to be protected from overcurrent at a level not exceeding 200% of either the maximum rated supply current or the rated primary current, whichever is given on the nameplate. The conductor is not permitted to be protected at a level greater than the ampacity of the supply conductor. It is permitted to round up to the next standard rating of overcurrent device if this value does not correspond to a standard rating overcurrent device as listed in *240.6*. The minimum permitted rating of supply conductor is determined from *Table 310.16* based upon the nameplate value of the effective rated primary current or the rated primary current and a multiplier found in *630.11(A)*. The supply conductors are likely to have an ampere rating less than the maximum rated supply current. Therefore, the overcurrent device sized to protect the conductors will frequently have a rating less than the maximum required to protect the welder. If a single overcurrent device is used to protect both the conductors and the welder as shown in Figure 12.7, the

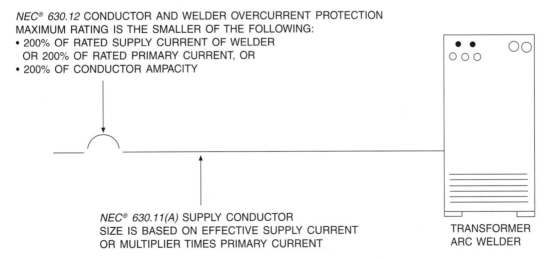

NEC® 630.12 CONDUCTOR AND WELDER OVERCURRENT PROTECTION
MAXIMUM RATING IS THE SMALLER OF THE FOLLOWING:
• 200% OF RATED SUPPLY CURRENT OF WELDER
OR 200% OF RATED PRIMARY CURRENT, OR
• 200% OF CONDUCTOR AMPACITY

NEC® 630.11(A) SUPPLY CONDUCTOR
SIZE IS BASED ON EFFECTIVE SUPPLY CURRENT
OR MULTIPLIER TIMES PRIMARY CURRENT

TRANSFORMER
ARC WELDER

Figure 12.7   One overcurrent device is permitted to protect both the supply conductor and the arc welder if the overcurrent device rating is not more than 200% of the ampacity of the conductor or 200% of the maximum rated supply current of the welder.

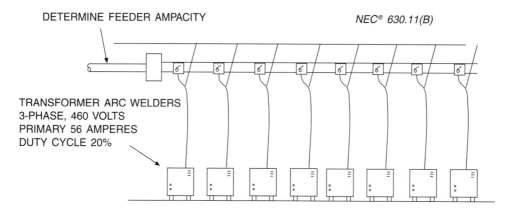

DETERMINE FEEDER AMPACITY   *NEC® 630.11(B)*

TRANSFORMER ARC WELDERS
3-PHASE, 460 VOLTS
PRIMARY 56 AMPERES
DUTY CYCLE 20%

Figure 12.8   **A method was specified for determining the minimum feeder ampacity for a group of arc welders.**

minimum rating is used unless the device opens under normal operation, in which case the next standard rating is permitted. If a feeder supplies a group of arc welders, the method for determining the minimum size of feeder conductor is found in *630.11(B)*. The following example will show how *630.11(B)* can be used to determine the minimum ampacity of the feeder supplying a group of arc welders. There cases where the duty cycle is fixed and a smaller feeder is permitted.

Example 12.2   Eight transformer arc welders are supplied by one feeder. The welders are 3-phase, 460 volts, with a primary supply current of 56 amperes marked on the rating plate, and the duty cycle is 20%, as shown in Figure 12.8. The conductors are copper with 75°C insulation and terminations. Determine the minimum ampacity required for the feeder supplying these welders.

Answer: First find the supply current rating of the conductors for each welder using *630.11(A)*. Multiply the rated primary current of the welder by the factor found in *Table 630.11(A)*, which is 0.45. This is a transformer-type arc welder. Therefore, use the nonmotor generator column. The individual welder current is 25.2 amperes (56 A × 0.45 = 25.2 A). Next use *630.11(B)* to determine the minimum ampacity of the feeder for all eight welders. The first two welders are taken at 100%, the next at 85%, the next at 70%, and all remaining at 60 % to obtain 149.9 amperes as the minimum feeder ampacity.
   [2 × 25.2 A + 0.85 × 25.2 A + 0.70 × 25.2 A + 0.60 × (4 × 25.2 A)]
   [50.4 A + 21.4 A + 17.6 A + 60.5 A] = 149.9 A

The secondary conductors of a welder are not considered as premises wiring, and the grounding requirements for these conductors are not those stated in *Article 250*. In particular, a welder is not to be considered as a separately derived premises system, and *250.30* containing grounding requirements for separately derived systems does not apply. A FPN in *630.15* points out the potential of parallel paths for the welder secondary current in cases where the return conductor is grounded to the building grounding system. The workpiece terminal of the welder should be connected to the workpiece or to the workpiece table but not to the grounding system of the building. The potential for objectionable currents on the building grounding conductors is illustrated in Figure 12.9.

Article 665 deals with equipment for induction heating and dielectric heating of materials in industrial processes and scientific applications. An electrical conducting material is heated by the induction heating process. The material is placed in an electromagnetic field operating at a frequency of a few kilohertz to several hundred kilohertz. Electrical current is induced into the material and with the impedance of the material results in heating.

The dielectric process is used when the material to be heated is not an electrical conductor. The material to be heated is placed between two electrical plates to which is applied a varying electric field in the range of a few megahertz to over 100 megahertz. The varying electric field vibrates the molecules of the material, thus producing heat. The high-frequency electromagnetic field or the electric field may be produced by a motor-generator, or it may be produced by some other type of field-producing equipment.

With the use of electromagnetic and electric fields of an output circuit at frequencies in the kilohertz and megahertz range to produce heat, there are specific requirements of the various systems to protect personnel from exposure to these fields or from electrical shock. Also, it is important that requirements be followed to prevent unintended heating of components and wiring of the system. Requirements for guarding and grounding are covered in *Part II*.

Article 668 deals with the wiring to electrolytic cells for the production of a particular metal, gas, or chemical compound. Definitions important for the application of the article are found in *668.2*. *NEC® 668.1*

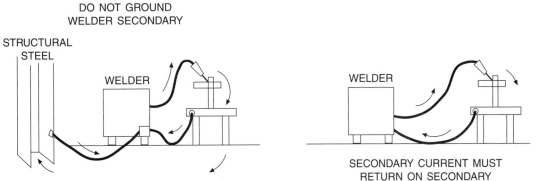

Figure 12.9   **A welder is not considered to be a separately derived premises wiring system, and therefore, is not required to be grounded as a separately derived system. If it is improperly grounded, objectionable current paths can be established, and welder current can flow on equipment grounding conductors throughout the building.**

states that this article does not apply to the production of hydrogen, electroplating, or electrical energy. This article provides some general wiring requirements for electrolytic cells, but each process is unique and requires engineering design for the specific material and process. This article allows for individual process design. A FPN in *668.1* refers to the IEEE standard *463-1993* for *Electrical Safety Practices in Electrolytic Cell Line Working Zones.* Electrolytic cell line is defined in *668.2,* and the cell line conductors shall meet the provisions of *668.3(C)* and *668.12.*

Article 669 deals with the installation of wiring and equipment for electroplating processes. *NEC® 669.9* requires overcurrent protection to be provided for the direct current conductors. Bare conductors are permitted to be run to the electroplating cells. *NEC® 669.6(B)* requires that when bare cell conductors operate at more than 50 volts, the conductors shall be guarded against accidental contact. When there is more than one power supply to the electroplating tanks, there shall be a means of disconnecting the direct current output of the power supply from the conductor to the electroplating tanks, as required by *669.8.*

Article 670 covers the wiring requirements for industrial machinery, the definition of industrial machinery, and required nameplate information. Industrial machinery is defined in *670.2. NEC® 670.3(A)* states the nameplate information to be provided on industrial machinery. The full-load current of the machine is required to be provided on the nameplate, as well as the full-load current of the largest motor or load. There is a separate NFPA standard for electrical wiring of the actual industrial machinery (*NFPA 79 2002*). If the industrial machine is provided with overcurrent protection, as permitted in *670.3(B)* and *670.4(C)* at the supply terminals, the supply conductor to the machine is permitted to be considered a feeder or tap from a feeder. When overcurrent protection is provided at the supply terminals, *670.3(B)* requires that a label be placed on the machine stating that overcurrent protection is provided at the machine supply terminals, as shown in Figure 12.10. *NEC®*

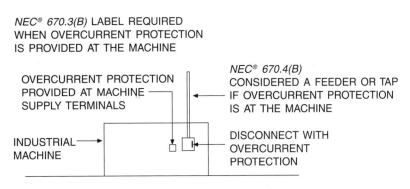

Figure 12.10   **When overcurrent protection is provided at the supply conductor terminals of the industrial machine, the supply conductors are permitted to be a feeder or a tap from a feeder.**

670.4(A) specifies the minimum permitted ampacity of supply conductors to industrial machinery. The supply conductor ampacity shall not be less than the sum of 125% of the full-load current of resistance heating loads, 125% of the full-load current of the largest motor, and the full-load current of all other loads.

Article 727 provides specifications for the use and installation of instrument tray cable. It is permitted to be used in industrial facilities per *727.2*. Instrument tray cable is not permitted to be installed with power, lighting, and non-power-limited circuits, as stated in *727.5*. There is an exception where the Type ITC Cable is permitted with other types of circuits when the Type ITC Cable has an approved outer metal sheath. Spacing requirements and other installation specifications are not provided to define what the Code means by Type ITC Cable not being installed with power, lighting, and non-power-limited circuits.

Type ITC Cable is not permitted to be used for circuits operating at more than 150 volts or more than 5 amperes, as stated in *727.5*. The 150 volts is between conductors as well as to ground. The conductor material for Type ITC Cable is only permitted to be copper or a thermocouple alloy. A thermocouple is a junction of two dissimilar metals that produce a small voltage that changes as the temperature changes. In industry and research, thermocouples are used to measure the temperature of materials and processes. *NEC® 727.7* requires that the cable be marked as Type ITC on the outer nonmetallic sheath. If the cable has a metal sheath, the marking is permitted to be on the nonmetallic sheath beneath the outer metal sheath.

Article 770 covers the markings and installation of optical fiber cable. Conductive optical fiber cable contains a noncurrent-carrying material, which may be a metallic vapor barrier, or a metal member may be present to add mechanical strength. Composite cables contain current-carrying electrical conductors. Nonconductive optical fiber cable is permitted to occupy the same raceway or cable tray with conductors for light, power, and heating circuits. Conductive optical fiber cable is not permitted to occupy the same cable tray or raceway with conductors of electric light, power, or Class 1 circuits, as stated in *770.133(A)*.

The markings on optical fiber cable run in buildings shall be as stated in *770.113*. The listing requirements, which are essentially the uses permitted of the various types, are summarized in *Table 770.113* and described in *770.154*. Conductive optical fiber cable for general-purpose use is marked as Type OFC and nonconductive optical fiber cable for general-purpose use is marked as Type OFN. When the letter **P** is included at the end of the type marking, such as Type OFCP, the cable is suitable for installation in air-handling spaces, as described in *770.154(A)*. The letter **R** at the end of the type marking, such as Type OFNR, designates that the cable is permitted to be run as cable from one floor to another of a building. Cable Types OFC and OFN are permitted to be run from one floor to another of a building if contained in metal conduit or in a shaft with fire stops at each floor. In the case of a one- or two-family dwelling, as stated in *770.154(B)(3)* Types OFC and OFN are permitted to be installed between floors without metal raceway protection. Cable substitutions are permitted, as stated in *Table 770.154*.

CABLE TRAY WIDTH AND CONDUCTOR AMPACITY

When cables are supported by cable tray, the allowable ampacity of the conductors depends upon whether the installation allows for cooling of the conductors when they are carrying current. If a solid cover for a length more than 6 ft (1.8 m) is placed over a ventilated channel or ladder type cable tray, the conductor ampacity will be lower than if the top of the cable tray is open. Spacing of the conductors in the cable tray is also important. If a minimum width of space between the cable is maintained as described in *392.11(A)(3)*, *392.11(B)(3)*, and *392.11(B)(4)*, the conductor ampacity will be higher. In all cases where the conductor spacing is maintained, the minimum width of the cable tray depends upon the diameters of the cables laid out in a single layer. In some cases, a single layer of cables is required but space is not required to be provided between the cables. In other cases, the cables are permitted to be placed in multiple layers. These rules are found in *392.9* for multiconductor cables and in *392.10* for single-conductor cables. The following several examples will illustrate the relationship between ampacity of conductor and width of cable tray.

Example 12.3　A feeder consisting of four single-conductor cables size 500 kcmil copper with Type XHHW insulation and 75°C terminations is run in aluminum ladder-type cable tray as shown in Figure 12.11. The cables are placed in the cable tray without maintaining a space between the cables. Determine the ampacity of the conductors and the minimum permitted width of cable tray.

Answer:　The minimum width of cable tray is determined by the method in *392.10(A)(2)* for single conductor cables. The cross-sectional area of the conductors is not permitted to exceed the area given in column 1 of *Table 392.10* for each width of cable tray. Look up the cross-sectional area of a 500 kcmil Type XHHW conductor in *Table 5, Chapter 9* and find 0.6984 sq. in. (450.6 mm²) The total area of the conductors is 2.7936 sq. in. (1802 mm²). According to column 1 of *Table 392.10*, a cable tray

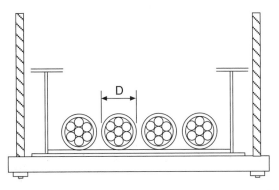

Figure 12.11 **Four size 500 kcmil Type XHHW copper conductors are supported by aluminum ladder-type cable tray with no maintained spacing between the conductors.**

with a 6 in. (150 mm) width is permitted to have a single-conductor cable fill of 6.5 sq. in. (4200 mm²). The minimum width, then, is 6 in. (150 mm).

4 conductors × 0.6984 in.² = 2.7936 in.²
4 conductors × 450.6 mm² = 1802 mm²

Next, determine the ampacity of the conductors according to *392.11(B)(2)*. The ampacity is permitted to be 0.65 times the ampacity found in *Table 310.17*, which is 403 amperes.

0.65 × 620 A = 403 A

Example 12.4 Determine the minimum width of cable tray and ampacity of the conductors when the conductors are installed in a single layer with a space between the conductors not less than the diameter (d) of the conductors as prescribed by *392.11(B)(3)* and shown in Figure 12.12. Four single-conductor cables size 500 kcmil copper with Type XHHW insulation and 75°C terminations are run in aluminum ladder-type cable tray.

Answer: *NEC® 392.11(B)(3)* specifies that this method of determining conductor ampacity requires a maintained space between the conductors placed in a single layer. First look up the diameter of the conductor in *Table 5, Chapter 9* and find 0.943 in. (23.95 mm). The minimum width of cable tray is then seven times the cable diameter, which is 6.60 in. (168 mm). Choose a 9 in. (225 mm) minimum cable tray width. The width of cable trays is listed in the left-hand two columns of *Table 392.9* and *Table 392.10*.

7 × 0.943 in. = 6.60 in.
7 × 23.95 mm = 168 mm

NEC® 392.11(B)(3) simply specifies the minimum conductor ampacity can be found in *Table 310.17*, which is 620 amperes. Note the big difference in conductor ampacity between the previous example and this example when a space is maintained between the conductors.

Example 12.5 Three Type TC cables with three size 250 kcmil XHHW copper conductors and an equipment grounding conductor are to be supported by aluminum ladder-type cable tray where a space

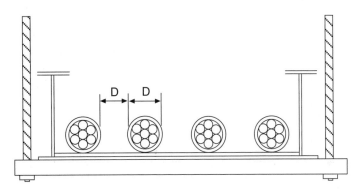

Figure 12.12 **Four size 500 kcmil Type XHHW copper conductors are run in aluminum ladder-type cable tray with a space maintained between the conductors equal to the cable diameter.**

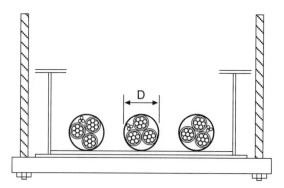

Figure 12.13   Three Type TC multiconductor cables, with size 250 kcmil Type XHHW copper insulated conductors and a bare equipment grounding conductor, are supported by an aluminum ladder-type cable tray without a maintained spacing between the cables.

is not maintained between the individual cables, as illustrated in Figure 12.13. All conductor terminations are rated 75°C, and the cable diameter is 1.76 in. (44.7 mm). Determine the minimum width of cable tray permitted, and the ampacity of the insulated conductors in the cable.

Answer: The minimum width of the cable tray is determined according to *392.9(A)(1)*. The cables are required to be in a single layer, but no space is required between the cables. The minimum width of cable tray is not permitted to be less than the sum of the diameters of the cables in the cable tray, which is 5.28 in. (134 mm). Therefore, a 6 in. (150 mm) cable tray is adequate.

3×1.76 in = 5.28 in
3×44.7 mm = 134 mm

Next determine the ampacity of the size 250 kcmil copper conductors in the cables according to *392.11(A)*. Note in *392.11(A)(1)* that the derating factors do not apply unless there are more than three current-carrying conductors in any one cable, and then they apply only to that particular cable. In this case, conductor ampacity is determined according to *Table 310.16*, and it is 255 amperes.

Example 12.6   For the cable tray installation of Figure 12.14, the three Type TC 3-conductor cables are installed with a space maintained between them of one cable diameter. Each cable contains three XHHW insulated conductors size 250 kcmil copper and has a diameter of 1.76 in. (44.7 mm). All conductor terminations are rated 75°C. Determine the minimum permitted width of aluminum ladder-type cable tray and the ampacity of the insulated conductors in the cable.

Answer: There is no rule for determining the minimum width of cable tray. It is simply a matter of adding the cable diameters. There are three cables of the same diameter and there are two spaces between the cables of the same diameter. The minimum width of cable tray is not permitted to be less than 8.8 in. (224 mm); therefore, the minimum standard width is 9 in. (225 mm).

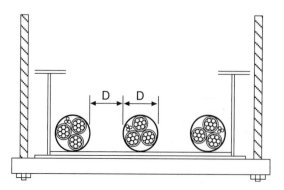

Figure 12.14   Three Type TC multiconductor cables, with size 250 kcmil Type XHHW copper insulated conductors and a bare equipment grounding conductor, are supported by an aluminum ladder-type cable tray with a space maintained between the cables equal to the diameter of the cable.

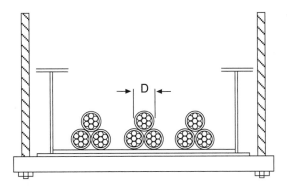

Figure 12.15 Three triplexed single conductor cable bundles with size 4/0 AWG Type THWN copper conductors are installed in aluminum ladder-type cable tray with no maintained space between the conductors.

5 × 1.76 in. = 8.8 in.

5 × 44.7 mm = 224 mm

The conductor ampacity is determined according to *392.11(A)(3)*, which specifies the method of *310.15(C)*. That section is one that requires engineering supervision, but the ampacity is usually determined from *Table B.310.3* in *Annex B*, and in this case is 274 amperes.

Example 12.7   Three sets of triplexed conductors are installed in an aluminum ladder-type cable tray as shown in Figure 12.15. The individual conductors are type THWN, size 4/0 AWG copper, and all terminations are rated 75°C. *NEC® 392.8(E)* permits these conductors to be bundled rather than installed in a single layer. Determine the minimum width of cable tray permitted and ampacity of the conductors.

Answer: The minimum width of cable tray is specified in *392.10(A)(4)*. The width is not permitted to be less than the sum of the diameters of all of the conductors in the cable tray. Look up the diameter of a size 4/0 AWG Type THWN conductor in *Table 5, Chapter 9*; it is 0.642 in. (16.3 mm). Then multiply by nine conductors to get the minimum width of 5.78 in. (147 mm). A 6 in. (150 mm) wide cable tray is permitted.

9 conductors × 0.642 in. = 5.78 in.
9 conductors × 16.31 mm = 147 mm

The ampacity of the individual conductors is determined according to *392.11(B)(2)*. The ampacity found in *Table 310.17* is multiplied by 0.65 to get 234 amperes.

0.65 × 360 A = 234 A

Example 12.8   Three triplexed bundles of Type THWN copper conductors are installed in aluminum ladder-type cable tray with a space maintained between each bundle equal to 2.15 times the individual cable diameter, as shown in Figure 12.16. All conductor terminations are rated 75°C. Determine the minimum permitted width of cable tray and the ampacity of the conductors.

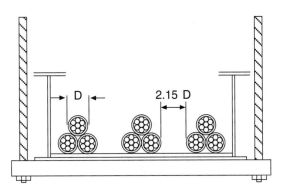

Figure 12.16 Three triplexed single conductor cable bundles with size 4/0 AWG Type THWN copper conductors are installed in aluminum ladder-type cable tray with a space not less than 2.15 times the individual cable diameter maintained between the cable bundles.

Answer: The width of the cable tray is simply the physical dimension necessary to accommodate the conductors when installed as required by *392.11(B)(4)*. There are three bundles of cables; therefore, there will be two spaces with a minimum dimension of 2.15 times the individual conductor diameter. The bundles will lay in the cable tray with two conductors touching the cable tray. Therefore, space must be allowed for a minimum of six conductors, as shown in Figure 12.16. Look up the diameter of a size 4/0 AWG Type THWN conductor in *Table 5, Chapter 9*; it is 0.642 in. (16.31 mm). The calculated minimum width permitted is 6.612 (168 mm), which requires at least a 9 in. wide (225 mm) cable tray.

$$
\begin{aligned}
\text{Width} \quad &= 2 \times (2.15 \times \text{D}) + 6 \times \text{D} \\
&= 2 \times (2.15 \times 0.642 \text{ in.}) + 6 \times 0.642 \text{ in.} \\
&= 2 \times 1.380 \text{ in.} + 3.852 \text{ in.} \\
&= 2.760 \text{ in.} + 3.852 \text{ in.} = 6.612 \text{ in.} \\
&= 2 \times (2.15 \times 16.31 \text{ mm}) + 6 \times 16.31 \text{ mm} \\
&= 70 \text{ mm} + 98 \text{ mm} = 168 \text{ mm}
\end{aligned}
$$

Because a space between the triplexed bundles is maintained for cooling, the ampacity is permitted to be looked up in *Table 310.20* and, in this case, is 287 amperes.

When multiconductor cables of different sizes are placed in a cable tray together, the method of determining the width of cable tray required depends upon the size of conductors in the cables. There is one simple method when all cables contain conductors size 4/0 AWG and larger. *NEC® 392.9(A)(1)* requires that the cables be arranged in a single layer. The width of the cable tray is equal to the sum of the diameters of the cables.

When the conductors in the cables are size 3/0 AWG and smaller, a single layer of conductors is not required. The cables can be placed in multiple layers as long as they are not stacked on top of any cables in the same cable tray that have conductors size 4/0 AWG and larger. Cables with conductor size 3/0 and smaller are not permitted to have a cross-sectional area greater than approximately 40% of the cross-sectional area of a standard usable depth cable tray, which is approximately 3 in. (75 mm). If a cable tray contains only cables with conductors that are size 3/0 and smaller, then the method of determining the cable tray width is specified in *392.9(A)(2)*. The total cross-sectional area of the conductors is not permitted to exceed the value given in column 1 of *Table 392.9*. The values given in column 1 of *Table 392.9* are approximately 40% of the usable cross-sectional area of a standard cable tray.

Example 12.9   A ladder-type cable tray contains 10 Type TC multiconductor cables that each have three copper conductors size 2/0 AWG. The total cross-sectional area of the cables is 1.35 sq. in. (870 mm²). Remember that how the cables are arranged in the cable tray determines the allowable ampacity of the cables. In this case assume the cables are installed in multiple layers in the cable tray. Determine the minimum width of the cable tray required for these cables.

Answer: The method of sizing the cable tray is described in *392.9(A)(2)*. First determine the total cross-sectional area of the cables in the cable tray, which is 13.6 sq. in. (8700 mm²). Next select the minimum size cable tray from column 1 of *Table 392.9* and find 12 in. (300 mm).

When cables with conductors larger and smaller than size 4/0 AWG are placed in the same cable tray, two methods are combined to determine the minimum width of cable tray. The rule is found in *392.9(A)(3)*, and utilizes column 2 of *Table 392.9*. An example is shown in Figure 12.17 which shows a cable tray arranged into two sections; one with cables with conductor sizes 4/0 AWG and larger, and the other section with cables with conductor sizes 3/0 AWG and smaller. The cables with conductor sizes 4/0 AWG and larger are required to be arranged in a single layer. The cables with conductor sizes 3/0 and smaller are permitted to be in multiple layers. In either case, the cable tray is not permitted to have a conductor cross-sectional area that exceeds 40% of the usable cross-sectional area of the cable tray. Cable trays have a standard depth of approximately 3 in. (75 mm). Therefore, the total cable cross-sectional area is not permitted to exceed 40% of the cable tray width times 3 in. (75 mm). Those values are given in column 1 of *Table 392.9*. The formulas in column 2 of *Table 392.9* essentially subtract the diameters of the single layer of cables from the total width, and then calculates 40% of the remaining area of the cable tray. This process is illustrated in Figure

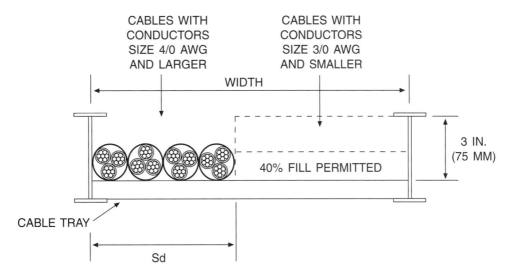

Figure 12.17 **After the diameters of the cables with size 4.0 AWG and larger conductors are subtracted from the width of the cable tray, the cables with conductors size 3/0 AWG and smaller are permitted to fill only 40% of the remaining portion of the cable tray.**

12.17. This is accomplished by determining the total width of the cables with conductors size 4/0 and larger (this value is **Sd**) and multiplying by 1.2, which is actually 1.2 in. assuming the usable depth of a cable tray is 3 in. The value 1.2 in. is 40% of 3 in. When worked out in metrics, the usable depth of the cable tray is 75 mm, of which 30 mm is 40%. The metric formulas in column 2 of *Table 392.9* multiplies **Sd** by 30, while the in.-pound formula multiplies **Sd** by 1.2. The result of the calculation using the formula in column 2 of *Table 392.9* is compared with the actual cross-sectional area of the cables with conductors size 3/0 AWG and smaller. If the value from the calculation is larger than the cross-sectional area of the conductors, then the cable tray is wide enough for the application.

Example 12.10 A cable tray contains two Type TC cables with three size 350 kcmil conductors with a cable diameter of 1.98 in. (50.3 mm), and six Type TC cables with three size 2/0 AWG conductors with a cable cross-sectional area of 1.35 sq. in. (870 mm²) as shown in Figure 12.18. Determine the minimum permitted width of ventilated-trough cable tray for this installation.

Answer: The method is described in *392.9(A)(3)*. The cross-sectional area and diameter of multiconductor cables are not given in the Code. The best source for this information is to contact the cable manufacturer. Some manufacturers provide cable technical data on their web sites. The symbol **Sd** in column 2 of the *Table 392.9* is the sum of the diameters of all cables with conductors size 4/0 AWG and larger, which is multiplied by 1.2 to convert to sq. in. (multiply by 30 to convert to square millimeters). In this example, there are two cables, so multiply the diameter by two to get 3.96 in. (100 mm).

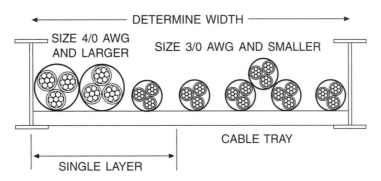

Figure 12.18 **The cable tray contains two Type TC cables with size 350 kcmil conductors and size Type TC cables with size 2/0 AWG conductors.**

Sd = (sum of cable diameters with conductors size 4/0 AWG and larger)
Sd = 2 × 1.98 in. = 3.96 in.
Sd = 2 × 50.3 mm = 100.6 mm

Next determine the cross-sectional area of all conductors size 3/0 AWG and smaller. There are six that are size 2/0 AWG, so multiply their area, which is 1.36 sq. in. (870 mm²), by six to get the total cross-sectional area of the conductors which is 8.10 sq. in. (5225 mm²).

6 cables × 1.35 in.² = 8.10 in.²
6 cables × 870 mm² = 5225 mm²

The method is described in the text is a trial and error method. Choose one of the standard cable tray widths believed to be wide enough and do the calculation. Start with a 12 in. (300 mm) width for this example. The formula is in column 2 of *Table 392.9*.

14 − (1.2 × Sd) = cross-sectional area of cables with conductors size 3/0 and smaller

The result from this calculation must not be smaller than the cross-sectional area of the cables with conductors size 3/0 AWG and smaller as determined earlier.

14 in.² − (1.2 × 3.96 in.) = 14 in.² − 4.75 in.² = 9.25 in.²
9000 mm² − (30 × 100 mm) = 9000 mm² − 3000 mm² = 6000 mm²

The cables with conductor size 3/0 AWG are only 8.10 in.² (5225 mm²); therefore, the 12 in. (300 mm) wide cable tray is adequate for the application.

MAJOR CHANGES TO THE 2005 CODE

These are the changes to the 2005 *NEC*® that correspond to the Code sections studied in this unit. The following analysis explains the significance of the changes from the 2002 to the 2005 Code only, and this analysis is not intended to be used in place of the Code. Refer to the actual section of the 2005 Code for the exact wording and meaning of each section discussed. Changes are indicated in the Code with a vertical line in the margin. If material was deleted or moved to another location in the Code, the location of the deletion is indicated with a dark dot in the margin.

Article 368 **Busways**

This article was renumbered to be consistent with the numbering system used in other similar articles.
368.56(A): This was *368.8(A)* in the previous edition of the Code. There is no actual change in this section, but rather than list the article numbers or the types of materials that are permitted to be used as branches from busway, this section now lists fifteen different wiring methods by name that are permitted to serve as branches from busway.

Article 392 **Cable Trays**

392.8(D): This section provides instructions as to how to install single-conductor cables when the conductor cables are run as parallel sets of conductors as permitted in *310.4*. To minimize the resistive effect of inductive reactance, the conductors are to be bundles into sets consisting of one of each phase conductor and neutral if a neutral is present. Inductive reactance is only an issue for ac circuits and feeders. When direct current conductors are run as parallel sets in cable tray, inductive reactance is not an issue. The change is the addition of the words "alternating current" to this section. This rule no longer applies to direct current circuits.
Table 392.9: This table is used to determine the minimum width of a cable tray permitted when multiconductor cables are installed in the cable tray. The SI metric formula given in *Column 2* and *Column 4* of the table were incorrect in the previous edition of the Code. That error was corrected. The formula was correct when cable area was in sq. in. and the cable diameter was given as in.
392.10(A)(1): It was not clear in the previous edition of the Code when single-conductor cables were required to be installed in a single layer in a cable tray and when they were permitted to be installed in multiple layers. There was a new sentence added to this section that applies to single-conductor cables size 1000 kcmil and larger. It is now clearly stated that these conductors are to be installed in a single layer unless permitted to be bundles to form a set of conductors.

Table 392.10: This table is used to determine the minimum width of a cable tray permitted when only single-conductor cables are installed in the cable tray. The SI metric formula given in *Column 2* of the table was incorrect in the previous edition of the Code. The error was corrected.

392.11(B)(3) Exception: This is a new exception that applies to single-conductor cables installed in an uncovered solid bottom cable tray where a space is maintained between the cables of not less than one cable diameter. The previous edition of the Code permitted the conductor ampacity to be determined using *Table 310.17* for single conductors in free air. With a solid bottom in the cable tray, free air movement between the conductors is restricted as shown in Figure 12.19. Now the ampacity of the conductors is required to be determined by an engineer using the provisions of *310.15(C)*.

Applying logic to this situation for field evaluation of conductor ampacity, if the conductors were installed in raceway, the ampacity would be determined using *Table 310.16*. Even though the cable tray has a solid bottom, with the conductors spaced a minimum on one cable diameter apart, there will be air circulation and conductor cooling with the open top. This would mean the ampacity of the conductors in the solid bottom cable tray would not be greater than the value given in *Table 310.16*, butt less than the value given in *Table 310.17*. A complicating factor to determination of conductor ampacity for solid bottom cable trays is dirt accumulation on the conductors and in the spaces between the conductors. According to *392.3(B),* single conductors are permitted to be installed in solid bottom cable trays in industrial areas.

Article 427   **Fixed Electric Heating Equipment for Pipelines and Vessels**

427.1 FPN: A new fine print note was added that provides references for the installation of industrial heat tracing, induction heating and skin-effect heating.

427.27 Exception: This section sets a maximum voltage for the secondary of a two-winding transformer that is used to pass current through a pipeline to apply heat. This process is called impedance heating. The new exception permits the secondary of the transformer to operate at up to 132 volts to ground for industrial applications only where the listed safety criteria are met. The previous edition of the Code limited the transformer secondary output to a maximum of 80 volts when ground-fault circuit-interrupter protection was provided.

Article 490   **Equipment Over 600 Volts Nominal**

490.46: This is a new section that requires a grounding bus to be provided as a part of metal enclosed high-voltage service equipment for the termination of high-voltage cable shields and other safety grounds. This is a requirement that must be met by the manufacturer of high-voltage service equipment.

Article 610   **Cranes and Hoists**

610.61: This section deals with the grounding of noncurrent-carrying metal parts of cranes, hoists, and similar equipment. The change deals with the method of bonding between the frame of the trolley and the bridge. The previous edition of the Code permitted contact between the trolley wheels and the track mounted to the bridge to provide bonding between the trolley and the bridge. If reliable metal-to-metal

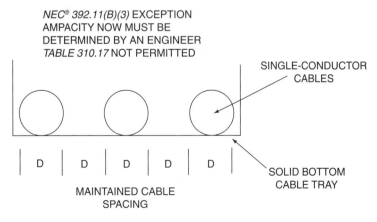

Figure 12.19   The ampacity of single-conductor cables installed in solid bottom cable tray where there is a maintained spacing of one cable diameter between the cables is now required to be determined by an engineer and is no longer determined using *Table 310.17* **or** *Table 310.19.*

contact is prevented by paint, corrosion, or other insulating substance, then a separate bonding jumper was required to be installed between the frame of the trolley and the bridge. Now that bonding jumper must be installed in all cases.

Article 630   **Electric Welders**

630.1: Plasma cutting equipment was added to the scope of this article that deals with arc welders and resistance welders, but no rules or additional reference were made to plasma cutting equipment installations. The article only provides rules for circuits and feeders for arc welders and resistance welders.

Article 670   **Industrial Machines**

670.1 FPN: A new sentence was added to the fine print note that now makes reference to *110.26* for workspace clearances at industrial machinery. This reference was added because *670.5* was deleted that provided working clearance requirements in front of openings that provided access to live parts and terminals. The clearance requirement was 21/2 ft (750 mm) unless a tool was required to open the equipment for only testing and diagnostics and then the clearance was permitted to be less than 2/12 ft (750 mm). The deletion of *670.5* and the addition of the reference in the fine print note means the minimum workspace clearance is found in *110.26,* which is 3 ft (900 mm) in most cases, as illustrated in Figure 12.20.

670.3(A)(4): Instead of just providing the short-circuit current rating of the industrial machine overcurrent device on the nameplate, now it is required to provide the short-circuit current rating of the machine industrial control panel. The machine control enclosure or assembly is also required to be listed and labeled.

670.3(A)(5): The previous edition of the Code required the diagram number or index to the electrical drawings for the industrial machine to be provided on the nameplate if available. Now this number must be provided.

670.4(C): This was part of *670.4(B)* in the previous edition of the Code, which was broken into two parts, one dealing with the rule for disconnecting means and the other dealing with the rules for overcurrent protection. There is no change in intent, but there was an important clarification in the overcurrent part of this section. It is made clear that the rules for determining the maximum rating overcurrent protection is for each feeder or branch-circuit supplying the machine. It seemed to imply in the past that there was only one overcurrent device protecting the entire machine.

Article 770   **Optical Fiber Cables and Raceways**

The sections in this article were renumbered to create a consistency with the numbering of similar articles.

770.12(C): The term "inner duct" is used without a definition. Optical Fiber Raceway is manufactured with a square cross-section such as 2 in. by 2 in. (50 mm by 50 mm), 4 in. by 4 in. (100 mm by 100 mm), 4

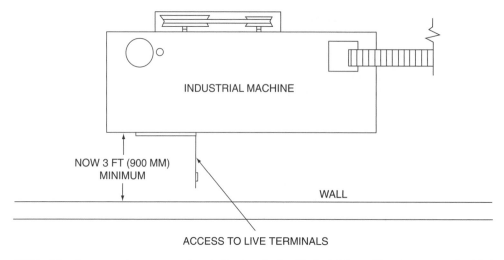

CLEARANCE TO LIVE PARTS NO LONGER 2 1/2 FT (750 MM)
NEC® 670.5 DELETED—NOW NEC® 110.26 APPLIES

INDUSTRIAL MACHINE

NOW 3 FT (900 MM) MINIMUM

WALL

ACCESS TO LIVE TERMINALS

Figure 12.20   The clearance at access openings to live terminals of industrial machines is now required to be determined using *110.26*, which will result in an increase in clearance, usually from 2¹/₂ ft (750 mm) to 3 ft (900 mm).

in. by 8 in. (100 mm by 200 mm), and 4 in. by 12 in. (100 mm by 300 mm). It is also manufactured with a round cross-section corrugated for flexibility similar to Electrical Nonmetallic Tubing. This Optical Fiber Raceway with a round cross-section can be run inside other raceways and is commonly called inner duct. This section now recognizes listed flexible Optical Fiber Raceway and permits it to be installed inside any type of raceway system recognized by the Code that is suitable for the application.

770.12(D): Plastic innerduct of the unlisted underground type or the outside plant construction type is required to be terminated at the point of entrance to a building. Firestopping is now required to be installed at the point of entrance for these materials. These materials are considered to be highly combustible.

770.24: This is the section that specifies how Optical Fiber Cable is to be installed and supported. This section supplied few specific requirements in the past. Now Optical Fiber Cable is required to be installed according to the rules in *300.11*. There is no specific spacing requirement for supports, but there is now a rule that the cable must be supported by the structure of the building. If run above suspended ceilings, it is not permitted to be placed on the ceiling or supported by the ceiling support wires. It is permitted to add support wires specifically for the support of wiring such as Optical Fiber Cable. An installation above a suspended ceiling is shown in Figure 12.21.

770.113 Exception 2: The previous edition of the Code permitted unlisted Optical Fiber Cable to be run through a building if installed in any acceptable raceway. Now only specific raceways are permitted to be used with unlisted Optical Fiber Cable, which are Rigid Metal Conduit, Intermediate Metal Conduit, Rigid Nonmetallic Conduit, and Electrical Metallic Tubing. Other types of raceways are not recognized for this application.

770.133(A): The list of electrical cables that are permitted to be run in the same cable tray or raceway with Optical Fiber Cable now includes Type ITC Instrument Tray Cable.

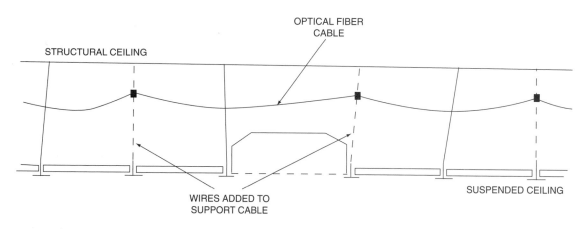

NEC® 770.24 OPTICAL FIBER CABLE PERMITTED
TO BE SUPPORTED TO SEPARATE WIRES

Figure 12.21 **The previous edition of the Code did not specifically state that Optical Fiber Cable was permitted to be supported by tie wires above a suspended ceiling that were specifically installed to support wiring.**

WORKSHEET NO. 12—BEGINNING INDUSTRIAL ELECTRICAL APPLICATIONS

Mark the single answer that most accurately completes the statement based upon the 2005 Code. Also provide, where indicated, the Code reference that gives the answer or indicates where the answer is found, as well as the Code reference where the answer is found.

1. A machine in a factory is supplied power from an overhead busway by a bus drop cable. A tension take-up device connected to the structure of the building supports the bus drop cable as shown in Figure 12.22. The distance from the busway plug-in device to the tension take-up device where no intermediate supports are provided is not permitted to exceed:

    A. 6 ft (1.8 m).   C. 12 ft (3.7 m).   E. 25 ft (7.5 m).
    B. 10 ft (3 m).   D. 15 ft (4.5 m).

                                Code reference _____

2. A cablebus is used as the service-entrance conductor from a transformer located outside a building with the cablebus running from the transformer to the service equipment inside the building, as shown in Figure 12.23. The cable bus is not specifically designed for extra-long spans. The maximum distance permitted between supports is:

    A. 5 ft (1.5 m).   C. 8 ft (2.5 m).   E. 12 ft (3.7 m).
    B. 6 ft (1.8 m).   D. 10 ft (3 m).

                                Code reference _____

3. When single-conductor cables size 4/0 and smaller are installed in ladder-type cable tray, the rung spacing is not permitted to be greater than:

    A. 6 in. (150 mm).   D. 16 in. (400 mm).
    B. 9 in. (230 mm).   E. 18 in. (450 mm).
    C. 12 in. (300 mm).

                                Code reference _____

4. Conductors are permitted to be run through the open provided they are not exposed to physical damage. The conductors are permitted to pass from one cable tray section to another, from the cable tray to equipment, or from the cable tray to a raceway

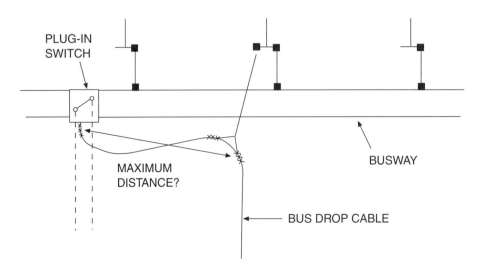

Figure 12.22   Determine the maximum length of bus drop cable permitted between the take-up support and the plug-in switch.

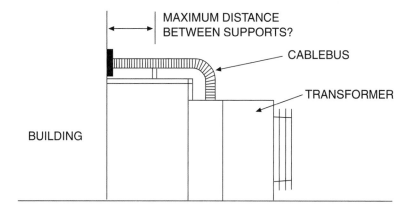

Figure 12.23 Determine the maximum distance between supports of an outside installation of cablebus that is not specifically designed for a longer support distance.

as shown in Figure 12.24. The maximum distance permitted for open spans of conductors from cable trays is:

A. 18 in. (450 mm). C. 4^1/2 ft (1.4 m). E. 8 ft (2.5 m).
B. 3 ft (900 mm). D. 6 ft (1.8 m).

Code reference _____

5. The cable tray shown in Figure 12.24 is aluminum ladder-type and contains three feeder circuits rated 200, 400, and 800 amperes. Two sections of the cable tray do not join and a copper conductor is used to bond the two sections of cable tray. The minimum size bonding jumper for the installation is:

A. 4 AWG. C. 1/0 AWG. E. 4/0 AWG.
B. 2 AWG. D. 3/0 AWG.

Code reference _____

6. Ventilated-trough cable tray contains only single copper conductor cables installed in a single layer with an air space maintained between the cables not less than the

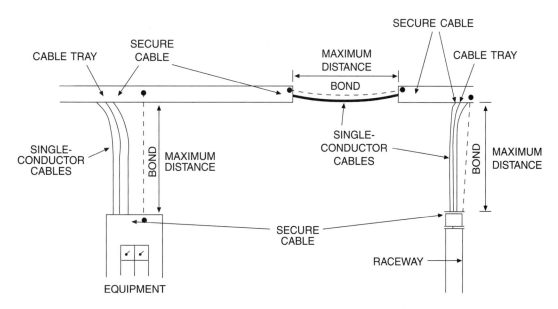

Figure 12.24 Cable tray installation in an industrial building where two sections of cable tray do not join and conductors are in the open between the cable tray and equipment or conduits.

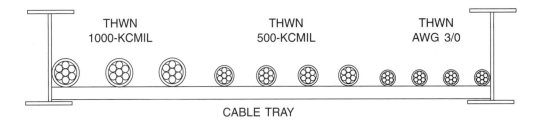

THWN
1000-KCMIL

THWN
500-KCMIL

THWN
AWG 3/0

CABLE TRAY

Figure 12.25   A cable tray in an industrial building that contains single-conductor cables arranged in a single layer with an air space between the cables with a width equal to the diameter of the largest adjacent cable. The cables are tied to the cable tray so they will not move after installation.

width of the largest adjacent cable, as shown in Figure 12.25. There are three 1000-kcmil conductors, four 500-kcmil conductors, and four size 3/0 AWG conductors all with THWN insulation. The cables are arranged in this manner to achieve maximum ampacity from the conductors. The cables are tied down to the cable tray so they will not move. The minimum width of cable tray required for this installation is:

A. 9 in. (225 mm).   D. 24 in. (600 mm).
B. 12 in. (300 mm).   E. 30 in. (750 mm).
C. 18 in. (450 mm).

Code reference _____

7. The size 500 kcmil, Type THWN conductors run in the cable tray of Figure 12.25 have a maintained spacing of one cable diameter between the cables for the entire circuit run in ventilated-trough cable tray with no cover. The allowable ampacity of the conductors is:

A. 380 amperes.   C. 500 amperes.   E. 620 amperes.
B. 420 amperes.   D. 560 amperes.

Code reference _____

8. The nonheating lead wires provided with an electric resistance type pipeline heating cable are longer than required to make the connections to the wiring system. The non-heating leads are:

A. required to be kept at the original length.
B. permitted to be shortened to the desired length provided the lead markings are not removed.
C. required to be shortened only to the length needed to make the connections.
D. permitted to be shortened to not less than 24 in. (600 mm).
E. permitted to be shortened but required to be 4 ft (1.2 m) in length to prevent conduction of heat to the terminal box.

Code reference _____

9. If a cable tray is to be used as an equipment grounding conductor for the circuit and feeder conductors that it supports, the cable tray shall be:

A. constructed of aluminum.
B. of the solid bottom or ventilated type.
C. marked to show the cross-sectional area of the cable tray or side rails, which is sufficient for the circuit rating.
D. mounted not less than 12 ft (3.7 m) above the floor
E. installed in lengths not to exceed 100 ft (30 m).

Code reference _____

10. A resistance welder makes repetitive spot welds and has a duty cycle of 4%. The actual primary supply current during the weld is 620 amperes at 460 volts. If the supply conductors for the welder are copper, Type THWN with 75°C terminations, the minimum permitted size conductor is:
    A. 3 AWG.   C. 1 AWG.   E. 2/0 AWG.
    B. 2 AWG.   D. 1/0 AWG.

    Code reference _____

11. An industrial machine is to be installed where only qualified persons will service the equipment. An area of the machine contains control wiring, operating at under 150 volts, that may require servicing while energized. The minimum clearance in front of the access to the live parts is required to be not less than:
    A. $2^1/2$ ft (750 mm).   D. 4 ft (1.2 m).
    B. 3 ft (900 mm).   E. $4^1/2$ ft (1.4 m).
    C. $3^1/2$ ft (1 m).

    Code reference _____

12. The secondary output conductors of an industrial arc welder consist of an electrode conductor and a workpiece conductor. The workpiece conductor:
    A. shall be grounded to the metal frame of the building in the area of the welder.
    B. shall be grounded to the equipment grounding conductor of the supply to the welder.
    C. is not required to be grounded to any premises electrical system grounding conductor or grounding electrode.
    D. is required to be connected to a ground rod in the area of the welder.
    E. is considered to be a separately derived system and is to be grounded accordingly.

    Code reference _____

13. In an existing building, optical fiber cable is to be installed where the most convenient way of running the necessary cables is across the upper side of a permanent suspended ceiling in a hallway. There is one communications cable and no other cables run across the same ceiling. The maximum number of optical fiber cables permitted to be installed by running them across the suspended ceiling from one end of the hallway to the other is:
    A. one.
    B. two.
    C. three.
    D. zero because this practice is not permitted.
    E. as many as desired.

    Code reference _____

14. A new optical fiber cable is installed in a building to replace an existing cable that will be abandoned. The abandoned cable:
    A. shall be removed where accessible.
    B. shall be removed only in places where it is visible.
    C. is permitted to be left in place provided it does not limit access to equipment.
    D. shall be completely removed from the building.
    E. is permitted to remain in place provided the ends are properly sealed.

    Code reference _____

15. For new construction, optical fiber cables are required to be secured to the structure of the building.
    A. within 12 in. (300 m) of terminations and at intervals not exceeding $4^1/2$ ft (1.4 m).
    B. within 12 in. (300 m) of terminations and at intervals not exceeding 5 ft (1.5 m).
    C. at intervals not exceeding 10 ft (3 m).
    D. only in areas that will not be accessible.
    E. with straps, staples, hangers, or similar fittings designed and installed so as not to damage the cable.

Code reference _____

WORKSHEET NO. 12—ADVANCED INDUSTRIAL ELECTRICAL APPLICATIONS

Mark the single answer that most accurately completes the statement based upon the 2005 Code. Also provide, where indicated, the Code reference that gives the answer or indicates where the answer is found, as well as the Code reference where the answer is found.

1. A set of integrated gas spacer cables, trade size 4500 kcmil are run underground for the service entrance of an industrial building. Assuming that the cable is installed such that derating of the ampacity is not required, the allowable ampacity of the Type IGS Cable is:

   A. 505 amperes.   D. 600 amperes.
   B. 519 amperes.   E. not specified in the Code.
   C. 550 amperes.

   Code reference_____

2. An 800-ampere busway suspended from the ceiling of an industrial area is to be extended to serve an additional load. A 400-ampere busway is adequate to supply the load and the 400-ampere section is added to the 800-ampere busway without over-current protection provided for the 400-ampere section of busway as shown in Figure 12.26. This practice is permitted provided the busway of reduced ampacity does not extend beyond the point of ampacity reduction more than:

   A. 25 ft (7.5 m).   C. 75 ft (22.5 m).   E. 200 ft (60 m).
   B. 50 ft (15 m).   D. 100 ft (30 m).

   Code reference_____

3. A cablebus system protected by a 1200-ampere circuit breaker is used as feeder conductors from a disconnect at a transformer to the main distribution panelboard. The cablebus will contain one set of 3-phase, 480/277-volt conductors, as shown in Figure 12.27. The calculated demand load on the conductors is 960 amperes. The minimum size Type THWN copper conductors permitted to be installed for this cablebus is:

   A. 1000 kcmil.   C. 1500 kcmil.   E. 2000 kcmil.
   B. 1250 kcmil.   D. 1750 kcmil.

   Code reference_____

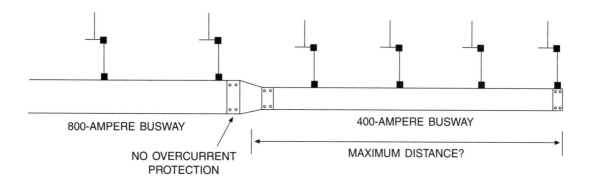

Figure 12.26 A 400-ampere rated busway is added to an 800-ampere busway to supply a load within the rating of the 400-ampere busway.

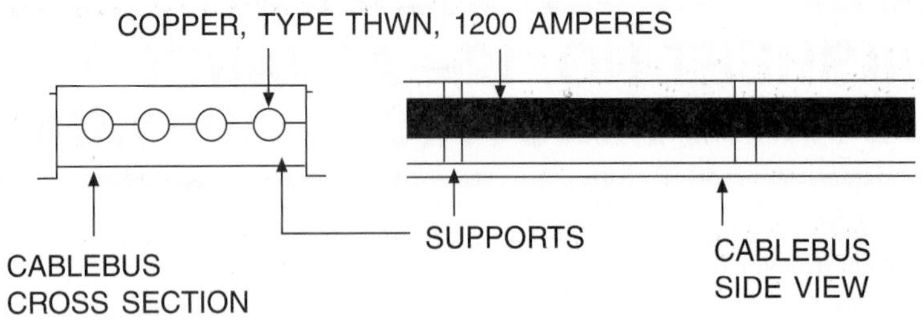

COPPER, TYPE THWN, 1200 AMPERES

CABLEBUS
CROSS SECTION

SUPPORTS

CABLEBUS
SIDE VIEW

Figure 12.27   **A 1200-ampere cablebus is used as a feeder from a transformer to a distribution panelboard.**

4. The smallest size ungrounded or grounded single-circuit conductors permitted to be installed in ladder-type cable tray is:
   A. not specified as long as rung spacing does not exceed 6 in. (150 mm).
   B. 6 AWG.
   B. 4 AWG.
   C. 2 AWG.
   D. 1/0 AWG.

   Code reference _____

5. A feeder is run with three size 500 kcmil copper conductors in raceway for a portion of the circuit and in aluminum ladder-type cable tray for a portion of the circuit. The conductors have 75°C insulation and terminations and are arranged in a single layer but a space is not provided between the conductors. The maximum allowable ampacity of the conductors is:
   A. 380 amperes.   C. 540 amperes.   E. 650 amperes.
   B. 403 amperes.   D. 620 amperes.

   Code reference _____

6. Single-conductor cables are permitted to be run in cable tray:
   A. in any type of building.
   B. only in commercial buildings.
   C. only in industrial buildings.
   D. in commercial buildings where there is a permanent maintenance staff to maintain the wiring system.
   E. only in approved industrial buildings with an engineer available to supervise maintenance.

   Code reference _____

7. Type TC multiconductor cables with three copper Type XHHW insulated power conductors are run in uncovered ladder-type cable tray in a single layer with no maintained spacing between the conductors, as shown in Figure 12.28. There are two other cables—one cable with copper size 3/0 AWG conductors. All conductor terminations are 75°C rated. The minimum size copper conductor in the other cable required for a 225-ampere feeder is:
   A. 2/0 AWG.   C. 4/0 AWG.   E. 300 kcmil.
   B. 3/0 AWG.   D. 250 kcmil.

   Code reference _____

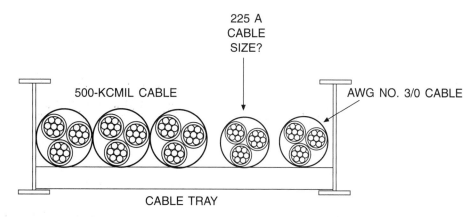

Figure 12.28 **A cable tray contains several multiple conductor cables arranged in a single layer but without a maintained air space between the cables. One of the cables is to be sized as a 225-ampere feeder circuit.**

8. The cable tray in Figure 12.28 is aluminum ladder-type and it is installed in an industrial building where only qualified maintenance staff will service the wiring. The highest rated circuit in the cable tray is 400 amperes. The cable tray is permitted to serve as the equipment grounding conductor for the circuits if it is listed as suitable for equipment grounding and it has a total cross-sectional area of the side rails of not less than:

   A. 0.40 in.² (258 mm²).   D. 1.00 in.² (645 mm²).
   B. 0.60 in.² (387 mm²).   E. 1.50 in.² (967 mm²).
   C. 0.70 in.² (451 mm²).

   Code reference _____

9. A ladder-type cable tray supports TC multiconductor power cables with Type XHHW copper conductors. There are two 3-conductor cables size 500 kcmil with a diameter of 2.26 in. (57.4 mm), three 4-conductor cables size 250 kcmil with a diameter of 1.93 in. (49.0 mm), and five 4-conductor cables size 3/0 AWG with a diameter of 1.58 in. (40.1 mm) and a cross-sectional area of 1.96 sq. in. (1264 mm²). An air space between the conductors, as shown in Figure 12.29, is not maintained. The minimum width of cable tray permitted for the installation is:

   A. 9 in. (225 mm).   D. 24 in. (600 mm).
   B. 12 in. (300 mm).   E. 30 in. (750 mm).
   C. 18 in. (450 mm).

   Code reference _____

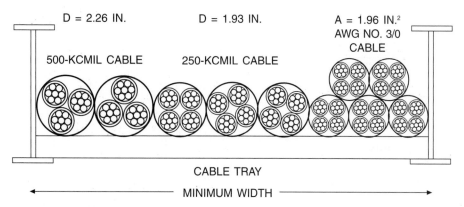

Figure 12.29 **A cable tray contains only Type TC multiconductor cables, several of which are larger than size 4/0 AWG and several of which are size 3/0 AWG. Determine the minimum width permitted for the cable tray.**

10. A 460-volt, 3-phase transformer arc welder with a duty cycle of 50% has a primary current listed on the rating plate as 40 amperes. The conductors supplying the welder are size 8 AWG copper with THWN insulation and conductor terminations are rated 75°C. A single overcurrent device protects both the welder and the circuit. The maximum rating permitted for the overcurrent device for this welder circuit is:

    A. 50 amperes.   C. 70 amperes.   E. 100 amperes.
    B. 60 amperes.   D. 80 amperes.

                                            Code reference _____

11. Six transformer arc welders are supplied by one feeder. The welders are 3-phase, 460-volt, with a primary supply current of 46 amperes marked on the rating plate, and the duty cycle is 30% as shown in Figure 12.30. The minimum recommended size of THWN copper feeder conductors to supply these welders, if all terminations are rated 75°C and no other load pattern information is available, is:

    A. 1 AWG.   C. 3 AWG.   E. 2/0 AWG.
    B. 2 AWG.   D. 1/0 AWG.

                                            Code reference _____

12. The branch-circuit conductors for an electroplating installation supply a load of 320 amperes and the conductors are solid copper busbars with a rectangular cross-section. The minimum busbar cross-sectional area permitted for this electroplating load is:

    A. 0.25 in.² (161 mm²).   D. 0.36 in.² (232 mm²).
    B. 0.30 in.² (194 mm²).   E. 0.40 in.² (258 mm²).
    C. 0.32 in.² (206 mm²).

                                          Code reference _____

13. Type ITC Cable, without a metallic sheath, that is installed as open wiring up to 50 ft (15 m) in length between a cable tray and equipment and protected from physical damage is required to be supported at intervals not to exceed:

    A. 3 ft (900 mm).   D. 10 ft (3 m).
    B. 4¹/2 ft (1.4 m).   E. 12 ft (3.7 m).
    C. 6 ft (1.8 m).

                                            Code reference _____

14. An industrial machine has a full-load current rating of 42 amperes, 3-phase, 480 volts, and the largest motor of the machine is 15 horsepower with a full-load current of 21 amperes. One additional motor has a full-load current rating of 11 amperes. The remainder of the load is 10 amperes of resistance heating. Copper wire in conduit with overcurrent protection at the supply end of the conductor is properly

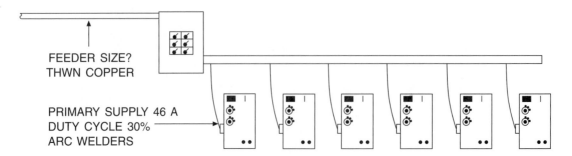

FEEDER SIZE?
THWN COPPER

PRIMARY SUPPLY 46 A
DUTY CYCLE 30%
ARC WELDERS

Figure 12.30   Six transformer type arc welders in an industrial plant are supplied by a common feeder, and the welders have a primary supply current of 46 amperes and a duty cycle of 30%.

sized for the machine. All conductor terminations are 75°C rated. The minimum size of Type THWN supply conductors permitted for the industrial machine is:

A. 10 AWG.   C. 6 AWG.   E. 3 AWG.
B. 8 AWG.   D. 4 AWG.

Code reference _____

15. When installing optical fiber cables in raceway with no electrical conductors in the raceway:
    A. only one cable is permitted for each raceway.
    B. the cross-sectional area of the cables is not permitted to exceed 20% of the area of the raceway.
    C. the cross-sectional area of the cables is not permitted to exceed 40% of the area of the raceway.
    D. a maximum of three cables are permitted in a raceway.
    E. the fill requirements that apply to electrical conductors do not apply to optical fiber cables.

Code reference _____

UNIT 13

Commercial Wiring Applications

OBJECTIVES

After completion of this unit, the student should be able to:

- state the requirements when multioutlet assembly passes through a partition.

- give a general description of underfloor raceway, cellular metal floor raceway, and cellular concrete floor raceway.

- explain the purpose of an insert for a floor raceway.

- explain what shall be done with wires remaining in use when a receptacle outlet is removed for cellular metal or concrete floor raceway.

- explain how wires are to be connected to outlet devices for underfloor raceway installations.

- describe a Flat Cable Assembly.

- state the maximum branch-circuit rating for a circuit of Flat Cable Assembly.

- explain a method used to connect building section wiring when assembling a manufactured building where the connection will be concealed.

- describe the wiring materials from which a manufactured wiring system is constructed.

- explain the requirements for conductor insulation when used for elevators, dumbwaiters, escalators, and moving walks.

- explain the minimum wire size and the number of circuits required for elevator car lighting.

- answer wiring installation questions from *322, 372, 374, 380, 384, 390, 518, 545, 600, 604, 605, 620, 645,* and *647.*

- state at least three changes that occurred from the 2002 to the 2005 Code for *Articles 322, 372, 374, 380, 384, 390, 518, 545, 600, 604, 605, 620, 645, 647,* and *Examples D9* and *D10.*

CODE DISCUSSION

The Code articles studied in this unit deal with installations most frequently encountered in commercial areas and buildings. These applications are not necessarily limited to commercial areas and buildings. Obviously, other articles of the Code also apply to commercial areas and buildings.

Article 322 covers the use and installation requirements for Type FC, Flat Cable Assembly installed in surface metal raceway identified for use with Type FC Cable. Figure 13.1 shows Type FC Cable installed in channel to supply lighting fixtures. This type of cable is permitted only for branch-circuits rated at not more than 30 amperes.

Article 372 deals with the use of the hollow cells of cellular concrete floor slabs as electrical raceways. Once the cellular concrete slabs are in place, a header is installed to provide access to the desired cells. A separate header and cells are used when communications wires are to be run in the floor. Splices and taps are only permitted to be made in header access units and floor junction boxes. When an outlet is abandoned, the

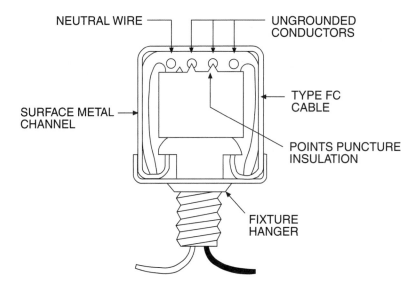

Figure 13.1   Cross-sectional view of Type FC Cable in channel-type surface metal raceway.

wire supplying the outlet shall be removed. Reinsulation of the wire is not permitted. If the wire is needed to supply other outlets, it will have to be removed and replaced with a new wire.

The raceway in cellular concrete floor slabs is not metal; therefore, an equipment grounding conductor is required to be run from insert receptacles to the header and connected to an equipment grounding conductor. This requirement is found in *372.9*. Figure 13.2 is a cross-sectional view of a cellular concrete floor slab and a header making access to two of the cells.

Article 374 covers the use of the cells of cellular metal floor as electrical raceways. Cellular metal floor decking is sometimes used as the structural floor support between main support beams. Concrete is then poured on this decking to form the finished floor. The Code permits these metal floor cells to be used as electrical raceways. A header is installed before the concrete is poured to provide access to the desired cells. Figure 13.3 is a cross-sectional view of a cellular floor used as an electrical raceway with the header shown connected to two of the cells. The header may connect directly to the supply panelboard, or this connection may be made with a suitable raceway. Splices and taps in wires are only permitted to be made in a header or in a floor-mounted junction box. Wiring requirements for cellular metal floor raceway are similar to those for underfloor raceway. An important requirement of this type of installation is that when an outlet is abandoned, the wire supplying the outlet shall be removed. Reinsulation of the wire is not permitted. If the wire is needed to supply other outlets, it will have to be removed and replaced with a new wire.

Article 380 covers the installation of multioutlet assemblies, which are raceway assemblies containing wiring and outlets. Multioutlet assembly is available factory-assembled, or it may be assembled on location. This wiring method is permitted only in dry locations. *NEC® 380.2* provides a list of uses permitted and not

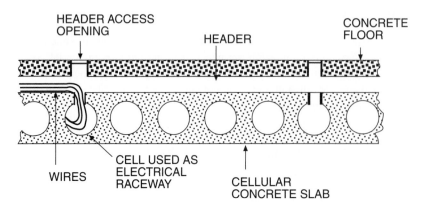

Figure 13.2   Cross-sectional view of header connecting to two cells of a cellular concrete floor slab.

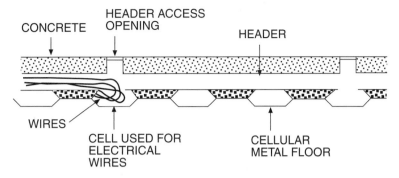

Figure 13.3 Cross-sectional view of cellular metal floor used as an electrical raceway.

permitted. Metal multioutlet assembly is permitted to pass through a dry partition provided no outlet is within the partition and caps or coverings on all exposed portions of the assembly can be removed.

Article 384 provides installation rules for strut-type channel raceway. This is a common support system material used both as a support system, and if listed as such, as a raceway for conductors. It is ideal for the support of luminaires (lighting fixtures) and acting as the raceway for the lighting circuit conductors. It is also permitted to be fastened to the floor and installed vertically as a power pole. Receptacle units are available that attach to the strut channel, and there are receptacles that are made to fit into the channel opening to allow the finished installation to have flush mounted receptacles. *NEC® 384.30(A)* requires that the strut-type channel be mounted to a surface or suspended from a structural member using straps and hangers that are external from the channel. Fixture support hangers are also available that are external to the channel. A strut-type channel showing the internal conductors and support hanger is shown in Figure 13.4. *NEC® 384.21* leaves the maximum permitted wire size up to the individual strut-channel listing. Generally, the maximum wire size for the larger area strut channels is 6 AWG.

The strut channel, when used as a raceway, can have the individual sections connected with joiners that are internal or external. This determines the percentage conductor fill. If the joiners are external, the maximum fill is 40% of the strut channel inside cross-sectional area. If the joiners are internal, the maximum fill is 25% of the inside cross-sectional area. *Table 384.22* gives values for the internal cross-sectional area for the common strut-channel trade sizes. These are trade sizes given with in. dimensions and do not have a practical metric equivalent. The ampacity adjustment factors of *310.15(B)(2)(a)* apply when there are more than three current-carrying conductors in the strut channel. If the fill does not exceed 20%, the number of conductors does not exceed 30 and the internal cross-sectional area is greater than 4 sq. in. (2500 mm²), then the adjustment factors of *310.15(B)(2)(a)* do not apply. The temperature adjustment factors at the bottom of the ampacity tables may apply, depending upon the installation. Luminaires (lighting fixtures) will produce heat.

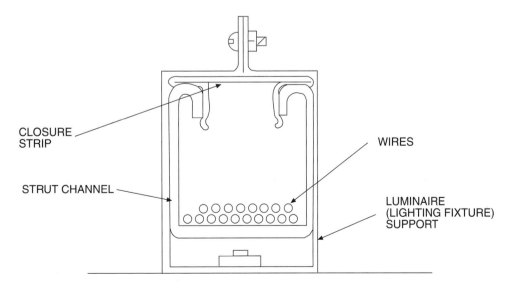

Figure 13.4 Cross-sectional view of strut-type channel raceway.

Therefore, follow the strut-channel manufacturer's specifications for mounting fixtures. In some cases a 1/2 in. (13 mm) spacing from the back of the luminaire (lighting fixture) to the strut channel is required. In most cases using 75°C conductors and terminations, a 1/8 in. (3 mm) spacing is considered adequate for fluorescent luminaires (lighting fixtures) to avoid being required to apply a temperature adjustment factor. The following example will show how to determine the maximum number of conductors permitted for an installation. The size of strut is generally determined based upon required strength for the application rather than the minimum required for the number of conductors.

Example 13.1   A 1⁵/₈ by 1⁵/₈ in. strut-type channel is used to support a row of fluorescent luminaires (lighting fixtures) and serve as the raceway for the circuit conductors. The multiwire branch-circuits rated 120/240-volt single-phase supply the luminaires (lighting fixtures). Boxes listed for the purpose and containing a receptacle are mounted to the strut-channel to supply the plug and cord luminaires (lighting fixtures) as shown in Figure 13.5. Determine the maximum number of size 12 AWG copper, Type THHN conductors permitted to be installed in this strut-channel raceway where the strut-channel joiners are internal.

Answer: First look up the cross-sectional area of a size 12 AWG, Type THHN conductor in *Table 5, Chapter 9* and find 0.0133 sq. in. (8.581 mm²). Use the formula given in *384.22* to determine the maximum number of conductors using a 25% fill. The footnote at the bottom of *Table 384.22* directs the use of the 25% column when the joiners fit inside the strut channel. The maximum number of conductors is 38.

$$\text{Number of wires} = \frac{0.507 \text{ in.}^2}{0.0133 \text{ in.}^2} = 38 \text{ wires}$$

$$\text{Number of wires} = \frac{327 \text{ mm}^2}{8.581 \text{ mm}^2} = 38 \text{ wires}$$

Splices and taps are permitted to be made in the strut channel provided the cross-sectional area at any point does not exceed a 75% fill. *NEC® 384.30* requires the strut channel to be supported at intervals not to exceed 10 ft (3 m) and within 3 ft (900 mm) of a termination. A strut channel is permitted to serve as an equipment grounding means. At the transition between other types of wiring such as EMT to strut channel, a means of effective bonding for equipment grounding must be provided.

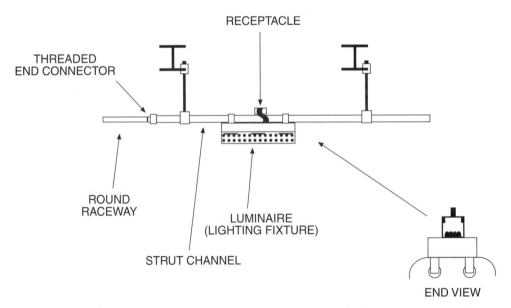

Figure 13.5   A 1⁵/₈ in. by 1⁵/₈ in. strut-type channel used as a raceway for the conductors supplig a row of fluorescent luminaries (lighting fixtures) also serving as support for the luminaries (lighting fixtures).

Article 390 deals with the installation of raceways installed under the floor. A raceway listed for the purpose is permitted to be installed in concrete floors. Installation is not limited to concrete floors. Underfloor raceway is permitted to be installed in a floor with a 3/4-in. (20-mm) minimum thickness wood covering, provided the raceway is not more than 4 in. (100 mm) wide. There is a minimum covering for underfloor raceway depending on the width and type of installation. More than one raceway may be run parallel in the floor, with one containing power wires and another containing communications wires. Underfloor raceway listed for the purpose is permitted to be run flush with the floor of an office building and covered with linoleum or an equivalent surface.

When an outlet is to be installed on the floor and have access to the underfloor raceway, an insert is installed. Splices are not permitted to be made at outlets. Connection to a receptacle outlet, for example, is done by means of stripping the insulation from the wire and looping the wire around the terminal of the device, as shown in Figure 13.6. Splices are permitted only in junction boxes, except in the case of flush-mounted underfloor raceway with removable covers, *390.6*.

Article 518 specifies the type of wiring method that is permitted in buildings or portions of structures where 100 or more persons are likely to assemble. Examples of assembly occupancies are given in the article. Extra care is taken in these areas to make the wiring system more resistant to fire ignition, fire spread, or the production of toxic vapors. The critical factor for emergency egress is the number of people in an area. It takes time to get a large number of people out of a building during an emergency. If less than 100 persons are in the building or area, egress can be expedited rapidly with a lesser danger to human life. The primary intent here is to buy extra time to get the people out of the building. Electrical Nonmetallic Tubing and Rigid Nonmetallic Conduit are permitted to be installed as concealed wiring in walls, ceilings, and floors of some types of areas where separated from the public by a listed 15-minute finish fire-rated surface, *518.4(C)*.

Article 545 covers the installation of wiring components in manufactured buildings. Building components, such as complete wall sections, are constructed (including the wiring) in a manufacturing facility and assembled at a building site. The components are of closed construction, which means the wiring is not accessible for inspection at the building site without disassembling the building components. All raceway and cable wiring methods in the Code are permitted for use in manufactured buildings. Other wiring methods shall be listed for the purpose. Receptacle outlets and switches with integral boxes are permitted if listed for the purpose. At the point where building sections are joined, fittings listed for the purpose and intended to be concealed at the time of on-site assembly shall be permitted to join the conductors of one section to the conductors of another section.

Article 600 deals with electric signs and outline lighting. This article is broken up into two parts, general information given in *Part I* and field-installed skeleton tubing in *Part II*. Skeleton lighting is actually neon tubing that is not mounted in a support structure or enclosure. Each commercial building that is accessible to patrons is required to have at least one outlet that is dedicated for signs or outline lighting, *600.5(A)*. When a building or structure houses more then one commercial establishment, each occupancy must be provided with their own outlet. The minimum rating of this circuit is 20 amperes but the Code also places maximum ratings on these circuits. In *600.5(B)*, the maximum permitted branch-circuit ratings for these signs and outline lighting circuits is 30 amperes for neon and 20 amperes for incandescent and fluorescent forms of lighting.

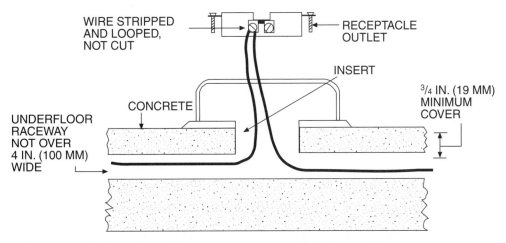

Figure 13.6   A receptacle outlet installed on the floor and supplied with underfloor raceway.

Signs and outline lighting are also required to be provided with a disconnect that is within sight of the equipment. If the disconnect is placed in a location that is not within sight from the sign, a disconnect that is capable of being locked in the open position is required, *600.6(A)*. The disconnecting means is permitted to be a switch, circuit breaker, or may consist of an attachment plug.

NEC® *600.32* provides the requirements for the installation and protection of the neon secondary conductors for field-installed skeleton tubing, operating above 1000 volts. The materials permitted for these conductors are listed in *600.32(A)(1)*. However, it is important to note that there are length restrictions depending on the material selected. This section also specifies that only one secondary conductor is to be installed in each run of raceway, as shown in Figure 13.7. The maximum length of these secondary conductors from the output of the power supply to the electrode is dependent on the power supply and the material used to protect the secondary conductors, *600.32(J)*.

Article 604 covers the use and construction of manufactured wiring systems. These are assemblies of sections of Type AC Cable, Type MC Cable, or wires in Flexible Metal Conduit or Liquidtight Flexible Conduit with integral receptacles and connectors for connecting electrical components in exposed locations, as shown in Figure 13.8. These manufactured wiring systems are permitted to extend into hollow walls for direct termination at switches and other outlets. The conductors in the manufactured wiring system are not permitted to be smaller than size 12 AWG copper, except for taps to single lighting fixtures. A common application has been the connection of lay-in fluorescent fixtures in suspended ceilings. Manufactured wiring systems can be listed for installation in outdoor locations as illustrated in Figure 13.9.

Article 620 deals with the wiring of elevators, dumbwaiters, escalators, and moving walks. A major issue is that wiring associated with this equipment shall have flame-retardant insulation, even when installed in a raceway, *620.11(C)*. This is of particular importance to minimize the spread of fire from one floor to another. *NEC*® *620.21* specifies wiring methods for electrical conductors and optical fiber cables permitted to be installed in an elevator machine room, machine space, and hoistways. It also specifies wiring methods for wellways for escalators and moving walks and for wheelchair and stairway lift runways. The required wiring method is Rigid Metal Conduit (RMC), Intermediate Metal Conduit (IMC), Electrical Metallic Tubing (EMT), Rigid Nonmetallic Conduit (RNC), and wireway. Type MC, AC, and MI Cables are also permitted. Some flexible conduits are permitted where flexibility is needed. *NEC*® *620.21* gives specific rules for the use of flexible conduits. Sometimes these materials are only permitted for specific applications. All flexible conduits are limited to lengths not greater than 6 ft (1.8 m) except in some applications such as machine room spaces. Liquidtight Flexible Nonmetallic Conduit (LFNC) in some situations is permitted to be installed in lengths greater than 6 ft (1.8 m).

When wireways are installed for conductors supplying elevator circuits, the current flow is generally intermittent and not continuous. Conductor heating takes time. Therefore, in the case of wireways, the fill

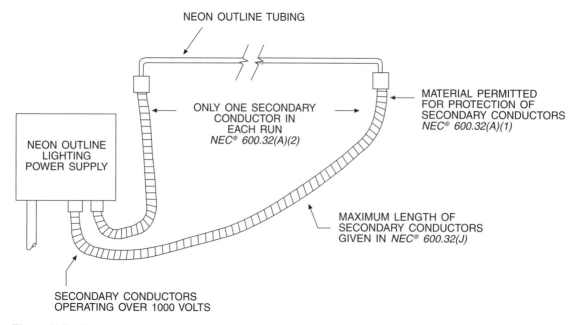

Figure 13.7   Field-installed skeleton tubing with secondary circuit conductors operating in excess of 1000 volts.

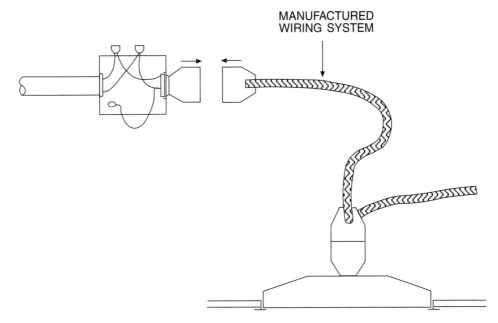

MANUFACTURED
WIRING SYSTEM

Figure 13.8   A manufactured wiring system is a prewired assembly of conductors and connectors in Type AC or Type MC Cable, Flexible Metal Conduit, or Liquidtight Flexible Conduit.

for a wireway is a higher percentage than for other installations. *NEC® 620.32* permits the total cross-sectional area of the conductors to be up to 50% of the cross-sectional area of the wireway. For other raceways, such as conduit and tubing, the percent fill is 40% as is the case with most other installations.

NEC® 620.61(B) states that elevator and dumbwaiter driving motors shall be classed as intermittent duty. In the case of escalator and moving walk driving motors, they shall be considered as continuous duty. Another code contains the requirement that a 3-phase drive motor be prevented from starting in the event there is a phase rotation reversal. If there is a phase rotation reversal, the drive motor will reverse direction. This can cause the elevator to run in reverse. In the case of a hydraulic drive, overheating will occur. Interchanging any two phase conductors of a 3-phase system of the building wiring or on the primary electrical system supplying power to the building will result in a phase rotation reversal.

Requirements for the wiring of wheelchair lifts and stairway chair lifts are also contained in *Article 620*. One complex task involved in the wiring of an elevator system is the determination of the minimum size of feeder required in the system. *Example D9* and *Example D10* show how the feeder conductor is sized

PERMITTED IN OUTDOOR LOCATIONS

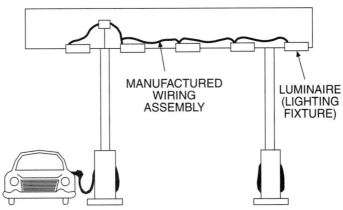

MANUFACTURED
WIRING
ASSEMBLY

LUMINAIRE
(LIGHTING
FIXTURE)

Figure 13.9   Manufactured wiring systems listed for the purpose are permitted to be installed in an outdoor location.

in two different situations. Several factors concerning elevator installations are not obvious by reading *Article 620*. Example 13.2 will help illustrate how the feeder conductor selection current is determined.

Example 13.2   Four passenger elevators are powered with motor-generator sets with the motor-generator set powered by 30-horsepower, 3-phase, 460-volt motors. In addition to this load, 12 amperes continuous load per elevator is required for operation control equipment. The motor-generator sets have a continuous-duty rating. Determine the minimum feeder conductor selection current.

Answer: *NEC® 620.13* states that for a motor-generator set, the feeder conductor shall be determined based on the ampacity of the drive motor. *NEC® 620.13(B)* can be misleading because passenger elevator motors are actually rated as intermittent duty motors. Start by looking up the 30-horsepower, 3-phase, 460-volt motor full-load current in *Table 430.250*, which is 40 amperes. The load consists of several motors plus other load; therefore, *430.24* is used to determine the feeder current rating. The feeder current is the sum of the full-load current of all the motors, plus other load, plus 25% of the full-load current of the largest motor. If all motors are the same size, then take one of them as the largest. But it is not quite so easy. *Exception 1* to *430.24* states that if the motors are used for short-time, intermittent, periodic, or varying duty, the motor current shall be as determined by *430.22(E)*. *Table 430.22(B)* states that a passenger elevator motor is considered to be intermittent duty. If the actual motor is a continuous-duty motor, then it can be overloaded without damage if operated intermittently. Therefore, the continuous full-load current rating of the motor is multiplied by 1.4 for use in the feeder calculation because it will be overloaded for short periods of time. In this case, the current to use for the motors is 56 amperes.

$$40 \text{ A} \times 1.4 = 56 \text{ A}$$

The feeder selection current calculation is determined according to *430.24, Exception 1* as follows:

$$4 \times 56 \text{ A} = 224 \text{ A}$$

According to *620.14*, a demand factor can be applied to that portion of the feeder current determined according to *620.13*. The demand factor for four elevators as determined in *Table 620.14* is 0.85. To get the actual feeder selection current, multiply the motor load calculated by this demand factor and then add the operation control load according to *Sections 430.24* and *215.2(A)*.

$$
\begin{aligned}
0.85 \times 224 \text{ A} &= 190 \text{ A} \\
4 \times 12 \text{ A} \times 1.25 &= \underline{60 \text{ A}} \\
& 250 \text{ A feeder selection current}
\end{aligned}
$$

An elevator machine room or machine space is required to be provided with at least one separate branch-circuit that supplies both the lighting for equipment, and at least one duplex receptacle. This rule is found in *620.23*. The switch for the lighting is required to be at the point of entry to the machine room or space. *NEC® 620.85* requires that each of these receptacles be of the ground-fault circuit-interrupter type. *NEC® 620.23(A)* requires that the lighting be tapped to the supply side of the ground-fault circuit interrupter. The same rules apply to an elevator hoistway pit. A separate circuit is required to supply a light and duplex receptacle in an elevator hoistway pit. *NEC® 620.24* requires that each hoistway pit area light be controlled by a switch at the entrance to the pit. The hoistway pit is also required to be provided with one ground-fault circuit-interrupter type receptacle. The light in the pit is required to be tapped from the supply side of the ground-fault circuit-interrupter receptacle, as illustrated in Figure 13.10.

Article 647 provides rules for establishing a unique type of separately derived electrical system called **technical power**. In the past, it was only permitted to be installed in motion picture and television studios. Technical power systems are now permitted to be installed in any commercial or industrial building where there is concern for electrical noise that affects audio and video signals. Much of the sound and video processing, recording, and reproduction equipment operates at 120 volts with one circuit conductor grounded. The reality is that grounds are sources of electrical noise. Even a properly installed and well-maintained electrical system can have differences in potential between different points on the grounding system that give rise to current flow. A technical power system operates at 120 volts with both conductors ungrounded. The system is grounded by connecting the midpoint of a 120-volt secondary winding of an insulating transformer

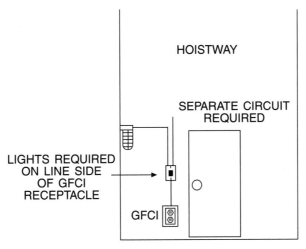

Figure 13.10 The lighting serving an elevator hoistway is required to be tapped on the line side of any ground-fault circuit-interrupter protecting receptacles.

to ground. This creates a system that has 60 volts to ground from either ungrounded conductor as shown in Figure 13.11. This is sometimes called a balanced 120-volt system.

NEC® 647.4(A) requires the technical power to be distributed with a single-phase, 3-wire panelboard. The two ungrounded lines with 120 volts between them are fed into the main circuit breaker. The center-tap grounded conductor is fed into the neutral terminal block. All branch-circuit breakers are 2-pole common-trip. The output to each piece of equipment will be 120 volts with two ungrounded conductors. *NEC® 647.4(C)* requires that all branch-circuit and feeder conductors be color coded or marked in some effective manner that the conductors can be identified as to the originating panelboard and the conductor at that panelboard. *NEC® 647.4(B)* requires that all junction boxes be identified as to the voltage of the conductors, and the panelboard from which the conductors originate.

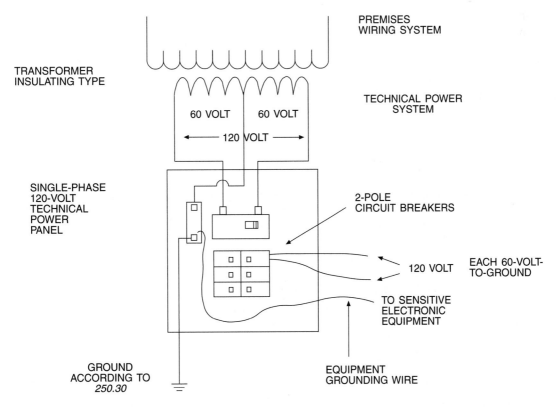

Figure 13.11 A technical power system is a separately derived grounded system operating at 60/120 volts with 60 volts measured line-to-ground and 120 volts line-to-line.

NEC® 647.6(A) clearly states that a technical power system is a separately derived system. This means the rules of *250.30* will apply with respect to grounding the system. Refer to *Unit 8* for a detailed discussion of grounding separately derived systems. *NEC® 647.6(B)* explains that the grounded conductor from the center tap of the transformer is required to be grounded on the supply side of the first disconnecting means of the separately derived system. That would mean the neutral terminal of the technical power panelboard is connected to a grounding electrode, as shown in Figure 13.11.

NEC® 647.7(A)(1) requires that all 15- and 20-ampere receptacles supplied from the technical power system be ground-fault circuit-interrupter protected. The concern is that there is no grounded neutral circuit conductor. What may be identified as a neutral conductor in connected equipment is actually 60 volts to ground. *NEC® 647.7(A)(4)* requires receptacles for technical power to have a unique configuration. The configuration is not specified. This same paragraph does permit standard 15- and 20-ampere rated 125-volt receptacles to be used in areas where only qualified personnel will work on the equipment. This essentially describes all areas. Therefore, standard receptacles are used for technical power systems, as shown in Figure 13.12. The circuit conductors are required to be color coded or identified but no specific colors are to be used. It is important to be consistent with respect to which conductor will be connected to the silver screw on receptacles and which will be connected to the brass-colored screw. These are both ungrounded conductors. Therefore, white is not used. There will be an equipment grounding conductor run to the grounding terminal of the receptacle and that wire will be either bare or green. Grounding of these circuits will be done in the same manner as grounding of other circuits. If there is concern about mixing receptacle and equipment grounds with box and raceway grounds, it is permitted to use insulated ground receptacles. In this case, a green insulated conductor is run to the grounding terminal of every receptacle and the boxes and raceways are grounded separately.

NEC® 647.7(A)(2) requires labeling of all receptacles that are supplied by the technical power system. It is important these receptacles not be used for any purpose other than supplying sensitive electronic equipment. To help ensure this is not violated, *647.7(A)(3)* requires a standard 15- or 20-ampere, 125-volt receptacle with a grounded neutral conductor be located within 6 ft (1.8 m) of every technical power receptacle. This is illustrated in Figure 13.12. One regular receptacle, if properly placed, can fill this requirement for many technical power receptacles in a given area.

If lighting equipment is also supplied from the technical power system, there are some special requirements because there is no grounded-circuit conductor. *NEC® 647.8(C)* requires that in the case of screw shell lamps, the screw shell of a lamp is not permitted to be exposed to contact while it is making contact with the screw shell of the lamp socket. This will prevent a person changing lamps from making contact with a 60-volt conductor. There will be cases where lighting units are a part of electronic sound and video processing, recording, or reproduction equipment. Some equipment may have electric discharge lighting with a ballast. The ballast is required to be one that is identified for use with a separately derived system. Avoiding connecting lighting equipment to the technical power system is desirable if practical.

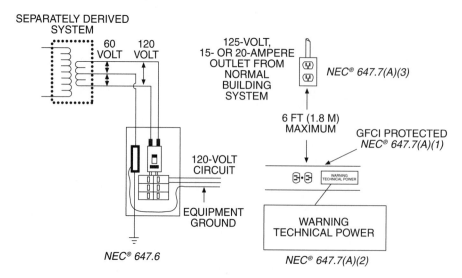

Figure 13.12   Spacing requirements between receptacles supplied by a technical power system and 125-volt, 15- and 20-ampere receptacles served by the normal building electrical system.

In facilities with 3-phase power available, it may be desirable to supply the technical power system from a 3-phase supply. This is permitted by *647.5*, which specifies using a wye 3-phase system. The Code refers to this as a 6-phase separately derived system. An example of how this can be accomplished is shown in Figure 13.13. The 208-volt output of each set of phases from the wye transformer is supplied to the primary of a single-phase transformer that has a 120-volt output with a center tap. The single-phase output of each transformer is fed to a separate 3-wire, single-phase panelboard in a manner similar to Figure 13.11. It would be necessary to keep the conductors from each technical power panel uniquely identified to avoid mixing up the conductors. Each panelboard is grounded to the same separately derived system grounding electrode, and therefore, a circuit will be completed if the conductors are not kept separated.

Voltage drop, especially on grounded conductors, creates the conditions that lead to objectionable ground currents. With technical power systems, it is important to make sure the source of the technical power is *grounded only at one point*. Some electronic equipment is extremely voltage sensitive and will not function properly if input voltage deviates from sometimes very narrow tolerances. Maximum voltage-drop requirements are placed on the conductors of a technical power system. *NEC® 647.4(D)* restricts voltage drop on technical power branch-circuits to specific fixed equipment to not more than 1.5%. The current draw of the equipment would need to be known in order to determine the voltage drop. The total voltage drop on feeders and branch-circuits is not permitted to exceed 2.5% There is no specific voltage drop

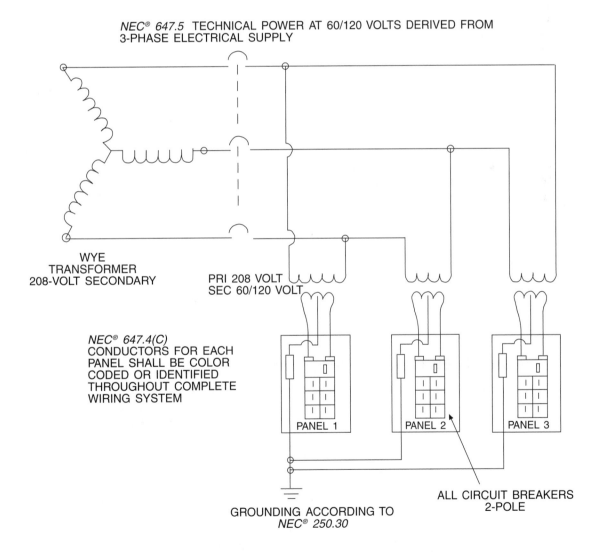

NEC® 647.5 TECHNICAL POWER AT 60/120 VOLTS DERIVED FROM 3-PHASE ELECTRICAL SUPPLY

WYE TRANSFORMER 208-VOLT SECONDARY

PRI 208 VOLT SEC 60/120 VOLT

NEC® 647.4(C) CONDUCTORS FOR EACH PANEL SHALL BE COLOR CODED OR IDENTIFIED THROUGHOUT COMPLETE WIRING SYSTEM

PANEL 1   PANEL 2   PANEL 3

ALL CIRCUIT BREAKERS 2-POLE

GROUNDING ACCORDING TO NEC® 250.30

Figure 13.13   A 60/120-volt technical power system derived from a 3-phase electrical system.

required on the feeder. Therefore, if the feeder voltage drop exceeds 1%, then less than 1.5% is permitted on the branch-circuits.

In the case where the actual equipment load is not known, and several receptacles are placed to supply the equipment, only 1% voltage drop is permitted on those circuits. The reason for this is that there is a power cord at equipment that will also result in some voltage drop. The voltage drop on a circuit supplying receptacles where the load is not known is based upon 50% of the rating of the branch-circuit. For a 20-ampere branch-circuit, the load is based on 10 amperes. Voltage-drop calculations were discussed in detail in *Unit 6*. These are single-phase circuits operating at 120 volts. Use Equation 6.4 to determine the voltage drop if the load, wire size, and circuit length are known. Use Equation 6.6 to determine the cross-sectional area of the conductor necessary to limit the voltage drop to a specific percentage such as 1% when the load and length of circuit are known. Then look up the wire size in *Table 8* in the Code. The following example will show how the minimum wire size is determined for a receptacle circuit supplied from a technical power system. Values of resistivity (**K**) of copper and aluminum will be needed for a voltage-drop calculation using Equation 6.4 or Equation 6.6. Values of **K** are given in Table 1.2 of *Unit 1*. The technical power circuits will most likely be copper. Therefore, it is recommended for these calculations to use a value of 12 ohm cmil/ft (0.02 ohm mm²/m) in the equations. These values are for a conductor operating temperature of approximately 50°C which, for a technical power circuit, is most likely higher than it will ever operate.

Example 13.3   A 20-ampere, technical power, branch-circuit supplies several receptacles located in a small area near where electronic equipment will be connected. The distance from the transformer to the technical power panelboard is short and the voltage drop on the feeder will not exceed 1%. The distance from the panelboard to the receptacle location is 40 ft (12.19 m). Determine the minimum size copper conductor with 75°C insulation and terminations required for this circuit.

Answer: *NEC® 647.4(D)(2)* only permits a 1% voltage drop on a circuit supplying receptacles. That same section specifies for receptacles that the load to be used in the calculation is 50% of the rating of the circuit, which in this case is 10 amperes. Choose a value of 12 for resistivity of the conductor and use Equation 6.6 to determine the wire cross-sectional area needed to limit voltage drop to 1%. The wire must have a cross-sectional area of not less than 8000 circular mils (4.06 mm²). Then look up the minimum conductor size in *Table 8, Chapter 9* of the Code and find size 10 AWG.

$$\text{Cross-sectional area of wire} = \frac{2 \times 12 \times 10\,\text{A} \times 40\,\text{ft}}{0.01 \times 120\,\text{V}} = 8000\ \text{cmil}$$

$$\text{Cross-sectional area of wire} = \frac{2 \times 0.02 \times 10\,\text{A} \times 12.19\,\text{m}}{0.01 \times 120\,\text{V}} = 4.06\ \text{mm}^2$$

MAJOR CHANGES TO THE 2005 CODE

These are the changes to the 2005 *NEC®* that correspond to the Code sections studied in this unit. The following analysis explains the significance of the changes from the 2002 to the 2005 Code only and this analysis is not intended to be used in place of the Code. Refer to the actual section of the 2005 Code for the exact wording and meaning of each section discussed. Changes are indicated in the Code with a vertical line in the margin. If material was deleted or moved to another location in the Code, the location of the deletion is indicated with a dark dot in the margin.

Article 372   **Cellular Concrete Floor Raceways**

372.17: This is a new section that specifies adjustment factors must be applied when detrmining the ampacity of conductors run in Cellular Concrete Floor Raceway if there are more than three current-carrying conductors. The adjustment factors are found in *310.15(B)(2)*. In the past, there have been differences in interpretation as to whether the adjustment factors applied to conductors in this case.

Article 374 Cellular Metal Floor Raceways

374.11: The connection to Cellular Metal Floor Raceway is usually partially embedded in concrete. The previous edition of the Code did not permit Liquidtight Flexible Metal Conduit (LFMC) to be used to connect a box or enclosure to Cellular Metal Floor Raceway. Now LFMC is permitted to be connected to Cellular Metal Floor Raceway. The grounding requirement of *250.118(6)* must be met. In many cases, in order to use LFMC for this application, there must be provisions within the Cellular Metal Floor Raceway to terminate a bonding conductor. Liquidtight Flexible Nonmetallic Conduit (LFNC) is also permitted to be used to make a connection to Cellular Metal Floor Raceway, but there is required to be a means provided to terminate a bonding jumper. A new last sentence to this section points out that LFMC and LFNC listed and identified for direct burial is also listed to be encased in concrete. It is not required to state on the raceway that it is suitable to be encased in concrete. The marking only needs to say suitable for direct burial.

374.17: This is a new section that specifies adjustment factors must be applied when detrmining the ampacity of conductors run in Cellular Metal Floor Raceway if there are more than three current-carrying conductors. The adjustment factors are found in *310.15(B)(2)*. In the past, there have been differences in interpretation as to whether the adjustment factors applied to conductors in this case.

Article 384 Strut-Type Channel Raceway

384.2: The word cables was added to the definition. Multiconductor cables as well as single conductors are permitted to be installed in Strut-Type Channel Raceway.

Article 390 Underfloor Raceway

390.17: This is a new section that specifies adjustment factors must be applied when determining the ampacity of conductors run in Underfloor Raceway if there are more than three current-carrying conductors. The adjusement factors are found in *310.15(B)(2)*. In the past there were differences in interpretaion as to whether the adjustment factors applied to conductors in this case.

Article 518 Assembly Occupancies

518.1: The scope of this article in previous editions of the Code was somewhat vague. The scope was completely reworded, giving examples of the types of facilities that are considered to be assembly occupancies. Live performance theaters and audience areas of television studios are not covered since they are covered in *Article 520*.

518.2(A): This section contains a list of the type of facilities that are to be included as assembly occupancies. Drinking facilities was added to the list. The term "church chapel" was considered to be too narrow description, and it was replaced by places of religious worship.

518.2(B): A new first sentence was added that makes it clear that in buildings of multiple use where a portion of the building is considered an assembly occupancy, the other portions of the building are not considered a part of the assembly occupancy. This is only true if there is a fire barrier between the two types of occupancies.

518.4(C): Rigid Nonmetallic Conduit (RNC) and Electrical Nonmetallic Tubing (ENT) are no longer permitted to be installed concealed in walls and ceilings that have a 15-minute finish fire rating in college and university classrooms, drinking establishments, or passenger stations and terminals for air, ground

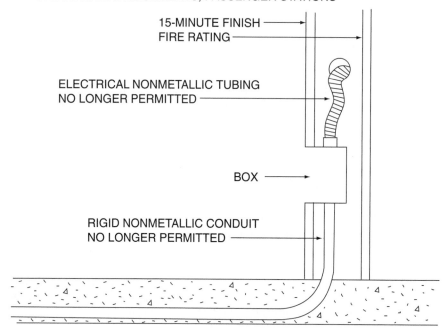

NEC® 518.4(C) COLLEGE AND UNIVERSITY CLASSROOMS
DRINKING ESTABLISHMENTS, PASSENGER STATIONS

15-MINUTE FINISH
FIRE RATING

ELECTRICAL NONMETALLIC TUBING
NO LONGER PERMITTED

BOX

RIGID NONMETALLIC CONDUIT
NO LONGER PERMITTED

Figure 13.14   It is no longer permitted to install ENT and RNC in walls and ceilings with a 15-minute finish fire-rating for college and university classrooms, drinking establishments, and passenger areas of transportation facilities.

transportation, subways, or marine transportation. An example is illustrated in Figure 13.14. Assembly occupancies in which RNC and ENT is permitted to be installed concealed behind a 15-minute finish fire-rated surface are listed in this section.

Article 600   Electric Signs and Outline Lighting

600.1: The scope was expanded beyond just electric signs and outline lighting to include neon tubes used as art forms and decorative elements.

600.2: A new definition of a **section sign** was added. It is a sign that consists of sections that must be electrically connected when assembled at the intended location.

600.3: The individual sections of a section sign are required to be listed.

Part II: These rules, according to the previous edition of the Code, only applied to the field installation of skeleton tubing systems. Now these rules apply to all field-installed wiring associated with a sign utilizing skeleton tubing.

Article 604   Manufactured Wiring Systems

604.6(A): The maximum conductor size permitted for a manufactured wiring system was increased from size 10 AWG to size 8 AWG. This increase in size was permitted so that larger size conductors would be available in locations where there were long circuit runs and voltage drop might be excessive if larger conductor sizes were not available.

604.6(A)(3): Hard-usage flexible cord not smaller than size 12 AWG and not longer than 6 ft (1.8 m) is permitted to be provided as a manufactured wiring system to make connections to utilization equipment that is not permanently attached to the structure of the building. This may be some piece of utilization equipment that is hook-and-eye attached. The change is that this technique of using flexible cord as a manufactured wiring system is not permitted for connection of power to luminaires. The equipment must be something other than a luminaire (lighting fixture).

604.6(F): This is a new subsection that permits luminaires (lighting fixtures) to be supplied with flexible cord-type manufactured wiring systems provided that installation meets the requirements of *410.30(C),* which

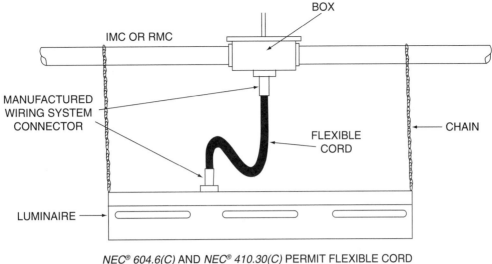

NEC® 604.6(C) AND NEC® 410.30(C) PERMIT FLEXIBLE CORD
SIZED SMALLER THAN CIRCUIT WIRES WITH MANUFACTURED
WIRING SYSTEM CONNECTORS

Figure 13.15 It is now permitted to supply luminaires (lighting fixtures) with a flexible-cord manufactured wiring system that contains wires smaller than size 12 AWG.

permits a luminaire (lighting fixture) to be mounted directly below an outlet, and the flexible cord is visible for the entire length as shown in Figure 13.15. There was also a change made in *410.30(C)(1)(2)* that recognizes a flexible cord with manufactured wiring system connectors. The previous edition of the Code did not have this provision. The real change is that it is now permitted to supply a luminaire (lighting fixture) with a flexible cord-type manufactured wiring system that has wires smaller than size 12 AWG. The wires to the luminaire (lighting fixture) must be sized at not less than 125% of the luminaire (lighting fixture) full-load current and meet the circuit tap requirements of *240.5* for flexible cords. Cord-type manufacured wiring systems with size 16 AWG wire will most likely be common.

Article 605 Office Furnishings

605.6: Prewired office partitions that are fixed in place or secured to building surfaces are required to be connected to power with a permanent wiring method. If the fixed office partition is supplied by a multiwire branch-circuit, it is now required that disconnect power to all ungrounded conductors can be disabled simultaneously. This means a 2-pole or a 3-pole circuit breaker or single-pole with a listed handle tie at the supply panelboard is required when the fixed office partition is supplied with a multiwire branch-circuit.

605.7: Prewired office partitions that are free standing are permitted to be supplied power with a cord and plug or by a permanent wiring method. If a permanent wiring method is used to supply a freestanding office partition, and the office partition is supplied by a multiwire branch-circuit, it is now required that disconnect power to all ungrounded conductors can be disabled simultaneously. This means a 2-pole or a 3-pole circuit breaker or single-pole with a listed handle tie is required at the panelboard.

Article 620 Elevators, Dumbwaiters, Escalators, Moving Walks, Wheelchair Lifts, and Stairway Chair Lifts

620.2: **Control Room** is now defined as a space that can be entered by personnel and contains controls and associated equipment, but not elevator driving machinery. This space is not permitted to be in the hoistway.

620.2: **Control Space** is now defined as an area that is not required to be capable of being entered by personnel and contains controls and associated equipment, but not driving machinery. This space is permitted to be part of the hoistway.

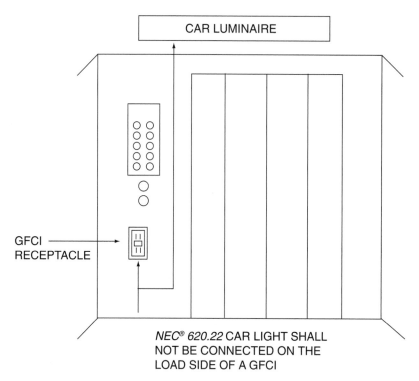

Figure 13.16   The required lighting for an elevator car is required to be connected on the supply side of any ground-fault circuit interrupter.

620.2: **Machine Room** is now defined as a space separate from the hoistway and intended for the driving machinery. This room must be capable of being entered by personnel.

620.2: **Machinery Space** is now defined as a space that is permitted to be a part of the hoistway that contains the driving machinery. This space is not required to permit full entry by personnel.

620.22: Required elevator car lighting now is not permitted to be supplied from the load side of a ground-fault circuit interrupter. This is illustrated in Figure 13.16

Article 645   **Information Technology Equipment**

645.17: This is a new section that recognizes power distribution units for information technology equipment. A convenient way to obtain individual control of equipment is by means of circuit breakers at a panelboard. It is easy to exceed the circuit-breaker capacity of an individual panelboard, thus creating the need for multiple panelboards. This new section permits multiple panelboards to be installed in a single cabinet provided no panelboard has more than 42 circuit breakers and the panelboards are listed for information technology equipment applications.

Article 647   **Sensitive Electronic Equipment**

647.4(A): Technical power systems provide 120 volts to sensitive electronic equipment with two ungrounded conductors and no grounded neutral conductor. Each ungrounded conductor is 60 volts to ground. A standard receptacle is used with an ungrounded conductor to each terminal. There is a new rule that both ungrounded conductors for a circuit or feeder are required to be common trip and disconnected simultaneously. This means that individual single-pole circuit breakers with handle ties are not permitted. A 2-pole circuit breaker is required as illustrated in Figure 13.17. Handle ties do not necessarily result in common trip on an overcurrent condition.

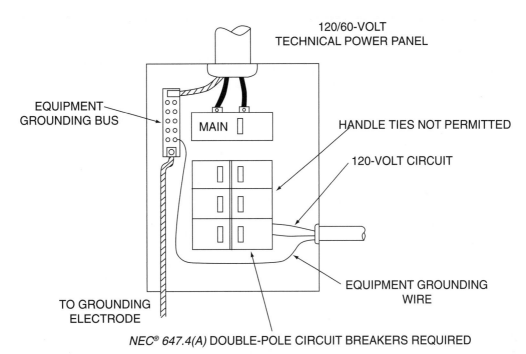

120/60-VOLT
TECHNICAL POWER PANEL

EQUIPMENT
GROUNDING BUS

MAIN

HANDLE TIES NOT PERMITTED

120-VOLT CIRCUIT

EQUIPMENT GROUNDING
WIRE

TO GROUNDING
ELECTRODE

NEC® 647.4(A) DOUBLE-POLE CIRCUIT BREAKERS REQUIRED

Figure 13.17   A technical power branch-circuit or a feeder is required to have a disconnect that simultaneously opens all ungrounded conductors.

WORKSHEET NO. 13—BEGINNING COMMERCIAL WIRING APPLICATIONS

Mark the single answer that most accurately completes the statement based upon the 2005 Code. Also provide, where indicated, the Code reference that gives the answer or indicates where the answer is found, as well as the Code reference where the answer is found.

1. A portion of a building has a precast cellular concrete floor that is being used for an electrical raceway. The cells in the concrete floor that are being used as a raceway are 3 in. (76.2 mm) in diameter with a cross-sectional area of 7.07 sq. in. (4560 mm²). Four Type THWN copper conductors are to be run through the concrete cells as shown in Figure 13.18. The maximum size conductors permitted is:
    A. 2 AWG.   D. 250 kcmil.
    B. 1/0 AWG.   E. 500 kcmil.
    C. 3/0 AWG.

    Code reference_____

2. Multioutlet assemblies are permitted to be installed:
    A. in damp, but not wet, locations.
    B. in an environment where corrosive vapors are present.
    C. in hoistways.
    D. through walls that are dry, provided the cover can be removed on all exposed assembly.
    E. where exposed to physical damage.

    Code reference _____

3. A strut-type channel raceway is surface mounted to a ceiling and used to support luminaires (lighting fixtures) as well as serve as the raceway for the conductors. The sections of strut channel are secured within 3 ft (900 mm) of ends and terminations and shall be secured at intervals not greater than:
    A. 3 ft (900 mm).   D. 8 ft (2.5 m).
    B. 4¹/2 ft (1.4 m).   E. 10 ft (3 m).
    C. 5 ft (1.5 m).

    Code reference_____

4. Grounding of receptacles and luminaires (lighting fixtures) attached to a strut-type channel raceway is:
    A. required to be accomplished using a bare equipment grounding wire.
    B. required to be accomplished using an insulated equipment grounding wire.
    C. only permitted to be accomplished by the strut channel if made of aluminum.

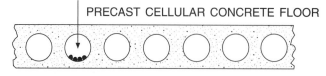

CELL WITH PROPER FITTINGS
USED AS A RACEWAY

PRECAST CELLULAR CONCRETE FLOOR

Figure 13.18   Determine the maximum size conductors permitted to be installed in a 3 in. (75 mm) diameter cells used as raceway in precast cellular concrete floor.

D.   required to be by means of a copper equipment grounding wire if the circuit operates at more than 150 volts to ground.

E.   permitted to be accomplished by the strut channel.

Code reference_____

5.   An insert is installed into underfloor raceway of an office building for electrical outlets at a workstation as shown in Figure 13.19. The connection of the individual wires to the receptacle at the insert:

A.   is permitted to be spliced with a pigtail connected to the receptacle.

B.   shall be made with the main conductor stripped, but not cut, with a short pigtail soldered to the main conductor and connected to the receptacle.

C.   is permitted to be the main conductor stripped and looped around the terminal screw of the receptacle.

D.   is permitted to cut the main circuit wire and connect one to each terminal screw of the receptacle.

E.   is permitted to cut the main circuit wire and push the wire into the receptacle provided a screw terminal is provided at the receptacle to ensure a tight connection.

Code reference_____

6.   A restaurant will be classified as an assembly occupancy if it is ruled to have a seating capacity of not less than:

A.   75 persons.   D.   200 persons.

B.   100 persons.   E.   500 persons.

C.   150 persons.

Code reference_____

7.   An area of a motel is used for conferences and meetings and as required by the local jurisdiction is classified as an assembly occupancy and is to be of fire-rated construction. A wiring method not permitted to be run as exposed wiring on the surface of walls and ceilings is:

A.   Type AC Cable with an insulated equipment grounding conductor.

B.   Flexible Metal Conduit.

C.   Type MC Cable.

D.   Electrical Metallic Tubing.

E.   Rigid Nonmetallic Conduit.

Code reference_____

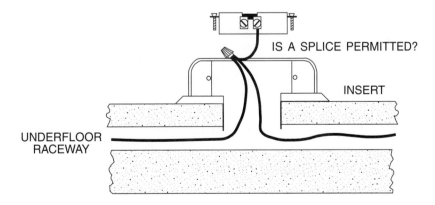

Figure 13.19   What type of connection is required for the wires to a receptacle outlet at an insert installed in underfloor raceway?

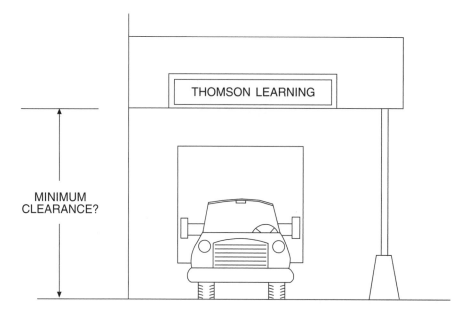

Figure 13.20 Determine the minimum height of a sign permitted when the area under the sign is open to vehicular traffic.

8. An electric sign is not protected from physical damage, and it is located above an area accessible to vehicles, as shown in Figure 13.20. The minimum clearance from the bottom of the sign to the paved vehicle surface shall not be less then:

   A. 10 ft (3 m).
   B. 12 ft (3.7 m).
   C. 14 ft (4.3 m).
   D. 18 ft (5.5 m).
   E. 20 ft (6 m).

   Code reference _____

9. An electric sign that contains either incandescent or fluorescent luminaires (lighting fixtures) shall be supplied from a branch-circuit that has a rating not to exceed:

   A. 15 amperes.
   B. 20 amperes.
   C. 30 amperes.
   D. 40 amperes.
   E. 50 amperes.

   Code reference _____

10. A neon sign on the outside of a building has the transformer installed inside the building with only a short distance between the transformer and the sign. The transformer is to be installed in a space:

    A. that is 3 ft wide (900 mm), 3 ft high (900 mm), and 3 ft deep (900 mm).
    B. 3 ft (900 mm) of clear space in all directions from the transformer.
    C. 30 in. wide (750 mm), 4 ft high (1.2 m), and 3 ft (900 mm) clearance in front.
    D. 24 in. (600 mm) of clear space in all directions from the transformer.
    E. 12 in. wide (300 mm) and 3 ft high (900 mm) and completely open in front.

    Code reference _____

11. A manufactured wiring system is installed in the ceiling of an industrial building to supply high-intensity discharge (HID) luminaires (lighting fixtures). Type SO flexible cord with size 12 AWG conductors, completely visible for the entire length, is part of a listed factory-made assembly for connecting the luminaires (lighting fixtures) to the branch-circuit as shown in Figure 13.21. The flexible cord is not permitted to be longer than:

    A. 24 in. (600 mm).
    B. 3 ft (900 mm).
    C. 4^1/2 ft (1.4 m).
    D. 6 ft (1.8 m).
    E. 8 ft (2.5 m).

    Code reference _____

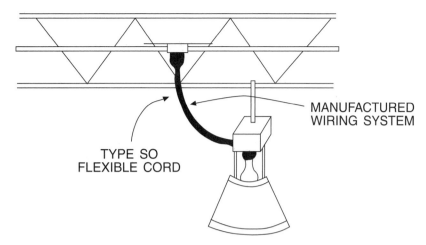

Figure 13.21 Determine the maximum length of flexible cord when a high-intensity discharge fixture is connected to a manufactured wiring system using Type SO flexible cord.

12. A required 125-volt, 15- or 20-ampere branch-circuit serves the lights and receptacles in an elevator machine room. The lights in the room are required to be:
    A. connected to the supply side of the ground-fault circuit interrupter.
    B. arc-fault protected.
    C. incandescent with at least two luminaires (lighting fixtures).
    D. ground-fault circuit-interrupter protected.
    E. grounded with an insulated equipment grounding conductor.

    Code reference _____

13. In the wellway of an escalator, trade size $^3/_8$ (12) Liquidtight Flexible Metal Conduit is used to protect wires supplying control equipment. The length of Liquidtight Flexible Metal Conduit is not permitted to exceed:
    A. 2 ft (600 mm).   C. $4^1/_2$ ft (1.4 m).   E. 10 ft (3 m).
    B. 3 ft (900 mm).   D. 6 ft (1.8 m).

    Code reference _____

14. A computer room has a raised floor as shown in Figure 13.22. A circuit is run through the raised floor space to supply field-installed equipment and receptacles. A wiring method not permitted to be installed in the raised floor space is:
    A. Type UF Cable.
    B. Electrical Nonmetallic Tubing (ENT).
    C. Type AC Cable.
    D. Flexible Metal Conduit (FMC).
    E. Rigid Nonmetallic Conduit (RNC).

    Code reference _____

15. In an area of a commercial building where sensitive electronic equipment is used, and electrical noise is a major concern, a technical power system is installed. The sensitive electronic equipment is supplied power from receptacles of the technical power system which has two ungrounded conductors operating at 120 volts. A 125-volt, 15- or 20-ampere rated receptacle from the premises wiring system, with a grounded neutral conductor, is required to be installed in the same area with the technical power receptacles (Figure 13.23), and located a distance from any technical power receptacle not more than:
    A. 3 ft (900 mm).   D. 10 ft (3 m).
    B. $4^1/_2$ ft (1.4 m).   E. 15 ft (4.5 m).
    C. 6 ft (1.8 m).

    Code reference _____

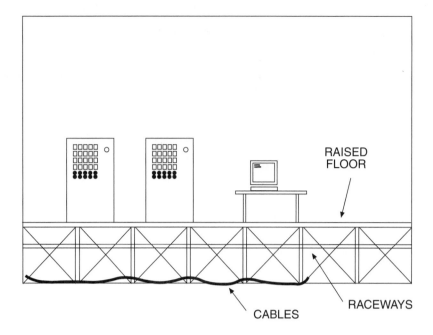

RAISED FLOOR

RACEWAYS

CABLES

Figure 13.22   What are the wiring methods permitted for a power circuit supplying receptacles for information technology equipment and installed through the raised underfloor space of a computer room?

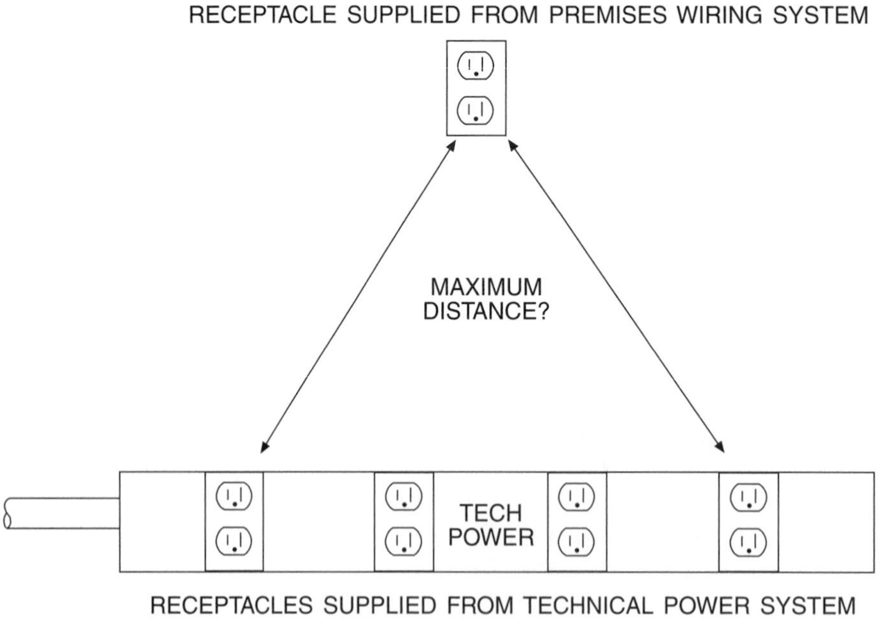

RECEPTACLE SUPPLIED FROM PREMISES WIRING SYSTEM

MAXIMUM DISTANCE?

TECH POWER

RECEPTACLES SUPPLIED FROM TECHNICAL POWER SYSTEM

Figure 13.23   A 15- or 20-ampere, 125-volt receptacle supplied from the premises wiring system with a neutral circuit conductor is required to be installed within a maximum of what distance from all receptacles of the technical power system?

WORKSHEET NO. 13—ADVANCED COMMERCIAL WIRING APPLICATIONS

Mark the single answer that most accurately completes the statement based upon the 2005 Code. Also provide, where indicated, the Code reference that gives the answer or indicates where the answer is found, as well as the Code reference where the answer is found.

1. A flat cable assembly consists of a Type FC Cable installed in strut that is attached to a ceiling. Taps are made for the installation of luminaires (lighting fixtures) by using terminal blocks that are installed into the strut and pierce the flat conductor cable as shown in Figure 13.1. If a protective cover is not installed to prevent access to the conductors, the minimum mounting height of the flat conductor assembly above the floor is:

A. 7 ft (2.1 m).   D. 10 ft (3 m).

B. 7^{1}/$_{2}$ ft (2.3 m).   E. 12 ft (3.7 m).

C. 8 ft (2.5 m).

Code reference _____

2. An insert is installed into a cellular concrete floor raceway for a receptacle outlet at a workstation. An equipment grounding conductor shall be installed and shall:

A. be bare.

B. be bonded to the equipment grounding conductor in the header for the cell.

C. be insulated.

D. be bonded directly to the equipment grounding conductor in the adjacent cell.

E. not exceed 100 ft (30 m) in length.

Code reference _____

3. A receptacle outlet is removed at an insert in cellular metal floor raceway and the outlet is abandoned. The conductors that were connected to the receptacle are:

A. permitted to be insulated with tape and placed into the raceway.

B. required to be removed.

C. permitted to remain in the raceway and reinsulated if the conductor is continuous without any nicks or breaks.

D. permitted to remain if the reinsulation is listed for underground use and is watertight.

E. required to be labeled as abandoned if reinsulated and left in the raceway.

Code reference _____

4. A strut-type channel raceway, trade size 1^{5}/$_{8}$ by 1, is installed across the ceiling of a commercial storage area to support wiring and luminaires (lighting fixtures). Internal joiners are used to connect the strut channel sections. The conductors installed in the strut channel are size 12 AWG copper, Type THWN with 75°C terminations, as shown in Figure 13.24. The maximum number of conductors permitted to be installed in the strut channel is:

A. 9.   C. 21.   E. 34.

B. 16.   D. 30.

Code reference _____

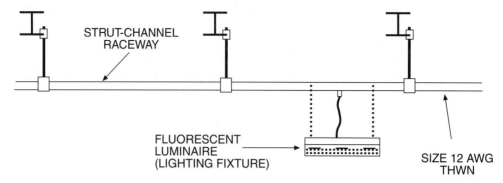

Figure 13.24   Determine the maximum number of size 12 AWG Type THWN conductors permitted to be installed in trade size 1⁵/₈ by 1 strut-channel raceway.

5. The suspended ceiling of a restaurant classified as a place of assembly has a 15-minute finish fire rating, as shown in Figure 13.25. The space above the ceiling is not used as a plenum for environmental air. A wiring method not permitted to be used in the space above the suspended ceiling is:
   A. Type MC Cable.
   B. Electrical Nonmetallic Tubing (ENT).
   C. Flexible Metal Conduit.
   D. Type AC Cable not containing an insulated equipment grounding wire.
   E. Electrical Metallic Tubing (EMT).

   Code reference _____

6. In an exhibition hall ruled as an assembly occupancy, a cable tray is installed for the purpose of placing cords used for temporary wiring for the floor displays, as shown in Figure 13.26. Only qualified personnel will service and maintain the installation. The cords placed in the cable tray and used for the temporary wiring and shall be rated for:
   A. hard or extra-hard usage and must be installed with a space not less than one cord diameter between each cord.
   B. extra-hard usage and placed in a single layer.
   C. extra-hard usage and limited to 100 ft (30 m) lengths.
   D. hard or extra-hard usage and must be placed in a single layer.
   E. hard usage.

   Code reference _____

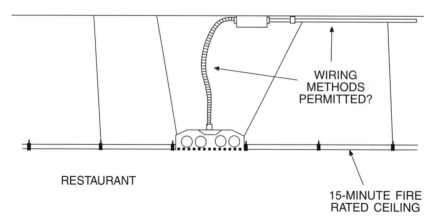

Figure 13.25   What wiring methods are permitted to be installed in a suspended ceiling of a restaurant that has a 15-minute finish rating.

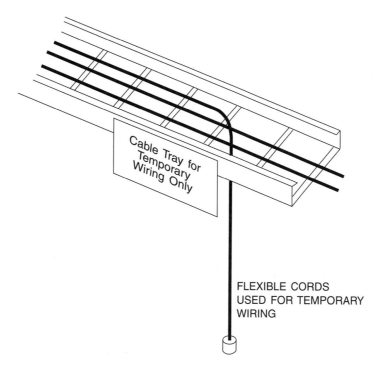

Cable Tray for Temporary Wiring Only

FLEXIBLE CORDS
USED FOR TEMPORARY
WIRING

Figure 13.26 **Cable tray is installed in an exhibition area for temporary wiring to exhibition booths has flexible cable provided to exhibit spates using what rating of flexible cord?**

7. A commercial building with grade level access to pedestrians has only one entrance intended for the general public. This building is required to provide for the purpose of supplying an electric sign:
   A. one outlet located near the entrance on a 30-ampere circuit that serves no other loads.
   B. two outlets located near the entrance each on a separate 20-ampere branch-circuit.
   C. one outlet locate near the entrance on a 20-ampere branch-circuit and permitted to serve other general use receptacle loads.
   D. one outlet located near the entrance on a branch-circuit rated not less than 15 amperes.
   E. one outlet located near the entrance on a 20-ampere circuit that serves no other loads.

   Code reference_____

8. A sign for a commercial building is supplied by a single branch-circuit and is operated by an electronic controller that is located separate from the sign. The disconnecting means:
   A. is required to be located within sight of the sign and capable of being locked in the open position if not in line of sight of any energized part of the sign.
   B. is required to be within the controller, disconnects all power to the controller, and capable of being locked in the open position.
   C. is required to be within sight of the controller but not necessarily within sight of the sign.
   D. for the sign is required to be located on the sign, which then requires an additional disconnecting means for the controller.
   E. is permitted to be a switch or circuit breaker, capable of being locked in the open position, and located within sight of the controller.

   Code reference_____

9. A receptacle is situated so that it may be used to supply power to a free standing office furnishing partition having electrical wiring and supplied power with a size 12 AWG extra-hard usage flexible cord. The cord is plugged into a receptacle that serves no other load. The maximum permitted distance from the partition to the receptacle is:

A. 12 in. (300 mm).
B. 24 in. (600 mm).
C. 36 in. (900 mm).
D. 4¹/2 ft (1.4 m).
E. 6 ft (1.8 m).

Code reference _____

10. A single 125-volt, 20-ampere receptacle is installed in the hoistway pit of an elevator to supply a permanently installed sump pump. This receptacle:

A. shall not require ground-fault circuit-interrupter protection.
B. shall be arc-fault interrupter protected.
C. is required to be of the ground-fault circuit-interrupter type.
D. is required to be ground-fault circuit-interrupter protected.
E. is not permitted to be located in an elevator hoistway pit for this purpose.

Code reference _____

11. The minimum number of duplex receptacles required to be installed in an elevator machine room or space is:

A. none because a single receptacle is permitted.
B. as many as desired, because no minimum requirement is established.
C. one.
D. two.
E. three.

Code reference _____

12. Eight passenger elevators are powered with motor-generator sets with generator field control and rated 30-horsepower, 3-phase, 460-volt continuous duty. In addition to this load, the motion-operation controller for each elevator has a continuous load of 9 amperes. The feeder conductors for the elevators are THWN copper conductors with 75°C terminations. The minimum size feeder conductors required to supply this elevator load is:

A. 600 kcmil.
B. 700 kcmil.
C. 750 kcmil.
D. 800 kcmil.
E. 900 kcmil.

Code reference _____

13. The control for the disconnecting means for the computer room dedicated air-handling equipment that is used to ventilate the area under the raised floor of an information technology equipment room is required to be located:

A. at the door considered to be the principal exit for the information technology equipment room.
B. at a centrally located location within the information technology equipment room.
C. within sight of the information technology equipment room.
D. located on the air-handling equipment.
E. located within 6 ft (1.8 m) of each exit.

Code reference _____

14. Sensitive electronic equipment in a room is supplied 120-volt, single-phase power with a technical power system. The equipment is connected to receptacle outlets supplied by 15-ampere rated circuits. The voltage drop on the feeder from the supply transformer to the technical power panel is less than 1% when all equipment is operating. The distance from the technical power panel to the receptacles in the electronics room is 75 ft (22.86 m). The actual load on the circuits is not known. The minimum size copper conductors with 75°C insulation and terminations permitted for these technical power receptacle circuits is: (assume K = 12 (0.02))

   A. 16 AWG.   C. 12 AWG.   E. 8 AWG.
   B. 14 AWG.   D. 10 AWG.

Code reference_____

15. A 15-kVA single-phase insulating transformer supplies 120-volt technical power to a single-phase panel with a 100-ampere main circuit breaker, as shown in Figure 13.27. The conductors from the transformer to the panelboard are size 1/0 copper. This separately derived electrical system is grounded to the structural steel of the building near the location of the panelboard. The minimum size grounding electrode conductor permitted for this technical power system is:

   A. 10 AWG.   C. 6 AWG.   E. 3 AWG.
   B. 8 AWG.   D. 4 AWG.

Code reference_____

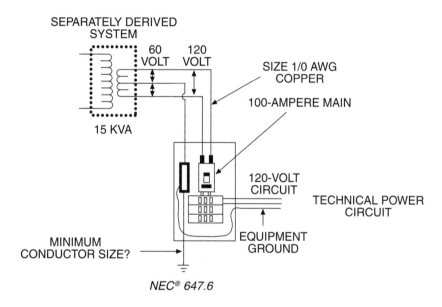

Figure 13.27   Determine the minimum size grounding electrode conductor for a technical power system where the conductors supplying the panel are size 1/0 AWG copper and the main in the panel is rated 100 amperes.

UNIT 14

Special Applications Wiring

OBJECTIVES

After completion of this unit, the student should be able to:

- state the wiring methods permitted for a theater or similar location.

- explain the location of the controls and overcurrent protection for stage lighting outlets and receptacles.

- explain the special requirements for switching lighting and receptacles in a theater dressing room.

- explain the wiring methods permitted to be used in a motion picture or television studio.

- state the minimum size of wire permitted to supply an arc or xenon projector in a motion picture projection room.

- define an equipotential plane, as related to agricultural building wiring.

- explain the grounding requirements for water pump on a farm.

- describe which article of the Code applies to the various types of communications circuits that can be installed within a building.

- describe the separation requirements of various types of communications wires in relation to circuit wires for normal light and power.

- explain the grounding of various types of communications circuits and equipment.

- answer wiring installation questions relating to *Articles 520, 525, 530, 540, 547, 553, 555, 625, 640, 650, 682, 800, 810, 820,* and *830.*

- state at least five significant changes that occurred from the 2002 to the 2005 Code for *Articles 520, 525, 530, 540, 547, 553, 555, 625, 640, 650, 682, 800, 810, 820, or 830.*

CODE DISCUSSION

This unit deals with special applications where conditions exist resulting in special requirements for wiring installations. Generally, the requirements of the Code apply except for specific requirements of these articles.

Article 520 covers wiring installations in buildings or portions of buildings used as theaters for dramatic and musical presentations and musical and dramatic presentations both indoors and outdoors. This article also applies to portions of motion picture and television studios used as assembly areas. *Parts I, III* and *VI* apply to the fixed wiring in the auditorium stage, dressing rooms, and main corridors leading to the auditorium. The fixed wiring for lighting and power shall be in metal raceway, nonmetallic raceway encased in at least 2 in. (50 mm) of concrete, or Type MI, Cable, Type MC Cable, or Type AC Cable with an insulated equipment grounding conductor. Many runs of wires are required for control of lighting. Many wires will be subjected to continuous load at times, but not all wires will be subjected to continuous load at the same time. For this reason, the derating factors of *310.15(B)(2)* do not apply when there are more than 30 wires in wireway or auxiliary gutter, as stated in *520.6.*

Dressing rooms of theaters and studios are areas where portable equipment is used, and there are usually high lighting levels. The potential for fire is great; therefore, special precautions are taken to minimize the chances of fire. Permanently attached open-ended guards are required to be installed around all incandescent lamps that are located less than 8 ft (2.5 m) above the floor, as indicated in *520.72*. According to *520.73*, all receptacles located adjacent to a mirror or serving a dressing table countertop are required to be

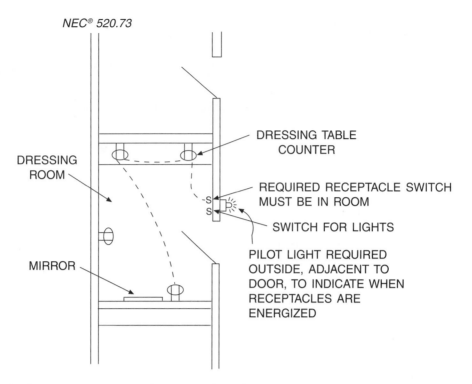

NEC® 520.73

DRESSING TABLE
COUNTER

DRESSING
ROOM

REQUIRED RECEPTACLE SWITCH
MUST BE IN ROOM

-S
S

SWITCH FOR LIGHTS

MIRROR

PILOT LIGHT REQUIRED
OUTSIDE, ADJACENT TO
DOOR, TO INDICATE WHEN
RECEPTACLES ARE
ENERGIZED

Figure 14.1   **Only the receptacles serving the dressing table counter top and the receptacles located adjacent to the mirror are required to be controlled by a switch and the pilot light to indicate when the receptacles are energized is located outside the dressing room adjacent to the door.**

controlled by a switch located in the dressing room. In addition, a pilot light must be installed outside of the dressing room adjacent to the door indicating when the dressing room receptacles are energized. These requirements are shown in Figure 14.1. Other receptacle outlets installed in the dressing room, not adjacent to a mirror or not serving a dressing table countertop, are not required to be controlled by a switch.

Part II applies to fixed stage switchboards and the feeders for switchboards. The requirements for portable switchboards are covered in *Part IV. NEC® 520.23* requires that receptacles intended to supply cord-and-plug stage lighting equipment, whether the receptacles are on the stage or located elsewhere, shall have their circuit overcurrent protection at the stage lighting switchboard location. *Part V* applies to portable stage equipment, but the receptacles for portable equipment are considered to be a fixed part of the building wiring and are covered in *520.45*. Any receptacle on a branch-circuit serving performance area may be required to carry current at the full rating of the circuit, therefore *520.9* requires all receptacles to have an ampere rating not less than the rating of the branch-circuit. The receptacle current limitation of *Table 210.21(B)(2)* does not apply to the receptacles serving performance areas. Conductors supplying receptacles shall be sized and protected for overcurrent in accordance with the provisions of *310.15* and *400.5*. The receptacles for connection of equipment shall supply continuous loads not in excess of 80% of the ampere rating of the receptacles since the receptacle has the same rating as the circuit overcurrent device. A receptacle outlet is permitted to supply a noncontinuous load rated at 100% of the receptacle ampere rating.

The definition of performance area in *520.2* makes it clear that temporary stage productions, both indoors and outdoors, are covered by the rules in *Article 520*. The rules in *Article 525* do not apply because they do not consider the unique nature of much of the equipment used for performances. An example of a portable performance stage and the equipment commonly used is shown in Figure 14.2. These temporary installations, according to *520.10,* are required to be supervised and operated by qualified personnel.

The power supply for temporary performance equipment is covered in *520.53(H)* and frequently consists of color-coded individual single-conductor flexible cables with pin-and-sleeve connectors. Portable utilization equipment generally requires a 120-volt nominal supply. Power supply feeders are generally 3-phase, 208Y/120-volt, 4-wire plus an equipment grounding cable. Good grounding is necessary for safety and proper operation of electronic equipment. The power supply may also be 208/120-volt, 3-wire plus and equipment grounding cable supplied from a 208Y/120-volt source. In this case, it is important to note that neutral current can be significant according to *310.15(B)(4)(b)*. Feeders can also be connected to a single-phase source and supplied 120/240-volt, 3-wire plus an equipment grounding cable.

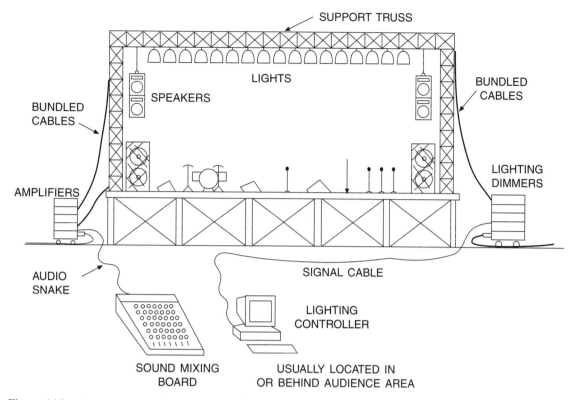

NEC® ARTICLE 520 INCLUDES TEMPORARY STAGE AND AUDIENCE
SETUPS FOR INDOOR AND OUTDOOR PERFORMANCES

Figure 14.2 A temporary performance stage is generally supplied power with single-conductor feeder cables with the main power run to amplifiers and lighting dimmers where the main power-carrying conductors are confined to the stage area.

Article 525 deals with wiring of a temporary nature, installed in adverse conditions, where the public is exposed to the temporary wiring and equipment supplied. This article deals with electrical power for equipment at carnivals, circuses, fairs, and similar events. *Article 590* does not adequately cover the conditions that exist at these events. *NEC® 525.5(B)* specifies the clearances of overhead wires from amusement rides and attractions. For example, a horizontal clearance of not less than 15 ft (4.5 m) must be maintained from amusement rides or attractions to overhead conductors as illustrated in Figure 14.3. *NEC® 525.21(A)* requires a disconnecting means for the power to all rides to be located within sight and not more than 6 ft (1.8 m) from the oper-

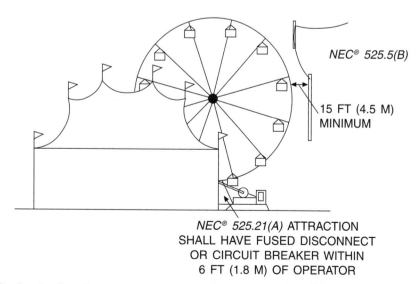

Figure 14.3 Overhead conductors are required to be kept at least 15 ft (4.5 m) horizontally from rides and attractions and a disconnect is to be located within sight and not more than 6 ft (1.8 m) from the operator's location.

ator's location. Flexible cords and cables shall be listed for extra-hard usage, wet locations, and be sunlight-resistant, *525.20(A)*. Where accessible to the public, flexible cords and cables shall be covered by approved nonmetallic mats. Receptacles shall have overcurrent protection that does not exceed the rating of the receptacle, *525.22(C)*. Receptacles rated 15 or 20 ampere, 125 volts and used by personnel shall have ground-fault circuit-interrupter protection. Grounding and bonding are covered in *Part IV*, which essentially complies with *Article 250*. It is pointed out that the grounded-circuit conductor and the equipment grounding conductor are to be maintained separate on the load side of the service disconnecting means or the separately derived system, *525.31*. In *Article 590*, 90 days is generally defined as the maximum length of time permitted for temporary wiring. There is no time limit for wiring to equipment at carnivals, circuses, fairs, and similar events.

Article 530 applies to buildings or portions of buildings used as studios, using motion picture film or electronic tape more than $^7/_8$ in. (22 mm) in width. This article also applies to areas where film and tape are handled, or where personnel are working with the film or tape for various purposes. An area for an audience is not present in the facilities covered by this article. The same requirements as for a theater, as covered in *Article 520*, should be applied. This article contains special requirements for areas where quantities of highly flammable materials, such as motion picture film and electronic recording tape, are present. The permanent wiring in stage and set areas shall be Type MC Cable, Type MI Cable, Type AC Cable with an insulated equipment grounding conductor, or approved raceway. The authority having jurisdiction will decide what wiring methods are approved. Generally, metal raceway wiring is the approved method for raceway wiring.

Article 540 deals with the wiring for motion picture projector rooms. NFPA standard, number 40-2001 provides additional information about the storage and handling of cellulose nitrate motion picture film. This material is highly flammable; therefore, special requirements are placed on the wiring of projection rooms. Local building codes will provide additional information as to which areas are considered part of the projection room. For example, a film rewinding room would be considered part of the projection room because it opens directly into the projection room.

Article 547 covers the wiring of agricultural buildings where excessive dust, dust with water, or corrosive atmospheres are present. Buildings in agricultural areas in which these conditions are not present are permitted to have the wiring installed according to the requirements elsewhere in the Code. Conditions on farms are extreme. Wiring is generally exposed to a wide variation in temperature which causes expansion and contraction as well as temperature differences that can lead to condensation. The wiring may be exposed to severe physical abuse from impact, rubbing, and chewing. The conditions are often corrosive. High humidity is common, and frequently portions of the wiring is exposed to standing water and animal waste. *NEC® 547.5(A)* covers the wiring methods permitted in an agricultural building that preferably are Rigid Nonmetallic Conduit or Type UF Cable. Other materials frequently used are Liquidtight Flexible Nonmetallic Conduit, copper SE Cable, and jacketed Type MC Cable. Equipment grounding in damp and wet locations that often exist on farms requires effective equipment grounding. *NEC® 547.5(F)* requires a separate copper equipment grounding conductor run with the circuit conductors to all noncurrent-carrying metal parts of equipment, raceways, and other enclosures requiring grounding. Whenever this copper equipment grounding conductor is run underground, it is required to be insulated or covered.

Boxes and enclosures installed in areas where there is excessive dust or damp and wet conditions or corrosive conditions are required to be corrosion resistant, weatherproof, and designed to prevent the entrance of dust as stated in *547.5(C)*. Nonmetallic boxes are generally ideal for these applications such as the weatherproof switchbox shown in Figure 14.4. The connector is of a type that is made for Type UF Cable and is watertight and dusttight. *NEC® 547.5(C)(1)* requires boxes to be mounted in such a manner that holes are not drilled in the back of the box. The box in Figure 14.4 is constructed for external mounting. Motors used in these areas shall be totally enclosed or designed to minimize the entrance of dust particles and moisture. Luminaires (lighting fixtures) in agricultural buildings shall be a type that minimizes the entrance of dust, moisture, and other foreign material.

Receptacle installed in agricultural areas that are rated 15 or 20 amperes, 125 volts and intended for general use are required to be ground-fault circuit-interrupter protected. This rule applies to all buildings and areas that have equipotential planes, all outside locations, and all damp or wet locations.

Farms consist of a group of buildings frequently including one or more dwellings. The typical manner for distributing power to a farm is to a centrally located distribution point, which is defined in *547.2*. From that point, power is extended to the various buildings, either overhead or underground, or a combination of both. *NEC® 547.9(A)* sets a minimum requirement for the distribution point. A means of disconnecting all ungrounded conductors to the buildings is required. The grounded service conductor is required to be connected to a grounding electrode at the distribution point in accordance with the rules in *Article 250*. The disconnecting means can be provided by the serving utility or the property owner. Overcurrent protection at the distribution point is optional unless supply to the building is underground.

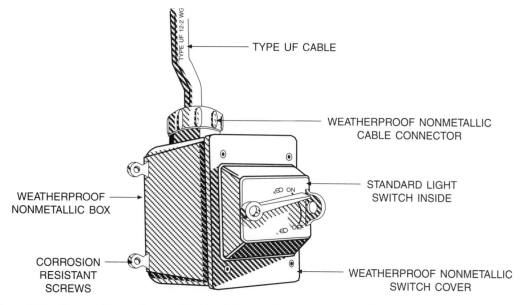

TYPE UF CABLE

WEATHERPROOF NONMETALLIC
CABLE CONNECTOR

STANDARD LIGHT
SWITCH INSIDE

WEATHERPROOF
NONMETALLIC BOX

CORROSION
RESISTANT
SCREWS

WEATHERPROOF NONMETALLIC
SWITCH COVER

Figure 14.4   Nonmetallic weatherproof boxes with watertight connectors and Type UF Nonmetallic Cable will help to prevent corrosion of the wiring and moisture from entering the wiring. Courtesy of Consumers Energy, Jackson,

There is permitted to be no overcurrent protection at the distribution point. There can also be one large disconnect with a single overcurrent device sized to the entire farm load. In this case, the conductors to the individual buildings are run as outside feeders according to *240.21(B)(5)*. Overhead conductors to the individual buildings are permitted to be installed according to *547.9(B)(1)*. It is permitted to ground the neutral conductor at each building and treat the installation as a service. It is also permitted to run a separate equipment grounding conductor to the building, in which case the neutral is not grounded at the building. This procedure is covered in *250.32*. This procedure is described in *Unit 5* and illustrated in Figure 5.5.

NEC® 547.9(B)(2) deals with the case where each set of conductors supplying the buildings is provided with an individual disconnecting means and overcurrent protection. In this case, the installation is treated as described in *250.32*. It is permitted to ground the neutral conductor at each building and treat the installation as a service. It is also permitted to run a separate equipment grounding conductor to the building, in which case the neutral is not grounded at the building. This procedure is covered in *547.9(B)(3)(b)* and illustrated in Figure 14.5.

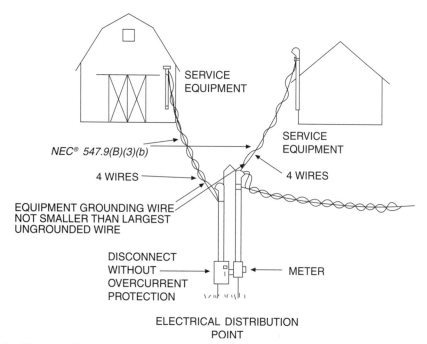

SERVICE
EQUIPMENT

SERVICE
EQUIPMENT

NEC® 547.9(B)(3)(b)

4 WIRES

4 WIRES

EQUIPMENT GROUNDING WIRE
NOT SMALLER THAN LARGEST
UNGROUNDED WIRE

DISCONNECT
WITHOUT
OVERCURRENT
PROTECTION

METER

ELECTRICAL DISTRIBUTION
POINT

Figure 14.5   Disconnecting means must be provided at the electrical distribution point on a farm.

The neutral conductor of a grounded electrical system carrying electrical current will usually have a measurable voltage from the neutral conductor to a ground rod driven into the earth in an isolated location. This is called neutral-to-earth voltage. The term **stray voltage** describes the condition where a neutral-to-earth voltage can be measured between points with which livestock may make contact, such as from a watering device in the barn to the concrete floor. A source of this stray voltage is the neutral wire of the feeder to a farm building. This article of the Code permits the neutral of the feeder to a building to be run separate from and insulated from, the equipment grounding conductor to the building. This procedure will prevent the neutral-to-earth voltage from the feeder neutral from getting to the areas where an animal may make contact. This is accomplished by running four wires to a farm building in the case of a single-phase, 120/240-volt feeder. An equipment grounding terminal is installed at the building service panel, which is separate from the neutral terminal, as shown in Figure 14.6.

It is extremely important to note that when the neutral conductors and the equipment grounding conductors are separated in a building, an equipment grounding conductor must be run from that building back to the source of power. That source of power may be another building or it may be a center distribution pole. Simply grounding the equipment grounding bus to the earth and not running the equipment grounding bus back to the main source of power is a violation of *250.4(A)(5)*. The earth is not permitted to be the only equipment grounding conductor.

Electrical wiring in agricultural buildings often is subject to physical damage and corrosive conditions. *NEC® 547.5(F)* requires that a copper equipment grounding conductor be run for all circuits even when run in metal conduit. This is to ensure that noncurrent-carrying parts of equipment will remain grounded even in the harsh environments found in agricultural areas. *NEC® 250.112(L)* and *(M)* require that a water pump installed on a farm be grounded. There is an additional requirement that in the case of a submersible water pump, as shown in Figure 14.7, a metal well casing shall be bonded to the pump equipment grounding conductor.

Another requirement of some agricultural buildings is that an equipotential plane shall be installed. Rules for installing an equipotential plane are covered in *547.10*. This equipotential plane shall be bonded to the grounding electrode system of the building. By bonding all metal objects in an animal confinement area together and then bonding the metal to the service grounding electrode system, an animal is not exposed to stray voltage in that confinement area. This article requires that an equipotential plane shall be bonded to the building grounding electrode system. An equipotential plane in a milking parlor is illustrated in Figure 14.8.

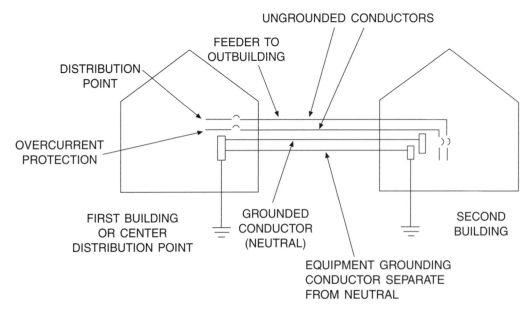

NEC® 547.9(C) AND *250.32(B)(1)* THE EQUIPMENT GROUNDING CONDUCTOR SHALL BE NOT SMALLER THAN *TABLE 250.122*

UNGROUNDED CONDUCTORS

FEEDER TO OUTBUILDING

DISTRIBUTION POINT

OVERCURRENT PROTECTION

FIRST BUILDING OR CENTER DISTRIBUTION POINT

GROUNDED CONDUCTOR (NEUTRAL)

EQUIPMENT GROUNDING CONDUCTOR SEPARATE FROM NEUTRAL

SECOND BUILDING

Figure 14.6   The equipment grounding conductor for a feeder on a farm is permitted to be run separated from the grounded circuit conductor and is required when there is a metal connection between buildings such as a water pipe.

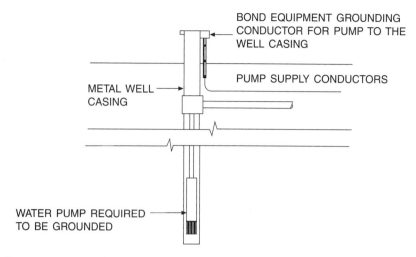

NEC® 250.112(L) AND (M) A WATER PUMP AT AN AGRICULTURAL
LOCATION SHALL BE GROUNDED

BOND EQUIPMENT GROUNDING
CONDUCTOR FOR PUMP TO THE
WELL CASING

PUMP SUPPLY CONDUCTORS

METAL WELL
CASING

WATER PUMP REQUIRED
TO BE GROUNDED

Figure 14.7 A water pump on a farm is required to be grounded and a metal well casing is required to be bonded to the equipment grounding conductor for a submersible pump.

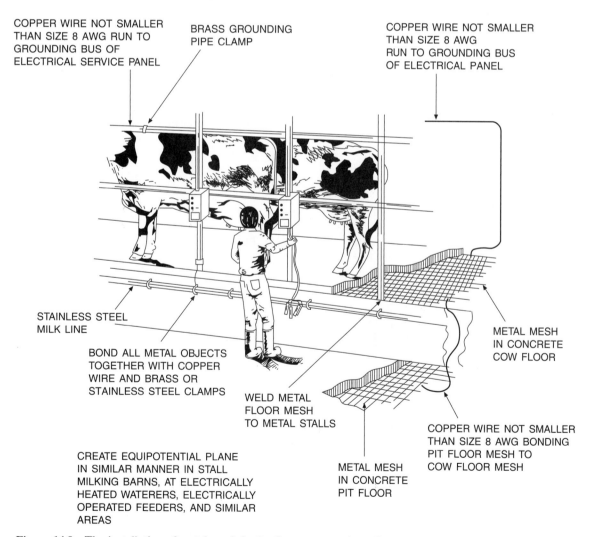

COPPER WIRE NOT SMALLER
THAN SIZE 8 AWG RUN TO
GROUNDING BUS OF
ELECTRICAL SERVICE PANEL

BRASS GROUNDING
PIPE CLAMP

COPPER WIRE NOT SMALLER
THAN SIZE 8 AWG
RUN TO GROUNDING BUS
OF ELECTRICAL PANEL

STAINLESS STEEL
MILK LINE

BOND ALL METAL OBJECTS
TOGETHER WITH COPPER
WIRE AND BRASS OR
STAINLESS STEEL CLAMPS

WELD METAL
FLOOR MESH
TO METAL STALLS

METAL MESH
IN CONCRETE
COW FLOOR

METAL MESH
IN CONCRETE
PIT FLOOR

COPPER WIRE NOT SMALLER
THAN SIZE 8 AWG BONDING
PIT FLOOR MESH TO
COW FLOOR MESH

CREATE EQUIPOTENTIAL PLANE
IN SIMILAR MANNER IN STALL
MILKING BARNS, AT ELECTRICALLY
HEATED WATERERS, ELECTRICALLY
OPERATED FEEDERS, AND SIMILAR
AREAS

Figure 14.8 The installation of metal mesh in the floor connected to all exposed metal forms an equipotential plane that helps prevent animals from being exposed to stray voltage. Courtesy of Consumers Energy, Jackson, MI.

Article 553 covers the wiring of an electrical service to floating buildings. A floating building is a structure moored in a permanent location, provided with electrical wiring, and permanently connected to an electrical system not associated with the floating building. The service equipment is not permitted to be located on or in the floating building. It shall be located adjacent to the building location. The equipment grounding conductor shall be run separate from the insulated neutral conductor. The equipment grounding conductor and the neutral conductors shall be maintained separate throughout the wiring system and equipment of the floating building. This is similar to the separation of neutral and equipment grounding conductor illustrated in Figure 14.6.

Article 555 covers the wiring of marinas and boatyards where electrical power is used on fixed or floating piers, wharfs, and docks, and where power from these structures supplies shore power to boats and floating buildings. A receptacle outlet intended to supply shore power to a boat shall be of the grounding and locking type. A receptacle outlet with a minimum rating of 30 amperes shall be provided for boats. If a receptacle rated at 15 or 20 amperes supplies 120-volt power for general use on the pier, ground-fault circuit-interrupter protection shall be provided. Wiring methods and grounding are covered. Extra-hard usage cord is permitted to be used as a feeder along the pier provided it is listed for use in wet locations and is marked as sunlight-resistant. Extra-hard usage cord also can be used as a feeder to floating docks and similar structures. But the electrical service is not permitted to be located on the floating dock. The limits of the classified hazardous location near gasoline dispensers are found in *Article 514*. The wiring within these classified areas shall be as required in *Articles 501* and *514*.

It is important to determine the electrical datum plane in order to locate receptacles and other electrical enclosures and to make connections. The electrical datum plane is defined in *555.2*. For a shore area on a body of water where there are tides, the electrical datum plane is a horizontal plane that is 2 ft (600 mm) higher than the normal high tide for the area. If the body of water is not subject to tides, then the electrical datum plane is a horizontal plane that is 2 ft (600 mm) higher than the normal water level. In the case of a floating dock or pier, the electrical datum plane of the dock or pier is 30 in. (750 mm) above the water level as illustrated in Figure 14.9.

Article 625 deals with the wiring and equipment external to a vehicle used for charging of electric vehicles. *NEC® 625.5* specifies that all materials and equipment shall be listed for the purpose. Specifications for the connection to the electric vehicle are covered in *Part II*. Electric vehicle charging loads are to be considered continuous loads, *625.14*. Electric vehicle supply cable shall be a type with the

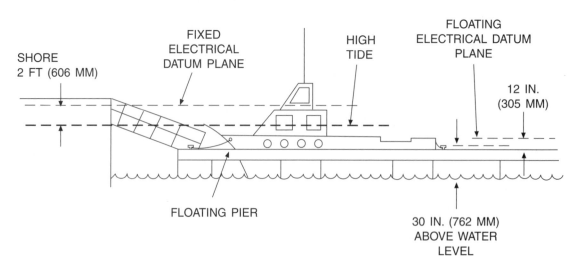

Figure 14.9 The datum plane must be established for a marina before the location of receptacles and other equipment can be determined.

letters EV as stated in *626.17* or a type listed as being suitable for the purpose. Basically this is a cable suitable for hard-usage and wet locations. The electric supply equipment for the vehicle shall have a listed shock protection system for personnel. It shall also have a ground-fault protection system that will disconnect power to the equipment at a level less than required to operate the circuit or feeder overcurrent device, *625.22*. A minimum ground-fault level is not specified. *NEC® 625.23* requires a disconnecting means capable of being locked in the open position when the equipment is rated more than 60 amperes or more than 150 volts to ground.

Attached and detached residential garages are included in the rules for supply to electric vehicle charging equipment. *NEC® 625.29(C)* provides specifications for ventilation. The batteries will give off vapors during charging that can create a hazard if not vented. Ventilation is not required where nonvented storage batteries are utilized. When vented batteries are used, such as deep discharge lead-acid batteries, ventilation is required when charging is in a closed garage. *NEC® 625.29(D)(2)* gives a formula for determining the required ventilation rate. The gas expelled from the batteries during charging is hydrogen, which will rise to the ceiling. A ventilation system for an electric vehicle garage with a charging system is illustrated in Figure 14.10. Systems for charging electric vehicles shall have provisions to disconnect power in the event of loss of normal power as stated in *625.25*.

Article 640 applies to the wiring of sound recording and sound reproduction systems, such as public address systems, and it even applies to sound reproduction systems of electronic organs. The power wiring to the components of a system follow the rules of the Code that apply to the particular type of wiring or type of conditions. The wiring of the amplifier output and similar wiring shall follow the requirements of *Article 725* for Class 2 or Class 3 wiring, whichever applies. When not more than 30 wires are installed in wireways or auxiliary gutters, ampacity derating factors do not apply, and the cross-sectional area of the wires is permitted to fill up to 20% of the cross-sectional area of the wireway or auxiliary gutter.

Audio distribution cables are required to be secured to the structure of the building in such a manner that they will not obstruct normal building operations. These cables are required to be installed according to *300.4* and *300.11*. They are not permitted to be placed on a suspended ceiling. These cables are permitted to be supported by ceiling tie wires that have been installed in addition to the ceiling grid support wires. Accessible portions of abandoned audio distribution cables, according to *640.3(A),* are required to be removed.

Article 650 applies to the circuit wiring for the keyboard of an electrically operated pipe organ and to the controls of the sounding devices. The wiring of electronic organs is to be done following the requirements of *Article 640*. The energy source for electrically operated pipe organs is not permitted to exceed 30 volts and shall originate from a self-excited generator, a rectifier supplied from a two-winding transformer, or a battery. When a motor-generator set is used, the bonding is important to prevent the 120- or 240-volt supply voltage from getting onto the generator output circuit wires which are limited to not more than 30 volts. Either the generator shall be effectively insulated from the motor or they shall be bonded together. The

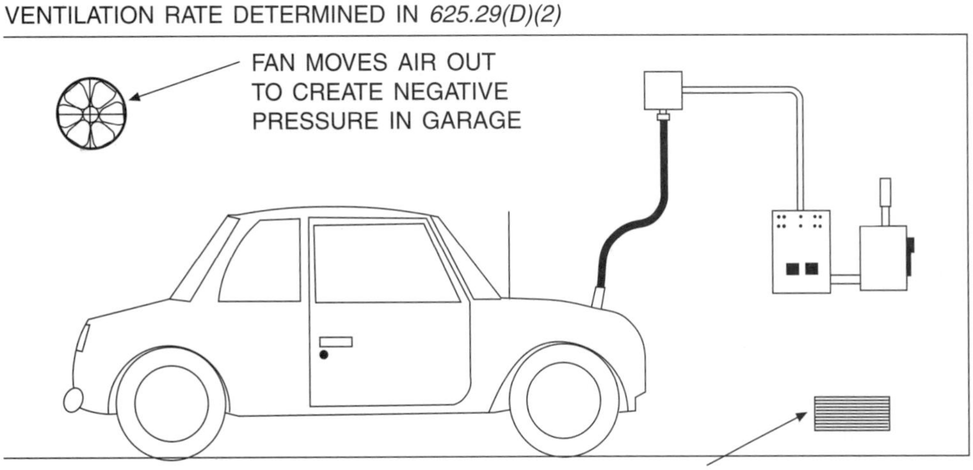

VENTILATION RATE DETERMINED IN *625.29(D)(2)*

Figure 14.10   Vented batteries in electric vehicles require ventilation when charging occurs in a closed garage.

conductor size, insulation, overcurrent protection, and installation of the wiring operating not over 30 volts are specified in this article.

Article 682 provides rules for the safe installation of electrical wiring and equipment in or associated with natural and artificially made bodies of water. Other than electrical equipment installed associated with natural and artificially made bodies of water. Other than electrical equipment installed in lakes, ponds, rivers, and streams, this article includes electrical equipment installed in such artificially made bodies of water as aeration ponds, fish farm ponds, storm retention basins, treatment ponds, and irrigation channels. Key factors are location of equipment and connections above a level where water may enter enclosures. The datum plane must be determined in order to locate equipment. The rules for establishing the datum plane are obvious in many situations, and not so obvious for other bodies of water. The rules for establishing the datum plane are the same as used in _Article 555_ for marinas. The disconnect required for the equipment in or on the water is required to be not closer than 5 ft (1.5 m) from the shore or edge of the water. Equipment and connections not rated to be submersible are required to be located not less than 12 in. (300 mm) above the datum plane.

Another aspect of electrical installations associated with electrical equipment in bodies of water is the potential for a person experiencing voltage differences between grounded equipment and the earth. This equipment ground-to-earth voltage is experienced by personnel in the form of step and touch voltages. A person standing on the wet earth and touching metal equipment can experience a touch voltage high enough to receive a mild shock. If the equipment is damaged and is faulting to the earth or water, a serious shock hazard can result. Ground-fault circuit interrupter protection is required in _682.33(B)_ for single-phase equipment operating at 120 volts, 208 volts, or 240 volts and protected by a circuit rated not over 60 amperes. For other circuits, GFCI protection is not required. To protect personnel from potentially dangerous step and touch voltages, an equipotential plane is required to be installed at exposed metal equipment such as the disconnect as shown in Figure 14.11. Specifications are not given in the Code for the illustration of the equipotential plane. The equipotential plane can be bare copper wires forming a grid beneath the ground surface, or concrete reinforcing steel mesh can also be used. Covering the area with crushed stone helps be keeping the standing surface dry. The equipotential plane is required to extend 3 ft (900 mm) beyond the normal standing area of personnel operating or maintaining the equipment. The equipotential plane is required by _680.33(C)_ to be connected to the equipment grounding conductor of the equipment using a copper wire not smaller than size 8 AWG. The connection to the equipotential plane is required to be by means of exothermic welding or a listed pressure connector.

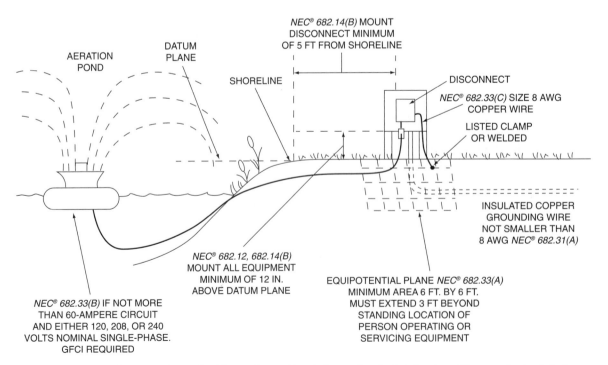

Figure 14.11 A disconnect is required for equipment located on or in a body of water with an equipotential plane installed in the earth around the disconnect to protect personnel from touch and step voltages.

NEC® 800.100(B) COMMUNICATIONS
MINIMUM GROUNDING ELECTRODE REQUIREMENT

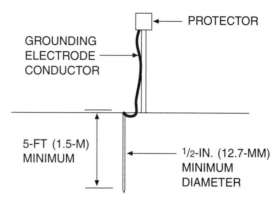

Figure 14.12 The grounding electrode for communication equipment is permitted to be a rod or pipe not less than ¹/₂ in. (12.7 mm) in diameter and a minimum of 5 ft (1.5 m) in length where there is no grounding means in the building.

Article 800 deals with the installation of communication circuits, such as telephone wiring within a building. *NEC® 800.1* provides a description of the types of communications circuits covered by this article. Some important terms are described in the FPN to *800.90(A)*. *Part III* of the article deals with the requirements for the installation of a primary protector on the communications circuit at or near the point of entry to a building or structure. Point of entry is defined in *800.2*. The minimum grounding requirements at the primary protector are covered in *800.100*. In the case of a mobile home, the primary protector is permitted to be grounded at the grounding electrode of the mobile home service or disconnecting means, which may be up to 30 ft (9 m) from the mobile home. The grounding requirements at the primary protector when a grounding means is not available are illustrated in Figure 14.12.

Part V provides the requirements for the installation of communications cables on the inside of buildings. It should be noted the requirements of *300.22* are followed when these cables pass through areas where environmental air is contained. The installation of communication conductors outside and entering buildings is covered in *Part II*. A local jurisdiction may have special rules different from the Code. Communications cables used for wiring in buildings are required to be marked as communications cable or a substitute listed in *800.154(G)* or *Figure 800.154*. The listings and markings of cables for inside installation are covered in *800.113*. The grounding of the communication equipment, communication cable sheath, and the protector is covered in *Part IV*. There is a provision in the Code that requires that communications systems be installed in a neat and workmanlike manner. Also, raceway is not permitted to be used as means of support for communications cables. This is illustrated in Figure 14.13.

Article 810 deals with radio and television receiving equipment and receiving and transmitting equipment used in amateur radio. A point of emphasis is the avoidance of contact between the normal electric circuits for light and power and the antenna and wiring of the radio or television lead-in circuits. The normal

NEC® 800.24 AND 300.11(B) COMMUNICATIONS CABLE NOT
PERMITTED TO BE SUPPORTED BY RACEWAYS

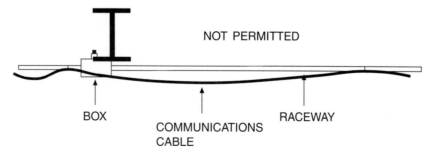

Figure 14.13 Communication cables are not permitted to use raceways as means of support.

power circuits for equipment shall follow the provisions of the appropriate portions of the Code. Specific clearances are given in the article for inside and outside installations. These requirements must be studied carefully before making an installation.

Article 820 covers the installation of coaxial cable for community antenna distribution systems for radio and television. Cables used for these purposes shall be listed for use as community antenna television cable, Type CATV. *NEC® 820.154* lists the specific cable markings required for particular installations. The requirements for installation of the various cable types are covered in *820.113* and *820.179*. The clearances of outside conductors are given in *Part II*. The grounding of the cable, protector, and equipment shall be done following the provisions of *Part IV*.

Article 830 covers wiring requirements for network-powered broadband communications systems. The FPN to *830.1* provides some insight into these systems. Broadband data communication uses a modem to introduce carrier signals onto the transmission system. The carrier signals are modulated by a digital signal. Broadband systems can be subdivided into multiple carrier signals, each carrying a different digitized signal. Broadband operates at the high end of the radio range with frequencies generally between 10 megahertz and 400 megahertz. Multiple carrier signals each transporting a digital signal can operate simultaneously, thus allowing large quantities of digital data to be transported. Teleconferencing including audio, video, and data communication is made possible with a broadband communication system.

MAJOR CHANGES TO THE 2005 CODE

These are the changes to the 2005 *NEC®* that correspond to the Code sections studied in this unit. The following analysis explains the significance of the changes from the 2002 to the 2005 Code only and this analysis is not intended to be used in place of the Code. Refer to the actual section of the 2005 Code for the exact wording and meaning of each section discussed. Changes are indicated in the Code with a vertical line in the margin. If material was deleted or moved to another location in the Code, the location of the deletion is indicated with a dark dot in the margin.

Article 520   **Theaters, Audience Areas of Motion Picture and Television Studios, Performance Areas, and Similar Locations**

520.68(A)(2): The flexible cord that supplies a portable stand lamp not subjected to physical abuse is now required to be listed as hard usage. The previous edition of the Code required the cord to be of a reinforced type.

Article 525   **Carnivals, Circuses, Fairs, and Similar Events**

525.11: If amusement rides and public attractions are supplied power from different sources, the grounded conductor of each source is to be bonded to the same grounding electrode system if the rides or attractions are separated by a distance of less than 12 ft (3.7 m). This is illustrated in Figure 14.14.

525.23(A): It is now made very clear which receptacles are required to be ground-fault circuit-interrupter protected. All 125-volt, 15- and 20-ampere nonlocking receptacles used for assembly or disassembly, or are otherwise readily accessible to the general public. Now other receptacles are not required to be GFCI protected.

525.23(B): Locking-type receptacles on 125-volt, 15- and 20-ampere circuits that are used for the connection of components of equipment are not required to be ground-fault circuit-interrupter protected.

Article 547   **Agricultural Buildings**

547.2: Site-isolating device is now defined as a disconnecting means located at the distribution point for the purpose of disconnecting the ungrounded conductors to a group of buildings for maintenance and emergencies, or to facilitate the connection of optional standby power. The requirements for the site isolating device are given *547.9(A)*. The grounded-circuit conductor is required to be connected to a grounding electrode and the rating is to be determined according to *220.103*. Overcurrent protection is not required to be provided at the site-isolating device. Figure 14.15 shows a site-isolating device that also serves as a transfer switch for an optional standby generator.

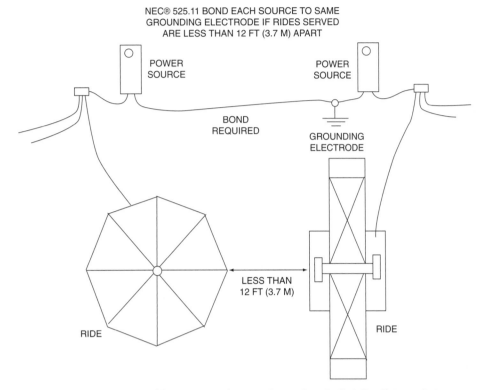

Figure 14.14   If two amusement rides or attractions are located such that the distance between exposed metal parts is less then 12 ft (3.7 m) and they are supplied from different power sources, the grounding electrodes for those power sources are required to be bonded.

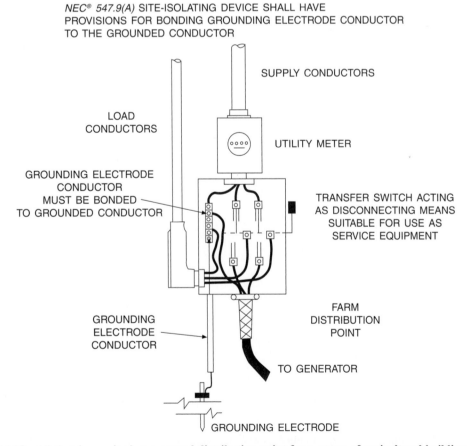

Figure 14.15   All that is required at a central distribution point for a group of agricultural buildings supplied power overhead is a site-isolation device such as a standby power transfer switch rated as suitable for use as service equipment.

547.5(G): All 125-volt, 15- and 20-ampere general-purpose receptacles installed in dirt confinement areas of agricultural buildings are required to be ground-fault circuit-interrupter protected. It is assumed that confinement area means livestock confinement. This is not really a change since it is hard to imagine any livestock confinement area that is not considered a damp or wet location, which was included as a part of this rule by the previous edition of the Code.

547.9: This new sentence places a restriction on underground feeders from a central distribution point to individual agricultural buildings. If the feeders are run overhead only a site-isolation device is required at the central distribution point. Providing individual overcurrent protection is optional. If the feeders are run underground, it is now required that overcurrent protection be provided at the central distribution point. A single disconnecting means with overcurrent protection can be installed wit the feeders run as outside taps according to *240.21(B)(5)*. Figure 14.16 is an example of a pole-top transfer switch acting as a site-isolating device and each feeder to the buildings has individual overcurrent protection.

547.9(A)(2): This is a new sentence that requires the site-isolating device to be pole mounted. There is a reference to meeting the requirements of *230.24*, which contains clearance requirements for service drop conductors.

547.9(A)(8): This paragraph places requirements on the location of the operating handle for a pole-top disconnecting means or transfer switch. The handle when in the up-position is not permitted to be more than 6 ft 7 in. (2 m) above the ground. An example of a pole-top transfer switch with a ground level operating handle is shown in Figure 14.16.

547.9(D): An equipment grounding conductor run underground was required to be insulated or covered copper according to the previous edition of the Code when supplying a building or structure housing livestock. The word "underground" was replaced with "direct-burial," which means that a bare equipment grounding conductor or an aluminum grounding conductor is permitted to be run underground in a cable or raceway. Only when the conductors are direct-burial is the equipment grounding conductor required to be insulated or covered copper.

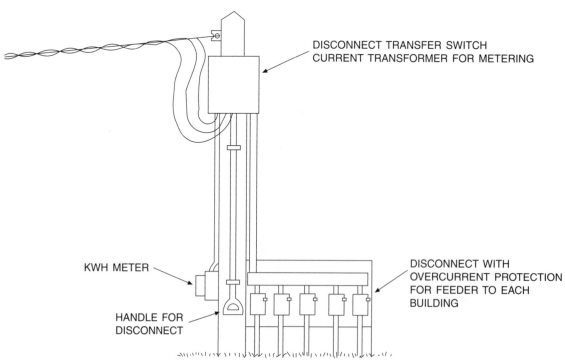

NEC® 547.9(C) DISCONNECT AND OVERCURRENT PROTECTION
FOR EACH FEEDER LOCATED AT THE DISTRIBUTION POINT

DISCONNECT TRANSFER SWITCH
CURRENT TRANSFORMER FOR METERING

KWH METER

HANDLE FOR
DISCONNECT

DISCONNECT WITH
OVERCURRENT PROTECTION
FOR FEEDER TO EACH
BUILDING

INSTALL FEEDERS ACCORDING TO NEC® ARTICLE 225, PART I AND II
AND *250.32*

Figure 14.16 Agricultural buildings supplied power underground from the central distribution point are required to be provided overcurrent protection at the supply end of the feeders.

547.10(B): This paragraph in the previous edition of the Code was deleted. Equipment likely to be energized and installed in a dirt confinement area, such as an electrically heated livestock watering device, was required to be protected with a ground-fault circuit interrupter if an equipotential plane was not installed in the first floor. With the deletion of the subsection, it is no longer required to provide GFCI protection for direct wired livestock watering devices and other electrical operated equipment installed in dirt confinement areas.

Article 553   **Floating Buildings**

553.8(C): The feeder from the service for a floating building located on shore or pier to the panelboard at the floating building is required to have an insulated equipment grounding conductor run with the feeder conductors. The equipment grounding conductor is now required to be copper. Rules for color coding this equipment grounding conductor are now the same as for all other similar cases in the Code. Sizes 4 AWG and larger are permitted to be field identified the same as other equipment grounding conductors except stripping of insulation as a means of identification is not permitted.

Article 555   **Marinas and Boatyards**

555.21: This rule change deals with the situation of a motor fuel-dispensing station location on a wharf, pier, or dock that is supplied by means of fuel piping installed along the wharf, pier, or dock. All wiring for power and lighting is now required to be installed on the opposite side of the wharf, pier, or dock from the fuel pipes. A new fine print note points out NFPA standard 303-2000 for guidance.

555.22: This is a new section that requires repair facilities for marine craft powered with combustible liquids or gases to also meet the rules of *Article 511* for commercial repair garages.

Article 625   **Electric Vehicle Charging System**

625.26: This is a new section that permits an on-vehicle electric power source to be used as an optional standby power source for a premises wiring system or be connected to operate interactively with the power grid. The equipment for this purpose is required to be listed and installed according to the rules in *Article 702* if used as an optional standby power system, and the rules in *Article 705* if connected as an interactive power source. Hybrid electric vehicles are under development that are capable of operating as an optional standby power source for loads not associated with the vehicle.

Article 682   **Natural and Artificially Made Bodies of Water**

682.1: This is a new article that provides specifications for the installation of equipment in or on natural bodies of water such as lakes, ponds, rivers, and streams, and artificially made bodies of water such as aeration ponds, fish farm ponds, storm retention basins, treatment ponds, and irrigation channels.

682.2: An **equipotential plane** is defined as wire mesh or other metal elements placed under the surface of a walk and bonded to metal equipment to prevent a voltage difference from developing between the walk surface and the metal equipment. This definition states the wire mesh or conductive elements in the earth are located at a depth of not more than 3 in. (75 mm) below the surface.

682.2: **Shoreline** is defined as the perimeter of the water line when the water is at the highest point. This would be the elevation of the datum plane. This the the same datum plane that is used in *Article 555*.

682.12: Electrical equipment and enclosures that are not rated as submersible are required to be located not less than 12 in. (300 mm) above the datum plane.

682.14: Electrical equipment located in the water or on the surface of the water is required to have a disconnect located in sight of the shore and located not closer than 5 ft (1.5 m) from the shoreline. It is necessary to determine the elevation of the datum plane before the shoreline can be located.

682.31(A): Electrical equipment with exposed metal that is likely to become energized is required to be grounded with an insulated copper equipment grounding conductor sized using *Table 250.122,* and not smaller than size 12 AWG.

682.33(A): An equipotential plane is required to be installed in the earth or floor about all outdoor service equipment and disconnecting means. The equipotential plane is required to have an area that extends a minimum of 3 ft (900 mm) beyond the perimeter of the normal standing area of a person operating or maintaining the service or disconnect. This is illustrated in Figure 14.17.

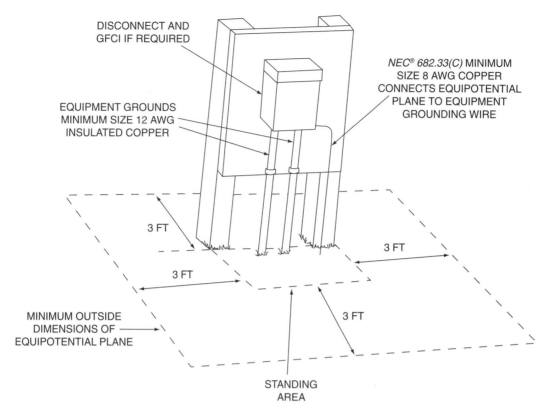

DISCONNECT AND
GFCI IF REQUIRED

NEC® 682.33(C) MINIMUM
SIZE 8 AWG COPPER
CONNECTS EQUIPOTENTIAL
PLANE TO EQUIPMENT
GROUNDING WIRE

EQUIPMENT GROUNDS
MINIMUM SIZE 12 AWG
INSULATED COPPER

3 FT

3 FT

3 FT

3 FT

MINIMUM OUTSIDE
DIMENSIONS OF
EQUIPOTENTIAL PLANE

STANDING
AREA

Figure 14.17   An equipotential plane is required to be provided around a disconnect or service supplying power to equipment installed on or in a body of water.

682.33(B): Ground-fault circuit-interrupter protection is required for equipment in or on the water that is supplied by a single-phase circuit rated not over 60 amperes, and operating at either 120 volts, 208 volts, or 240 volts.

682.33(C): The equipotential plane is required to be bonded to the equipment grounding conductor in the disconnecting means or service equipment with a solid copper wire not smaller than size 8 AWG. The copper wire is to be connected to the equipotential plane by exothermic welding or by a pressure connector listed for the application.

Article 800   Communications Circuits

800.2: **Circuit Integrity Cable** is now defined for communications cables as one that will continue operation for a specified period of time such as 2 hours. This cable is marked with the designation (CI).

800.24: This section specifies the installation of communications cable in buildings. There was an added reference to *300.11*, which specifies how communications cable is to be installed above suspended ceilings. It is no longer permitted to be placed on the ceiling grid. it is permitted to be attached to support wires that are added to support wiring.

800.100(A)(4) FPN: This section specifies that for one- and two-family dwellings the length of the grounding electrode conductor for the primary protector is not permitted to exceed 20 ft (6 m). A new fine print note was added that suggests the primary protector grounding electrode conductor length be as short as possible to limit possible differences in voltage from developing between the communications ground and the electrical system ground during a lightning event for other types of buildings.

Article 820   Community Antenna Television and Radio Distribution Systems

820.24: This section specifies the installation of CATV cable in buildings. There was an added reference to *300.11*, which specifies how CATV cable is to be installed above suspended ceilings. It is no longer permitted to be placed on the ceiling grid. it is permitted to be attached to support wires that are added to support wiring.

820.100(A)(4) FPN: This section specifies that for one- and two-family dwellings, the length of the grounding electrode conductor for the primary protector is not permitted to exceed 20 ft (6 m). A new fine print note was added that suggests the primary protector grounding electrode conductor length be as short as possible to limit possible differences in voltage from developing between the CATV system ground and the electrical system ground during a lightning event for other types of buildings.

Article 830   **Network-Powered Broadband Communications Systems**

830.24: This section specifies the installation of broadband cable in buildings. There was an added reference to *300.11*, which specifies how broadband cable is to be installed above suspended ceilings. it is no longer permitted to be placed on the ceiling grid. It is permitted to be attached to support wires that are added to support wiring.

830.100(A)(4) FPN: This section specifies that for one- and two-family dwellings, the length of the grounding electrode conductor for the primary protector is not permitted to exceed 20 ft (6 m). A new fine print note was added that suggests the primary protector grounding electrode conductor length be as short as possible to limit possible differences in voltage from developing between the broadband system ground and the electrical system ground during a lightening event for other types of buildings.

WORKSHEET NO. 14—BEGINNING SPECIAL APPLICATIONS WIRING

Mark the single answer that most accurately completes the statement based upon the 2005 Code. Also provide, where indicated, the Code reference that gives the answer or indicates where the answer is found, as well as the Code reference where the answer is found.

1. The minimum copper conductor size permitted to supply an outlet for an arc or xenon projector in the projection room of a theater is:
   A. 14 AWG.
   B. 12 AWG.
   C. 10 AWG.
   D. 8 AWG.
   E. not specified in the Code.

   Code reference _____

2. A 30-ampere, locking-type grounding receptacle outlet is used to supply shore power a boat at a marina. A disconnecting shall be readily accessible and located so that the distance from the disconnect to the receptacle it controls is not more than:
   A. 6 ft (1.8 m).   C. 24 in. (600 mm).   E. 50 ft (15 m).
   B. 10 ft (3 m).   D. 30 in. (762 mm).

   Code reference _____

3. A dimmer is to be installed for a bank of lights for a theater stage. The circuit originates from a fixed stage switchboard. Unless listed for a higher voltage, a solid-state dimmer shall not be permitted to be used on a circuit where:
   A. the voltage between conductors exceeds 150 volts.
   B. the voltage to ground exceeds 150 volts.
   C. the voltage between conductors exceeds 300 volts.
   D. the voltage to ground exceeds 300 volts.
   E. higher than 200 watt lamps are to be dimmed.

   Code reference _____

4. The permanent power wiring for a commercial television studio where there is no provisions for an audience may include all the wiring methods listed below except:
   A. Type MI Cable.   D. Type MC Cable.
   B. Type AC Cable with insulated equipment ground.   E. approved raceway
   C. Type UF Cable.

   Code reference _____

5. A metal grid of welded steel rods is installed in the concrete floor of a barn where dairy cows are to be milked and the grid is welded to the metal support poles for the equipment in the barn. This metal grid in the concrete floor is required to be bonded to the building electrical grounding system with copper wire not smaller than size:
   A. 10 AWG.   C. 4 AWG.   E. 2/0 AWG.
   B. 8 AWG.   D. 2 AWG.

   Code reference _____

6. In an area of an agricultural building having an equipotential plane installed, general use receptacles installed on 15- and 20-ampere, 125-volt circuits are required to be:
   A. only grounding-type receptacles.
   B. tamper-proof.
   C. arc-fault protected.
   D. listed for hazardous locations.
   E. ground-fault circuit-interrupter protected for personnel.

   Code reference_____

7. As shown in Figure 14.18, a 125-volt, 20-ampere, ground-fault circuit-interrupter receptacle is mounted on a floating pier. The purpose of this receptacle is for general use, not for shore power to boats. The minimum mounting height of the receptacle is:
   A. 30 in. (762 mm) above the high water level.
   B. 24 in. (600 mm) above the datum plane.
   C. 24 in. (600 mm) above the deck of the pier.
   D. 12 in. (305 mm) above the deck of the pier.
   E. 12 in. (305 mm) above the datum plane.

   Code reference_____

8. Unless listed otherwise, the coupling means for an electrical vehicle supply equipment, installed indoors, shall be located at a height above the floor not to exceed:
   A. 24 in. (600 mm).
   B. 4 ft (1.2 m).
   C. 5 ft (1.5 m).
   D. 6 ft (1.8 m).
   E. 8 ft (2.5 m).

   Code reference_____

9. Telephone cable run from outlet to outlet in a one-story commercial building. The type of cable permitted for this application, but not permitted to be used as a riser or in air handling spaces without being run in raceway is Type:
   A. CM.   C. CMP.   E. CMP-50.
   B. CMX.   D. CMR.

   Code reference_____

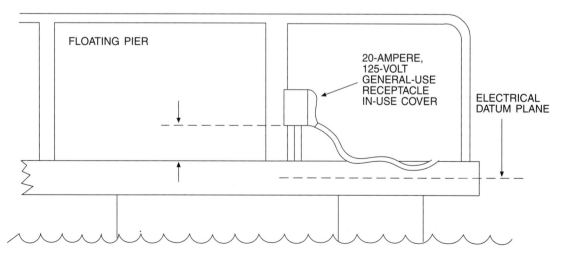

Figure 14.18   A general-use, 125-volt 20-ampere, ground-fault circuit-interrupter receptacle is mounted above a floating pier.

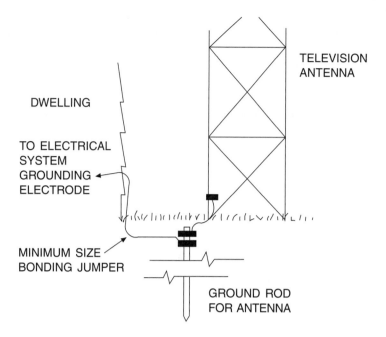

Figure 14.19 Bonding a television antenna grounding electrode to the electrical system grounding electrode at a dwelling.

10. A television antenna is placed on a tower for a single-family dwelling. It is grounded to an 8 ft (2.5 m) ground rod placed at the base of the footings for the tower. A bonding jumper is required to connect the antenna's grounding electrode to the buildings electrical grounding system as shown in Figure 14.19. This copper bonding jumper is not permitted to be smaller than size:
   A. 10 AWG.
   B. 8 AWG.
   C. 6 AWG.
   D. 4 AWG
   E. the size based upon *Table 250.66*.

   Code reference_____

11. A minimum spacing between community antenna system coaxial cables and conductors used for lightning protection where practical shall not be less than:
   A. 10 ft (3 m).   D. 12 in. (300 mm).
   B. 6 ft (1.8 m).   E. 4 in. (100 m).
   C. 5 ft (1.5 m).

   Code reference_____

12. Direct buried network powered broadband cable is run underground from one building to another at a commercial site. The conductors are installed in a trench without raceway protection nor is it protected with any concrete placed above the cable. If the cable does not pass under a driveway the minimum depth of burial is:
   A. 6 in. (150 mm).
   B. 12 in. (300 mm).
   C. 18 in. (450 mm).
   D. 24 in. (600 mm).
   E. not specified in the Code.

   Code reference_____

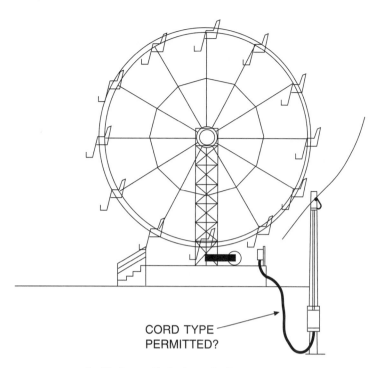

CORD TYPE
PERMITTED?

Figure 14.20 An amusement park ride is supplied with a flexible cord for temporary power.

13. An amusement park ride is temporally placed on a site for a week during a carnival. From the disconnecting means to the ride a flexible cord is installed, as shown in Figure 14.20. This flexible cord that supplies temporary power to the ride shall be sunlight-resistant, listed for wet locations, and be:
   A. listed for the purpose and may only be used as permanent wiring on the ride.
   B. extra-hard usage, unless arc-fault circuit-interrupter protected, then hard-usage cords are permitted.
   C. hard-usage and must be sunlight-resistant.
   D. extra-hard usage, unless ground-fault circuit-interrupter protected, then hard-usage cords are permitted.
   E. extra-hard usage, unless located in an area not subjected to physical damage, then hard-usage cords are permitted.

   Code reference _____

14. A group of control wires to electric pipe organ electromagnetic sounding devices mounted on a common frame is permitted to have a common copper return conductor not smaller than size:
   A. 14 AWG.   C. 20 AWG.   E. 28 AWG.
   B. 18 AWG.   D. 26 AWG.

   Code reference _____

15. When installed in an agricultural building that houses livestock, the wiring material that is not permitted be installed is:
   A. jacketed Type MC Cable.
   B. Rigid Nonmetallic Conduit.
   C. Liquidtight Flexible Nonmetallic Conduit.
   D. Type UF Cable.
   E. Type AC Cable with an insulated equipment ground.

   Code reference _____

WORKSHEET NO. 14—ADVANCED SPECIAL APPLICATIONS WIRING

Mark the single answer that most accurately completes the statement based upon the 2005 Code. Also provide, where indicated, the Code reference that gives the answer or indicates where the answer is found, as well as the Code reference where the answer is found.

1. A metal auxiliary gutter contains only Class 3 wires of a public address system for a building and no power supply wires, and it is permitted to be grounded to a common ground for the public address system with a copper wire not required to be larger than size:

   A. 14 AWG.
   B. 12 AWG.
   C. 10 AWG.
   D. 8 AWG.
   E. not specified in the Code.

   Code reference _____

2. A group of control wires to electric pipe organ sounding devices mounted on a common frame are permitted to have a common copper return conductor that:

   A. shall have an overcurrent device that is rated not more than 6 amperes.
   B. shall be ground-fault circuit-interrupter protected.
   C. shall have an overcurrent device that is rated not more than 2 amperes.
   D. requires raceway protection.
   E. is not required to have overcurrent protection.

   Code reference _____

3. Egress lighting at a carnival shall:

   A. not be ground-fault circuit-interrupter protected.
   B. have a disconnecting means within 6 ft (1.8 m) of each lighting source.
   C. be ground-fault circuit-interrupter protected.
   D. be connected to two sources of power, emergency and normal, through the use of a transfer switch.
   E. be arc-fault circuit-interrupter protected, when accessible.

   Code reference _____

4. At an outdoor location, a temporary stage is assembled for a concert. The wiring methods for this performance area shall be according to the requirements of:

   A. *Article 527.*
   B. *Article 525.*
   C. *Article 520.*
   D. *Article 518.*
   E. *Article 590.*

   Code reference _____

5. Cables used to supply stage set lighting for a television studio shall be protected with an overcurrent device set at no higher than the ampacity of the conductors taken at:

   A. 100%.
   B. 125%.
   C. 175%.
   D. 200%.
   E. 400%.

   Code reference _____

6. A motion picture projector that is installed in a permanently constructed projection room shall have a working clearance on the sides and rear of the projector of not less than:

   A. 12 in. (300 mm).
   B. 18 in. (450 mm).
   C. 24 in. (600 mm).

   D. 30 in. (750 mm).
   E. 36 in. (900 mm).

   Code reference _____

7. An agricultural building is supplied single-phase, 120/240-volt power from a 400-ampere rated pole-top disconnect that does not contain overcurrent protection. The conductors to the building are overhead aluminum with 75°C insulation and terminations, as shown in Figure 14.21. The conductors supply a panelboard that contains a 200-ampere main circuit breaker. The overhead ungrounded conductors are size 3/0 AWG aluminum. The overhead conductors consist of two ungrounded conductors, an insulated neutral, and a bare equipment grounding conductor. The neutral is not grounded at the agricultural building. The minimum size overhead equipment grounding conductor permitted is:

   A. 6 AWG.
   B. 3 AWG.
   C. 2 AWG.

   D. 3/0 AWG.
   E. not specified in the Code.

   Code reference _____

8. A disconnecting means that is capable of being locked in the open position is required for an electric vehicle charging system with a rating of more than 150 volts-to-ground or rated at more than:

   A. 30 amperes.
   B. 50 amperes.
   C. 60 amperes.

   D. 70 amperes.
   E. 100 amperes.

   Code reference _____

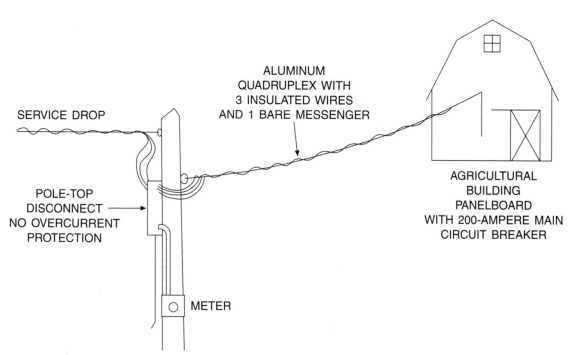

Figure 14.21 **Aluminum quadruplex supplies an agricultural building from a pole-top disconnect that contains no overcurrent protection.**

9. An equipotential plane:
   A. is required to have a voltage transition ramp where animals enter and exit from the plane.
   B. is required in all concrete floors in livestock confinement areas, where metal electrical equipment is accessible to the livestock.
   C. requires the bonding of slats for a slatted flooring system.
   D. must be installed wash rooms where milking equipment is washed.
   E. is required in all areas where the surface the livestock are confined on is dirt.

   Code reference _____

10. The receptacles installed adjacent to the mirror and above the dressing tables of a dressing room of a theater are required to be:
    A. arc-fault interrupter protected.
    B. ground-fault circuit-interrupter protected.
    C. controlled by a wall switch located in the dressing room.
    D. controlled by a wall switch located outside and adjacent to the door.
    E. provided with a pilot light at each receptacle.

    Code reference _____

11. A wiring method permitted to be used for power circuits in a movie theater, which is run as surface wiring and in wall and ceiling cavities, is:
    A. Type AC Cable with an insulated equipment grounding conductor.
    B. Type UF Cable with an insulated equipment grounding conductor.
    C. Electrical Metallic Tubing.
    D. Rigid Nonmetallic Conduit encased in 2 in. (50 mm) of concrete.
    E. Type MC Cable.

    Code reference _____

12. A mast or metal supporting structure for a television antenna or a satellite receiving dish is required to be grounded to limit the damaging effects of high voltage surges. When grounding this equipment it is:
    A. not required to protect the grounding conductors when exposed to physical damage.
    B. permitted to ground this equipment by connecting onto the grounding electrode conductor for the service to the building.
    C. only permitted to install the grounding electrode conductor outdoors.
    D. required to use either insulated or covered conductors.
    E. necessary to install an individual grounding electrode, such as a ground rod, for this equipment.

    Code reference _____

13. CATV Cable from a community antenna system is installed underground to a single-family dwelling. An additional grounding electrode is required to be installed when the grounding electrode conductor connecting the shield of the cable to the electrical system grounding electrode is longer than:
    A. 6 ft (1.8 m).   C. 20 ft (6 m).   E. 50 ft (15 m).
    B. 10 ft (3 m).   D. 25 ft (7.5 m).

    Code reference _____

14. Rigid Nonmetallic Conduit is used to supply a motor in an agricultural building as shown in Figure 14.22. The raceway is not installed in the earth. The equipment grounding conductor for the motor.
    A. is permitted to be aluminum.
    B. is required to be covered or insulated.
    C. is not required because the Code permits the motor to be grounded to a ground rod at the motor.
    D. shall be copper.
    E. is required to be the same size as the branch-circuit conductors serving the motor.

    Code reference _____

15. A network-powered broadband communications medium-power cable that is permitted to be installed underground without additional protection is:
    A. CLU.   C. BLX.   E. BMU.
    B. BL.   D. UBM.

    Code reference _____

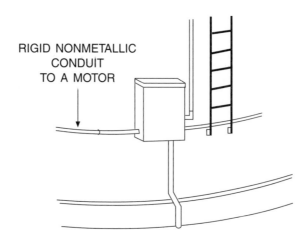

RIGID NONMETALLIC
CONDUIT
TO A MOTOR

Figure 14.22   Rigid Nonmetallic Conduit is installed above ground and used to supply a motor in an agricultural building where physical strength is needed.

UNIT 15

Review

OBJECTIVES

After completion of this unit, the student should be able to:

- evaluate ability to solve basic electrical fundamentals calculations.
- evaluate ability to find answers to questions from the Code.
- evaluate ability to make electrical wiring calculations.
- determine which articles of the Code require further work to increase the student's level of understanding.

EVALUATION PROCESS

This review is designed to serve as a self-evaluation. Mark only the answer on the answer sheet. It is best to do this evaluation in a specified period of time. A suggested time interval is two hours. If the review test is taken in a two-hour time limit, it will generally not be possible to look up the answer to every question in the Code. Usually, the answer can be narrowed to two possible answers by elimination. The following example of an electrical fundamentals problem will illustrate how the possible answers can be reduced to only two by using basic understanding of the concept.

Example 15.1   Three resistors with the values 4 ohms, 6 ohms, and 12 ohms are connected in parallel. The total resistance of the circuit is:
- A.   1 ohm.
- B.   2 ohms.
- C.   6 ohms.
- D.   22 ohms.

Answer: The circuit resistance is less than the smallest resistor in the group; therefore, the answer must be either response A or B. Working out the answer shows that the answer is response B.

$$\frac{1}{R_T} = \frac{1}{4} + \frac{1}{6} + \frac{1}{12} = \frac{3}{12} + \frac{2}{12} + \frac{1}{12} = \frac{6}{12}$$

$$R_T = \frac{12}{6} = 2 \text{ ohms}$$

WORKSHEET NO. 15—BEGINNING REVIEW

Mark the single answer that most accurately completes the statement based upon the 2005 Code. Also provide, where indicated, the Code reference that gives the answer or indicates where the answer is found, as well as the Code reference where the answer is found.

1. A 260 ft (79.25 m) length of size 4 AWG uncoated copper wire operating at a temperature of 75°C has a resistance of:
   A. 0.0801 ohm.   C. 0.308 ohm.   E. 1.010 ohm.
   B. 0.0792 ohm.   D. 0.524 ohm.

   Code reference _____NA_____

2. An extension cord with size 16 copper conductors operating at 50°C has a length of 150 ft (45.72 m) and is supplying a 120-volt load that is drawing 8 amperes. If the total resistance for the neutral and ungrounded wire is 1.375 ohms, the voltage drop caused by the resistance of the cord is:
   A. 5.2 volts.   C. 11.0 volts.   E. 18.3 volts.
   B. 7.6 volts.   D. 14.0 volts.

   Code reference _____NA_____

3. A chandelier has eight 60-watt incandescent lamps and operates at 120-volts. The total current drawn by the chandelier is:
   A. 0.5 amperes.   C. 2.5 amperes.   E. 8.0 amperes.
   B. 2.0 amperes.   D. 4.0 amperes.

   Code reference _____NA_____

4. A circuit has three resistors connected in parallel. One resistor is 4 ohms, the next is 6 ohms, and the last is 12 ohms. The total resistance of the circuit is:
   A. 1.5 ohms.   C. 4 ohms.   E. 22 ohms.
   B. 2 ohms.   D. 11 ohms.

   Code reference _____NA_____

5. A circuit has two resistors connected in series—one with 3 ohms and the other with 21 ohms. The two resistors are connected to a 120-volt ac electrical supply. The voltage across the 21-ohm resistor will be:
   A. 15 volts.   C. 60 volts.   E. 120 volts.
   B. 45 volts.   D. 105 volts.

   Code reference _____NA_____

6. A feeder supplies 76 amperes of continuous load and 45 amperes of noncontinuous load. The feeder consists of three Type THHN copper current-carrying conductors in raceway. The minimum rating overcurrent device permitted for this feeder is:
   A. 100 amperes.   D. 150 amperes.
   B. 110 amperes.   E. 175 amperes.
   C. 125 amperes.

   Code reference _____

7. A raceway contains six size 2/0 AWG copper current-carrying conductors with THHN insulation. The maximum allowable ampacity of these conductors in this raceway is:

    A. 116 amperes.
    B. 123 amperes.
    C. 136 amperes.
    D. 140 amperes.
    E. 156 amperes.

    Code reference _____

8. A feeder consisting of three current-carrying conductors supplies a continuous load of 94 amperes and is protected with a 125-ampere circuit breaker. All conductor terminations in the circuit are rated at 75°C. If the circuit conductors are THWN copper, the minimum size permitted for the circuit is:

    A. 6 AWG.
    B. 4 AWG.
    C. 3 AWG.
    D. 2 AWG.
    E. 1 AWG.

    Code reference _____

9. A 20-ampere, 120-volt circuit for lighting in the yard of a dwelling is not protected with a ground-fault circuit interrupter, although there is one GFCI type receptacle installed on the circuit. The circuit is run as Direct Burial Type UF Cable that passes under the dwelling drive that is not paved, and is not subject to other than light vehicle traffic. The minimum depth of burial permitted for the cable is:

    A. 6 in. (150 mm).
    B. 12 in. (300 mm).
    C. 18 in. (450 mm).
    D. 24 in. (600 mm).
    E. 30 in. (750 mm).

    Code reference _____

10. Type NM-B Cable is used in a dwelling as a switch loop from a ceiling box at a luminaire (lighting fixture) to a single-pole switch on the wall. A 120-volt black insulated conductor and a white insulated neutral conductor are run using Type NM-B Cable to the ceiling box. The white insulated conductor from the ceiling box to the switch box:

    A. is only permitted to be the 120-volt feed to the switch and must be marked to identify it as an ungrounded conductor.
    B. is only permitted to be the return to the light and must be marked to identify it as an ungrounded conductor.
    C. is permitted to be either the 120-volt supply to the switch or the return to the light.
    D. if marked with black tape is permitted to be the return to the light.
    E. is only permitted to be the return to the light.

    Code reference _____

11. Type NM-B Cable enters a metal device box and is secured by a cable clamp at the bottom of the box. The cable is installed so the cable jacket is flush with the cable clamp so free conductors begin at the cable clamp. If the device box has a depth of 2³/4 in. (70 mm), the minimum length of free conductor permitted in this box is required to be not less than:

    A. 3 in. (75 mm).
    B. 6 in. (150 mm).
    C. 6¹/2 in. (163 mm).
    D. 8 in. (200 mm).
    E. 12 in. (300 mm).

    Code reference _____

12. Four Type THWN size 500 kcmil copper conductors are run in Electrical Metallic Tubing (EMT). The minimum trade diameter EMT for this run is:

    A. trade size 2 (53).
    B. trade size 2¹/2 (63).
    C. trade size 3 (78).
    D. trade size 3¹/2 (91).
    E. trade size 4 (103).

    Code reference _____

13. A trade size 4 (103) Intermediate Metal Conduit (IMC) enters one side of a pull box and two trade size 3 (78) Rigid Metal Conduits enter the adjacent side. If the conductors running from the trade size 4 (103) conduit to either of the trade size 3 (78) conduits is larger than size 4 AWG, the minimum distance permitted between the conduit entries inside the box is:

A. not specified.
B. 8 in. (200 mm).
C. 12 in. (300 mm).
D. 18 in. (450 mm).
E. 24 in. (618 mm).

Code reference _____

14. Wiring in a dwelling uses single-gang nonmetallic boxes without cable clamps. A 20-ampere circuit is wired with Type NM-B Cable size 12 copper. A box on the wall has a three-way switch and a Type NM-B Cable with three insulated conductors and an equipment ground entering one end of the box. It also has a Type NM-B Cable with two insulated conductors and an equipment ground entering the other end of the box. The internal volume of the box is:

A. 15 in.³ (246 cm³).
B. 16.25 in.³ (266 cm³).
C. 18 in.³ (295 cm³).
D. 20 in.³ (328 cm³).
E. 22.5 in.³ (369 cm³).

Code reference _____

15. Fluorescent lay-in luminaires (lighting fixtures) are mounted in a suspended ceiling and supplied from 20-ampere branch-circuits. Flexible Metal Conduit supported only by a listed connector at a solidly mounted junction box and at the luminaire is permitted to be installed to supply a luminaire in lengths not to exceed:

A. 18 in. (450 mm).
B. 24 in. (600 mm).
C. 3 ft (900 mm).
D. 4 ft (1.2 m).
E. 6 ft (1.8 m).

Code reference _____

16. Trade size 1 (27) PVC type Rigid Nonmetallic Conduit is run as exposed surface wiring in a building and securely fastened within a distance of not more than 3 ft (900 mm) of each box and cabinet. The maximum distance between supports in a run of this RNC is not permitted to be greater than:

A. 3 ft (900 mm).
B. 5 ft (1.5 m).
C. 6 ft (1.8 m).
D. 8 ft (2.5 m).
E. 10 ft (3 m).

Code reference _____

17. The minimum branch-circuit rating permitted for a 12-kW, 120/240-volt electric range in a dwelling is:

A. 30 amperes.
B. 35 amperes.
C. 40 amperes.
D. 50 amperes.
E. 60 amperes.

Code reference _____

18. An unbroken length of kitchen counter along a wall in a dwelling is 10 ft (3.05 m) long. The counter extends from the edge of the sink for 4 ft (1.22 m), then makes a corner and continues for 6 ft (1.83 m) to the end of the counter at the refrigerator. The minimum number of receptacle outlets required on this length of wall space is:

A. not specified and the number is up to the owner of the dwelling.
B. one.
C. two.
D. three.
E. four.

Code reference _____

19. A single-family dwelling has three bathrooms. The minimum number of 20-ampere, 125-volt branch-circuits required for the dwelling is:
   A. three.
   B. four.
   C. five.
   D. six.
   E. not specified in the Code.

   Code reference _____

20. A single-family dwelling has two wall-mounted ovens and one counter-mounted cooking unit supplied by a single branch-circuit. The total rating of the two ovens and the cooking unit is 22 kW. The minimum permitted demand rating for this range branch-circuit is:
   A. 8 kVA.
   B. 8.4 kVA.
   C. 9.2 kVA.
   D. 11 kVA.
   E. 12 kVA.

   Code reference _____

21. A service entrance in a single-family dwelling has a 150-ampere main circuit breaker. The service-entrance conductor is aluminum with 75°C insulation and terminations. The calculated demand load for the dwelling is 115 amperes. The minimum size ungrounded conductors for this service is:
   A. 1 AWG.
   B. 1/0 AWG.
   C. 2/0 AWG.
   D. 3/0 AWG.
   E. 4/0 AWG.

   Code reference _____

22. Service-Entrance Cable Type SE style U is installed on the outside of a building for a service entrance. The weather head is fastened to the side of the building and the cable is supported within 12 in. (300 mm) of the weather head. The maximum distance permitted between supports for the cable is:
   A. 8 in. (200 mm).
   B. 12 in. (300 mm).
   C. 18 in. (450 mm).
   D. 24 in. (600 mm).
   E. 30 in. (750 mm).

   Code reference _____

23. The service drop to a building is required to meet minimum clearance requirements depending upon the use of the area beneath the service drop. In no case is the point of attachment of the service drop permitted to be less than:
   A. the minimum required to prevent personal contact.
   B. 10 ft (3 m).
   C. 12 ft (3.7 m).
   D. 15 ft (4.5 m).
   E. 18 ft (5.5 m).

   Code reference _____

24. A short section of Liquidtight Flexible Nonmetallic Conduit connects the circuit conductors to a machine. The conductors are size 1 AWG copper protected by 125-ampere time-delay fuses. The copper equipment grounding conductor required for this circuit is not permitted to be smaller than size:
   A. 6 AWG.
   B. 4 AWG.
   C. 3 AWG.
   D. 2 AWG.
   E. 1 AWG.

   Code reference _____

25. A metal underground water-piping system is used as a grounding electrode for a service of a dwelling and it is supplemented with an additional grounding electrode consisting of a ground rod driven 8 ft (2.44 m) into the earth. If the supplemental ground rod is not 25 ohms or less to ground:
    A. a second ground rod driven 6 ft (1.8 m) away and bonded to the first ground rod meets the service grounding requirement.
    B. nothing additional is required because the underground water pipe acts as the additional grounding electrode.
    C. more ground rods are required to be driven until a resistance of less than 25 ohms is achieved.
    D. soil treatment is required to lower the resistance to less than 25 ohms.
    E. a bare copper wire at least 20 ft (6 m) in length and not smaller than size 2 AWG is required to be used in place of the ground rod.

    Code reference _____

26. The service-entrance conductors in a commercial building are size 250 kcmil copper and the main circuit breaker is rated 225 amperes. The grounding electrode for the service is the structural steel of the building that is effectively grounded by bonding to reinforcement steel in the building footings. The minimum size copper grounding electrode conductor from the service panel to the structural steel is not permitted to be smaller than:
    A. 8 AWG.   C. 4 AWG.   E. 1/0 AWG.
    B. 6 AWG.   D. 2 AWG.

    Code reference _____

27. The equipotential bonding grid for a permanent swimming pool is formed by connecting together all metal parts associated with the pool such as reinforcing steel in the pool wall and deck, metallic parts of the pool structure, forming shells and mounting brackets of lighting fixtures, metal ladders and other metal fixtures attached to the pool, metal equipment associated with the water circulating system, pool covers and similar metal parts. The conductor used for this bonding is:
    A. permitted to be solid aluminum if insulated and not smaller than size 8 AWG.
    B. required to be insulated copper, stranded, and not smaller than size 8 AWG.
    C. permitted to be bare, insulated, or covered copper if not smaller than size 6 AWG.
    D. required to be copper not smaller than size 2 AWG.
    E. required to be solid copper not smaller than size 8 AWG.

    Code reference _____

28. A standard rating of an overcurrent device as recognized by the Code is:
    A. 55 amperes.   D. 750 amperes.
    B. 110 amperes.   E. 1500 amperes.
    C. 130 amperes.

    Code reference _____

29. When terminating neutral conductors and equipment grounding conductors to the neutral terminal bus in a service panelboard:
    A. only a single neutral conductor is permitted to be connected to a terminal.
    B. only one neutral and one equipment grounding conductor of the same circuit are permitted to be connected to a terminal.
    C. any number of neutral and equipment grounding conductors are permitted to be connected to a terminal if they will fit the space available.
    D. not more than two neutral conductors are permitted to be connected to a terminal.
    E. there is no limit to the number of conductors for any terminal provided a minimum torque is applied to the terminal.

    Code reference _____

30. A mobile home served with 120/240-volt, 3-wire service that has a calculated load greater than 50 amperes is required to be supplied power using a permanent wiring method from the adjacent power supply pole to the mobile home. The power supply feeder is required to consist of:
    A. three insulated conductors and an equipment ground permitted to be bare.
    B. three insulated conductors and an equipment grounding conductor with the insulated conductors permitted to be identified with colored tape.
    C. only three insulated conductors if the mobile home panel is grounded to the earth.
    D. two insulated and color coded ungrounded conductors with the neutral permitted to be bare.
    E. four insulated and color coded conductors—one of which is an equipment grounding conductor.

    Code reference _____

31. According to *430.102(B)* in the Code, a disconnecting means is required to be located "in sight from" a motor and driven machinery location. In addition to being directly in the line of sight, the term "in sight from" means the disconnect is not permitted to be located from the motor and driven machine a distance greater than:
    A. 20 ft (6 m).   D. 50 ft (15 m).
    B. 25 ft (7.5 m).   E. 75 ft (22.5 m).
    C. 30 ft (9 m).

    Code reference _____

32. A 10-horsepower single-phase, 240-volt induction motor has a nameplate current of 46 amperes and a service factor of 1.0. If the motor is protected from overload by a current-sensing device in the supply conductors, the maximum current setting of the overload device is:
    A. 46 amperes.   C. 53 amperes.   E. 63 amperes.
    B. 50 amperes.   D. 58 amperes.

    Code reference _____

33. A 3-phase, 20 horsepower, 460-volt, design B induction motor is supplied with copper conductors with 75°C rated insulation and terminations. The minimum size branch-circuit conductors permitted for this circuit is:
    A. 10 AWG.   C. 6 AWG.   E. 3 AWG.
    B. 8 AWG.   D. 4 AWG.

    Code reference _____

34. A 3-phase, 25-horsepower, 460-volt, design B induction motor with a service factor of 1.15 is supplied with copper conductors with 75°C insulation and terminations. The motor is powering an easy starting load. The maximum rating inverse time circuit breaker permitted on the circuit to provide branch-circuit, short-circuit and ground-fault protection is:
    A. 50 amperes.   C. 70 amperes.   E. 90 amperes.
    B. 60 amperes.   D. 80 amperes.

    Code reference _____

35. A 3-phase transformer is rated 37.5 kVA and supplies a panelboard with a 100-ampere main circuit breaker with 208/120-volt power from a 480/277-volt supply. The conductors from the transformer to the panelboard are size 3 AWG copper and the input conductors to the transformer are size 8 AWG copper. The maximum standard rating overcurrent device permitted on the primary side of the transformer is:
    A. 35 amperes.   C. 50 amperes.   E. 125 amperes
    B. 40 amperes.   D. 110 amperes.

    Code reference _____

36. A transformer is installed in a building and is considered to be a separately derived system. There is a metal underground water-piping system entering the building but no metal water pipe in the area of the transformer. The structural steel of the building is effectively grounded. The required grounding electrode for the transformer secondary electrical system is the:
    A. metal underground water-piping system connected within 5 ft (1.5 m) of the point where it enters the building.
    B. metal underground water-piping system connected within 5 ft (1.5 m) of the point where it enters the building and supplemented with one additional electrode.
    C. ground rod driven next to the transformer.
    D. structural steel of the building with a connection as close to the transformer as possible.
    E. bare copper wire not smaller than size 2 AWG buried under the floor near the transformer.

    Code reference _____

37. A duplex receptacle is installed on a masonry wall in the service area of a commercial garage to provide power for portable tools. The commercial garage is used only for the repair of gasoline and diesel powered vehicles. The receptacle is installed in a surface-mount masonry box and supplied with EMT using set screw connectors. This installation is permitted provided the distance from the service area floor to the bottom of the box is not less than:
    A. 12 in. (300 mm).   C. 24 in. (600 mm).   E. 5 ft (1.5 m).
    B. 18 in. (450 mm).   D. 3 ft (900 mm).

    Code reference _____

38. Wiring using field-threaded and installed Rigid Metal Conduit in a Class I, Division 1 hazardous location is required to be made up with the conduit having a minimum number of threads fully engaged of:
    A. not less than 5 threads.
    B. not less than 6 threads.
    C. not less than 8 threads.
    D. not less than 4 and sealed with a sealing compound.
    E. any number as long as the connection is tight.

    Code reference _____

39. The extent of the Class I, Division 2 classified hazardous location at a gasoline dispenser is measured up from the surface 18 in. (450 mm) and out from the edge of the dispenser a minimum of:
    A. 10 ft (3 m).   D. 25 ft (7.5 m).
    B. 15 ft (4.5 m).   E. 50 ft (15 m).
    C. 20 ft (6 m).

    Code reference _____

40. Even though circuit wiring is run in metal raceway of a hospital, an insulated copper equipment grounding conductor is required to be run in the raceway and connected to the grounding terminal of:
    A. all equipment, receptacles, and lighting fixtures in the hospital.
    B. all receptacles in patient care areas.
    C. receptacles installed in basements and other below grade areas.
    D. every circuit protected with a ground-fault circuit interrupter.
    E. ceiling lighting fixtures in patient care areas.

    Code reference _____

41. Receptacles rated 20 amperes at 125 volts in an operating room of a hospital are supplied from an isolated power system. The orange conductor of the isolated power system is to be:
    A. connected to the equipment grounding terminal of the receptacles.
    B. connected only to the silver-colored screw terminal of the receptacles.
    C. run without splices and looped at each receptacle terminal connection.
    D. connected only to the brass-colored screw terminal of the receptacles.
    E. connected to either the silver- or brass-colored terminal of the receptacles.

    Code reference _____

42. A receptacle in the unfinished portion of the basement of a dwelling that is installed to provide power to a non-power-limited fire alarm system is:
    A. not permitted and the installation is required to be supplied by a circuit near an exit door on the first level.
    B. required to be ground-fault circuit-interrupter protected.
    C. not permitted to be ground-fault circuit-interrupter protected.
    D. required to be protected by an arc-fault circuit interrupter.
    E. required to be controlled by a switch.

    Code reference _____

43. The circuit conductors from an emergency panelboard to exit signs in a building are:
    A. permitted to be run in the same raceway with normal power conductors provided all conductors have 600-volt insulation.
    B. required to be run in Rigid Metal Conduit.
    C. required to be run only in Type MI, MC, or AC Cable, or metallic raceway.
    D. required to be run in separate raceways even when two separate emergency circuits originate from the same panelboard.
    E. not permitted to be run in the same raceway with other power or lighting circuit conductors.

    Code reference _____

44. A fire pump installation consists of only one motor that is rated 60-horsepower, 3-phase, 460-volt, design B. Conductors are copper with 75°C rated terminations. The minimum size copper Type THWN conductors permitted to be run from the controller to the fire pump motor is:
    A. 8 AWG.   C. 4 AWG.   E. 2 AWG.
    B. 6 AWG.   D. 3 AWG.

    Code reference _____

45. In an existing building, optical fiber cable is to be installed where the most convenient way of running the necessary cables is across the upper side of a permanent suspended ceiling in a hallway. There are no other cables run across the suspended ceiling. The maximum number of optical fiber cables permitted to be installed by running them across the suspended ceiling from one end of the hallway to the other is:
    A. two.
    B. three.
    C. zero, because this practice is not permitted.
    D. only one.
    E. as many as desired.

    Code reference _____

46. Ventilated-trough cable tray contains only single-conductor cables installed in a single layer with an air space maintained between and secured to the cable tray so they will not move. The width of the air space is equal to the cable diameter. There are eleven Type XHHW size 500 kcmil copper single-conductor cables. The minimum width cable tray required for this installation is:

    A. 9 in. (225 mm).   D. 24 in. (600 mm).
    B. 12 in. (300 mm).   E. 30 in. (750 mm).
    C. 18 in. (450 mm).

    Code reference _____

47. A required 125-volt, 15- or 20-ampere branch-circuit serves the lights and receptacles in an elevator machine room. The lights in the room are required to be:

    A. connected to the supply side of any ground-fault circuit interrupter.
    B. arc-fault protected
    C. incandescent with at least two luminaires (lighting fixtures).
    D. grounded with an insulated equipment grounding conductor.
    E. ground-fault circuit-interrupter protected.

    Code reference _____

48. Strut-type channel raceway is surface mounted to a ceiling and used to support luminaires (lighting fixtures) as well as serve as the raceway for the conductors. The sections of strut channel are secured within 3 ft (900 mm) of ends and terminations, and shall be secured at intervals not to exceed:

    A. 3 ft (900 mm). C. 5 ft (1.5 m). E. 10 ft (3 m).
    B. 4.5 ft (1.4 m). D. 8 ft (2.5 m).

    Code reference _____

49. A restaurant is classified as a place of assembly when the seating capacity of the building is:

    A. 50 or more persons.   D. 150 or more persons.
    B. 75 or more persons.   E. 200 or more persons.
    C. 100 or more persons.

    Code reference _____

50. A wiring method not permitted to be used for light and power circuits in a movie theater and run as surface wiring, and in wall and ceiling cavities is:

    A. Type AC Cable with an insulated equipment grounding conductor.
    B. Type UF Cable with an insulated equipment grounding conductor.
    C. Electrical Metallic Tubing.
    D. Rigid Nonmetallic Conduit encased in 2 in. (50 mm) of concrete.
    E. Type MC Cable.

    Code reference _____

WORKSHEET NO. 15—ADVANCED REVIEW

Mark the single answer that most accurately completes the statement based upon the 2005 Code. Also provide, where indicated, the Code reference that gives the answer or indicates where the answer is found, as well as the Code reference where the answer is found.

1. A fluorescent luminaire (lighting fixture) has a ballast that operates two 40-watt lamps and draws 1.32 amperes at 120 volts with a power factor of 0.58. The power drawn by the luminaire (lighting fixture) is approximately:

   A. 80 watts.   C. 115 watts.   E. 273 watts.
   B. 92 watts.   D. 158 watts.

   Code reference _____ NA _____

2. A 3-phase resistance-type electric heater has a rating of 15 kW at 240 volts. The current drawn by the electric heater is approximately:

   A. 20.8 amperes.   D. 45.1 amperes.
   B. 28.9 amperes.   E. 62.5 amperes.
   C. 36.1 amperes.

   Code reference _____ NA _____

3. A circuit has three resistors connected in parallel. One resistor is 4 ohms, the next is 14 ohms, and the last is 28 ohms. The total resistance of the circuit is:

   A. 1.5 ohms.   C. 4 ohms.   E. 46 ohms.
   B. 2.8 ohms.   D. 23.0 ohms.

   Code reference _____ NA _____

4. A circuit supplying twelve 150-watt, 120-volt incandescent lamps is not turned off for a week. If the average cost of electrical power for the building is 10.2 cents per kilowatt-hour, the approximate cost of operating the lamps for one week is:

   A. $12.38.   C. $23.64.   E. $30.84.
   B. $18.22.   D. $26.47.

   Code reference _____ NA _____

5. A 3-phase electric motor running at full-load and developing exactly 10 horsepower draws 28 amperes at 230 volts with a power factor of 71.2%. The efficiency of the motor is approximately:

   A. 62%.   C. 71%.   E. 94%.
   B. 66%.   D. 82%.

   Code reference _____ NA _____

6. A 3-phase, 208/120-volt feeder consists of three ungrounded conductors and a neutral, and all four conductors count as current-carrying conductors because of the type of load being supplied. The conductors are copper with THHN insulation. The only terminations are rated 75°C and they are at panelboards on each end of the feeder.

The feeder passes through an area of the building where the ambient temperature is typically 125°F (51.7°C). If the conductors are size 300 kcmil, the allowable ampacity of these conductors for this feeder is:

A. 195 amperes.   D. 335 amperes.
B. 204 amperes.   E. 526 amperes.
C. 320 amperes.

Code reference _____

7. A feeder consisting of three current-carrying conductors supplies a continuous load of 90 amperes, and is protected with a 125-ampere circuit breaker. All conductor terminations in the circuit are rated at 75°C. If the circuit conductors are THWN copper, the minimum size permitted for the circuit is:

A. 6 AWG.   C. 3 AWG.   E. 1 AWG.
B. 4 AWG.   D. 2 AWG.

Code reference _____

8. Two sets of 3-phase feeders are run in the same raceway for a total of six current-carrying conductors. Each circuit supplies a 42-ampere continuous load. The circuit conductors are copper with THHN insulation and all circuit conductor terminations are rated 75°C. A portion of the raceway run goes through a room that is 30 ft wide and runs typically at a temperature of 130°F (54.4°C). The minimum size conductors permitted for these feeders is:

A. 6 AWG.   C. 3 AWG.   E. 1 AWG.
B. 4 AWG.   D. 2 AWG.

Code reference _____

9. An 480/277-volt electrical panelboard is mounted to the surface of a poured concrete wall and extends out from the wall surface 10 in. (250 mm). The minimum clearance in front of the panelboard from the front to the opposite concrete block wall is not permitted to be less than:

A. 24 in. (600 mm).   C. 3 ft (900 mm).   E. 4 ft (1.2 m).
B. 30 in. (750 mm).   D. 3^1/2 ft (1 m).

Code reference _____

10. Receptacles, rated 125 volts, 20 amperes, are installed on a construction site for the purpose of supplying power for portable equipment. These receptacles are not permitted to be:

A. of the grounding type.
B. supplied with Type NM-B Cable if the building is more than three floors in height.
C. installed on branch-circuits that also supply temporary lighting.
D. supplied with Type NM-B Nonmetallic-Sheathed Cable except for dwellings.
E. ground-fault circuit-interrupter protected.

Code reference _____

11. Type SE-R Service-Entrance Cable with three insulated conductors and a bare equipment grounding conductor contained within the outer nonmetallic sheath is used as a feeder to provide power from the service panel to a panelboard located in another part of the building. The cable run along the flat surface of structural materials is required to be supported at intervals not to exceed:

A. 1 ft (300 mm).   C. 3 ft (900 mm).   E. 6 ft (1.8 m).
B. 2 ft (600 mm).   D. 4^1/4 ft (1.4 m).

Code reference _____

12. Electrical Nonmetallic Tubing (ENT) is permitted to be installed:
    A. as concealed wiring in buildings of not more than three floors unless the walls, floors, and ceilings provide a thermal barrier that has at least a 15-minute finish fire rating.
    B. as concealed wiring in buildings of not more than three floors unless the walls, floors, and ceilings provide a thermal barrier that has at least a 1-hour finish fire rating.
    C. for exposed work in buildings of any height.
    D. as concealed wiring only if the walls, floors, and ceilings provide a thermal barrier that has at least a 15-minute finish fire rating.
    E. as concealed wiring in buildings of any height with no requirement the walls, floors, or ceiling have a fire rating.

    Code reference _____

13. A run of Rigid Metal Conduit contains three size 3/0 AWG Type THWN conductors and six size 8 AWG Type THHN conductors. The minimum size RMC permitted for these circuits is:
    A. trade size $1^1/4$ (35).
    B. trade size $1^1/2$ (41).
    C. trade size 2 (53).
    D. trade size $2^1/2$ (63).
    E. trade size 3 (78).

    Code reference _____

14. A junction box has two Type NM-B cables size 12-3 AWG with ground and one Type NM-B Cable size 8-3 AWG with ground. The box has a blank cover. The minimum trade size box from the following list permitted for these conductors is:
    A. $4 \times 2^1/8$ in. (100×54 mm) octagonal.
    B. $4 \times 2^1/8$ in. (100×54 mm) round.
    C. $4 \times 1^1/2$ in. (100×38 mm) square.
    D. $4^{11}/16 \times 1^1/4$ in. (120×32 mm) square.
    E. $4^{11}/16 \times 2^1/8$ in. (120×54 mm) square.

    Code reference _____

15. Several runs of Electrical Metallic Tubing enter adjacent sides of a pull box. On one side of the box are trade size 4 (103) and two trade size 2 (53) runs of EMT, and from the adjacent size there are two trade size 4 (103) runs of EMT. All conductors are size 4 or larger. The pull box is required to have dimensions not less than:
    A. 24 in. $\times$ 24 in. (618 mm $\times$ 618 mm).
    B. 27 in. $\times$ 28 in. (696 mm $\times$ 724 mm).
    C. 24 in. $\times$ 32 in. (618 mm $\times$ 812 mm).
    D. 26 in. $\times$ 30 in. (660 mm $\times$ 762 mm).
    E. 24 in. $\times$ 36 in. (610 mm $\times$ 914 mm).

    Code reference _____

16. An 80 ft (24.38 m) horizontal run of trade size 2 (53) schedule 40 Rigid Nonmetallic Conduit in a parking garage will experience a change in temperature of 110°F (61.1°C) throughout the year. The change in length of this run of Rigid Nonmetallic Conduit throughout a year is approximately:
    A. 1.50 in. (38.1 mm).
    B. 1.94 in. (49.3 mm).
    C. 2.43 in. (61.7 mm).
    D. 3.57 in. (90.6 mm).
    E. 4.46 in. (113.3 mm).

    Code reference _____

17. A single-family dwelling has a living area of 3100 sq. ft (288.0 m²). All circuits for general illumination, in addition to those for small appliance, laundry and the bathroom receptacles, are rated at 15 amperes. The minimum number of general illumination branch-circuits permitted for this dwelling is:
    A. two.
    B. three.
    C. four.
    D. five.
    E. six.

    Code reference _____

18. An electric range rated 18.4 kW is installed in a single-family dwelling. The minimum demand load permitted to be used to determine the rating of the circuit is:
    A. 8.0 kVA.
    B. 8.8 kVA.
    C. 10.2 kVA.
    D. 10.56 kVA.
    E. 18.4 kVA.

    Code reference _____

19. In a dwelling, the light for a stairway is required to be controlled from each level:
    A. if there are six risers or steps or more.
    B. if there is an exit to the outside of the dwelling from both levels.
    C. if there are more than four steps.
    D. if there is a change in elevation in a step and not a ramp.
    E. unless the light is controlled by a circuit breaker.

    Code reference _____

20. A 208/120-volt, 4-wire feeder supplies power from one building to a separate building on the same property. The feeder is protected by a 100-ampere overcurrent device in the first building. The property management does not have a qualified electrician on site at all times. The disconnecting means for the second building is:
    A. required to be located at the first building.
    B. permitted to be located on the outside the second building.
    C. required to be located inside the second building at the nearest practical point where the conductors enter the building.
    D. permitted to be located inside the first building.
    E. required to be located on the outside of the second building near the point of entry of the conductors to the building.

    Code reference _____

21. A service has a calculated demand load of 405 amperes. The service-entrance conductors are copper with 75°C rated insulation and terminations. The service conductors terminate at six disconnecting means arranged in one location that add up to a combined rating of 600 amperes. The minimum size copper conductors permitted for this service is:
    A. 500 kcmil.
    B. 600 kcmil.
    C. 700 kcmil.
    D. 750 kcmil.
    E. 800 kcmil.

    Code reference _____

22. A single-family dwelling with a total living area of 2300 sq. ft (213.7 m²) is served with a single-phase 120/240-volt electrical system. Appliances in the dwelling are a 12-kW electric range and a 5-kW electric clothes dryer operating at 120/240 volts, a 4.5-kW, 240-volt electric water heater, a 1.2-kW, 120-volt dishwasher, a ¹/₂-horsepower, 120-volt garbage disposer, and a central air-conditioner with a nameplate rated load current of 17 amperes at 240 volts. The minimum service demand load for the dwelling using the optional method of *220.82* is:
    A. 101 amperes.
    B. 125 amperes.
    C. 130 amperes.
    D. 148 amperes.
    E. 160 amperes.

    Code reference _____

23. A multifamily dwelling has 18 living units all provided with an electric range. Six of the units have 14-kW electric ranges, eight have 10-kW electric ranges, and four units have 12-kW electric ranges. For the purpose of determining the demand load for the building, the average rating of electric range is:
    A. 11.8 kVA.   C. 12.7 kVA.   E. 14.0 kVA.
    B. 12.0 kVA.   D. 13.2 kVA.

                                    Code reference_____

24. A 480/277-volt, 4-wire service to a building consists of two parallel sets of 600 kcmil copper conductors run underground from a transformer in separate Rigid Nonmetallic Conduit. A metal water pipe from underground serves the building and is used as a grounding electrode for the service. The minimum size grounding electrode conductor permitted from the main disconnect to the water pipe is
    A. 4 AWG.   C. 2/0 AWG.   E. 4/0 AWG.
    B. 2 AWG.   D. 3/0 AWG.

                                    Code reference_____

25. One building on the same property is supplied 120/240-volt single-phase power from another building. The second building is supplied from a 4-wire feeder with two ungrounded conductors, a neutral conductor, and an equipment grounding conductor. The neutral conductor at the second building is:
    A. not permitted to be connected to a grounding electrode at either building.
    B. required to be bonded to the equipment grounding conductor at the load end in the second building.
    C. required to be bonded to the disconnect enclosure in the second building and connected to a grounding electrode.
    D. only permitted to be connected to a grounding electrode at the second building.
    E. only permitted to be connected to a grounding electrode and bonded to the disconnect enclosure at the supply end of the feeder in the first buildings.

                                    Code reference_____

26. A feeder in an industrial building is protected with 800-ampere fuses and has the ungrounded conductors run as two parallel sets of copper conductors in individual Rigid Nonmetallic Conduits. The conductors in each conduit are size 500 kcmil. The minimum size copper equipment grounding conductor required to be run in parallel in each conduit is:
    A. 3 AWG.   C. 1 AWG.   E. 2/0 AWG.
    B. 2 AWG.   D. 1/0 AWG.

                                    Code reference_____

27. A permanent swimming pool is installed in a large backyard at a dwelling and at least one general-purpose receptacle that is ground-fault circuit-interrupter protected for personnel is required to be installed not more than 20 ft (6 m) from the inside edge of the pool. This receptacle is not permitted to be installed closer from the inside edge of the pool than:
    A. 5 ft (1.5 m).   D. 12 ft (3.7 m).
    B. 8 ft (2.5 m).   E. 15 ft (4.5 m).
    C. 10 ft (3 m).

                                    Code reference_____

28. A tap is made to supply a load protected by a 50-ampere circuit breaker. The feeder conductors are size 250 kcmil copper and the feeder is rated 208/120-volts protected with a 250-ampere circuit breaker. All conductors are copper with insulation and terminations rated 75°C. The distance from the tap point on the feeder to the 50-ampere circuit breaker is 22 ft (6.7 m). The minimum tap conductor size permitted for this installation is:

A. 4 AWG. C. 2 AWG. E. 250 kcmil.
B. 3 AWG. D. 1 AWG.

Code reference _____

29. The service equipment or a disconnecting means listed as suitable for use as service equipment is required to be installed adjacent to a mobile home and within sight of the mobile home and located from the exterior wall of the mobile home not more than:

A. 25 ft (7.5 m). C. 35 ft (11 m). E. 50 ft (15 m).
B. 30 ft (9 m). D. 40 ft (12 m).

Code reference _____

30. A 200-ampere panelboard is supplied with a 120/240-volt, 3-wire electrical supply containing a neutral conductor. The 12-space panelboard contains only six two-pole circuit breakers, one rated at 100 amperes, three rated at 40 amperes, and two rated at 20 amperes. The neutral conductor is only used with the 100-ampere circuit breaker. This panelboard is rated as a:

A. lighting and appliance branch-circuit panelboard.
B. power panel.
C. sub-panel.
D. load center.
E. heavy-duty panelboard.

Code reference _____

31. A motor and driven machine are located within sight of the controller and a fusible switch capable of being locked in the open position is located within sight of the controller but not within sight of the motor or driven machine. This installation is permitted:

A. if there is an additional disconnect located between the controller and the motor and within sight of the motor and driven machine
B. in any type of occupancy.
C. only in a commercial occupancy where there are qualified personnel on duty to service the installation.
D. even if the disconnect cannot be locked in the open position as long as a warning label can be attached to the disconnect during servicing.
E. because the fuses can be removed from the disconnect during servicing.

Code reference _____

32. A 3-phase, design B, 50-horsepower induction motor is rated 60 amperes at 460 volts. Circuit conductors are copper with 75°C insulation and terminations. The minimum size copper conductors permitted to supply this motor is:

A. 8 AWG. C. 4 AWG. E. 2 AWG.
B. 6 AWG. D. 3 AWG.

Code reference _____

33. A 3-phase, design B, 40-horsepower induction motor is rated 48 amperes at 460 volts. The circuit conductors are size 6 AWG copper. The motor is powering an easy starting load. The maximum standard rating time-delay fuse permitted to provide short-circuit and ground-fault protection for the circuit is:

    A. 60 amperes.
    B. 70 amperes.
    C. 80 amperes.
    D. 90 amperes.
    E. 100 amperes.

    Code reference _____

34. Three induction motors are supplied by a single feeder. There are two 40-horsepower, 3-phase, design B, 460-volt motors and one 50-horsepower, 3-phase, design B, 460-volt motor. The feeder conductor is copper with 75°C insulation and terminations. The minimum size conductor permitted as a feeder to supply these motors is:

    A. 1 AWG.
    B. 1/0 AWG.
    C. 2/0 AWG.
    D. 3/0 AWG.
    E. 4/0 AWG.

    Code reference _____

35. A 75-kVA 3-phase transformer is tapped to a 480/277-volt feeder to supply a 200-ampere 208/120 volt panelboard. The distance from the feeder tap point through the transformer to the panelboard is less than 25 ft (7.5 m). It is permitted to tap the transformer directly to feeder without an overcurrent device protecting the transformer primary provided the feeder overcurrent device is not rated higher than:

    A. 200 amperes.
    B. 225 amperes.
    C. 250 amperes.
    D. 300 amperes.
    E. 400 amperes.

    Code reference _____

36. A 50-kVA single-phase transformer supplies 120/240-volt, 3-wire power to a panelboard with a 200-ampere main circuit breaker. The secondary conductors from the transformer to the panelboard are size 3/0 AWG copper. The structural steel of the building is effective grounded and is used as the grounding electrode for the separately derived system. The minimum size copper grounding electrode conductor permitted from the transformer secondary system to the structural steel is:

    A. 6 AWG.
    B. 4 AWG.
    C. 3 AWG.
    D. 2 AWG.
    E. 1/0 AWG.

    Code reference _____

37. A trade size 1 (27) Rigid Metal Conduit enters a NEMA 7 explosionproof motor starter installed in a Class I, Division 1 classified hazardous area. The seal placed in the conduit is not permitted to be located a distance from the motor starter enclosure more than:

    A. 6 in. (150 mm).
    B. 12 in. (300 mm).
    C. 18 in. (450 mm).
    D. 24 in. (600 mm).
    E. 3 ft (900 mm).

    Code reference _____

38. A gasoline dispenser at a service station is supplied with 120-volt power from a circuit breaker panelboard where the circuit breaker will act as the disconnecting means for the dispenser. The circuit breaker permitted to serve as the dispenser disconnect is:

    A. a single-pole circuit breaker.
    B. a single-pole, ground-fault circuit-interrupter circuit breaker.
    C. a single-pole, arc-fault circuit interrupter.
    D. an instantaneous trip circuit breaker.
    E. a 2-pole switched neutral circuit breaker.

    Code reference _____

39. An intrinsically safe circuit runs from a control panel containing electronic barriers located in a nonclassified area to control devices located in a Class I hazardous location. The intrinsically safe circuits are run in a cable and not in raceway through the nonclassified area. The minimum separation distance from the intrinsically safe circuit cable to other power circuit shall not be less than:

   A. $^1/_2$ in. (13 mm).   D. 1.97 in. (50 mm).
   B. $^3/_4$ in. (19 mm).   E. 6 in. (150 mm).
   C. 1 in. (25 mm).

   Code reference _____

40. A hospital has a demand load on the essential electrical system of more than 150 kVA. A single standby generator supplies the equipment system, the critical branch, and the life safety branch. Upon a loss of normal utility power to the hospital, the equipment system is:

   A. required to be energized in not more than 10 seconds.
   B. permitted to be completely or partially transferred to the generator manually.
   C. required to be started simultaneously along with the emergency system.
   D. permitted to be started simultaneously along with the emergency system.
   E. required to have power restored automatically with an appropriate time-lag.

   Code reference _____

41. For a critical care area in a hospital, each patient bed location is required to be provided with two branch-circuits:

   A. one of which shall originate from the emergency electrical system.
   B. both of which shall originate from the emergency electrical system.
   C. both of which shall originate from the normal electrical system.
   D. with not less than two receptacles connected to each branch-circuit.
   E. that are permitted to originate from any electrical system desired.

   Code reference _____

42. Fire alarm cables installed in a room with a non-fire-rated suspended ceiling grid of an existing building is not permitted to be supported by the ceiling grid:

   A. unless there are no more than three cables in any 10 ft by 10 ft (3 m by 3 m) area.
   B. except where there is a maintenance electrician.
   C. unless the cables have a diameter not more than $^1/_2$ in. (13 mm) and there are no more than three cables across any one ceiling tile.
   D. unless only one cable is across any one ceiling tile.
   E. in any type of building.

   Code reference _____

43. The authority having jurisdiction may rule in some situations that the electrical supply for an emergency panel is permitted to be:

   A. a tap to the normal service conductors entering the building provided the tap is made ahead of the main service disconnect.
   B. a circuit in the first panelboard of the normal power system.
   C. a tap to the normal service made at the main lugs of the disconnect.
   D. a separate and independent service to the building supplying only the emergency panelboard.
   E. any circuit from the normal power system in the building.

   Code reference _____

44. A 50-horsepower, 460-volt, 3-phase, design B fire pump motor in a building is supplied with copper conductors size 4 AWG with 75°C insulation and terminations. The fire pump circuit is protected with a fusible disconnect. The minimum rating fuse permitted for the circuit is:
    A. 90 amperes.
    B. 100 amperes.
    C. 300 amperes.
    D. 350 amperes.
    E. 400 amperes.

    Code reference _____

45. When installing optical fiber cables in raceway with no electrical conductors in the raceway:
    A. only one cable is permitted for each raceway.
    B. the fill requirements that apply to electrical conductors do not apply to optical fiber cables.
    C. the cross-sectional area of the cables is not permitted to exceed 20% of the area of the raceway.
    D. the cross-sectional area of the cables is not permitted to exceed 40% of the area of the raceway.
    E. a maximum of three cables are permitted in a raceway.

    Code reference _____

46. A ladder-type cable tray supports TC multiconductor power cables with Type XHHW copper conductors. There are four 3-conductor cables size 500 kcmil with a diameter of 2.26 in. (57.4 mm), a five 4-conductor cables size 3/0 AWG with a diameter of 1.58 in. (40.0 mm) and a cross-sectional area of 1.96 sq. in. (1264 mm²). There is no maintained air space between conductors. The 3-conductor cables are in a single layer and the 4-conductor cables are in a double layer. The minimum width of cable tray permitted for this installation is:
    A. 9 in. (225 mm).
    B. 12 in. (300 mm).
    C. 18 in. (450 mm).
    D. 24 in. (600 mm).
    E. 30 in. (750 mm).

    Code reference _____

47. The minimum number of duplex receptacles required to be installed in an elevator machine room or space is:
    A. one.
    B. two.
    C. three.
    D. as many as desired because no minimum requirement is established.
    E. none because a single receptacle is permitted.

    Code reference _____

48. A strut-type channel raceway, trade size 1⅝ by 1, is installed across the ceiling of a commercial storage area to support wiring and luminaires (lighting fixtures). Internal joiners are used to connect the strut channel sections. The conductors installed in the strut channel are size 12 AWG copper Type THWN with 75°C terminations. The maximum number of conductors permitted to be installed in the strut channel is:
    A. 11.
    B. 19.
    C. 21.
    D. 28.
    E. 34.

    Code reference _____

49. The suspended ceiling of a restaurant classified as a place of assembly has a 15-minute finish fire rating. The space above the ceiling is not used as a plenum for environmental air. A wiring method not permitted to be used in the space above the suspended ceiling is:

    A. Type MC Cable.
    B. Electrical Nonmetallic Tubing (ENT).
    C. Flexible Metal Conduit.
    D. Type AC Cable not containing an insulated equipment grounding wire.
    E. Electrical Metallic Tubing (EMT).

    Code reference_____

50. Conductors for border and pocket lighting in the stage area of an auditorium are run in metal wireway. The cross-sectional area of the conductors does not exceed 20% of the cross-sectional area of the wireway. Adjustments to the ampacity of conductors:

    A. shall be applied for more than 3 current-carrying conductors.
    B. shall be applied for more than 9 current-carrying conductors.
    C. shall be applied for more than 30 current-carrying conductors.
    D. is a fixed 0.8 whenever there are more than twelve conductors.
    E. do not apply in this application.

    Code reference_____

Annex A-1 Common English distances and the metric equivalent frequently used in the Code.

English Distance	Rounded Metric	Actual Metric	Difference
$^1/_4$ in.	6 mm	6 mm	
$^1/_2$ in.	13 mm	13 mm	
$^5/_8$ in.	16 mm	16 mm	
$^3/_4$ in.	19 mm	19 mm	
1 in.	25 mm	25 mm	
$1^1/_2$ in.	38 mm	38 mm	
2 in.	50 mm	51 mm	
$2^1/_2$ in.	65 mm	64 mm	
3 in.	75 mm	76 mm	
$3^1/_2$ in.	90 mm	89 mm	
4 in.	100 mm	102 mm	$-^1/_{16}$ in.
6 in.	150 mm	152 mm	$-^1/_{16}$ in.
8 in.	200 mm	203 mm	$-^1/_8$ in.
12 in.	300 mm	305 mm	$-^3/_{16}$ in.
18 in.	450 mm	457 mm	$-^1/_4$ in.
20 in.	500 mm	508 mm	$-^5/_{16}$ in.
24 in.	600 mm	610 mm	$-^3/_8$ in.
30 in.	750 mm	762 mm	$-^7/_{16}$ in.
3 ft	900 mm	914 mm	$-^9/_{16}$ in.
$3^1/_2$ ft	1 m	1.067 m	$-2^5/_8$ in.
4 ft	1.2 m	1.219 m	$-^3/_4$ in.
$4^1/_2$ ft	1.4 m	1.372 m	$-1^1/_{16}$ in.
5 ft	1.5 m	1.524 m	$-1^5/_{16}$ in.
6 ft	1.8 m	1.829 m	$-1^1/_8$ in.
7 ft	2.1 m	2.134 m	$-1^5/_{16}$ in.
8 ft	2.5 m	2.438 m	$-2^7/_{16}$ in.
10 ft	3 m	3.048 m	$-1^7/_8$ in.
12 ft	3.7 m	3.658 m	$-1^5/_8$ in.
14 ft	4.3 m	4.267 m	$-1^5/_{16}$ in.
15 ft	4.5 m	4.572 m	$-2^{13}/_{16}$ in.
16 ft	4.9 m	4.877 m	$^7/_8$ in.
18 ft	5.5 m	5.486 m	$^9/_{16}$ in.
20 ft	6 m	6.096 m	$-3^3/_4$ in.
25 ft	7.5 m	7.620 m	$-4^{11}/_{16}$ in.
30 ft	9 m	9.144 m	$-5^{11}/_{16}$ in.
35 ft	11 m	10.668 m	$13^1/_{16}$ in.
40 ft	12 m	12.192 m	$-7^9/_{16}$ in.
50 ft	15 m	15.240 m	$-9^7/_{16}$ in.
75 ft	22.5 m	22.860 m	$-14^3/_{16}$ in.
100 ft	30 m	30.480 m	$-18^7/_8$ in.

Length:
in. $\times$ 25.4 = millimeters (mm)   millimeters / 25.4 = inches (in.)
ft $\times$ 0.3048 = meters (m)   meters $\times$ 3.28 feet (ft)
miles $\times$ 0.62 = kilometers (km)   kilometers $\times$ 1.61 = miles (mi)

Area:
in.^{2} $\times$ 645 = mm^2 mm^2 / 645 = in.^{2}
mm^2 / 100 = cm^2 cm^2 $\times$ 100 = mm^2
ft^2 $\times$ 0.0929 = m^2 m^2 $\times$ 10.76 = ft^2
acre $\times$ 43,560 = ft^2 ft^2 / 43,560 = acres
hectares $\times$ 10,000 = m^2 m^2 / 10,000 = hectares
acres $\times$ 0.4047 = hectares   hectares $\times$ 2.471 = acres

Volume:
in.^{3} $\times$ 16.387 = cm^3 cm^3 / 16.387 = in.^{3}
gallons $\times$ 3.785 = liters   liters $\times$ 0.264 = gallons
gallons $\times$ 0.1336 = ft^3 ft^3 $\times$ 7.48 = gallons
liters $\times$ 0.001 = m^3 m^3 $\times$ 1000 = liters (l)
ft^3 $\times$ 28.32 = liters   liters $\times$ 0.0535 = ft^3
gallons $\times$ .003784 = m^3 m^3 $\times$ 264.2 = gallons (gal)

Weight:
pounds $\times$ 0.454 = kilograms (kg)   kilograms $\times$ 2.20 pounds (lb)

Force:
pounds $\times$ 4.448 = Newtons (N)   Newtons $\times$ 0.2248 = pounds

Torque:
pound-ft $\times$ 1.357 = Newton-meters (Nm) Newton-meters $\times$ 0.737 = pound-ft
pound-in $\times$ 0.1130 = Newton-meters   Newton-meters $\times$ 8.850 = pound-in.

Pressure:
psi $\times$ 6.9 = kilopascals (kPa)   kilopascals $\times$ 0.145 = psi
in. of water $\times$ 249.3 = pascals (Pa)   Pascals $\times$ 0.00401 = in. of water
psi $\times$ 27.68 = in. of water   in. of water 3 0.0361 = psi

Air or Vapor Flow Rate:
cubic ft per minute $\times$ 0.0283 = m^3/minute   m^3/minute $\times$ 35.3 = cubic ft per minute (cfm)
ft^3 per hour $\times$ 28,339 = cm^3 per hour   cm^3 per hour $\times$ 0.0000353 = ft^3 per hour

Power:
horsepower $\times$ 746 = watts (W)   kilowatts (kW) $\times$ 1.34 = horsepower (hp)
horsepower $\times$ 33,000 = ft-pounds per minute
kilowatts $\times$ 3,413 = Btu/hour

Annex A-3   Conduit and tubing trade sizes and equivalent metric designators.

^{3}/8 in.	12	2^1/2 in.	63
^{1}/2 in.	16	3 in.	78
^{3}/4 in.	21	3^1/2 in.	91
1 in.	27	4 in.	103
1^1/4 in.	35	5 in.	129
1^1/2 in.	41	6 in.	155
2 in.	53		

ANSWER SHEET

Name _____

Unit No._____   **Beginning or Advanced** (circle one)

	Answer	Code reference
1.	A B C D E	_____
2.	A B C D E	_____
3.	A B C D E	_____
4.	A B C D E	_____
5.	A B C D E	_____
6.	A B C D E	_____
7.	A B C D E	_____
8.	A B C D E	_____
9.	A B C D E	_____
10.	A B C D E	_____
11.	A B C D E	_____
12.	A B C D E	_____
13.	A B C D E	_____
14.	A B C D E	_____
15.	A B C D E	_____

ANSWER SHEET Name _____

No. 15 REVIEW—Beginning or Advanced (circle one)

	Answer	Code reference		Answer	Code reference
1.	A B C D E	NA	26.	A B C D E	_____
2.	A B C D E	NA	27.	A B C D E	_____
3.	A B C D E	NA	28.	A B C D E	_____
4.	A B C D E	NA	29.	A B C D E	_____
5.	A B C D E	NA	30.	A B C D E	_____
6.	A B C D E	_____	31.	A B C D E	_____
7.	A B C D E	_____	32.	A B C D E	_____
8.	A B C D E	_____	33.	A B C D E	_____
9.	A B C D E	_____	34.	A B C D E	_____
10.	A B C D E	_____	35.	A B C D E	_____
11.	A B C D E	_____	36.	A B C D E	_____
12.	A B C D E	_____	37.	A B C D E	_____
13.	A B C D E	_____	38.	A B C D E	_____
14.	A B C D E	_____	39.	A B C D E	_____
15.	A B C D E	_____	40.	A B C D E	_____
16.	A B C D E	_____	41.	A B C D E	_____
17.	A B C D E	_____	42.	A B C D E	_____
18.	A B C D E	_____	43.	A B C D E	_____
19.	A B C D E	_____	44.	A B C D E	_____
20.	A B C D E	_____	45.	A B C D E	_____
21.	A B C D E	_____	46.	A B C D E	_____
22.	A B C D E	_____	47.	A B C D E	_____
23.	A B C D E	_____	48.	A B C D E	_____
24.	A B C D E	_____	49.	A B C D E	_____
25.	A B C D E	_____	50.	A B C D E	_____

INDEX